Proceedings of the XVI Ibero-American Congress of Mechanical Engineering

Oscar Francisco Farías Fuentes ·
Emilio Enrique Dufeu Delarze ·
Juan Carlos García Prada
Editors

Proceedings of the XVI Ibero-American Congress of Mechanical Engineering

Selected Papers of CIBIM/CIBEM 2024

Editors
Oscar Francisco Farías Fuentes
Department of Mechanical Engineering
University of Concepción
Concepción, Chile

Emilio Enrique Dufeu Delarze
Department of Mechanical Engineering
University of Concepción
Concepción, Chile

Juan Carlos García Prada
Departamento Mecánica
E.T.S. de Ingenieros Industriales - UNED
Madrid, Spain

ISBN 978-3-032-22822-2 ISBN 978-3-032-22823-9 (eBook)
https://doi.org/10.1007/978-3-032-22823-9

This work was supported by University of Concepcion, Chile.

This Springer imprint is published by the registered company Springer Nature Switzerland AG
The registered company address is: Gewerbestrasse 11, 6330 Cham, Switzerland

Contents

Mechanical Engineering Topics

Thermofluids

Mechanical Engineering Topics

Vibration Transmissibility of a Passive-Mechanical Isolation System for RPAS Applications

Karen Riedel Hornig[1], Félix Leaman[2(✉)], Frank Tinapp Dautzenberg[1], and Cristián Vicuña[2]

[1] Laboratorio de Técnicas Aeroespaciales, Universidad de Concepción (LTA-UdeC), Concepción, Chile
[2] Laboratorio de Vibraciones Mecánicas, Universidad de Concepción (LVM-UdeC), Concepción, Chile
fleaman@udec.cl

Abstract. This work focuses on the vibration isolation of electronic and/or mechanical components and equipment that operate on board a Remotely Piloted Aircraft System (RPAS). Experimental tests are performed on a mechanical shaker to characterize the vibration response of an adjustable passive-mechanical isolation system incorporating mountable mechanical isolators in different spatial configurations and to determine the effect of the arrangement on vibration isolation. As a result, the transmissibility curves of the isolation system are determined for four suspension ball-type isolators in four configurations each over a frequency sweep of 5 to 220 Hz. This frequency range includes that at which the highest amplitudes are reached in a small commercial RPAS. It is concluded that the effectiveness of vibration isolation depends on the arrangement of the mechanical isolators, which must be selected correctly according to the operating frequency range.

Keywords: Vibration isolation · mechanical-passive isolation system · RPAS

1 Introduction

In the aerospace [1,2], automotive [3] and marine [4] industries, there are great challenges in developing new technologies and methods for vibration isolation of mechanical and electronic components. The machines used in these industries are exposed to adverse conditions and dynamic forces that generate vibrations, causing discomfort and performance losses. In some cases, these vibrations pose a latent danger to the operation. This is why it is essential to study the vibrations to which they are exposed and to develop isolation systems, such as mechanical isolators, to minimize transmitted vibrations.

An area of research in development is the application of mechanical vibration isolation systems for RPAS equipment and sensors onboard. In particular, due to

O. F. Farías Fuentes et al. (Eds.): CIBIM 2024, *Proceedings of the XVI Ibero-American Congress of Mechanical Engineering*, pp. 3–12, 2026.
https://doi.org/10.1007/978-3-032-22823-9_1

their variety of applications and uses in scientific, technological, agricultural and other areas, the technological advancement of RPASs is a topic of great interest nowadays [5,6]. RPASs can be equipped with a variety of equipment and sensors, with image acquisition systems being the most commonly used [7,8]. However, RPASs are exposed to different sources of vibration during flight due to flight conditions, maneuvers, moving mechanical components and external factors. If vibrations experienced during flight are transmitted to the onboard equipment and sensors, their performance may be negatively affected. In severe cases, they may even be physically damaged.

When vibrations are transmitted to the onboard cameras, motion blur and scene misframing occur, altering the definition of the resulting images [8,9]. This makes scene mapping, pattern recognition and target tracking difficult. To address this problem, different image stabilization methods have been proposed to mitigate the effect of vibrations on RPAS cameras. These methods can be categorized as digital, optical or mechanical, and when used in combination, better results can be achieved [8,10].

Digital image stabilization methods are used when images have already been processed and are blurred. These methods correct the motion blur effect in the captured images by estimating the displacement induced by the vibration and making the appropriate corrections through digital techniques and algorithms. However, the field of view captured by the camera is a limitation of this method. In case of high amplitude vibrations, the trace of a reference point may be lost, and consequently, digital corrections are not possible [11].

Optical stabilization methods correspond to internal camera mechanisms that control and modify the optical path of the lens [12]. The advantage of these methods is that the image projected on the lens is stabilized before being transformed into digital information. However, this technology is not available in all camera systems.

Mechanical stabilization methods are the first opportunity to eliminate image blurring before data processing, thus reducing the risk of damage to optical components. A mechanism is implemented to counteract vibration induced displacements and isolate vibrations. These methods have the advantage of operating over a wide frequency and amplitude range [9,10]. Depending on their mode of action, they are classified as active or passive. Active mechanical stabilization methods isolate vibrations by implementing active control systems, while passive methods correspond to conventional mechanisms designed with viscoelastic materials. Previous results suggest that passive mechanical systems allow the reduction of in-flight vibrations by up to 90 % [13].

This work focuses on the vibration isolation of electronic and mechanical components and equipment operating on board RPAS, such as cameras and sensors. This is an extremely relevant issue, since depending on the vibration severity experienced during flight, damage to the components and alterations in the acquired data can be generated, negatively compromising the flight mission. Therefore, it is important to determine the transmissibility of the system in the operational frequency ranges in order to select a suitable system and thus achieve a better quality of the images captured and the data obtained. Specifically, empirical laboratory tests are performed to record the vibration response of a passive-mechanical isolation system to known and controlled input excitation. The isolation characteristics of different configurations of passive-mechanical isolators are studied, being a highly reproducible experiment applicable to other more complex mechanical system problems. As reported in the literature [2,3], experimental studies have been performed on vibration isolation systems; however, there is a lack of detailed documentation on the influence of different mechanical isolator arrangements on isolation performance.

2 Method

The vibration response of an adjustable isolation system with mechanical-passive suspension ball isolators is characterized. These isolators are commercially available and commonly used in RPAS applications because of their characteristics: they are lightweight, small in size, low in cost, highly reliable and do not require a power supply. The procedures performed during the experimental tests, along with the processing of the acquired data, are documented in the following. The transmissibility curves of the vibration isolation systems are determined, and a suitable configuration is selected based on a selection criterion for implementation in small commercial RPAS.

2.1 Experimental Tests

The proposed insulation system is designed with two perforated carbon fiber plates into which the isolators are inserted in four different spatial configurations. A camera is assembled to the isolation system so that the isolators act in compression. Figure 1 shows the components of the passive-mechanical isolation system with four isolators and the mounted camera.

For the purposes of this study, only the assembly of four suspension ball isolators of different sizes is considered. These are visually differentiated by their color: yellow (Y), red (R), black (B) and gray (G). The four isolators under study are each mounted in four spatial configurations within the adjustable vibration isolation system. Figure 2 illustrates the four configurations studied, each of which is assigned an identification number. Figure 2 shows the location of the isolators in the isolation system as gray circles and the spaces where no isolator is located in this configuration as white circles. Additionally, Fig. 3 shows Configuration No. 3 and Configuration No. 4 with R and G isolators, respectively.

Vibration transmissibility is determined by performing experimental tests under laboratory conditions using a TIRA Vib S 51110 shaker. The vibrations experienced by the upper structure and the payload resting on the isolation system are then recorded. For the purposes of this study, the payload corresponds to an RPAS camera mounted on an adjustable mechanical-passive isolation system, as well as a vibration measurement system that operates in conjunction

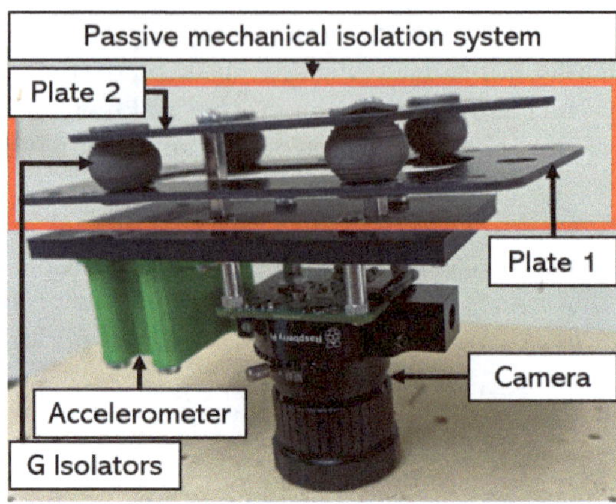

Fig. 1. Adjustable passive-mechanical vibration isolation system

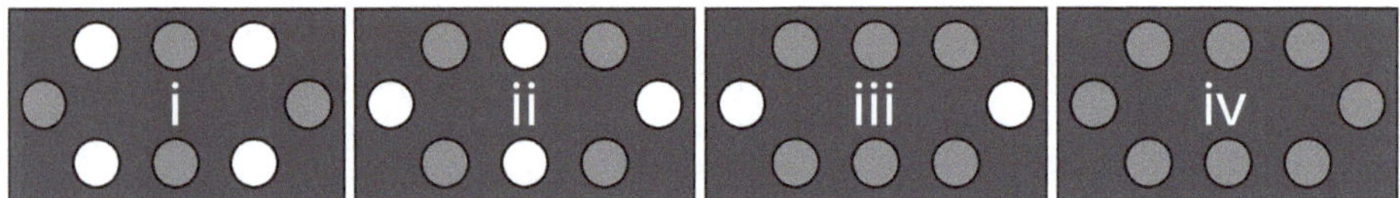

Fig. 2. Test configurations of the isolators in the passive mechanical isolation system: i) Configuration No. 1, ii) Configuration No. 2, iii) Configuration No. 3 and iv) Configuration No. 4.

with a Raspberry Pi 4B (RPi) microcomputer and two MPU6050 triaxial digital accelerometers. Figure 4 shows the experimental setup used for tests carried out in a controlled laboratory environment.

Specifically, excitation is applied in the form of a sinusoid over a sweep ranging from 5 to 220 Hz, as this is the operational range at which the highest amplitudes are achieved in small commercial RPAS. Accelerations are measured at two locations: the undamped structure (input) and the camera, which is mounted under the vibration isolation system (output). Subsequently, a frequency analysis of the measured vibratory responses is performed and transmissibility curves are plotted to identify frequency regions where isolation and amplification occur.

2.2 Transmissibility Curve

Transmissibility describes how motion is transmitted from the base or supporting structure to the mass body, which in this case corresponds to the camera. Transmissibility is defined as shown in Eq. 1, where T denotes transmissibility and A is amplitude. It is satisfied that: if $T > 1$, then there is an amplification in the amplitude of the vibrational signal measured at the camera location, if $T < 1$, a reduction in the amplitude of the measured signal is observed at the camera location, and if $T = 1$ both the vibrational amplitude of the signal at the base and at the camera location have the same magnitude.

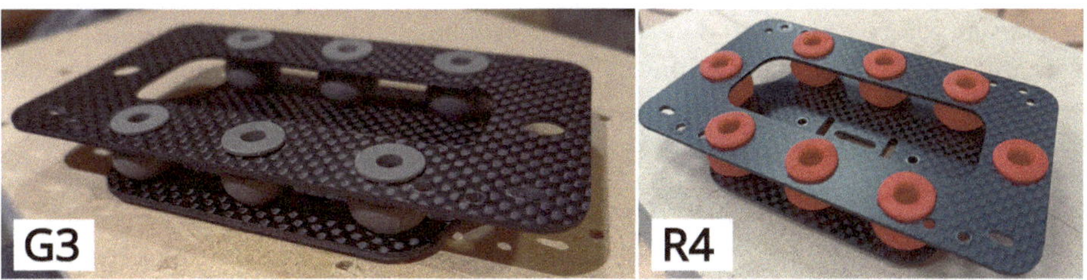

Fig. 3. Left: Configuration No. 3 with G isolators. Right: Configuration No. 4 with R isolators.

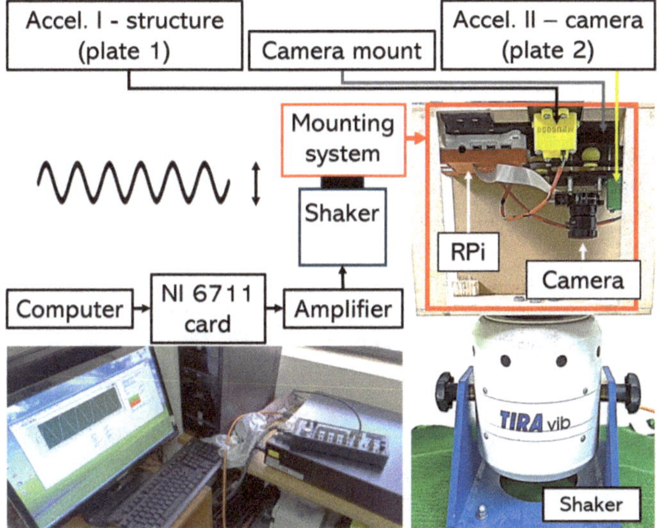

Fig. 4. Laboratory setup for experimental vibration measurement tests.

$$T = \frac{A_{output}}{A_{input}} = \frac{A_{camera}}{A_{structure}} \tag{1}$$

Vibrations can never be completely eliminated from the system; therefore, this ratio (Eq. 1) must be minimized within the desired frequency range of vibration isolation. For a vibration isolation system to be effective, it must exhibit low transmissibility throughout the entire frequency range contained in the Fourier spectrum of the structural displacement input. The following criteria should be considered when selecting a vibration isolation system in order for it to operate effectively.

I Low natural frequency of the vibration isolation system
II Low transmissibility at resonance regions
III The transmissibility should decrease rapidly at frequencies above the natural vibration frequency of the system

It is therefore crucial to determine the transmissibility curves of the isolation system in different configurations, taking into account the operational frequency ranges, in order to identify the isolation regions and select an appropriate isolation system based on the above mentioned criteria. The primary objective is to improve the quality of the images captured during flight, as well as the data acquired by onboard sensors.

The methodology used to evaluate the transmissibility response of the vibration isolation system is explained below. Once the experimental tests are completed, a database is compiled that contains the vibration measurements acquired from the two digital accelerometers, one mounted on the main structure and the other at the camera location, for the different configurations and corresponding isolators. The data processing procedure is described as follows.

The vibration measurement data obtained is processed using Fourier transform (FT), a mathematical process that converts a function in the time domain to the frequency domain. This is done to separate the frequency components of the original function. The amplitude spectra are then determined for each measurement test performed, and the amplitude values of the structure and the camera location at the corresponding excitation frequency are retrieved. Once the amplitude values in the structure and the camera are known, it is possible to determine the transmissibility for the frequency sweep. In this way, the transmissibility curves for each configuration are determined using the different isolators and evaluated along the three axes (transverse, longitudinal and vertical relative to the camera). The transmissibility curve is presented on the ordinate axis in a logarithmic scale [dB] to visualize the variations between the measurements, while the frequency is shown on the abscissa axis in [Hz]. The results are documented in Sect. 3 for the vertical Z axis only. Given that the results are expressed in a logarithmic scale, values above zero indicate signal amplification at the camera location, whereas values below zero represent signal attenuation. In the transmissibility curves obtained, a solid line is used to indicate the level where $T = 0$. Thus, regions where the values fall below this line correspond to vibration isolation at the camera location, while those above it indicate vibration amplification. For small commercial RPAS applications, maintaining vibration isolation within the 150–200 Hz frequency band is essential, as this range corresponds to operational conditions where rotor blade rotation may induce significant vibration amplitudes [14].

3 Results and Discussion

The results of the transmissibility curves obtained for the four isolators studied are presented in Figs. 5, 6, 7 and 8, corresponding to the isolators Y, R, G and B, respectively, each evaluated under the four test configurations.

It is important to note that the documented results correspond exclusively to the transmissibility curves obtained from vertical vibration measurements, which is the axis along which the mechanical shaker excites the entire system. From these transmissibility curves, it is possible to identify the isolation regions

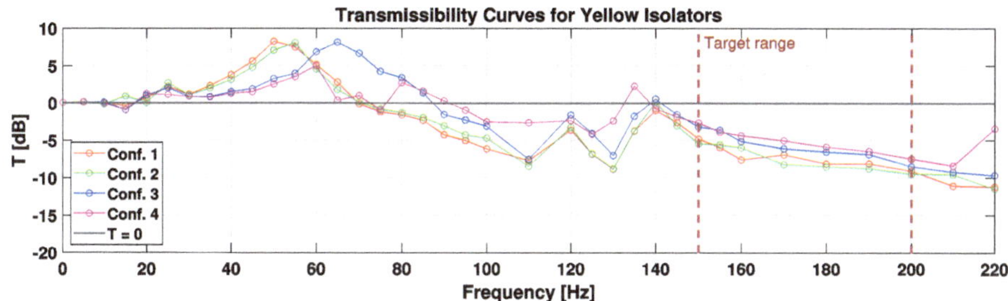

Fig. 5. Transmissibility curves of isolator Y in the four test configurations of the mechanical isolation system.

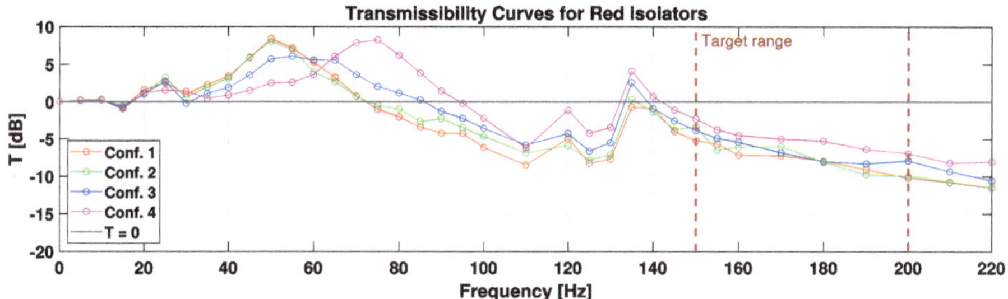

Fig. 6. Transmissibility curves of isolator R in the four test configurations of the mechanical isolation system.

($T < 0$) and amplification regions ($T > 0$) within the frequency sweep from 5 to 220 Hz, for the mechanical isolation system in the configurations studied.

From the results, the following observations can be made.

I. Regardless of the isolators implemented, there is a tendency in the behavior along the frequency sweep that both configurations with fewer isolators (Configurations No. 1 and 2) have lower transmissibility values. This is because in Configurations No. 1 and 2 the transmissibility value stays below the reference line ($T = 0$), which means that it is in the isolation region. This occurs in these two configurations for a greater frequency range, obtaining lower transmissibility values compared to configurations with more isolators.

II For configurations No. 3 and 4, the range over which vibration attenuation occurs is narrower. This is due to the increased number of isolators that make the mounting structure stiffer, which in turn leads to greater transmission of vibrations from the main structure to the camera.

III G isolators have a lower natural frequency compared to Y, B and R isolators.

IV Due to the design characteristics of the vibration isolation system, two natural vibration frequencies are observed.

Finally, based on the results obtained, an alternative to be implemented is selected. For this purpose, the criteria mentioned in Sect. 3 are considered. Therefore, the system with the lowest natural frequency, with ideally low transmissibil-

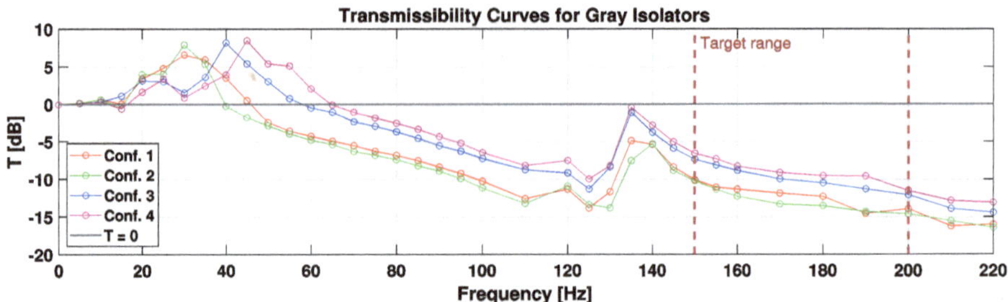

Fig. 7. Transmissibility curves of isolator G in the four test configurations of the mechanical isolation system.

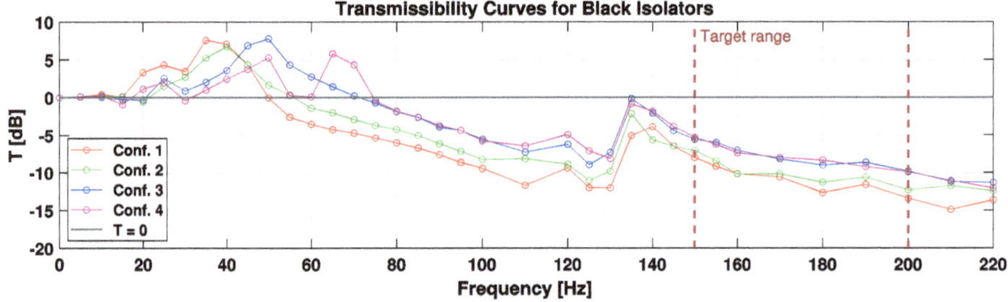

Fig. 8. Transmissibility curves of isolator B in the four test configurations of the mechanical isolation system.

ity in the zones where resonance occurs, and with transmissibility that decreases rapidly after a resonant zone, should be selected. To obtain a low value of the natural frequency, two possibilities can be considered: i) increase the mass of the system and/or ii) consider a system with low stiffness. Given the operational conditions and payload capacity limitations of RPAS, the use of lightweight equipment is essential, making alternative (i) unsuitable for these types of applications. The transmissibility curves reveal that the G isolators, Fig. 6, have lower natural frequencies than the other isolators analyzed in this study. Thus, Configuration No. 2 combined with G isolators represents the most appropriate solution for implementing a vibration isolation system in small RPAS, as it most effectively satisfies the criteria for efficient vibration isolation.

4 Conclusion

Based on the reviewed literature, it is concluded that the implementation of vibration isolation systems in RPAS applications is highly relevant to ensure mission effectiveness and to protect onboard optical and mechanical equipment. When vibrations during flight are transmitted to sensors such as onboard cameras, image degradation occurs in the form of motion blur and scene misframing. Therefore, the integration of an appropriate mechanical isolation system is essential and should be considered a primary measure to mitigate these effects. In addition, other image stabilization methods, such as digital and optical techniques, can be used in combination to further enhance image quality.

This study demonstrates through empirical testing in a controlled laboratory environment that an effective mechanical-passive vibration isolation system for small RPAS can be achieved using commercially available suspension ball isolators. Of the four spatial configurations studied, the setup using four low-stiffness gray isolators was found to be the most effective based on transmissibility curve analysis and isolation performance criteria. This configuration offers a practical, lightweight solution for isolating in-flight vibrations, thereby enhancing image quality and safeguarding sensors in RPAS applications.

These findings lead to the conclusion that both the configuration and quantity of mechanical isolators in a vibration isolation system are decisive factors in its effectiveness and must not be chosen arbitrarily. The selection of the isolation system should be guided by a detailed analysis of the operational frequency range relevant to the specific application and the corresponding transmissibility performance requirements.

5 Future Work

The next step in this research project is to implement the selected passive mechanical isolation system in a small commercial RPAS to verify the reduction in vibrations perceived by the camera during flight.

In future work, it is also important to consider the study of mechanical isolators made of different materials and shapes, such as: Sorbothane sheets, Kyosho Zeal sheets and silicone foam, among others.

Acknowledgments. This research was made possible thanks to the Department of Mechanical Engineering of the Universidad de Concepción.

References

1. Shi, H.T., Abubakar, M., Bai, X.T., Luo, Z.: Vibration isolation methods in spacecraft: a review of current techniques. Adv. Space Res. **73**(8), 3993–4023 (2024). https://doi.org/10.1016/j.asr.2024.01.020
2. Oh, H.-U., Lee, K.-J., Jo, M.-S.: A passive launch and on-orbit vibration isolation system for the spaceborne cryocooler. Aerosp. Sci. Technol. **28**(1), 324–331 (2013). https://doi.org/10.1016/j.ast.2012.11.013
3. Panda, K.C.: Dealing with noise and vibration in automotive industry. Procedia Eng. **144**, 1167–1174 (2016). https://doi.org/10.1016/j.proeng.2016.05.092
4. Li, Y., He, L., Shuai, C.-G., Wang, C.-Y: Improved hybrid isolator with maglev actuator integrated in air spring for active-passive isolation of ship machinery vibration. J. Sound Vib. **407**, 226–239 (2017). https://www.sciencedirect.com/science/article/pii/S0022460X17305230

5. Bolch, E. A., Hestir, E.L., Khanna, S.: Performance and feasibility of drone-mounted imaging spectroscopy for invasive aquatic vegetation detection. Remote Sens. **13**(4), 582 (2021). https://doi.org/10.3390/rs13040582
6. Abdel-Maksoud, H.: Combining UAV-LiDAR and UAV-photogrammetry for bridge assessment and infrastructure monitoring. Arab. J. Geosci. **17**(4), 144 (2024). https://doi.org/10.1007/s12517-024-11897-5
7. Dunbar, M.B., Caballero, I., Román, A., Navarro, G.: Remote sensing: satellite and RPAS (remotely piloted aircraft system). In: Blasco, J., Tovar-Sánchez, A. (eds) Marine Analytical Chemistry, pp. 389–417. Springer, Cham (2023). https://doi.org/10.1007/978-3-031-14486-8_9
8. Dahlin Rodin, C., de Alcantara Andrade, F.A., Hovenburg, A.R., Johansen, T.A.: A survey of practical design considerations of optical imaging stabilization systems for small unmanned aerial systems. Sensors **19**(21), 4800 (2019). https://doi.org/10.3390/s19214800
9. Windau, J., Itti, L.: Multilayer real-time video image stabilization. In: IEEE/RSJ International Conference on Intelligent Robots and Systems, San Francisco, CA, USA, pp. 2397–2402 (2011). https://doi.org/10.1109/IROS.2011.6094738
10. Verma, M., Collette, C.: Active vibration isolation system for drone cameras. In: Sapountzakis, E.J., Banerjee, M., Biswas, P., Inan, E. (eds.) Proceedings of the 14th International Conference on Vibration Problems, ICOVP 2019. Lecture Notes in Mechanical Engineering, pp. 1067–1084. . Springer, Singapore (2021). https://doi.org/10.1007/978-981-15-8049-9_67
11. Duric, Z., Rosenfeld, A.: Image sequence stabilization in real time. Real-Time Imag. **2**(5), 271–284 (1996). https://doi.org/10.1006/rtim.1996.0029
12. Park, R.Y., Pak, J.M., Ahn, C.K., Lim, M.T.: Image stabilization using FIR filters. In 15th International Conference on Control, Automation and Systems (ICCAS), Busan, Korea (South), pp. 1234–1237 (2015). https://doi.org/10.1109/ICCAS.2015.7364819
13. Imam, A., Bicker, R.: Design and construction of a small-scale rotorcraft UAV system (2014). https://api.semanticscholar.org/CorpusID:212569615
14. Riedel Hornig, K.: Sistema de medición de vibraciones basado en Raspberry Pi para estabilización de dispositivos de adquisición de imágenes en dron hexarotor F550 (2023). https://repositorio.udec.cl/handle/11594/11176

Acoustic Emissions Generated by the Contact Between Involute Spur Gear Teeth

Félix Leaman(✉) and Felipe Segura

University of Concepción, Edmundo Larenas 219, Concepción, Chile
fleaman@udec.cl

Abstract. This study aims to design a test bench capable of replicating the sliding-rolling contact condition to validate the hypothesis on the origin of acoustic emission (AE) bursts generated by the contact between involute spur gear teeth. AE signals were measured on the test bench using a piezoelectric AE sensor, and a tachometer was employed to determine the angular position at the pass point for accurate burst identification. Preliminary results showed that the continuous part of the signals is consistent with the asperity contact hypothesis, but the AE bursts showed variability in both amplitude and timing, revealing the need for improvements to the test bench setup, such as better alignment and more controlled motion. Furthermore, distinct bursts were observed at both the beginning and the end of contact, indicating that the hypothesis of tooth impact justifies further investigation. Consequently, these results neither confirm nor refute the original hypothesis. This work establishes a basis for further research on theoretical modeling and phenomenological validation of AE in gear contacts.

Keywords: Acoustic emissions · Gears · Test bench · Signal measurement

1 Introduction

Acoustic emissions (AE) have proven to be a valuable tool for detecting potential failures in rotating machinery and various types of mechanical components. Since AE "listens" to the internal condition of a system, it allows early detection of progressive failure modes such as bearing pitting, excessive friction, and gear tooth faults that could go unnoticed using conventional predictive maintenance techniques. For example, this technique has been commonly used to detect pitting in individual bearing balls at low rotational speeds [1].

Numerous authors have applied AE analysis to fault detection in different scenarios. Nirwan et al. [2] used AE to monitor the progression of a defect in the outer race of a cylindrical roller bearing in a rolling mill. From the measured signals, they were able to experimentally correlate the amplitude to the development of the fault, observing components at the ball pass frequency of the outer

O. F. Farías Fuentes et al. (Eds.): CIBIM 2024, *Proceedings of the XVI Ibero-American Congress of Mechanical Engineering*, pp. 13–26, 2026.
https://doi.org/10.1007/978-3-032-22823-9_2

race (BPFO) and its harmonics. A clear amplitude increase was observed as a crack in the outer race progressed, under constant operating conditions. Similarly, Hou et al. [1] developed a theoretical model capable of predicting various types of defects in tapered roller bearings of high-speed train wheels, obtaining predicted RMS values that closely matched experimental results across multiple fault scenarios.

Another successful application was reported by Yu and Li [3], who used AE to detect small gas leaks caused by loose threaded connections in pipelines, which were visually imperceptible. From the acquired signals, they trained machine learning models to automatically detect leaks using AE measurements from piezoelectric sensors attached to the pipe surface.

As illustrated by the previous cases, AE emerge as a valuable diagnostic alternative under varying speed and load conditions, particularly for faults that develop progressively. This is especially relevant for equipment that must operate continuously, since failure progression is directly tied to remaining useful life. However, despite its advantages, AE based condition monitoring still faces significant challenges. These include the large number and complexity of AE sources, combined with the high sensitivity of AE signals to operating conditions and machine-specific behavior, making it extremely difficult to develop a general theoretical model applicable to all machines.

A particularly important application is gear fault detection, given the extensive use of gear systems in critical industrial equipment such as wind turbines, mining trucks, and heavy machinery. These systems often operate continuously and tend to fail through mechanisms such as wear, pitting, surface fatigue, and fracture [4], all of which evolve progressively and could be effectively detected through AE analysis. However, the use of AE for condition monitoring in gears presents challenges due to the appearance of burst-type signals at tooth meshing frequencies; even under healthy conditions. This phenomenon was analyzed experimentally by Vicuña [5] in a planetary gear, highlighting the random nature of these bursts due to the non-stationary nature of AE signals. Understanding the origin of these bursts is critical for distinguishing normal operational AE from those indicating a fault. One of the most prominent hypotheses is that bursts appear during the tooth contact period, coinciding with the pitch point; where the contact transitions from mixed sliding/rolling to pure rolling, and where the direction of the friction force between teeth reverses. Specifically, Tan and Mba [6] proposed that sliding generates continuous AE signals, whereas pure rolling produces short-duration, high-amplitude bursts. They observed that inducing damage to the involute profile reduced the prominence of the bursts, supporting their hypothesis.

Although many experimental studies agree with the hypothesis proposed by Tan and Mba [7], the behavior of AE along the full contact path of a gear tooth pair has not been thoroughly investigated. Existing analyses have primarily focused on the observation of burst events occurring at the meshing frequency, without deeper exploration of AE activity throughout the entire engagement. The small size of the teeth and the rapid nature of gear tooth contact make it

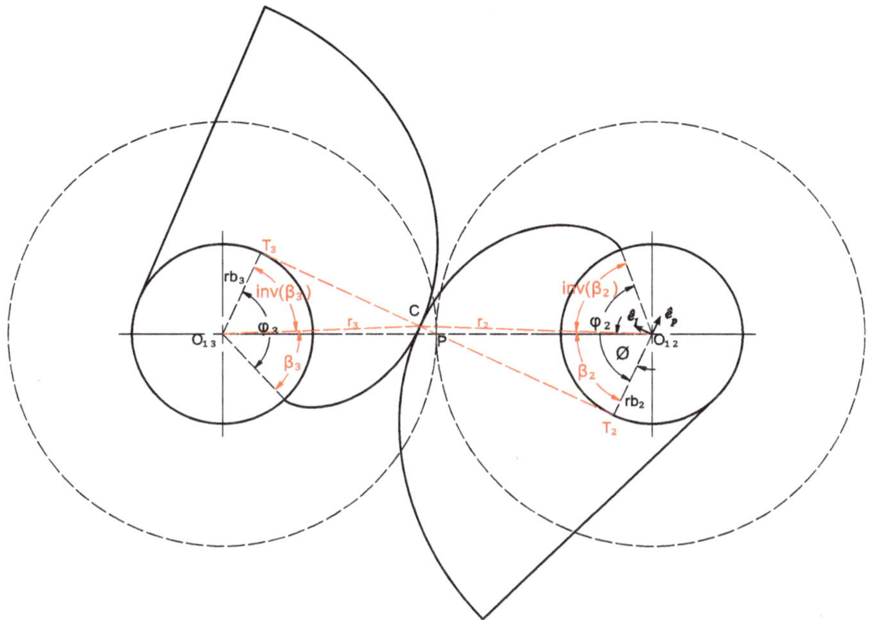

Fig. 1. Top view of the involute teeth mesh

particularly challenging to conclusively correlate the sliding-to-rolling transition with the onset of AE bursts in a standard gearbox. In light of this limitation, the present work proposes the design of a test bench capable of replicating the rollingsliding conditions occurring during gear tooth contact. The design aims to amplify the contact behavior to distinguish between sliding, the transition to pure rolling, and the return to sliding. To achieve this, a gear tooth with a sufficiently large involute profile is designed, allowing signal segmentation according to contact condition and correlation with shaft rotation angle. AE signals are recorded and synchronized using a tachometer and reflective tape markers placed at the start of contact, the pitch point, and the end of contact positions.

2 Test Bench Design

2.1 Geometric Conditions

The involute curve is defined parametrically by:

$$x = r_b(\cos\varphi + \varphi\sin\varphi) \tag{1}$$

$$y = r_b(\sin\varphi - \varphi\cos\varphi) \tag{2}$$

Where r_b is the base radii and φ denotes the angular position from the base circle to the current contact point. Figure 1 illustrates the meshing of two involute gear teeth. The angle β between the line of action and the contact point position relates to ϕ as follows:

$$\varphi = \beta + inv(\beta) \tag{3}$$

With inv denoting the involute function:

$$inv(\beta) = \tan\beta - \beta \tag{4}$$

The total line of contact is given by:

$$L_c = L_{ac} + L_{al} \tag{5}$$

where approach and recess segments are:

$$L_{ac} = \sqrt{r_{a3}^2 - r_{p3}^2\cos^2(\phi)} - r_{p3}\sin(\phi) \tag{6}$$

$$L_{al} = \sqrt{r_{a2}^2 - r_{p2}^2\cos^2(\phi)} - r_{p2}\sin(\phi) \tag{7}$$

And the full contact line is:

$$L_c = \sqrt{r_{a2}^2 - r_{p2}\cos^2(\phi)} + \sqrt{r_{a3}^2 - r_{p3}\cos^2(\phi)} - A_0\sin(\phi) \tag{8}$$

Let A_0 denote the distance between the centers of the two teeth. Equations 6 and 7 can then be related to the pinion approach and the recess angles as follows:

$$\gamma_{12} = \frac{L_{ac}}{r_{p2}\cos(\phi)} \tag{9}$$

$$\gamma_{22} = \frac{L_{al}}{r_{p2}\cos(\phi)} \tag{10}$$

These angles correspond to the angular distances that the tooth must travel to reach the pitch point. In the case of the approach angle, it refers to the angular distance from the initial contact to the pitch point. For the recess angle, it refers to the distance from the pitch point to the last point of contact. Thus, the total contact angle is defined as:

$$\gamma_{12} + \gamma_{22} = \sqrt{\left(\frac{r_{a3}}{r_{p2}}\right)^2 \sec^2(\varphi) - \left(\frac{r_{p3}}{r_{p2}}\right)^2} + \sqrt{\left(\frac{r_{a2}}{r_{p2}}\right)^2 \sec^2(\varphi) - 1} - \left(\frac{r_{p3}}{r_{p2}} + 1\right)\tan(\varphi) \tag{11}$$

Additionally, the position of the contact point with respect to both teeth and the pressure angle can be determined as follows:

$$\boldsymbol{r}_2 = -r_{b2}\,\hat{e}_\ell + r_{b2}\,\varphi_2\,\hat{e}_\ell \tag{12}$$

$$\boldsymbol{r}_3 = r_{b3}\,\hat{e}_p - r_{b3}\,\varphi_3\,\hat{e}_\ell \tag{13}$$

where $\hat{e}_p$ and $\hat{e}_\ell$ correspond to the normal and tangential directions with respect to the contact point, as illustrated in Fig. 1.

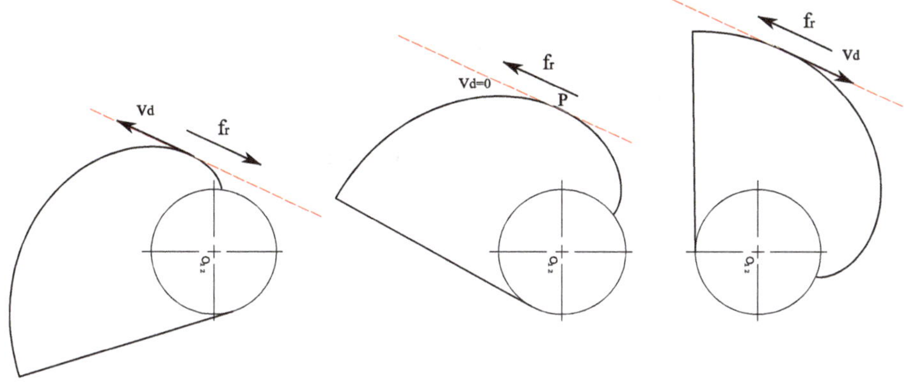

Fig. 2. Evolution of Sliding Velocity and Friction Force Along the Contact Path

2.2 Kinematic Conditions

The relevant kinematic conditions are illustrated in Fig. 2, where a linear decrease in sliding speed between teeth is expected as the contact progresses, reaching zero when the contact point coincides with the pitch point. Beyond this point, during the recess phase, the sliding speed should increase gradually but in the opposite direction. Since the friction force opposes the sliding speed, a corresponding change in friction force is also expected throughout the contact, which is hypothesized to be the underlying mechanism responsible for AE bursts.

The contact condition requires that the speeds in the normal direction be equal, leading to the following expression.

$$\frac{\omega_3}{\omega_2} = \frac{r_{b2}}{r_{b3}} \tag{14}$$

The speed in the tangential direction depends solely on the pressure angle. From the figure, the following relation can be deduced:

$$r_{b2}\varphi_2 + r_{b3}\varphi_3 = \overline{T_2C} + \overline{T_3C} \tag{15}$$

By developing the expression, one obtains:

$$r_{b2}\varphi_2 + r_{b3}\varphi_3 = A_0 \sin(\phi) \tag{16}$$

The rolling speed is defined as the speed of the teeth at the contact point projected onto the tangential direction:

$$v_{r2}(\widehat{ep}) = \omega_2 r_{b2}\varphi_2 \tag{17}$$

$$v_{r3}(\widehat{ep}) = \omega_3 r_{b3}\varphi_3 \tag{18}$$

On the other hand, the sliding speed corresponds to the relative velocity at the contact point between the rolling speed measured from gear 2 and that measured from gear 3. Thus, the following expression is obtained:

$$v_d(\widehat{ep}) = \omega_2 r_{b2} \varphi_2 - \omega_3 r_{b3} \varphi_3 \tag{19}$$

From Eqs. (16) and (19), the sliding speed can be expressed as

$$v_d = \omega_2 r_{b2} \left[\varphi_2 \left(1 + \frac{r_{b2}}{r_{b3}} \right) - \frac{A_0 \sin(\overline{\phi})}{r_{b3}} \right] \tag{20}$$

The sliding speed varies with changes in the pressure angle φ_2, while the other parameters in the expression remain constant.

Finally, the ratio between sliding and rolling is defined as the coefficient between the sliding speed at any contact point and the rolling speed of gear 1 or 2, depending on which is being analyzed. Considering the driving gear, this ratio is expressed as follows.

$$r_n = \frac{v_d}{v_{rn}} \tag{21}$$

3 Prototype Design and Dimensions

Based on the geometric conditions, an initial sizing of the teeth can be carried out by selecting the parameters listed in Table 1.

Table 1. Geometric parameters of the gear teeth

	Driving Tooth	Driven Tooth
r_p [mm]	118.31	118.31
r_a [mm]	182.89	182.89
r_b [mm]	50	50
α [deg]	65	65
ω [rpm]	25	25

In addition, the corresponding lengths used in the equations are obtained and listed in Table 2.

Table 2. Derived contact lengths

Parameter	Value [mm]
A_0	236.62
L_{ac}	68.69
L_{al}	68.69
L_c	137.39

Table 3. Calculated angular parameters

Parameter	Value [deg/rad]
γ_{12}	78.72/1.37
γ_{22}	78.72/1.37
Total Contact Angle	157.44/2.75
$\varphi_{2,min}$	44.15/0.77
$\varphi_{2,max}$	201.59/3.52

Lastly, Table 3 shows the calculated values for the approach, recess, and aperture angles:

It is noted that at an aperture of $\varphi_2 = 122.9°$, the contact point coincides with the pitch point.

From a sliding speed magnitude standpoint, there is no strict design criterion that limits the parameters. However, it is considered that a higher sliding speed will lead to greater amplitudes in AE bursts due to an increased rate of asperity contact over time. In addition, higher speeds are sought to enhance the sliding effect, making it easier to differentiate rolling and sliding phenomena in the recorded signals.

By substituting the corresponding values into the kinematic expressions developed in Sect. 2.2, Fig. 3 is obtained for different aperture angles. It highlights the change in direction of the sliding speed after passing the pitch point. This sudden transition from sliding to rolling and back to sliding could be the cause of the burst-like patterns observed in the AE signals, suggesting that the design meets the desired kinematic conditions.

Furthermore, the comparison between sliding speed and the sliding-to-rolling ratio reveals that at the ends of the contact path, the sliding effect increases drastically, while near the pitch point, the sliding effect is very small and increases at a much lower rate.

4 Test Bench Implementation

4.1 Fabrication

The gears, shafts, and masses were machined from AISI 1020 steel. The gear teeth were milled using a CNC machine, while the shafts were rough turned and finished on a lathe, where the necessary tolerances were applied for press fitting into the pulley and a close running fit with the gear. Additionally, the tooth support platform was machined from ultra high molecular weight polyethylene (UHMWPE) using the same CNC machine.

The manufacturing processes are illustrated in Fig. 4.

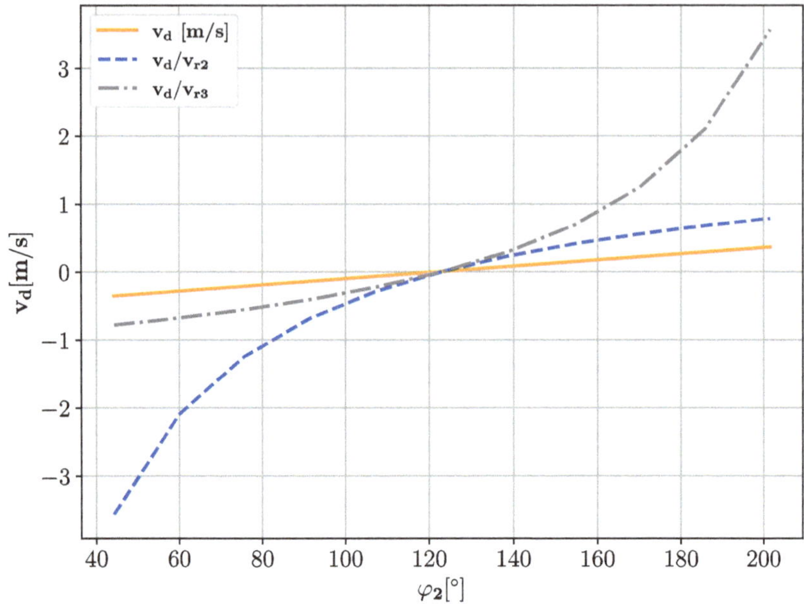

Fig. 3. Sliding speed and sliding/rolling ratio along contact

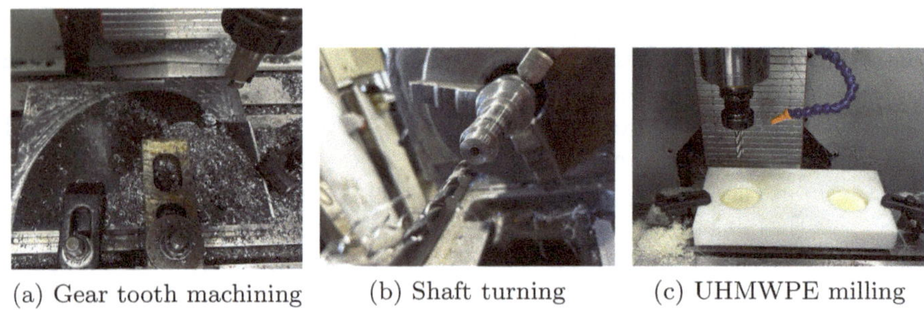

(a) Gear tooth machining (b) Shaft turning (c) UHMWPE milling

Fig. 4. Manufacturing of the test bench components: (a) teeth, (b) shafts, and (c) UHMWPE support platform

With the components manufactured, the test bench is ready for assembly. Figure 5 shows the assembled test bench mounted on the aluminum frame, incorporating a pulley system based on the Atwood machine concept, which enables the generation of a driving/resisting torque difference by means of weight imbalance.

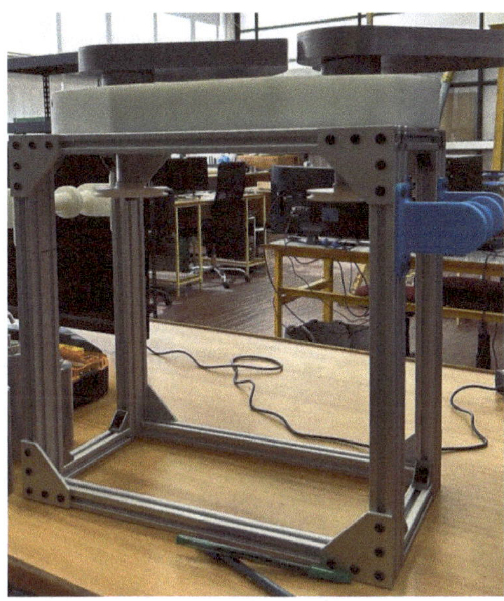

Fig. 5. Assembled test bench

4.2 Instrumentation

Data acquisition is carried out using the measurement chain schematized in Fig. 6. Two piezoelectric sensors (model VS375-M) are used; one mounted on each tooth. These sensors have a resonance frequency of 375 kHz and a frequency response range from 250 to 700 kHz. The captured signal is first amplified by a preamplifier with an adjustable gain ranging from 34 to 49 dB. This device includes a built-in high-pass filter at 95 kHz and a low-pass filter at 1000 kHz, providing an initial filtering stage to attenuate noise and mitigate aliasing effects during signal amplification. Subsequently, the signal passes through a DCPL2 decoupling box, which electrically isolates the circuits to prevent undesired signals or electrical disturbances from propagating between them, thereby reducing interference and noise. Finally, the NI-9223 data acquisition module is used to interface the measurement system with the computer. This module performs the analog-to-digital conversion, enabling the captured signals to be stored and processed digitally.

5 Preliminary Measurements and Discussion

For the preliminary measurements, the two piezoelectric sensors were mounted on the upper surface of each tooth, positioned sufficiently close to the engagement point in order to minimize potential signal attenuation that may occur if placed farther away.

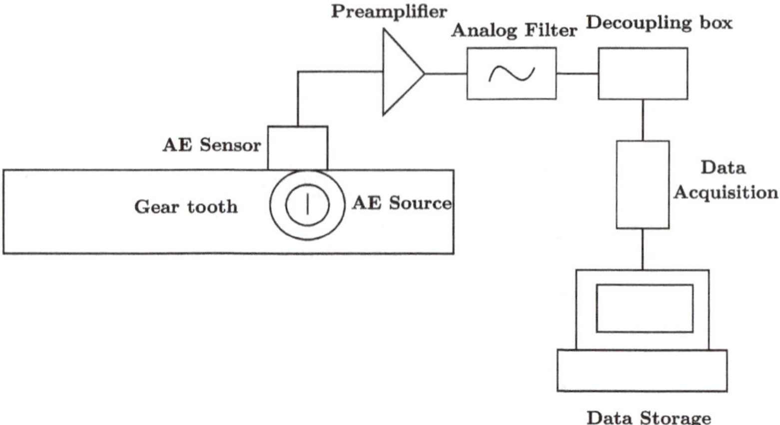

Fig. 6. Measurement chain overview

Fig. 7. Used weights for the AE measurements

Once the measurement chain was configured, the weights shown in Fig. 7 were used. These weights were selected based on preliminary tests, aiming to generate adequate driving and resisting torques to enable smooth and controlled motion.

Table 4 summarizes the torque configurations that yielded the best performance, assuming a counterclockwise rotation of the driving tooth and a clockwise rotation of the driven tooth. The left and right sides of each tooth are treated as separate reference systems, defined from the perspective of an observer located behind the corresponding tooth.

Table 4. Mass values and equivalent torques used for each measurement

Measurement	Measurement 1		Measurement 2	
	Driving	Driven	Driving	Driven
Right-side mass [g]	657	649	657	649
Left-side mass [g]	3138	1047	4125	1047
Equivalent torque [N.mm]	910.6	146.07	1272.8	146.07

5.1 Results

With the test bench fully assembled, AE measurements were conducted, yielding varied results. The teeth were in healthy condition; however, as previously mentioned, machining marks were present due to the gear profile milling process. Figures 8 and 9 show two representative samples of the measurements obtained from the test bench. These allow for the observation of several expected phenomena, such as an increase in continuous AE activity with increasing sliding speed.

On the other hand, the phenomenon of burst emission at the pitch point is not as consistent, exhibiting variability in both amplitude and time of occurrence.

Figure 8 presents one of the best measured cases, with a clear transient observed at the engagement point. In contrast, other measurements, such as that shown in Fig. 9, exhibit a much weaker transient. After the pitch point, due to the acceleration caused by the applied masses, the system becomes unstable, making the signal difficult to interpret. However, from the beginning of contact up to the engagement point, a repeatable pattern is observed, confirming the pulley system's effectiveness for motion control.

In addition, small signal bursts can be observed throughout the measurement, which may be attributed to minor impacts occurring along the contact path. An AE burst is observed at the start of the contact mesh, which is attributed to an impact generated as the gear teeth first engage.

6 Discussion

During the development of the test bench, several issues arose that may have contributed to the inconsistencies observed in the measurements. Marks from the milling tool were visible on the fabricated gear teeth, which is problematic as it

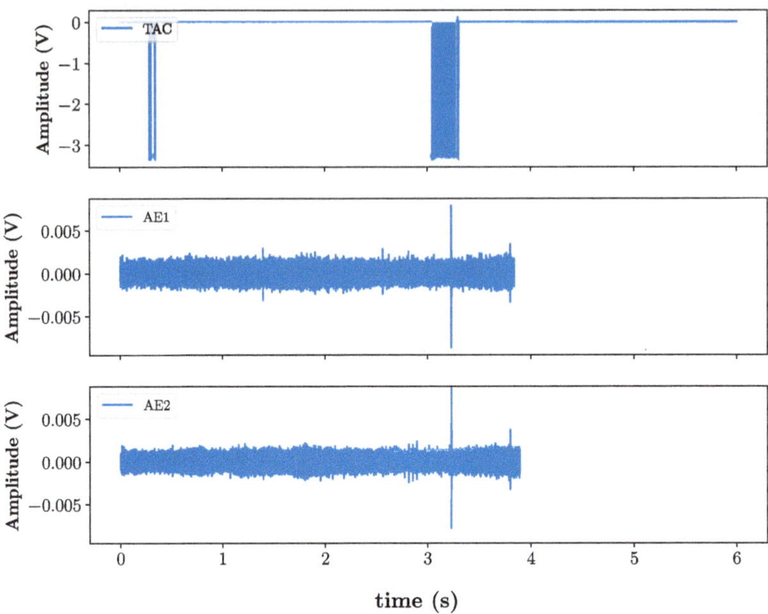

Fig. 8. Sample AE signal with clear burst at pitch point

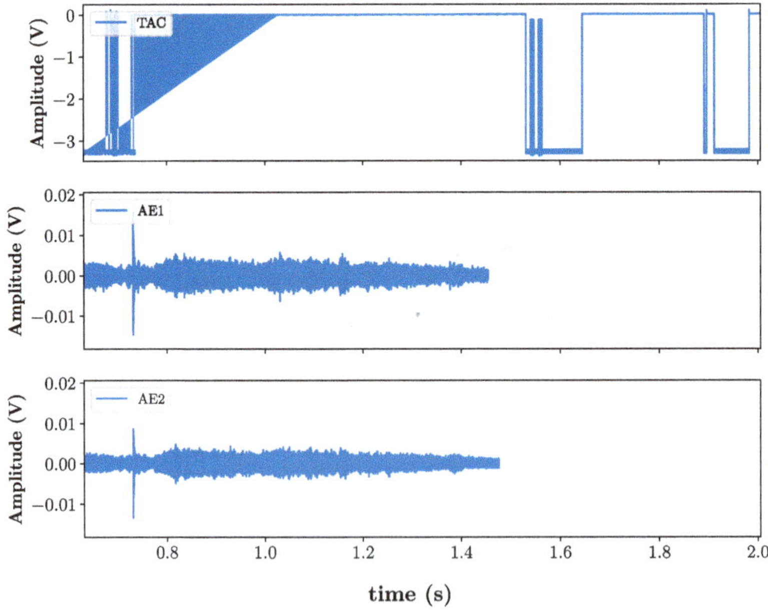

Fig. 9. Sample AE signal with weak burst visibility

disrupts the smoothness of the involute curve. This can lead to impacts or even loss of contact near the engagement point during meshing. Additionally, slight angular misalignment was observed, resulting in the driving tooth being slightly elevated at the beginning of contact and lower at the end. This misalignment can cause incomplete contact and axial sliding, both of which affect the measured signals.

To improve the performance of the test bench and obtain more consistent measurements, the following measures are proposed:

- Improve the smoothness of the involute curve, ensuring that contact is maintained across the entire surface during motion, thereby avoiding potential impacts and loss of contact.
- Enhance motion control to prevent initial and final impacts, aiming to obtain a more symmetrical signal suitable for post-processing and analysis. The normal force and the sliding speed should be independently adjustable in order to isolate their respective effects on AE generation.
- Improve alignment of the teeth to avoid incomplete contact and angular misalignment, which may result in unwanted axial sliding and distortion of the measured signals.
- Fabricate a new gear tooth with a modified involute profile such that no engagement point occurs during contact. This eliminates the occurrence of pure rolling, ensuring that the observed transient is due solely to the transition from sliding to rolling, and not caused by other factors.

The impacts occurring at the start of contact may be responsible for the transients observed in the AE signals of gear systems, especially considering that such transients also would appear at the gear mesh period. This observation underscores the need for further research aimed at distinguishing transients caused by mechanical impacts from those resulting from changes in contact conditions, such as the transition from sliding to rolling.

7 Conclusions

This study investigated the hypothesis regarding the origin of AE bursts in gears with involute profiles. Through the design and testing of an experimental setup, patterns of AE were identified that correspond to sliding contact conditions. However, inconsistencies in the amplitude and timing of the bursts suggest that further refinements in the experimental design and a deeper understanding of the influencing variables are required. Additionally, tooth impacts merit further investigation, as they might represent a potential source of the AE bursts observed during healthy gear meshing.

By implementing improvements such as enhanced surface finishing of the gear teeth, better alignment, more controlled motion, and comparisons with gear teeth featuring modified involute profiles, more consistent and conclusive results are expected.

The test bench developed in this work provides a foundation for generating representative acoustic emissions from involute gear contacts, which is valuable for future investigations. This represents a significant contribution to the development of more accurate theoretical models and more effective condition monitoring methodologies in gear systems.

References

1. Hou, D., et al.: High-speed train wheel set bearing fault diagnosis and prognostics: research on acoustic emission detection mechanism. Mech. Syst. Signal Process. **179**, 109325 (2022)
2. Nirwan, N., Ramani, H.: Condition monitoring and fault detection in roller bearing used in rolling mill by acoustic emission and vibration analysis. Mater. Today: Proc. **51**, 06 (2021)
3. Yu, L., Li, S.Z.: Acoustic emission (ae) based small leak detection of galvanized steel pipe due to loosening of screw thread connection. Appl. Acoust. **120**, 85–89 (2017)
4. Mohammed, O.D., Rantatalo, M.: Gear fault models and dynamics-based modelling for gear fault detection – a review. Eng. Fail. Anal. **117**, 104798 (2020)
5. Vicuña, C., Burgwinkel, P., Vaehsen, K., Wellhausen, J.: Analysis of the acoustic emissions generated during the meshing process of planetary gearboxes. In: 6th International Conference on Condition Monitoring and Machinery Failure Prevention Technologies 2009, vol. 2, pp. 1252–1263 (2009)
6. Keong, C., Mba, D.: Identification of the acoustic emission source during a comparative study on diagnosis of a spur gearbox. Tribol. Int. **38**, 469–480 (2005)
7. Leaman, F., Vicuña, C.M., Clausen, E.: A review of gear fault diagnosis of planetary gearboxes using acoustic emissions. Acoust. Aust. **49**(2), 265–272 (2021)

Biomechanical Evaluation of Custom Cranial Implants in PMMA and PEEK Using Finite Element Analysis: Bone Coupling and Fixation System

Freddy Patricio Moncayo-Matute[1,2](✉), Paúl Bolívar Torres-Jara[1,2], Diana Denisse Bohorquez-Vivas[2], Diana Patricia Moya-Loaiza[1,2], and Efrén Vázquez-Silva[1,2]

[1] Universidad Politécnica Salesiana (UPS), 010105 Cuenca, Azuay, Ecuador
[2] Grupo de Investigación en Nuevos Materiales y Procesos de Transformación (GIMAT), UPS, 010105 Cuenca, Azuay, Ecuador
fmoncayo@ups.edu.ec

Abstract. This study investigates the effect of the von Mises stress distribution caused by an external load of 50 N on a cranial implant and its fixation system for anchorage in the bone of a specific patient. Two simulations of the behaviour of the central zone of the implant, manufactured in PMMA and PEEK, were performed. Anatomical models were obtained from computed tomography scans, and the implant design was derived through reverse engineering. The anchorage systems were modelled in detail. The maximum stresses obtained for the external load state were 4.76 MPa for the PMMA-based device and 3.18 MPa for the PEEK-based device. Cranial reconstruction with the customised implant is safe since the stresses did not exceed the yield strength.

Keywords: Polymethylmethacrylate (PMMA) · Polyether-ether-ketone (PEEK) · finite element method · additive manufacturing · surgical planning

1 Introduction

Reconstruction of cranial defects caused by trauma or brain tumours is a well-established surgical procedure [1,2]. Autologous bone grafts are effective for small, simple-contour defects; but they are challenging for larger, more complex defects [3].

The objective of cranial reconstruction is to restore its protective function, improve the neurological system, mainly motor function, optimise cerebral blood flow, prevent disorders in the dynamics of cerebrospinal fluid and repair the patient's aesthetics [4].

Various materials have been used for implants and bone reconstruction, such as medical grade titanium alloys, Polyether-ether-ketone (PEEK), Polymethylmethacrylate (PMMA), composites and hybrid materials [5].

O. F. Farías Fuentes et al. (Eds.): CIBIM 2024, *Proceedings of the XVI Ibero-American Congress of Mechanical Engineering*, pp. 27–40, 2026.
https://doi.org/10.1007/978-3-032-22823-9_3

Current craniomaxillofacial implants are designed from digital images of the patient in Digital Imaging and Communications in Medicine (DICOM) format, commonly obtained from computed tomography (CT), cone beam computed tomography (CBCT), or magnetic resonance imaging (MR), using computational design assistance. In addition, they are manufactured using 3D additive manufacturing techniques such as Fused Deposition Modelling (FDM), Stereolithography (SLA), or Direct Metal Laser Sintering (DMLS) [6–8].

One of the most widely used materials for the reconstruction of cranial defects, since approximately 2000, has been PMMA. This material is a polymerised ester of acrylic acid, presented in powder form with benzoyl peroxide, and mixed with a liquid monomer. During the exothermic reaction that occurs during mixing, it slowly cools, converting into a translucent material with a strength similar to that of human bone. Methyl methacrylate, during the cooling phase, can be moulded to fit any complex cranial defect. These implants are biocompatible, chemically inert, nonconductive, radiolucent, and inexpensive. With the use of current technologies, such as tomography and 3D printing, it is possible to create customised PMMA implants, which significantly reduces surgical times and morbidity rate [9].

On the other hand, PEEK has emerged as a high-performance alternative in cranioplasty procedures due to its biocompatibility, thermal stability, and excellent mechanical properties, which make it comparable to cortical bone. Its radiolucent nature facilitates postoperative monitoring using imaging techniques, and its surface can be modified to improve integration with bone tissue. Furthermore, in finite element simulation studies, PEEK has shown uniform stress distribution, reducing the risk of concentration points that could compromise implant integrity. This makes it a promising material for clinical applications requiring anatomical precision, mechanical strength and long-term durability [3,10].

Advances in open source medical software for hard and soft tissue segmentation have facilitated the automatic or semi-automatic extraction of anatomical structures from computerised medical images [10–12]. Once the 3D model is obtained in STL format, it is possible to create the model of the missing bone using editing tools available in the software, applying the mirror effect for symmetrical anatomical damage, and reverse engineering to obtain a Computer Aided Design (CAD) model. The craniofacial region presents special challenges for tissue engineering due to the overall mechanical stresses and deformations it experiences under external loading conditions, which have been poorly studied. The Finite Element Method (FEM) is a useful tool to virtually evaluate the mechanical conditions of the interaction between the implant and the cranial bone [13,14].

Several finite element analysis software programs provide a design and analysis methodology, with tools for reverse engineering and repairing models exported from other modelling codes. Analyzing cranial implants coupled to the patient's bone interface, and the selected anchoring system, allows a better understanding of the mechanical system, obtaining von Mises stresses and directional deformations under an impact load to verify if there is damage to the bone in contact with the implant [15]. Some studies have reported on finite element analysis only on the cranial implant, considering unrealistic boundary conditions, and raising questions about how the von Mises stress distribution is generated when an external and internal load acts on the implant, the microplate anchoring system, screws and the cranial bone [16].

How is the von Mises stress distribution generated on the anchoring system of microplates, screws and bone, when a point and inclined external load acts on the implant?

This article reports on the design of a customized cranial implant for a patient with damage to the frontal cranial bone, and also on the analysis of the bone-implant interface and the fixation system with microplates and microscrews, to obtain the von Mises stresses, applying FEM, with point load conditions and external inclined of 50 N in the central area of the implant.

1.1 Methodology

The analysis of the custom cranial implant was based on a computed tomography scan and a finite element study to determine stress levels. This approach was carried out quantitatively. The applied research methodology includes several steps. First, a literature review was conducted to identify the most commonly used materials in cranial reconstruction and current finite element design and analysis technologies. A CT scan of the patient's affected area was then obtained. Using the CT data and computer modelling software, a 3D model of the skull and implant was generated. Mechanical properties of materials, such as density, modulus of elasticity, Poisson's ratio, and yield strength, were established for the materials used in the implant, screws, plates, and cranial bone. A controlled structural mesh of finite elements was subsequently generated in the 3D models.

Point and inclined external loads were applied to the implant model to perform the analysis, considering PMMA and PEEK polymers as raw materials for the medical device. A linear numerical simulation was performed using finite element analysis software to obtain the von Mises stress levels for the implant, screws, plates, and bone. The numerical simulation results were interpreted and evaluated, comparing the stress levels with reference values to ensure that the implant meets the necessary strength and stability requirements.

1.2 Clinical Case

A 19-year-old male patient presented with a wound caused by a steel-shot blast in the left fronto-orbital region. CT images revealed damage to the roof of the superior orbital rim and the frontal bone, as well as foreign bodies and bone fragments embedded in the frontal lobe. Following medical evaluation, a craniectomy and surgical debridement were performed. Vision in his left eye was severely compromised. Three and a half months after surgery, the trauma was assessed without primary reconstruction, and no medical or psychological complications were observed that would prevent or limit restorative treatment. Figure 1 shows, in a tomographic image, the damage suffered by the young man, highlighting that the greatest damage occurred in the upper edge of the orbit and the left frontal lobe.

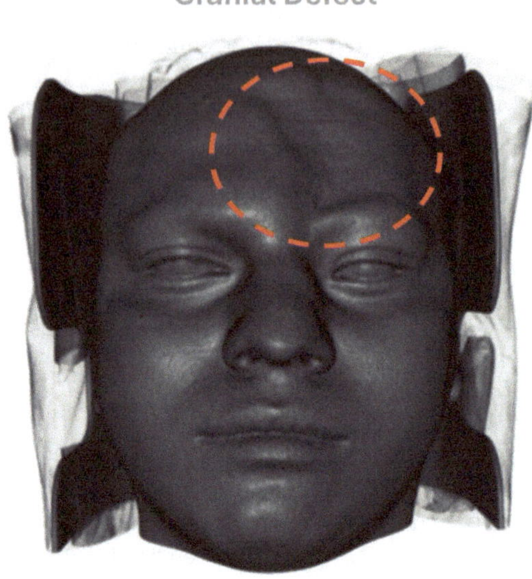

Fig. 1. CT scan performed shortly after surgery. Sunken tissue is observed in the upper left part of the face.

2 Data Acquisition

A computed tomography scan of the patient's affected area was performed, in DICOM format, high resolution, with voxels of $512 \times 512 \times Z$, Z varies from 25 to 670. The images were processed using the open-source software 3D Slicer to generate an STL file representing the anatomy of the study area.

Computed tomography image segmentation was performed by selecting specific intensities in Hounsfield Units (HU), which measure the grayscale attenuation coefficient for the different tissue types (bone, skin, muscle) of the anatomical region of interest. This segmentation was performed with the aid of a thresholding algorithm to delimit the anatomical area of interest. For this study, values ranging from 188 to 2000 HU were used to obtain a compact bone model.

Figure 2 shows the segmentation performed with the 3D Slicer software, showing the process and results of the anatomical modelling.

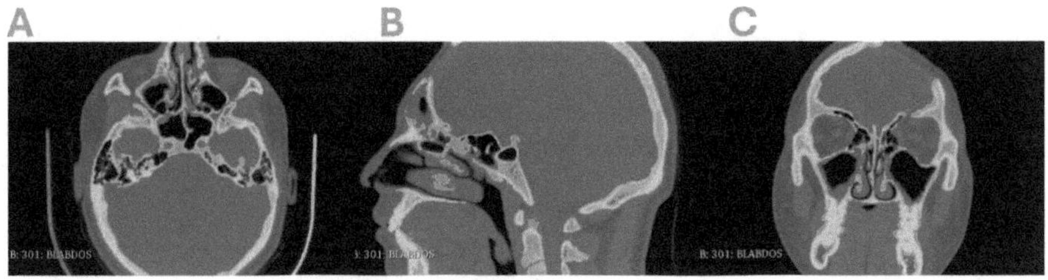

Fig. 2. Computed Tomography: Axial Section (A), Sagittal Section (B), Coronal Section (C).

2.1 Reverse Engineering Cranial Reconstruction

Reverse engineering is a process that allows creating 3D models of an existing physical object by capturing and processing data obtained from laser scanners, CT scans or medical images [17,18]. This method is applied to anatomical models of patients to obtain the most accurate representation of their anatomy, and can then be used for finite element analysis (FEM). To obtain the 3D geometry of the patient's cranial model, the STL file was processed using mesh and point cloud processing, applying reverse engineering tools provided by Autodesk Meshmixer and Ansys Workbench software. These programs allow the CAD model to be correlated with the stereolithographic reconstruction.

The post-processed model was used to reconstruct the missing bone area using Autodesk Meshmixer. A symmetrical reference plane was created in the sagittal plane of the cranial model, assuming the anatomical symmetry of the human body [19,20]. Using the software's editing tools, the healthy side of the structure was inverted, creating a mirror image that was superimposed on the area to be replaced (missing bone). Both parts of the structure were assembled to completely "fill" the affected cavity [21,22]. The Boolean subtraction tool was then applied to obtain the initial design of the customised implant, and refinements were made to the implant contours. The reconstructed model can be seen in Fig. 3.

The cranial anatomical model and the reverse-engineered implant show complex surfaces that have been processed and adapted to the defect, allowing modifications to the CAD models. The anchoring systems used during the surgical intervention, the titanium alloy microplates and screws [23], were also identified. The meshing process was then carried out. Figure 4 shows the fixing elements model prepared for the FEM analysis.

3 Coupling Details

The contours of the cranial defect interface profile were detailed for the adaptation of the customised implant, studying the peripheral area of damage and stress transfer under external loading conditions. According to the authors Pramana et al. [24], it is crucial to analyse the interface contacts between the bone

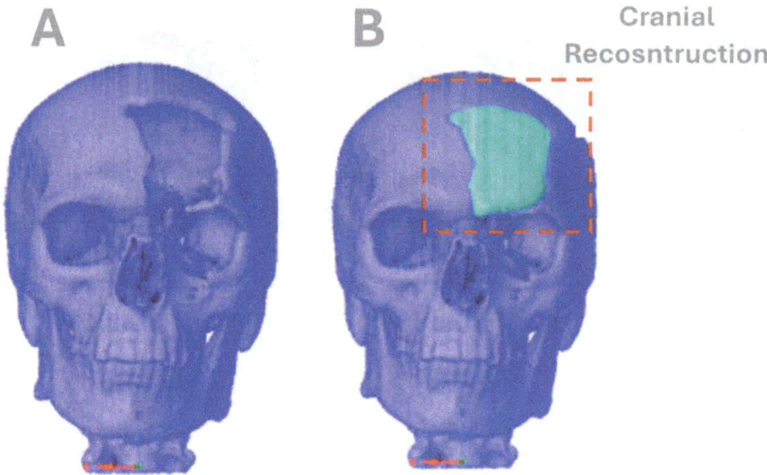

Fig. 3. Stereolithographic model of the skull with the defect (A). Cranial model with the implant adapted to the profile of the affected area (B).

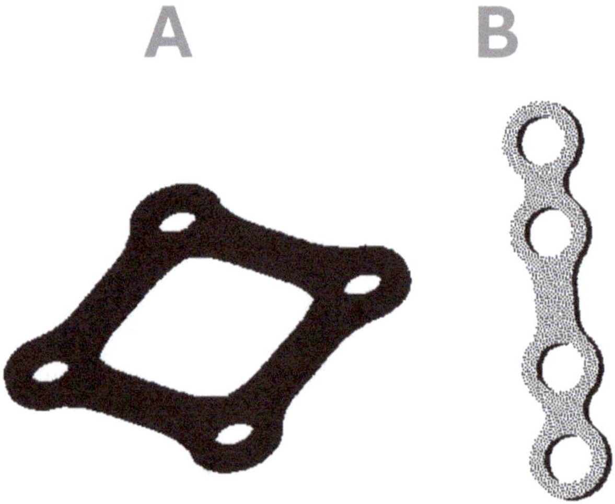

Fig. 4. FEM computational model: Anchoring system: 4-hole square plate, 12 mm (A). 4-hole medium straight plate, 17 mm (B).

and the implant. These authors conclude that the edges of the defect should have a surgical preparation with positive angles. Otherwise, if the implant does not have adequate support in the cranial bone, when an external load is applied, the stresses are transferred to the fixation elements. This can compromise the integrity of the implant and cause non-physiological stresses in the adjacent bone, affecting the bone morphology necessary for anchorage. Therefore, in the simulations, the trauma edge was analysed with a fixation contact at positive angles to ensure correct implant adaptation (see Fig. 5).

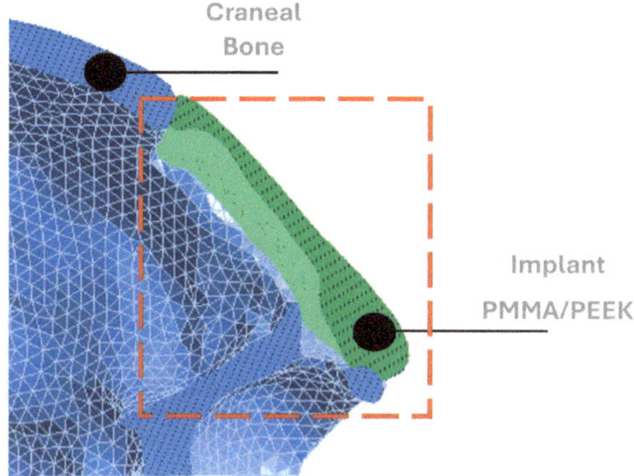

Fig. 5. Model setup to simulate the clinical scenario, showing the defect contour and implant fit. Sagittal section of the skull with the implant in the defect (red segmented lines). The shape and angle variation of the interface along the defect contour are also detailed.

3.1 Mechanical Properties

The materials used in the computational model were assumed to be isotropic, homogeneous and linearly elastic, as suggested by Ameen et al. [22]. The anchoring system, composed of screws and microplates, is made of medical-grade titanium ($Ti6Al4V$), while the custom implant was manufactured with PMMA. The mechanical properties of each material used in the simulation are presented in Table 1.

Table 1. Mechanical properties of the materials used

Property	Bone	Ti6AlV	PMMA	PEEK
Young's Modulus [MPa]	15000	110000	3000	4200
Poisson's ratio	0.3	0.3	0.38	0.39
Ultimate Stress [MPa]	130	950	72	120
Reference	[12]	[14]	[16]	[18]

For this study, a commercial surgical screw made of titanium alloy was selected. Its typical dimensions are: 37.5 mm total length, $\frac{1}{3}$ of which extends from the conical tip, and $\frac{2}{3}$ of which extends from the tip to the cylindrical head.

Once the mechanical properties of the components were defined, FEM simulation runs were performed to study the anatomical model's responses to different types of loads. These simulations are essential for understanding the behaviour of the human body in various situations, such as during physical activity or in cases of trauma. This process will be discussed in detail in the section on boundary conditions.

3.2 Boundary Conditions

Finite element analysis was performed using the static structural modulus. A loading condition was established on the external surface of the implant. An external load of 50 N was applied, according to Motherway et al. [25], which simulates the mass of the head when the person lies down. The cranial model was simplified in the axial plane to reduce meshing and simulation times. A fixed support condition was assigned at the base of the axial section of the skull, with zero displacements and rotations, as shown in Fig. 6.

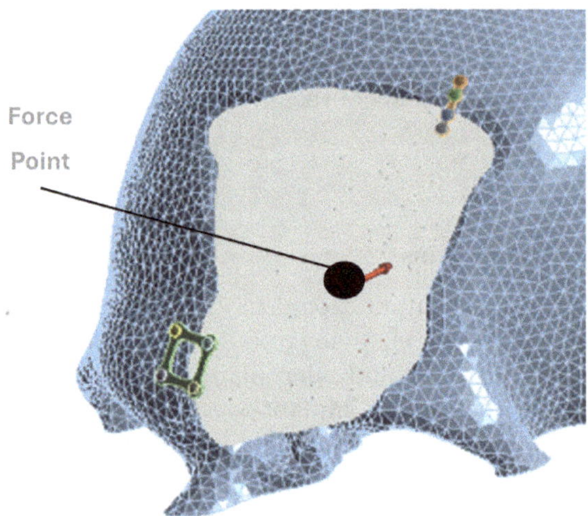

Fig. 6. Loading condition: Force 1, placed at the top left of the front view of the implant.

A finite element model of the 3D human cranial structure was used, using ANSYS WORKBENCH R21.1 software (ANSYS Inc., Canonsburg, Pennsylvania, USA). The selected mesh consists of tetrahedral elements (SOLID185), and refinement tests were performed to ensure a 5% convergence. The model comprised 251,520 elements of 0.5 mm, and 444,343 nodes (see Fig. 7). For the simulation, a bonded contact was applied to the boundary conditions of the skull, specifically at the interfaces between the screws and the cortical bone. According to Ameen et al. [22], this condition is considered the most realistic. The fixed support restriction was assigned to the base of the lower external cranial structure, restricting displacements and rotations.

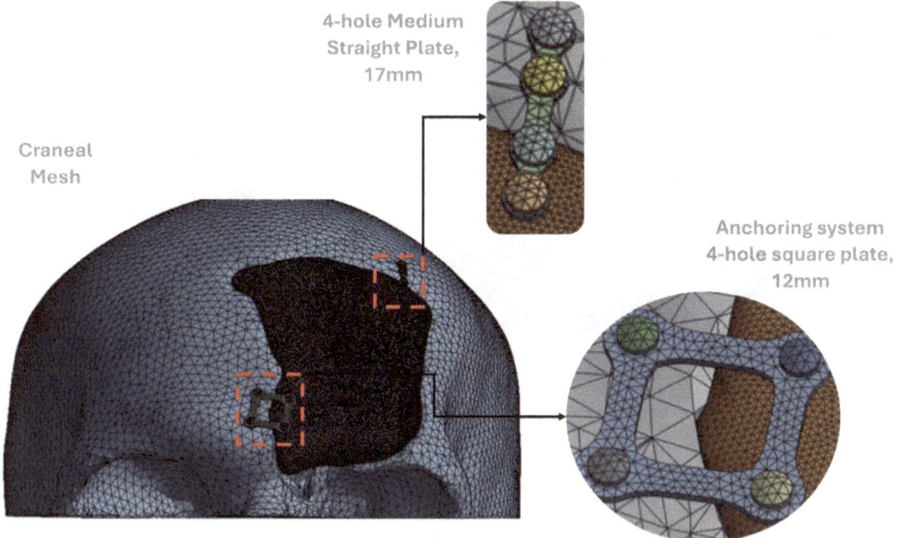

Fig. 7. Meshing of the implant with the fixation systems towards the skull.

4 Results

Computer numerical simulation was performed using PMMA and PEEK for the implant, and the von Mises stress distribution in the cranial bone and microplates for anchorage was obtained. For the external load state (Force 1), a maximum stress of 4.76 MPa was obtained in the PMMA implant, in the square plate for the lower anchorage, a maximum stress of 0.47 MPa, and in the upper middle plate a maximum stress of 2.84 MPa. The average stress distribution at the cranial trauma interface yielded a result of 1.08 MPa. Figure 8 shows the von Mises stress distribution for each element, according to the external loading state.

For the external loading condition (Force 1), a maximum stress of 3.18 MPa was obtained for the PEEK implant, a maximum stress of 1.21 MPa for the square plate for the lower anchorage, and a stress of 0.91 MPa for the upper middle plate. The average distribution at the head trauma interface yielded a stress of 0.88 MPa (see Fig. 9).

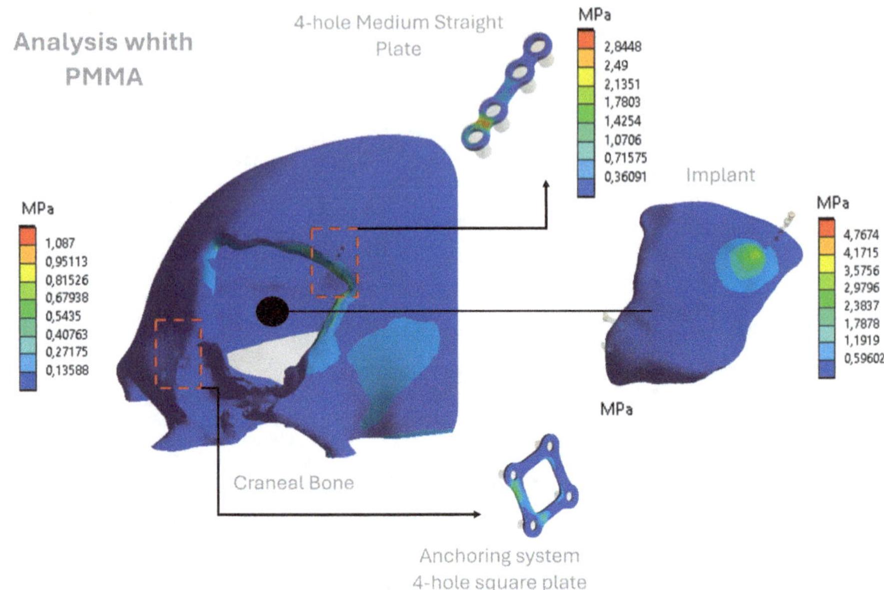

Fig. 8. Computational analysis of the PMMA implant-skull interface and fixation plates.

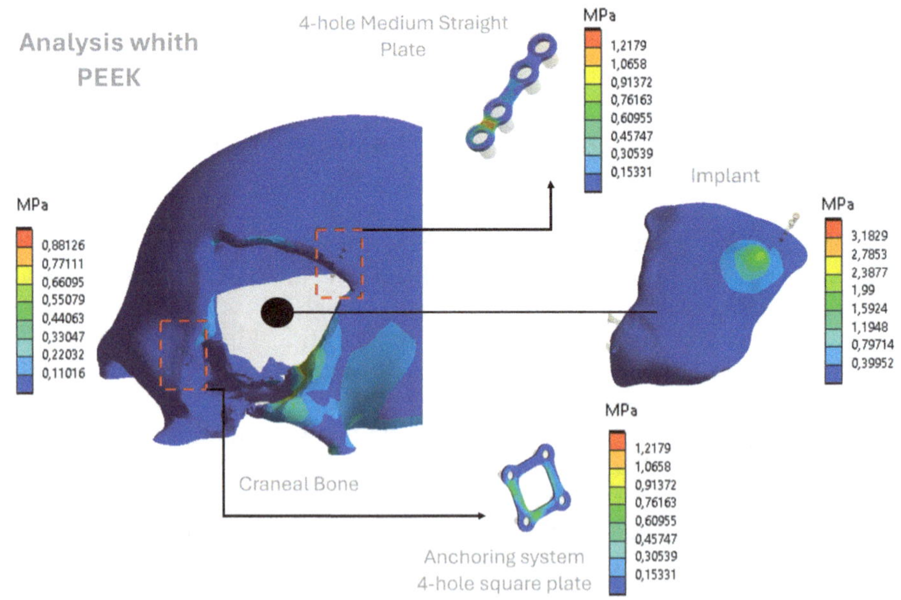

Fig. 9. Computational analysis of the PEEK-skull implant interface and fixation plates.

5 Conclusions

The results achieved in this study demonstrated that the methodology applied for the design, simulation, and evaluation of a customised cranial implant based on PMMA and PEEK constitutes a comprehensive, safe, and replicable approach in clinical and biomedical engineering settings. Using a high-resolution computed tomography scan in DICOM format, precise segmentation was performed using the open-source software 3D Slicer, enabling a three-dimensional reconstruction of the patient's cranial anatomy, including the bone defect caused by trauma.

Using reverse engineering and CAD editing tools (Autodesk Meshmixer and Ansys Workbench), an implant model adapted to the defect geometry was developed, respecting anatomical symmetry and considering the principles of structural biomechanics at the bone-implant interface. The structural meshing process using finite elements and the assignment of mechanical properties such as isotropy, homogeneity and linear elasticity, allowed a reliable numerical simulation with a biomechanical load of 50 N, equivalent to the head weight in the supine position.

The results showed that both the PMMA and PEEK implants resisted the von Mises stresses generated in the cranial environment, without exceeding their respective yield strengths or compromising the stability of the bone and the fixation system (titanium microplates and microscrews). Specifically, a maximum stress of 4.76 MPa was reached for PMMA, and a maximum stress of 3.18 MPa for PEEK. Both values are consistent with the implant's structural safety under physiological loading conditions. Biomechanical behaviour was also favourable in the bone-implant contact areas, reducing the risk of nonphysiological stresses or mechanical failure.

The study shows that the use of accessible technologies, such as 3D printing and CAD-CAM modeling, combined with advanced computational analysis (FEM), enables the development of effective, safe, and potentially low-cost personalized implants, contributing to functional, aesthetic, and psychological improvement for the patient.

Additive manufacturing has radically transformed the design of biomedical implants by enabling the creation of highly customised devices that precisely fit the patient's anatomy. In the context of cranioplasty, technologies such as 3D printing based on materials like PMMA and PEEK have proven to be efficient tools for reducing surgical times, improving aesthetic results, and reducing post-operative complications. The ability to create three-dimensional models from computed tomography (DICOM) scans and convert them into physical objects using additive manufacturing techniques has eliminated many of the limitations associated with standard prefabricated implants, providing safer and more effective solutions for bone reconstruction.

As future work, we propose validating the results obtained in the simulation through experimental tests, which will allow us to analyse the results of applying dynamic and cyclic loads to implants manufactured in PMMA and PEEK, to corroborate their structural behaviour under conditions closer to reality. Furthermore, it would be important to incorporate multi-scale biomechanical models that allow for the analysis of the biological response of the surrounding bone tissue, including bone integration and possible long-term adverse reactions. It is also suggested to extend the finite element analysis to multiaxial and nonlinear loading scenarios, considering different head movements and patient postures.

Another possible line of research would be to compare different anchorage configurations, evaluating the number and geometric arrangement of plates and screws based on stress distribution and implant stability. Furthermore, applying this methodology to more complex craniofacial defects could allow the development of implants with porous internal structures that promote osseointegration. Finally, it would be desirable to design longitudinal clinical studies that evaluate the efficacy, safety, aesthetics, and quality of life of patients undergoing this type of treatment, thus consolidating the applicability of this technology in the neurosurgical field.

However, recent advances in non-planar additive manufacturing (NPAM) further expand the horizon of possibilities. This technology enables material deposition on curved and complex surfaces, which is especially useful in irregular anatomical regions such as the skull. By eliminating planar layering constraints, the structural integrity of the implant is improved, layer bonding is optimised, and manufacturing defects are reduced. These options, combined with trajectory planning algorithms and curvature-adaptive non-solid fill techniques, offer a new paradigm for the design of lighter and stronger implants. Thus, additive manufacturing not only complements but enhances the principles of personalised medical engineering, bringing the technology closer to a truly patient-centred approach [26].

References

1. Klammert, U., Gbureck, U., Vorndran, E., Rödiger, J., Meyer-Marcotty, P., Kübler, A.C.: 3D powder printed calcium phosphate implants for reconstruction of cranial and maxillofacial defects. J. Cranio-Maxillofac. Surg. **38**(8), 565–570 (2010). https://doi.org/10.1016/j.jcms.2010.01.009
2. Lethaus, B., et al.: Cranioplasty with customized titanium and PEEK implants in a mechanical stress model. J. Neurotrauma **29**(6), 1077–1083 (2012). https://doi.org/10.1089/neu.2011.1794
3. El Halabi, F., Rodriguez, J.F., Rebolledo, L., Hurtós, E., Doblaré, M.: Mechanical characterization and numerical simulation of polyether-ether-ketone (PEEK) cranial implants. J. Mech. Behav. Biomed. Mater. **4**(8), 1819–1832 (2011). https://doi.org/10.1016/j.jmbbm.2011.05.039
4. Piazza, M., Grady, M.S.: Cranioplasty. Neurosur. Clin. **28**(2), 257–265 (2017). https://doi.org/10.1016/j.nec.2016.11.008
5. Vázquez-Silva, E., et al.: Composites and hybrid materials used for implants and bone reconstruction: a state of the art. Contemp. Eng. Sci. **15**, 105–135 (2022). https://doi.org/10.12988/ces.2022.91974
6. Tsouknidas, A., Michailidis, N., Savvakis, S., Anagnostidis, K., Bouzakis, K.D., Kapetanos, G.: A finite element model technique to determine the mechanical response of a lumbar spine segment under complex loads. J. Appl. Biomech. **28**(4), 448–56 (2012). https://doi.org/10.1123/jab.28.4.448

7. Moncayo-Matute, F.P., et al.: Description and application of a comprehensive methodology for custom implant design and surgical planning. Interdisc. Neurosurg. **29**, 101585 (2022). https://doi.org/10.1016/j.inat.2022.101585
8. Khader, B.A., Towler, M.R.: Materials and techniques used in cranioplasty fixation: a review. Mater. Sci. Eng., C **66**, 315–322 (2016). https://doi.org/10.1016/j.msec.2016.04.101
9. Unterhofer, C., Wipplinger, C., Verius, M., Recheis, W., Thomé, C., Ortler, M.: Reconstruction of large cranial defects with poly-methyl-methacrylate (PMMA) using a rapid prototyping model and a new technique for intraoperative implant modeling. Neurol. Neurochir. Pol. **51**(3), 214–220 (2017). https://doi.org/10.1016/j.pjnns.2017.02.007
10. Honigmann, P., Sharma, N., Okolo, B., Popp, U., Msallem, B., Thieringer, F.M.: Patient-specific surgical implants made of 3D printed PEEK: material, technology, and scope of surgical application. Biomed. Res. Int. **2018**(1), 4520636 (2018). https://doi.org/10.1155/2018/4520636
11. Bücking, T.M., Hill, E.R., Robertson, J.L., Maneas, E., Plumb, A.A., Nikitichev, D.I.: From medical imaging data to 3D printed anatomical models. PLoS ONE **12**(5), e0178540 (2017). https://doi.org/10.1371/journal.pone.0178540
12. Rubio-Pérez, I., Diaz Lantada, A.: Surgical planning of sacral nerve stimulation procedure in presence of sacral anomalies by using personalized polymeric prototypes obtained with additive manufacturing techniques. Polymers **12**(3), 581 (2020). https://doi.org/10.3390/polym12030581
13. Leal-Naranjo, J.A., Ceccarelli, M., Torres-San-Miguel, C.R., Aguilar-Perez, L.A., Urriolagoitia-Sosa, G., Urriolagoitia-Calderón, G.: Multi-objective optimization of a parallel manipulator for the design of a prosthetic arm using genetic algorithms. Lat. Am. J. Solids Struct. **15**(3), e26 (2018). https://doi.org/10.1590/1679-78254044
14. Torres-San-Miguel, C.R., Hernández-Gómez, J.J., Urriolagoitia-Sosa, G., Romero-Ángeles, B., Martínez-Sáez, L.: Design and manufacture of a customised temporomandibular prosthesis. Revista internacional de métodos numéricos para cálculo y diseño en ingeniería, **35**(1) (2019). https://doi.org/10.23967/j.rimni.2019.02.001
15. Cuc, N.T.K., Hung, P.D., Duc, B.M., Anh, N.H.: Mechanical evaluation of the large cranial implant using finite elements method. In: IFToMM Asian Conference on Mechanism and Machine Science, pp. 649–658. Springer, Cham (2021). https://doi.org/10.1007/978-3-030-91892-7_62
16. Fedorov, A., et al.: 3D slicer as an image computing platform for the quantitative imaging network. Magn. Reson. Imaging **30**(9), 1323–1341 (2012). https://doi.org/10.1016/j.mri.2012.05.001
17. Al-Ahmari, A., Ashfaq, M., Mian, S.H., Ameen, W.: Evaluation of additive manufacturing technologies for dimensional and geometric accuracy. Int. J. Mater. Prod. Technol. **58**(2–3), 129–154 (2019). https://doi.org/10.1504/IJMPT.2019.097665
18. Moncayo-Matute, F.P., et al.: Surgical planning and finite element analysis for the neurocraneal protection in cranioplasty with PMMA: a case study. Heliyon **8**(9) (2022). https://doi.org/10.1016/j.heliyon.2022.e10706
19. Lee, J.W., Fang, J.J., Chang, L.R., Yu, C.K.: Mandibular defect reconstruction with the help of mirror imaging coupled with laser stereolithographic modeling technique. J. Formos. Med. Assoc. **106**(3), 244–250 (2007). https://doi.org/10.1016/S0929-6646(09)60247-3
20. Arango-Ospina, M., Cortés-Rodriguez, C.J.: Engineering design and manufacturing of custom craniofacial implants. In: The 15th International Conference on

Biomedical Engineering: ICBME 2013, 4–7 December 2013, Singapore, pp. 908–911. Springer, Cham (2014). https://doi.org/10.1007/978-3-319-02913-9_234

21. Zhou, L.B., et al.: Accurate reconstruction of discontinuous mandible using a reverse engineering/computer-aided design/rapid prototyping technique: a preliminary clinical study. J. Oral Maxillofac. Surg. **68**(9), 2115–2121 (2010). https://doi.org/10.1016/j.joms.2009.09.033
22. Ameen, W., Al-Ahmari, A., Mohammed, M.K., Abdulhameed, O., Umer, U., Moiduddin, K.: Design, finite element analysis (FEA), and fabrication of custom titanium alloy cranial implant using electron beam melting additive manufacturing. Adv. Prod. Eng. Manage. **13**(3), 267–278 (2018). https://doi.org/10.14743/apem2018.3.289
23. Omidi, A., Jeannin, C., Nazari, M., Panahi, M.: Analysis of temporomandibular joint prosthesis using finite element method and a patient specific design. Eng. Solid Mech. **7**(1), 83–92 (2019)
24. Ridwan-Pramana, A., Marcián, P., Borák, L., Narra, N., Forouzanfar, T., Wolff, J.: Structural and mechanical implications of PMMA implant shape and interface geometry in cranioplasty–a finite element study. J. Cranio-Maxillofac. Surg. **44**(1), 34–44 (2016). https://doi.org/10.1016/j.jcms.2015.10.014
25. Yoganandan, N., Pintar, F.A., Zhang, J., Baisden, J.L.: Physical properties of the human head: mass, center of gravity and moment of inertia. J. Biomech. **42**(9), 1177–1192 (2009). https://doi.org/10.1016/j.jbiomech.2009.03.029
26. Guzman-Bautista, A., et al.: Quasi-uniform density non-solid infill strategy for axisymmetric non-planar additive manufacturing. Appl. Sci. **15**(11), 5899 (2025). https://doi.org/10.3390/app15115899

Synthesis and Characterization of Tricalcium Phosphate (β-TCP) for Potential Biomedical Applications

Gonzalo Fonseca Miranda[1], Ariel Nenen Huenchul[2], Mario Flores Flores[3], Loreto Troncoso Aguilera[1], and Caroline Silva Danna[1](✉)

[1] Laboratorio de Materiales, Facultad deCienciasdeLaIngeniería, Universidad Austral de Chile, Valdivia, Chile
caroline.silva@uach.cl
[2] Laboratorio de Biointerfaces, Libertas Laboratories, Santiago, Chile
[3] Laboratorio de Polímeros, FacultaddeCiencias, Universidad Austral de Chile, Valdivia, Chile

Abstract. β-TCP is a ceramic material of current interest for applications in bone tissue engineering. However, different synthesis methods and thermal treatments produce variations in their chemical composition, crystalline structure, and porosity, these factors being determining in its biocompatibility and bioactivity. The objective of this work was to study the characteristics of β-TCP synthesized by two methods (mechanical synthesis and sol-gel) and calcined at different temperatures, to establish standardized production protocols. Mechanical synthesis was carried out at 350 RPM in 24 h. For sol-gel synthesis, constant stirring was used for 30 min at pH 2. The samples, in powder and pellet form, were calcined for 3 h at 700, 800, 900, 1000, and 1100 °C. Material characterization was performed using XRD, SEM-EDS, FTIR, and TGA.

Keywords: β-TCP · bio-ceramics · sol gel process · mechanical milling · bone tissue engineering

Nomenclature

CPP	Calcium Pyrophosphate $Ca_2P_2O_7$ Ca/P = 1
HA	Hydroxyapatite $Ca_{10}(PO_4)_6(OH)_2$ Ca/P = 1.6
CDHA	Calcium-deficient HA $Ca_9(HPO_4)(PO_4)_5(OH)$ Ca/P = 1.5
TCP	Tricalcium Phosphate $Ca_3(PO_4)_2$ Ca/P = 1.5
TTCP	Tetracalcium Phosphate $Ca_4(PO_4)_2O$ Ca/P = 2
SM	Mechanical Synthesis
SG	Sol-Gel Synthesis

O. F. Farías Fuentes et al. (Eds.): CIBIM 2024, *Proceedings of the XVI Ibero-American Congress of Mechanical Engineering*, pp. 41–53, 2026.
https://doi.org/10.1007/978-3-032-22823-9_4

1 Introduction

β-TCP is a ceramic material with interesting potential for applications in bone tissue engineering, currently being one of the most widely used bone substitutes; however, bibliographic reviews have indicated that it possesses an inconsistent biological response [1]. The application of different synthesis methods can lead to variations in its chemical composition, forming intermediate compounds, mainly hydroxyapatite (HA) and calcium pyrophosphate (CPP) [2]. The former is formed when the Ca/P ratio is greater than 1.5, with approximately 10% by weight of HA forming due to a 1% variation in the Ca/P ratio. On the other hand, when the ratio is less than 1.5, CPP precipitates are obtained. Although currently, by regulation, a β-TCP bone substitute is considered pure if its content is greater than 95% (ISO 13175-3:2012), both HA and CPP negatively affect the reabsorption mechanisms and subsequent bone deposition. Thus, it is a challenge to decrease the content of both phases within the β-TCP.

The main limitation of mechanical synthesis is the precise control in the mixing of reactants, since, during the milling process, areas of concentration of one of them can be generated, altering the local Ca/P ratio. For its part, sol-gel synthesis demands strict pH control during the reaction; variations can lead to the precipitation of phases other than calcium-deficient hydroxyapatite (CDHA), which requires subsequent calcination for its transformation into β-TCP. Likewise, for both synthesis methods, a wide variation in treatment temperatures has been reported, with ranges from 600 °C [3] to 1300 °C. The consensus indicates that the transformation from CDHA to β-TCP occurs in the range of 650–750 °C, while above 1115–1200 °C, the transformation to α-TCP occurs. This phase has a much higher solubility than the β-phase and is therefore undesirable. The wide range of intermediate temperatures can affect the crystalline structure and porosity of the material [3]; these factors being determining in its biocompatibility and bioactivity. Standardizing β-TCP manufacturing methods is key for its safe clinical use and consistent results. This work presents a study of the characteristics of β-TCP synthesized by two methods (mechanical synthesis and sol-gel) and subsequently treated at different temperatures, at 100 °C intervals, to establish standardized production protocols.

2 Methodology

2.1 Synthesis

Mechanical Synthesis (MS)

The precursors, calcium carbonate ($CaCO_3$) and dibasic calcium phosphate ($CaHPO_4$), (both Sigma-Aldrich, purity $\geq 98\%$) were weighed and incorporated in the stoichiometric proportion of $Ca/P = 1.5$. They were then placed in a pair of tungsten carbide vials, along with the milling bodies in a 1:8 mass ratio (powder: balls) per vial [5, 6]. A planetary mill (FRITSCH Pulverisette 7) was used, where the first process was the homogenization of the mixture, for 15 min at 100 RPM, and immediately after, the synthesis condition was started, at a rotation speed of 350 RPM for a total of 32 work cycles, each consisting of 45 min of milling and 15 min of pause, for a total of 24 h of machining.

Sol-Gel (SG)

The reagents used were calcium nitrate tetrahydrate ($Ca(NO_3)_2$-$4H_2O$), citric acid monohydrate ($C_6H_8O_7$-H_2O), and diammonium phosphate ($(NH_4)_2HPO_4$), (all Sigma-Aldrich, purity > 99%). Additionally, nitric acid was used for pH control [3, 4]. The amounts of each reagent were calculated respecting the Ca/P = 1.5 ratio and were dissolved in 25 mL of deionized water with constant stirring for 30 min, maintaining a pH of 2. Once the reaction time was completed, the solution was heated to 80 °C until a semi-transparent gel appearance was obtained, then the temperature was raised to 100 °C to eliminate excess water.

Thermal Treatment

The obtained samples were calcined in 2 formats: powder and pellets. For the fabrication of the pellets, 500 mg were weighed, and using a steel die and hydraulic press, they were formed, applying pressures of 75, 187, and 375 MPa, respectively, isostatically for 40 s. Subsequently, they were subjected to thermal treatment at different temperatures, from 700 °C to 1100 °C. The heating rate was 3 °C per minute and the target temperature was maintained for 3 h (Table 1).

Table 1. Crystallite size of β-TCP samples. N/C = not calculated.

Sample	β (radianes)	θ (degrees)	D (ángstrom)
SM - 700 °C	N/C	N/C	N/C
SM - 800 °C	0.00715585	15.557	199.05
SM - 900 °C	0.00314159	15.485	453.24
SM - 1000 °C	0.00314159	15.520	453.31
SM - 1100 °C	0.00296706	15.503	479.94
SG - 700 °C	N/C	N/C	N/C
SG - 800 °C	N/C	N/C	N/C
SG - 900 °C	0.00610865	15.643	233.27
SG - 1000 °C	0.00663225	15.600	214.81
SG - 1100 °C	0.00331613	15.473	429.36

2.2 Sample Characterization

X-Ray Diffraction (XRD)

The purity of the phases and the crystalline structure of the uncalcined powder samples (precursors and CDHA) and post-thermal treatment powder (β-TCP) were analyzed using X-ray diffraction techniques. The equipment used was a Bruker D2 PHASER X-ray diffractometer (30 kV, 10 mA) with Cu Kα radiation ($\lambda = 1.5418$ Å). The scanning window for the diffraction patterns was between $10° < 2\theta < 60°$ at a scanning speed of 0.02 °/s. The obtained patterns were matched with the database available in the

X'Pert HighScore Plus software and then contrasted with the patterns mentioned in the literature for β-TCP (JCPDS No. 09-0169), α-TCP (JCPDS No. 09-0348), HA (JCPDS No. 9-0432), CPP (PDF 01-073-0440), CDHA (PDF 00-046-0905) and TTCP (PDF 00-025-1137).

Pellet Density
The pellets were weighed and dimensioned, based on the method mentioned in ISO 13175-3, before and after being subjected to the oven. With these data, the densities and porosity fractions for each sample were calculated.

Fourier Transform Infrared Spectroscopy (FTIR)
The study was performed with a JASCO FT/IR-4600 Spectrometer; the samples were analyzed in powder form. The obtained spectra were processed and analyzed using Spectra Analysis software and then plotted in Origin.

Thermogravimetric Analysis (TGA)
Tests were performed on a Perkin Elmer thermogravimetric analyzer (TGA 4000), in a nitrogen atmosphere (2 bar). The temperature ramp was 40 °C/min from 30 to 100 °C, followed by 20 °C/min from 100 to 995 °C. Data were processed with Pyris Manager and then plotted in Origin.

Scanning Electron Microscopy (SEM)
For the morphological analysis of the pellet surfaces, a Zeiss microscope, model AURIGA Compact Field emission scanning, transmission (FESEM, STEM) was used; surface scanning was performed at 1000 X, 5000 X, and 20000 X magnifications.

Energy Dispersive Spectroscopy (EDS)
The pellets were fractured to allow separate analysis of the external surface and the internal material, then they were carbon coated with 15 nm. The samples were analyzed by surface mapping and by selection of random points. For this study, a Zeiss EVO MA10 Variable Pressure Scanning Electron Microscope (VPSEM) with an Oxford x-act EDS detector was used. The analysis of the information collected by the detector was performed with Aztec 6.0 SP1 software.

3 Results and Discussion

X-ray diffraction results are presented in Fig. 1 and are contrasted with the characteristic patterns for different calcium phosphates to verify the formation of β-TCP, and, secondly, the approximate minimum crystallite size obtained for each sample was calculated using the Scherrer formula. MS alone did not provide enough energy to produce β-TCP directly, and it was not even formed after treatment at 700 °C, appearing as a mixture of precursors and different calcium phosphates [4–7]. With treatments at temperatures of 800 °C and above, β-TCP was produced, with similar results. In samples treated at 1000 and 1100 °C, some peaks indicating the presence of HA (18.78, 28.12, 39.20, and 58.07°) were found. This can be associated with local concentrations within the material due to the synthesis method used, and rather than being associated with the thermal treatment temperature, they suggest the existence of non-stoichiometric local Ca/P concentrations [2, 5, 7]. These suspicions were later confirmed by EDS studies. For SG, the transformation to β-TCP occurs at higher temperatures, with the transformation process completing from 900 °C upwards. As the temperature increases up to 1100 °C, more defined peaks can be observed, which may be associated with the increase in the crystallinity of the material. These changes mentioned would accompany the morphological changes observed by SEM. For the sample treated at 1100 °C, some peaks indicating the presence of CPP (20.85°) and HA (31.77°) were found.

The presence of HA was not quantified as a percentage, and it is proposed to perform this calculation as a complement for future work, especially due to the importance given in recent years to biphasic calcium phosphates (BCP) as a bone graft material [8–10].

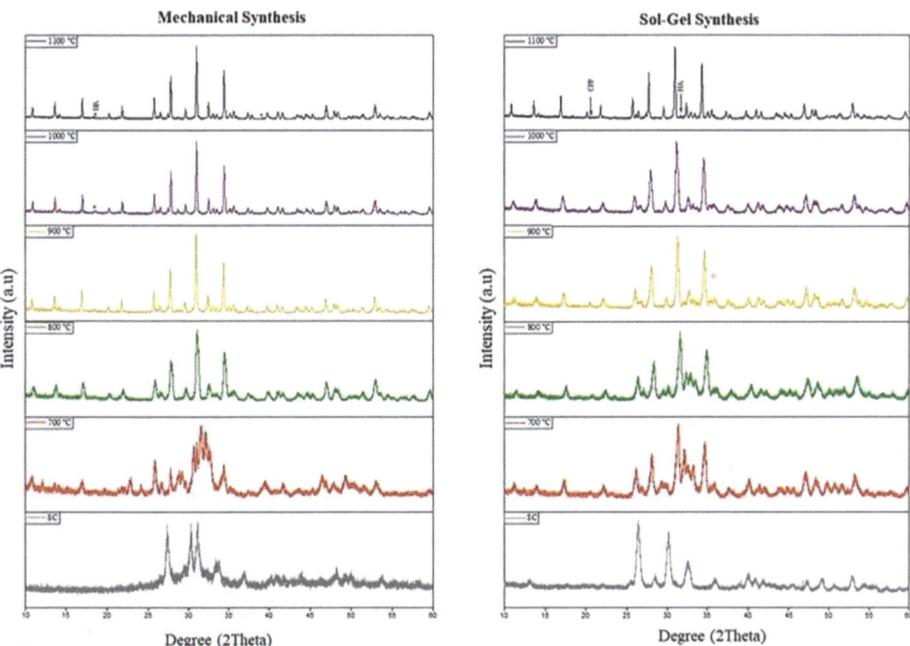

Fig. 1. XRD for the different β-TCP samples. Mechanical Synthesis (left) and Sol-Gel (right).

For both syntheses, the proposed maximum temperature (1100 °C) was reached without α-TCP formation. Furthermore, it is noteworthy that for MS, it was possible to lower the β-TCP formation temperature compared to SG [2]. For samples where β-TCP formation was confirmed, the crystallite sizes presented in Fig. 2 were calculated. It can be observed that the smallest sizes were calculated for samples obtained by sol-gel. Additionally, the most pronounced changes occur between 800 and 900 °C for MS, and between 1000 and 1100 °C for SG, in agreement with literature [3, 5–7].

The densities of the pellets presented in Fig. 2 increase proportionally to the pressures exerted during their compaction, and variations were also observed as the treatment

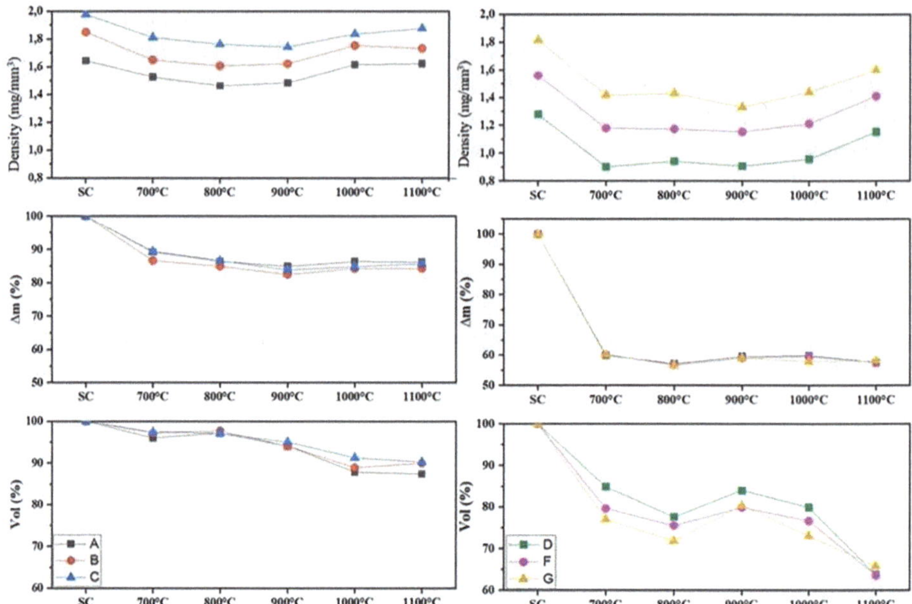

Fig. 2. Variations in density, mass, and volume of β-TCP pellets. A, B y C correspond to pellets obtained by MS, formed at pressures of 75, 187, and 375 MPa. D, F, and G represent the same pressures for sol-gel synthesis.

temperature increased. For both synthesis methods, an initial decrease in density associated with mass loss is observed (more details in the TGA section, Fig. 3). However, after 800 °C, this loss stabilizes [7, 10, 11], while a volumetric contraction of the material continues due to the increase in grain size and coalescence (see SEM section, Figs. 5 and 6). As a result, there is a densification of the material above 900 °C. Figure 3 summarizes the calculated porosity fractions for those samples in which β-TCP was formed. This is higher for those obtained by the SG technique and decreases with increasing pressure and/or temperature.

With the infrared spectroscopy analysis Fig. 4, it can be observed that for the sample synthesized via sol-gel, there are evident changes between the precursor and the samples treated at different temperatures. In contrast, for the β-TCP obtained by SM, the untreated sample and the treated ones are similar to each other. This appears to be a clear indication that the energy provided by the SM method to the β-TCP precursors is greater than that of sol-gel. Both graphs are limited to wavelengths between 400 and 2500 cm^{-1} because no significant peaks or absorbance bands were recorded between this value and 4000 cm^{-1}. The vertical dotted lines mark representative peaks of phosphate groups (PO_43^-) in β-TCP [5].

The samples obtained by SM are quite similar across the entire temperature range studied. Both samples prior to β-TCP phase formation exhibit characteristic HA peaks at 875 cm^{-1} and near 1420–1460 cm^{-1}. These peaks corresponding to HPO_42^- and CO_32^- usually appear in samples with low-temperature thermal treatment, where the

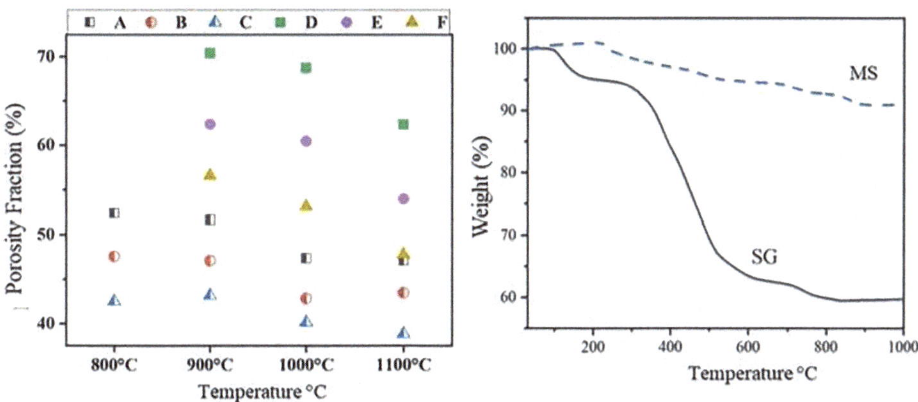

Fig. 3. Porosity Fraction (left), and TGA of "untreated" powder samples (right).

formed compounds are not stable and the material's surface interacts with environmental humidity, absorbing water and generating hydroxyapatite [6, 7, 11]. At 700 °C, there is also a low-intensity peak at 720 cm^{-1} indicating the presence of CPP impurities, which disappear upon reaching 800 °C. All samples between 800 and 1100 °C have a peak near 600 cm^{-1} (red circle), which marks the presence of HA impurities. The β-TCP treated at 900, 1000, and 1100 °C show no relevant changes, except that the phosphate group (PO_43^-) peaks are slightly accentuated. Finally, all samples obtained by SM exhibit a slight absorbance band between 1700 and 2200 cm^{-1} corresponding to humidity (H_2O).

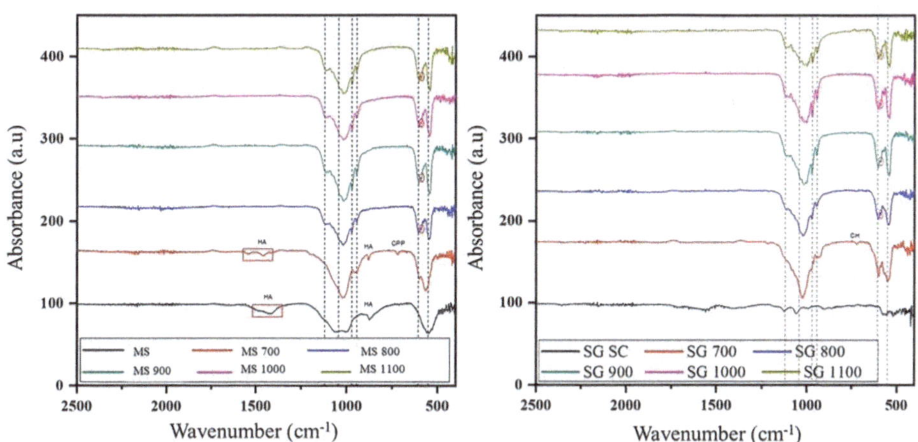

Fig. 4. FTIR spectra for β-TCP Mechanical Synthesis (right) and Sol-Gel (left).

In the case of SG, the untreated sample is clearly different from the others; a shallow band between 420 and 580 cm^{-1} corresponding to phosphate groups is observed, and other faint bands appear near 860 and 1330 cm^{-1} as residues of nitrogenous compounds from the synthesis. A band from 1500 to 1750 cm^{-1} indicates the presence of H_2O and

$CO_3 2^-$, evidencing the presence of CDHA and moisture absorption. With the 700 °C treatment, some peaks corresponding to $PO_4 3^-$ groups emerge, some of which can be attributed to β-TCP; there is a faint peak at 720 cm^{-1} corresponding to CH groups (organic residues from citric acid) [10, 11]. As the treatment temperature is increased, the distinct β-TCP phosphate groups ($PO_4 3^-$) become better defined. Similar to SM, a peak near 600 cm^{-1} (red circle) is observed, which indicates the presence of HA impurities.

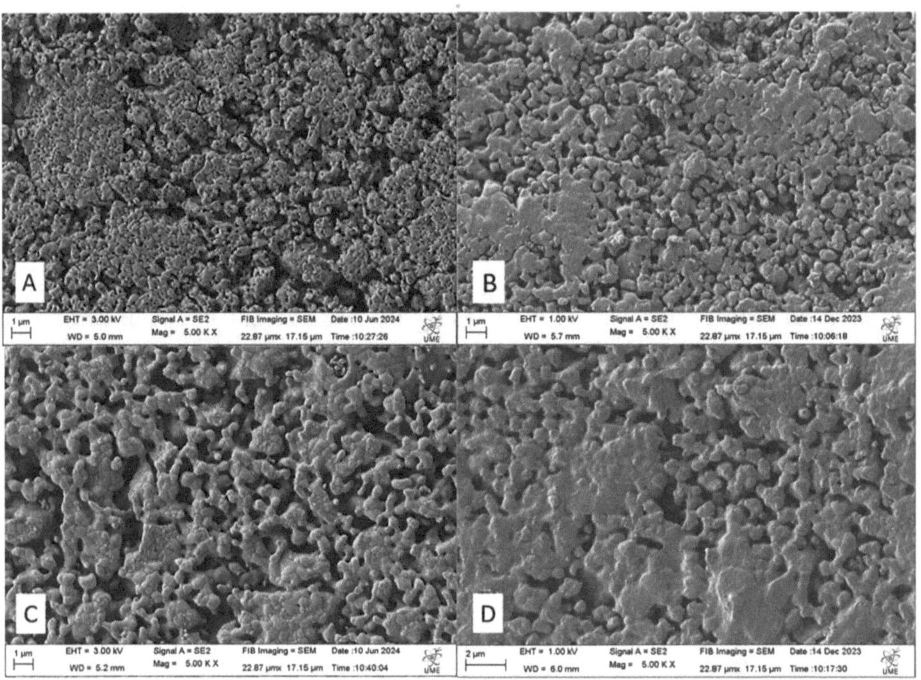

Fig. 5. SEM of β-TCP Mechanical Synthesis, 5,000X magnification (A) 800 °C, (B) 900 °C (C) 1000 °C and (D) 1100 °C.

Regarding the thermogravimetric analysis, significant variations can be observed depending on the chosen synthesis route. In the case of the sample obtained by SG, the thermal decomposition curve corresponds to the sum of several processes that occur sequentially and simultaneously. The first occurs between room temperature (30 °C) and 100 °C and is due to the loss of residual water within the precursor. This water loss continues during the process as hydroxyl groups are released. Between 170 and 310 °C, the decomposition and evaporation of residual citric acid, used as a reaction catalyst, occurred [11, 12]. Between 300 and 700 °C, changes are due to several factors: on one hand, the condensation of hydrogen phosphate ions occurs, accompanied by the subsequent evaporation of water [2]; furthermore, above 400 °C, the decomposition of organic residues occurs. Finally, near 750 °C, a last mass loss indicates the decomposition of $CaCO_3$ and intermediate compounds, marking the transformation from CDHA to β-TCP [8]. The total weight loss is close to 40%.

Meanwhile, for SM, the curve begins by showing a weight increase that reaches ~ 1% near 215 °C; this may correspond to the hydrolysis of calcium phosphates on the surfaces due to moisture absorption during material handling [4]. From this temperature, the weight begins to decrease due to the decomposition of hydroxyl groups, and near 850 °C, the decomposition of carbonates marks the complete transition to β-TCP. The total mass loss was approximately 10%.

As for the electron microscopy studies, the presented micrographs correspond to pellets obtained at 375 MPa, and only those obtained above 800 °C are shown. In the case of SM synthesis, Fig. 5, it was possible to verify that the increase in treatment temperature has an inversely proportional relationship with the amount of microporosities in the material. At 800 °C (A), multiple microporosities located at grain boundaries and within them can be easily distinguished. At 900 °C, a large part of the microporosities disappeared. At 1000 °C (C), the contact regions between neighboring grains continue to increase, such that at 1100 °C (D), the occurrence of large plates formed by multiple coalescing grains can be observed. This densification process occurs heterogeneously, creating some areas clearly marked by grain coalescence, where micropores disappear and give rise to denser particles, while in other areas of the material, fissures are generated due to the release of local stresses. In this way, in these sites, the proportion of macroporosities increases. In other words, macropores are not formed by the union of micropores, but quite the opposite: they form and grow in those areas that grains leave free as they coalesce elsewhere. This explains the inverse relationship between the presence of microporosities and macroporosities and highlights the importance of the balance that must be achieved to obtain a material capable of receiving cells that colonize its surfaces in the micropores and that allows adequate neoformation of blood vessels for proper irrigation of the deepest areas of the graft through interconnected macropores [3, 7, 11–14].

For SG Fig. 6, a similar process of grain growth and material densification occurs, but in this case, a change in grain shape can also be observed as precursors begin to form the β-TCP phase, which occurs simultaneously with an increase in its crystallinity (as shown previously in XRD studies). Thus, in (A), the predominant geometries are "flake-like" and relatively flat [11], which with increasing temperature begin to take on a more three-dimensional body (B and C) until well-defined crystalline grains are obtained at 1100 °C (D) (Table 2).

EDS mappings, shown in Fig. 7, reported Ca/P ratios close to 1.5 for both the outer surface and the internal areas (of the fractured pellets). Some authors point out that MS techniques generate materials with inconsistent compositions, with precipitation of contamination phases. This was verified by analyzing the results of EDS taken at random points where the local Ca/P ranges found were mostly between 1.3 and 1.6, but there were some instances that yielded extreme values above 1.8 and below 1.2 [4, 6, 13, 15]. All the mentioned cases occurred in the internal zone of the pellets obtained by SM, with the lowest value being for the sample treated at 900 °C. It is believed that these variations are not related to the treatment temperature but to the synthesis method. For the SG pellets, all mappings reported values close to a Ca/P of 1.5. Whereas for the point analyses, the only notable difference occurred in the sample treated at 1100 °C,

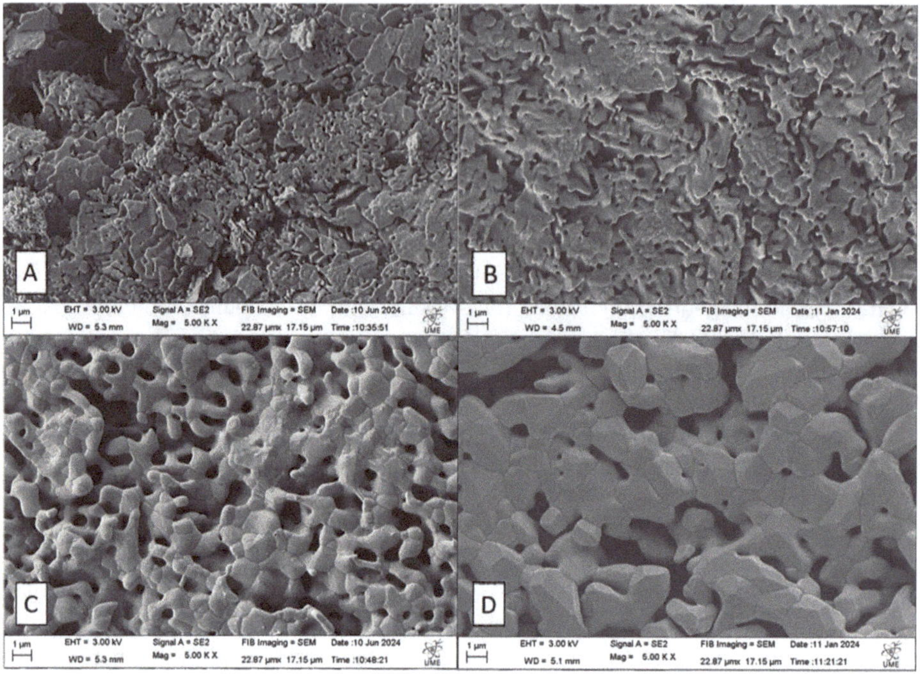

Fig. 6. SEM of β-TCP Sol-gel Synthesis, 5.000X magnification (A) 800 °C, (B) 900 °C, (C) 1000 °C y (D) 1100 °C.

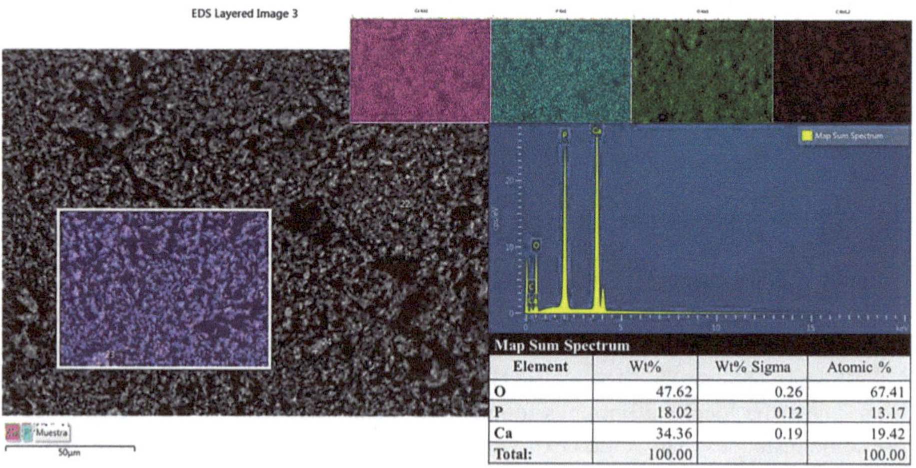

Map Sum Spectrum

Element	Wt%	Wt% Sigma	Atomic %
O	47.62	0.26	67.41
P	18.02	0.12	13.17
Ca	34.36	0.19	19.42
Total:	100.00		100.00

Fig. 7. EDS Mapping.

yielding an average Ca/P value of 1.34. There is no clear bias in any of the samples. The quasi-quantitative results for the different EDS measurements are detailed in Fig. 8.

Table 2. Composition of β-TCP samples studied by EDS. S = Surface, I = Interior.

Sample	Site	%Ca (at)	%P (at)	Ca/P	Site	%Ca (at)	%P (at)	Ca/P
SM - 700 °C	S	20,71	14,11	1,47	I	18,74	13,03	1,44
SM - 800 °C	S	18,71	13,86	1,35	I	14,79	11,46	1,29
SM - 900 °C	S	22,38	14,83	1,51	I	8,97	9,41	0,95
SM - 1000 °C	S	21,50	14,12	1,52	I	17,92	11,95	1,50
SM - 1100 °C	S	19,00	13,58	1,40	I	22,36	14,33	1,56
SG - 700 °C	S	19,77	13,91	1,42	I	20,24	12,96	1,56
SG - 800 °C	S	23,30	15,05	1,55	I	20,65	12,91	1,59
SG - 900 °C	S	21,11	14,36	1,47	I	21,27	13,75	1,54
SG - 1000 °C	S	21,04	13,77	1,53	I	19,03	12,98	1,47
SG - 1100 °C	S	25,06	15,75	1,59	I	19,27	14,30	1,34

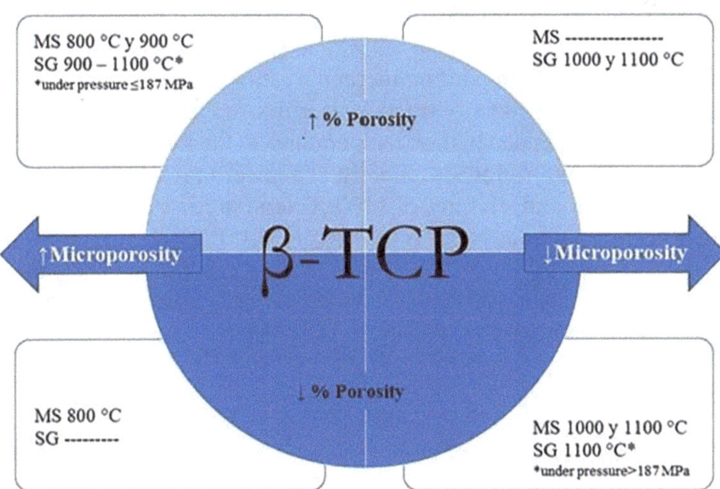

Fig. 8. Variation of porosity fraction and microporosity according to synthesis method and treatment temperature.

4 Conclusions

β-TCP was successfully synthesized by both SM and SG techniques. Evident morphological differences were observed depending on the method used.

SM takes significantly longer (32 h if pause times are included) than the sol-gel reaction. However, both techniques present their own complexities. Mainly, for SM, the adequate mixing of reactants, and for SG, pH control during the process. Thanks to XRD, TGA, and FTIR studies, it was demonstrated that there is a lower limit for β-TCP formation between 700 and 800 °C for SM, completing above 900 °C for SG. The decrease in temperature for β-TCP phase formation can be associated with the energy input to the precursors during the milling process. Up to the maximum studied treatment temperature (1100 °C), no transformation to the α phase was detected. From the above, the question arises for future studies to investigate modifications to thermal treatment times, potentially shortening them for SM to optimize resources and prolonging them for SG to obtain crystalline β-TCP at lower temperatures.

The morphological characterization performed by SEM demonstrates that, as treatment temperatures increase, particle sizes increase, agglomeration is more prevalent, the presence of microporosities decreases, and consequently, densification of the material occurs. EDS mapping characterization indicates acceptable Ca/P ratios close to 1.5 for the external surfaces and internal areas of the β-TCP pellets. However, some of the point analyses (at random sites) detected Ca/P ratios far from this proportion within the pellets obtained by SM. These results are consistent with what some authors have stated, who believe that SM may have difficulties in generating a homogeneous material. From the calculation of the porosity fraction of the pellets manufactured by both methods, and the information gathered by the different characterization techniques used in this work, it can be concluded that the most porous (porosity fraction > 50%) are those obtained by the sol-gel technique at temperatures between 900 and 1000 °C (at all applied pressures), and up to 1100 °C (when using 187 MPa or less). For SM, the highest porosities were found at 800 and 900 °C at pressures of 75 MPa. (*Compaction pressures of the pellets, not to be confused with pressing during thermal treatment).

If, in addition to a high porosity fraction, it is desired that these are predominantly microporous, it is recommended to limit temperatures to the ranges of 900 to 1000 °C for SG, and not to exceed 800 °C for SM. Finally, if a lower porosity fraction and material densification are sought, temperatures of 1100 °C (and pressures close to 375 MPa) are recommended for SG, and ranges between 900 and 1100 °C for SM.

Acknowledgments. The authors wish to thank the Escuela de Graduados de la Facultad de Ciencias de la Ingeniería and the al Magister en Ingeniería Mecánica y Materiales (Master in Mechanical and Materials Engineering) of the Universidad Austral de Chile for the support provided in financing this study.

References

1. Bohner, M., Santoni, B.L.G., Döbelin, N.: β-tricalcium phosphate for bone substitution: synthesis and properties. Acta Biomater **113**, 23–41 (2020)
2. Destainville, A., Champion, E., Bernache-Assollant, D., Laborde, E.: Synthesis, characterization and thermal behavior of apatitic tricalcium phosphate. Mater. Chem. Phys. **80**(1), 269–277 (2003)
3. Zhao, D., Zhang, Y., Dong, Y., Zhang, W., Chen, D., Zhang, H.: Low-temperature synthesis of highly pure β-tricalcium phosphate nanoparticles by a novel co-precipitation method. J. Mater. Sci. Mater. Med. **30**(7), 80 (2019)
4. Perera, F.H., Martinez-Vazquez, F.J., Miranda, P., Ortiz, A.L., Pajares, A.: Clarifying the effect of sintering conditions on the microstructure and mechanical properties of β-tricalcium phosphate. Ceramics International **36**(6), 1929–1935 (2010)
5. Ruiz-Aguilar, C., Olivares-Pinto, U., Aguilar-Reyes, E.A., López-Juárez, R., Alfonso, I.: Characterization of β-tricalcium phosphate powders synthesized by sol–gel and mechanosynthesis. Bol Soc Esp Cerám Vidrio **57**(5), 213–220 (2018)

6. Panda, S., Dash, S.S., Nayak, B.B.: Mechanochemically synthesized phase stable and biocompatible β-tricalcium phosphate from avian eggshell for the development of tissue ingrowth system. Appl. Phys. A **128**(1), 1–10 (2022)
7. Zhao, D., Zhang, Y., Dong, Y., Zhang, W., Chen, D., Zhang, H.: Low-temperature synthesis of highly pure β-tricalcium phosphate nanoparticles by a novel co-precipitation method. J. Mater. Sci. Mater. Med. **30**(7), 80 (2019)
8. León, B., Jansen, J.A.: Thin Calcium Phosphate Coatings for Medical Implants. Springer, New York (2009)
9. Fan, D., Chen, Z., Li, X., Wang, Q., Hu, Z., Zhang, J.: Recent advances in bone tissue engineering: from materials to clinical translation. Bioactive Materials **32**, 337–361 (2024)
10. Chen, Z., Cao, B., Wang, P., Zhou, B.: Microstructure evolution and mechanical properties of β-tricalcium phosphate bioceramics prepared by pressureless sintering. J. Eur. Ceram. Soc. **37**(10), 3295–3304 (2017)
11. Sun, L., Ma, W., Yang, L., Li, B., Chen, S., Deng, Z.: Sol–gel synthesis and characterization of Ca/P ratio adjustable β-tricalcium phosphate. Mater. Chem. Phys. **248**, 122971 (2020)
12. Rangavittal, N., Landa-Cánovas, A.R., González-Calbet, J.M., Vallet-Regí, M.: Structural study and stability of hydroxyapatite and tricalcium phosphate: two important bioceramics. J. Biomed. Mater. Res. **51**, 660–668 (2000)
13. Choi, D., Kumta, P.N.: Mechano-chemical synthesis and characterization of nanostructured β-TCP powder. Mater. Sci. Eng. C **27**(3), 377–381 (2007)
14. Grigoraviciute-Puroniene, I., et al.: A novel wet polymeric precipitation synthesis method for monophasic β-TCP. Adv. Powder Technol. **28**(9), 2325–2331 (2017)

Finite Element Analysis of Bimetallic Sheet Deep Drawing

Carlos Sánchez-Alvarracín[1], Marco Amaya-Pinos[2], Julio Loja-Quezada[2], and Luis López-López[3](✉)

[1] Environmental Risk Assessment Group in Production and Service Systems, Environmental Engineering, Universidad de Cuenca, Cuenca, Ecuador
carlos.sancheza@ucuenca.edu.ec

[2] Research Group on Simulation, Optimization, and Decision Making, Mechanical Engineering, Universidad Politécnica Salesiana, Cuenca, Ecuador
{mamaya,jlojaq}@ups.edu.ec

[3] New Materials and Transformation Processes Research Group, Mechanical Engineering, Universidad Politécnica Salesiana, Cuenca, Ecuador
llopez@ups.edu.ec

Abstract. The behavior of an aluminum-copper bimetallic sheet was evaluated in a single-phase cylindrical drawing process with a force applied by a plate press on the sheet to avoid deformations. The properties of aluminum compared to copper-clad aluminum were determined through laboratory tests and the theoretical analysis of drawing is based on the energy method, obtaining the mini-mum required force. With the theoretical results, FEM finite element simulations are carried out and experimental drawing tests were used to validate the results. Tensile tests indicate that the bimetallic sheet fails before peeling off and has better mechanical and formability properties than aluminum alone. A good relationship was observed between the theoretical, numerical and experimental results with deformation errors of less than 4%, therefore, the simulation can be used to analyse and predict failure in the drawing process with bimetallic sheets.

Keywords: Simulation · Bimetallic · Finite Elements · Stuffing

1 Introduction

The demands of modern industry and the metal forming market necessitate materials with enhanced physical and mechanical characteristics. Multilayered metallic sheets fabricated from dissimilar metals offer superior properties in terms of strength, plastic deformation, fracture resistance, electrical conductivity, and corrosion resistance, attributes not found in single-metal sheets. For this reason, composite materials are extensively employed across the aerospace, chemical, automotive, and electrical industries, as well as in the production of household appliances and other products re-quiring high mechanical strength [1, 2]. The most common process for joining sheets of dissimilar metals, such as Aluminum-Copper (Al-Cu), is cold roll bonding (CRB). This method allows for

O. F. Farías Fuentes et al. (Eds.): CIBIM 2024, *Proceedings of the XVI Ibero-American Congress of Mechanical Engineering*, pp. 54–72, 2026.
https://doi.org/10.1007/978-3-032-22823-9_5

the simultaneous achievement of excellent corrosion resistance, high deformability, and superior electrical and thermal properties, all at relatively low costs. In contrast, hot roll bonding often leads to the formation of brittle intermetallic compounds, which adversely affect the material's ductility and deformation behavior [3–6].

Deep drawing of metal sheets is one of the most widely applied industrial processes for manufacturing a vast array of products with diverse applications. Its high mechanical strength-to-weight ratio, uniform temperature distribution (due to differing thermal expansion coefficients of the sheets), reduced spring back, and cost effectiveness, among other advantages, have led to extensive analytical study of the process. This research spans from the design of the desired geometry to the operational para-meters of the press and tooling, all aimed at identifying optimal processing conditions. It's crucial to consider that the plasticity of the composite material is de-pendent on both strain rate and anisotropy [7–10]. The formation of wrinkles or folds is a primary failure mode in the deep drawing of thin, high-strength metal sheets. A key focus in this area is the prediction, prevention, and evaluation of these wrinkles, which are considered a form of local buckling caused by excessive compressive stresses [11, 12]. The finite element method (FEM) is currently the only feasible method capable of predicting such defects before the actual stamping operation takes place [11].

Yu and Johnson [13] developed a two-dimensional buckling model for an elastic-plastic annular plate, employing the energy method with critical conditions for both elastic and plastic buckling. Their work successfully improved the understanding of blank holder influence on sheet buckling and quantitatively predicted the number of generated waves or wrinkles. Morovvati M., et al. [14] applied the Yu and Johnson model through analytical methods, numerical simulations using finite elements, and experimental validation to determine the minimum blank holder force required to prevent wrinkle formation during the deep drawing of a composite sheet composed of two layers of aluminum and stainless steel. They concluded that the results were dependent on the sheet's geometry and the material of the upper layer in the deep drawing process. Samie S., Morovvati M., and Vaghasloo A. [15] investigated the wrinkling of a circular plate in the flange area of a deep-drawn part subjected to tension and large deformation. They applied a model based on Hill's bifurcation method, examining the friction between the forming plate and the blank holder, as well as strain hardening parameters. They used simulation software with finite elements, comparing their findings with an experimental study to validate the results and obtain the critical blank holder force associated with the critical wrinkling condition in the flange area.

Darendeliler and Kaftanoglu [16] developed a finite element method to obtain elastic-plastic information for sheet materials in the presence of large deformations and large displacements. Their work was based on a Lagrangian formulation and membrane theory, assuming the metal sheet to be isotropic and rate-insensitive, thereby adhering to flow theory. They incorporated material hardening characteristics and Coulomb friction between the sheet and forming tools to model deep drawing with appropriate boundary conditions. Their numerical solutions were compared with experimental results, revealing a small variation attributed to changes in friction conditions between the punch, sheet, and forming die. In his research, Flores F. [17] presented algorithmic aspects for handling the frictional contact problem between tools and the metal sheet in deep drawing simulations. His goal was an effective and computationally economical search for contact points. To achieve this, he applied an explicit integration scheme that processed tool surfaces defined through triangulations for mesh handling. Numerical simulations of deep drawing processes with large elastoplastic deformations have also been performed, where the material's behavior is based on a hyper elastic model. This model utilizes the explicit integration of the equations of motion through a dynamic model, considering element friction. This approach yielded a numerical model that accurately corresponds to displacements, stresses, and deformations [18, 19].

Optimizing a deep drawing process involves considering various parameters, from the device design to the material's properties, all aimed at preventing defects like wrinkling, rupture, or tearing in the drawn part. Therefore, it's essential to establish or predict the forces and deformations involved in cylindrical deep drawing using finite elements. In the present case study, a bimetallic material was used for forming. We considered different aspects of the deep drawing process to determine the percentage of error between simulated results and experimental tests, ultimately validating the forming of the bimetallic component.

2 Methodology

2.1 Material

For this investigation, a bimetallic sheet joined by explosion bonding and rolling was used. It consisted of a base of pure aluminum (Al 1060) clad with a copper sheet (Cu T2), with thicknesses of 2.5 mm and 0.5 mm, respectively. The chemical composition of these individual materials is presented in Table 1.

Table 1. Chemical composition.

Element	Al 1060 (%)	Al 1060 – Cu T2 (%)
Si	0.25	–
Fe	0.35	–
Cu	0.05	99.60
Mn	0.03	–
Al	99.60	–
Bi	–	0.005
Pb	–	0.005

2.2 Mechanical Properties of the Bimetallic Material

To determine the material's properties, test specimens were obtained in accordance with ASTM E517-00, as shown in Fig. 1.

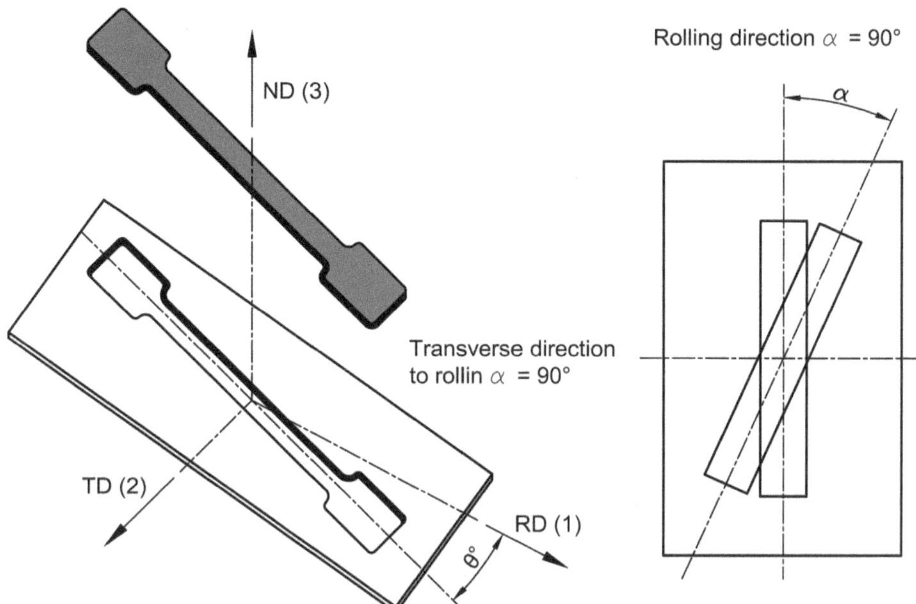

Fig. 1. Cutting direction of test pieces.

Twenty-one bimetallic specimens were tested, divided into three groups of seven each, oriented at 0°, 45°, and 90° with respect to the material's rolling direction. The properties obtained are presented in Table 2.

Table 2. Properties of the bimetallic sheet.

Al 1060 – CuT2 Bimetallic Material	
Elongation	19.22%
Ultimate Load or Maximum Load	4.51 kN
Breaking Load or Fracture Load	74.42 N/mm^2
Breaking Load or Rupture Load	0.82 kN

The strain hardening exponent (n) and the plastic anisotropic ratio (r) were calculated. These two properties are crucial for determining material formability. The bilayer properties used for these calculations were obtained from uniaxial tension tests conducted under displacement and strain control. The strain hardening exponent (n) was calculated following the procedures outlined in ASTM E646–16, 2016, while the plastic anisotropic ratio (r) was determined using the methodology of ASTM E517–00, 2010, with dependence on each cutting inclination. For this research, a rigid-plastic material is considered, where the strain hardening characteristics are defined by Eq. 1.

$$\sigma_v = K(\varepsilon_v)^N \tag{1}$$

where, συ is the true stress and ευ is the effective true strain. K is the strength coefficient (or the exponential hardening curve coefficient) applied to the sheet. It's obtained from the equation of a straight line in the form $y = a+bx$, as shown in Fig. 2, with an intercept $a = \ln(K)$ and a slope $b = N$.

The strain hardening exponent, N, was determined from the engineering stress-strain curve. This calculation utilized the engineering strain, $\varepsilon_{eng} = (L_1 - L_2)/L_0$, and the engineering stress, σeng, recorded at the onset of necking in the tensile specimen. Subsequently, the true stress, $\sigma true$, was computed using the relationship $\sigma_{true} = \sigma_{eng}\left(1 + \varepsilon_{eng}\right)$, while the true strain, ϵtrue, was calculated as $\ln(L_1/L_0)$.

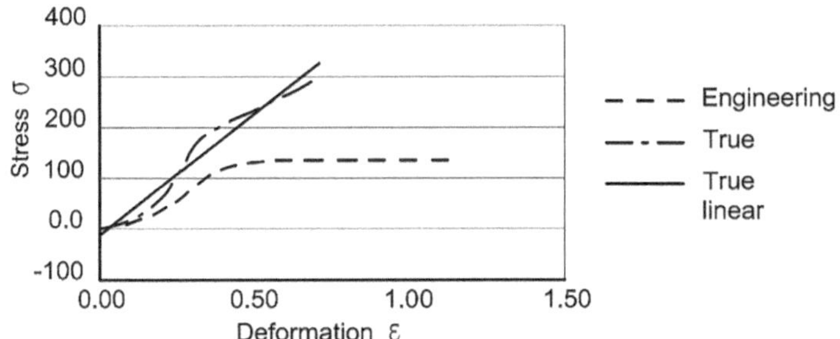

Fig. 2. Stress-strain diagram (True vs. Engineering).

The plastic strain ratio, often denoted as "r" or the Lankford parameter, which quantifies a metal sheet's aptitude to undergo thinning or thickening under tensile or compressive forces, was ascertained using Eq. 2. This equation incorporates the uniaxial strains.

$$r = \frac{\ln(w_0/w_f)}{\ln(l_f w_f/l_0 w_0)} \tag{2}$$

Materials often exhibit varying "r" values depending on their orientation relative to the rolling direction. Due to the inherent difficulty in precisely measuring thickness changes, an equivalent relationship, derived from strain measurements in both length and width, is commonly employed. Furthermore, we utilized rm, as defined by Equa-tion 3. This represents the weighted average of "r" and is also known as the normal anisotropy across the three analysed directions. Additionally, Δr, expressed in Equa-tion 4, indicates the propensity for wrinkling in the metal sheet, also referred to as planar anisotropy.

$$r_m = \frac{r_0 + 2r_{45} + r_{90}}{4} \tag{3}$$

$$\Delta_r = \frac{r_0 + r_{90} + 2r_{45}}{2} \tag{4}$$

Table 3 presents the values calculated from the stress-strain curve data.

Table 3. Mechanical properties of the bilayer sheet.

Property	Magnitude
Young's modulus of elasticity (E)	81 kN/mm^2
Poisson's ratio (υ)	0,34
Tensile strength σ_u	134,96 N/mm^2
Ductibility (% elongation)	30,95%
Density	3289 kg/m^3
Hardening exponent (n)	0,16
Lankord parameter r_0	1,66
Lankord parameter r_{45}	1,77
Lankord parameter r_{90}	1,96
Anisotropic coefficient (r_m)	1,84
Planar anisotropy (Δ_r)	0,04

Through the conducted tests and subsequent analysis of the obtained values, it was confirmed that the bimetallic sheet exhibits perfect adhesion between its two layers, experiencing tensile failure before delamination. Consequently, it was treated as a monolithic material for the deep drawing process.

2.3 Deep Drawing Parameters

The blank holder force exerted on the bilayer sheet is one of the most influential parameters in the deep drawing process. Insufficient force leads to wrinkling on the surface, while excessive force results in fracture. For this investigation, the model proposed by Morovvati, Mollaei-Dariani, and Asadian-Ardakani (Eq. 5) was applied to determine the minimum required blank holder force, Q.

$$Q = \frac{\frac{4D_{11}^e}{tY^2} - \frac{H(n,\rho)}{F_\rho^e(n,\rho)}}{4\frac{uw_{max}G(n,\rho)-4(1-\rho)^2 b}{\pi(tY)bw_{max}F_\rho^e(n,\rho)}} \tag{5}$$

where

$$G(n,\rho) = -\frac{1}{2}n^2\left(1+3\rho^2\right) - \frac{3}{2}\left(1-\rho^2\right) + 2n^2\left(p + \frac{1}{2}\rho^2 ln\rho\right)$$

$$H(n,\rho) = \left(n^2-3\right)\left(1-\rho^2+2\rho^2 ln\rho\right) - 2n^2\rho^2(ln\rho)^2$$

$$F_\rho^e(n,\rho) = \left[2+\left(n^2-1\right)^2\right]ln\frac{1}{\rho} + n^2(1-\rho)\left[\frac{1}{2}(1+\rho)\left(n^2+1\right) - 2\left(n^2-1\right)\right]$$

The constant ρ is determined by dividing the outer radius of the formed plate (defined by the drawing radius) by the radius of the blank or circular sheet, as observed in Fig. 3, where $\rho = a/b$.

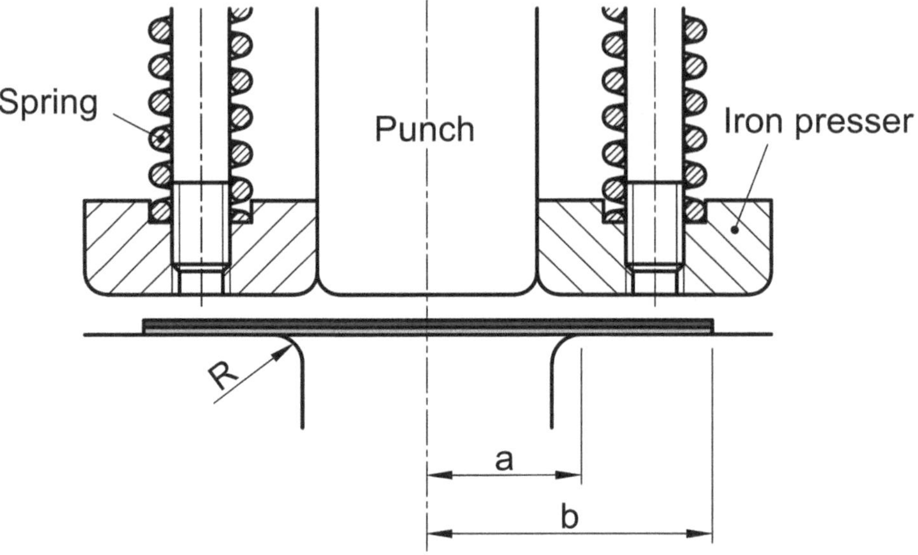

Fig. 3. Geometry of drawing radii and part radius.

The number of wrinkles (n) is calculated using the expression $n = 1.65\,\frac{r_m}{f}$, where r_m is the mean radius of the flange, and f is the flange distance and the maximum deflection, which is given by Eq. 6.

$$w_{max} = W|_{r=0,\theta=0} = 2c(b-a) \tag{6}$$

where c is a constant obtained by equating the circumference of the outer edge before and after the formation of drawing wrinkles.

$$c = \frac{2\sqrt{(b_0 - b)}}{n(b-a)}$$

The flexural rigidity D is

$$D_{11}^{e} = \frac{Et^3}{12(1-v^2)}$$

where:
v = Poisson's ratio
E = Young's Modulus or Modulus of Elasticity ($kN/mm^2 \approx MPa$).

Y is the material's yield stress

$$Y = \sigma_r - \sigma_\theta$$

The stress distribution shown in Fig. 4 is given by:

$$\sigma_r = \frac{\sigma \times a^2}{b^2 - a^2}\left(\frac{b^2}{r^2} - 1\right)$$

$$\sigma_\theta = -\frac{\sigma \times a^2}{b^2 - a^2}\left(1 + \frac{b^2}{r^2}\right)$$

where σ is the tensile strength

Finally, for the deep drawing process, t is the sheet thickness and μ is the coefficient of friction. Figure 4 illustrates the stress state in a section of the sheet, where the inner edge is subjected to tensile stresses (σ_r), circumferential compression (σ_θ) occurs, and the stress through the thickness is negligible.

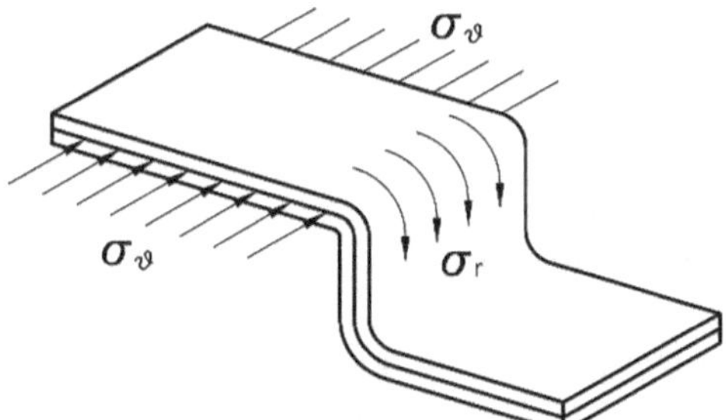

Fig. 4. Stress state in the flange zone.

The deep drawing force exerted on the sheet metal was established by applying the theory of plasticity proposed by Siebel–Beisswanger in Eq. 7.

$$F_{d.max} = \pi \times d_m \times t_0\left[1.1\frac{\sigma_{fml}}{n}\left(ln\frac{D_b}{d_p} - 0.25\right)\right] \quad (7)$$

where:

$F_{d.max}$ = maximum deep drawing force (N).
d_m = mean diameter of the drawn part (mm), determined by:

$$d_m = d_p - t0$$

d_p = punch diameter (mm)
D_b = blank diameter (mm)
t_0 = sheet thickness (mm)
n = a correction factor primarily due to friction; for this study, the value is 0.75.

f_{fmI} = average flow stress corresponding to the deformation between the inner and outer radii of the flange (N/mm^2), determined by:

$$\sigma_{fml} = 0.5 \times K(\varphi_1^N + \varphi_2^N)$$

φ_1 = true strain localized at the flange edge, obtained by:

$$\varphi_1 = ln\frac{D_b}{D}$$

φ_2 = true strain localized at the end of the flange or at the beginning of the die radius, obtained by:

$$\varphi_2 = ln\frac{\sqrt{D_b^2 - D^2 + (d_d + 2r_d)^2}}{d_d + 2r_d}$$

where:
d_d = die cavity diameter (mm)
D = instantaneous flange diameter (mm) where the maximum force will occur when D = 0,77D_b
r_d = die edge radius (mm)
N = exponent of the exponential hardening curve
K = strength coefficient or coefficient of the exponential hardening curve. For this research, a hardening coefficient K of 168 MPa was obtained. With the analysed equations and mathematical model, the deep drawing parameters presented in Table 4 are calculated.

Table 4. Deep drawing parameters.

Parameter	Magnitude
Blank Diameter (D)	71 mm
Maximum Drawing Force ($F_{m,max}$)	8 541.4 N
Punch Force (F_p)	4 207.0 N
Blank Holder Force (F_{p-p})	4 334.4 N

2.4 Numerical Simulation

Using the previously determined mechanical properties of the bimetallic sheet, the deep drawing process was simulated via finite elements using Ansys Workbench with the Explicit Dynamics module. The deep drawing die consists of a punch, blank holder, sheet metal, and die (given), all modelled as solid bodies, as shown in Fig. 5.

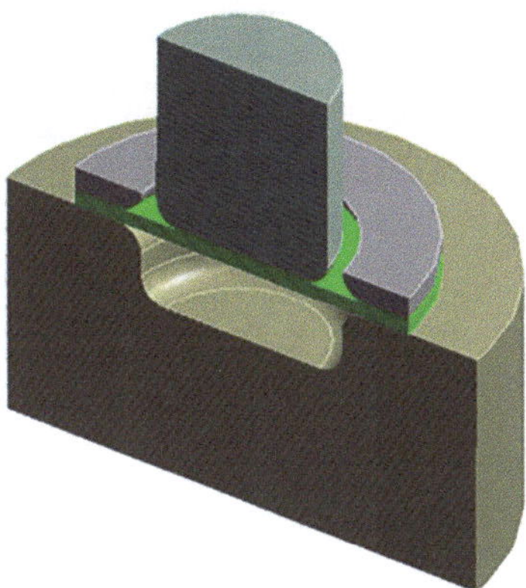

Fig. 5. Die drawing geometry.

The mechanical properties of the bimetallic material were calculated based on the tensile tests performed. Density was determined from the weight and volume of a material sample. The multilinear isotropic hardening values, as well as the specific heat, were obtained by interpolating the properties of aluminum and copper from the ANSYS explicit materials database with respect to the tensile strength of the individual and bimetallic materials obtained from the tensile test. The shear modulus was calculated u-sing Eq. 8.

$$G = \frac{E}{2(1 + v)} \tag{8}$$

For the Shock EOS Linear data in Fig. 6, the Rankine-Hugoniot equations were used for the shock conditions. These can be considered as defining a relationship between any pair of the variables: ρ (density), P (pressure), e (energy), vp (particle velocity), and U (shock velocity). The Gruneisen coefficient was taken from the tables by Harris and Avrami. The constant $C1$ corresponds to a strain rate hardening parameter of the material and is calculated using Eq. 9.

$$C_1 = \frac{\pi t^3 b(b_0 - b)[(n^4 - 2n^2 + 3)r^2 + 2n^2(1 - n^2)ar + n^4 a^2}{3n^4(b - a)^2 r^3} \tag{9}$$

The quadratic parameter S2 is zero for linear materials but can have a non-zero value for highly nonlinear materials. Figure 6 shows the input parameters for the bi-metallic sheet used in the simulation.

Properties of Outline Row 3: AlCu

	A
1	Property
2	Density
3	Multilinear Isotropic Hardening
4	Scale
5	Offset
6	Specific Heat
7	Shear Modulus
8	Shock EOS Linear
9	Gruneisen Coefficient
10	Parameter C1
11	Parameter S1
12	Parameter Quadratic S2

Fig. 6. Bilayer material properties.

For the material properties of the deep drawing die elements, data from an explicit structural steel within the Ansys software was utilized. The die was considered rigid and a fixed support, as shown in Fig. 7. The punch's displacement is 15 mm in the negative Y-axis direction and is constrained in the X and Z-axis directions. The punch is rigid, and a force of 4 207 N was applied to it. The force applied to the blank holder is in tabular form: at time zero, the pressure is also zero, and at a time of 1×10^{-3} s, the force is 4 334 N. Due to computational expense in the simulation, the punch's velocity is applied in the negative Y direction and set at –200 m/s. The time was selected by assumption and is 1×10^{-3} s.

For the contact surfaces between the sheet and blank holder (steel-aluminum), static and dynamic friction coefficients were set at 0.1 and 0.05, respectively. Meanwhile, at the sheet-die (copper-steel) contact surface, the static and dynamic friction coefficients were 0.1 and 0.08, respectively, considering that these surfaces are lubricated.

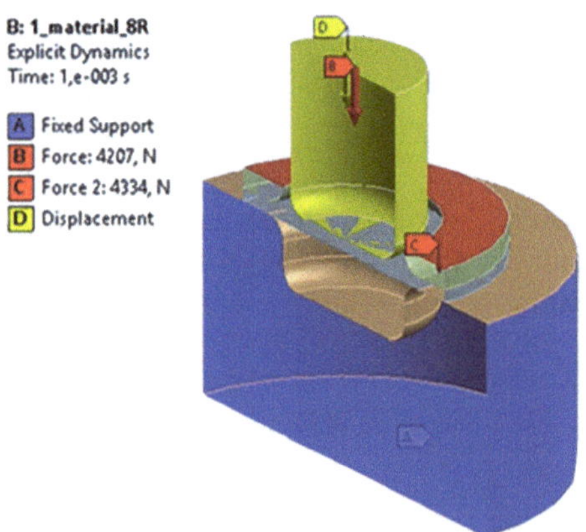

Fig. 7. Boundary conditions.

Figure 8 shows the meshing applied to the system. The element size for the blank to be drawn is 1 mm, while for the other components, ANSYS's default settings are used. Additionally, a medium mesh with curvature refinement is applied to all elements. The number of nodes obtained is 32 868, and the number of elements is 41 743.

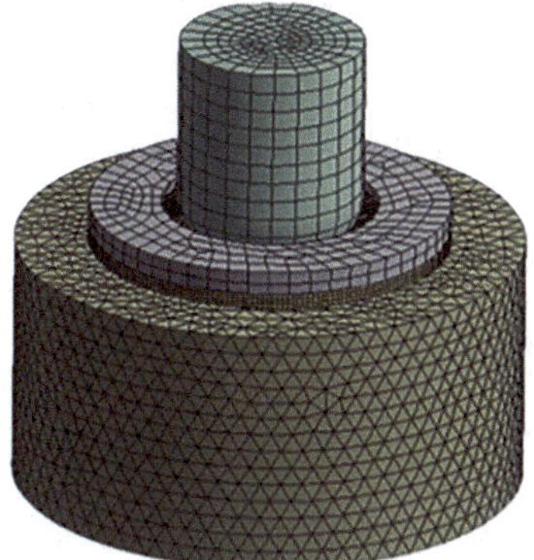

Fig. 8. Meshing applied to the deep drawing process.

Figure 9 displays the convergence plot for the deep drawing system, showing an 86.65% convergence rate, which confirms that the problem statement has a valid solution.

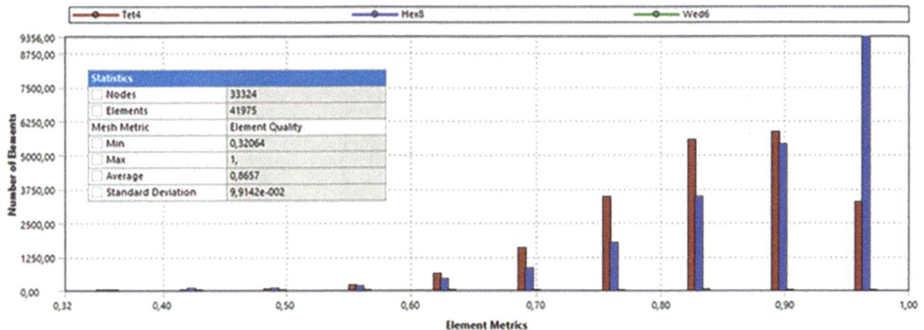

Fig. 9. Deep drawing process convergence diagram.

For post-processing, the simulation was performed with the calculated blank holder force of 4 334 N, which prevents wrinkling in the flange of the drawn part. To further investigate wrinkle formation, tests were also conducted with forces of 2 167 N and 1083 N, corresponding to one-half and one-quarter of the calculated blank holder force, respectively.

To complete the simulations, a run without a blank holder was performed, revealing maximum wrinkle deformation. Additionally, another test was conducted with a force of 5416 N, exceeding the calculated value, to confirm that under these conditions, no wrinkles form in the drawn part's flange; however, the part undergoes shearing at its bottom.

2.5 Experimental Procedure

Laboratory tests were conducted with the die both with and without a blank holder. The applied force was variable, utilizing two, four, eight, and ten equally spaced springs to ensure a uniform force on the blank holder. The springs comply with ISO 10243 G16–076 series standards, with a 20% compression corresponding to 15.2 mm, which generates a force of 518 N. This value is the closest approximation to the calculated 541 N for the current study.

The die designed for the deep drawing tests is shown in Fig. 10. Its operation was performed using a 12-ton hydraulic press, which allowed for pressure measurement during each process. A stopper was used in the die to ensure the punch's dis-placement was precisely 15 mm in all tests.

The deep drawing force, or total calculated force, required to obtain a drawn part without flange wrinkles is 8 541 N. This force comprises components such as the blank holder force, the punch force, the friction force, and other minor forces generated by the process. The calculated blank holder force is 4 334 N, making the sum of the remaining forces 4 207 N. These values were utilized in the simulation process.

Fig. 10. Experimental deep drawing die.

3 Results and Discussion

In the practical setup, the exact calculated blank holder force of 4334.4 N could not be applied due to the use of standardized springs; placing eight springs provides a force of 4 144 N. The remaining forces are attributed to the punch force, which was treated as a single force in the simulation. The calculated punch force is 4 207 N, whereas the experimental force is 4 000 N. With these results, the deep drawing force (total force) initially shows a 4.65% error between theoretical and practical values.

In subsequent tests, the number of springs was gradually reduced to demonstrate that the force applied to the blank holder causes flange wrinkling. Additionally, a test was conducted with a force exceeding the calculated value (ten springs) to induce fracture in the drawn part. As the force applied to the blank holder decreases, the percentage error in the deep drawing force increases, albeit not significantly.

Directional deformation using Finite Element Method (FEM) was exclusively performed along the Y-axis, which represents the height of the part. This analysis allowed for the determination of the flange's deformation, which was then compared against the deformation observed in the deep-drawn parts. The maximum error percentage recorded was 6.51%. This occurred during the deep drawing process utilizing 10 springs, indicating that the blank holder force (BHF) was greater than the calculated value. Consequently, the experimental force was lower in all instances, suggesting that the simulated deformation should also be greater than the experimental deformation. The relationship between the force and deformation errors was termed the "real error" (Table 5).

Table 5. Deep drawing parameters.

N°	Directional deformation (mm)			
Compression spring	Simulated	Experimental	% Error	% Simulated error
10	14.89	13.92	6.51	−1.91
8	15.12	15.50	2.51	2.14
4	15.94	16.08	0.88	3.86
2	16.23	16.45	1.35	3.45
0	18.91	18.15	4.01	0.91

Positive percentage values indicate that the experimental deformation exceeds the simulated deformation. However, for the blank holder utilizing 10 springs, the opposite is true: the experimental deformation is lower. This discrepancy arises because, in practice, the deep-drawn part shears before reaching the required depth, and deformations across all tests remain below 4%. The FEM and experimental results are presented in Figs. 11 and 12, respectively.

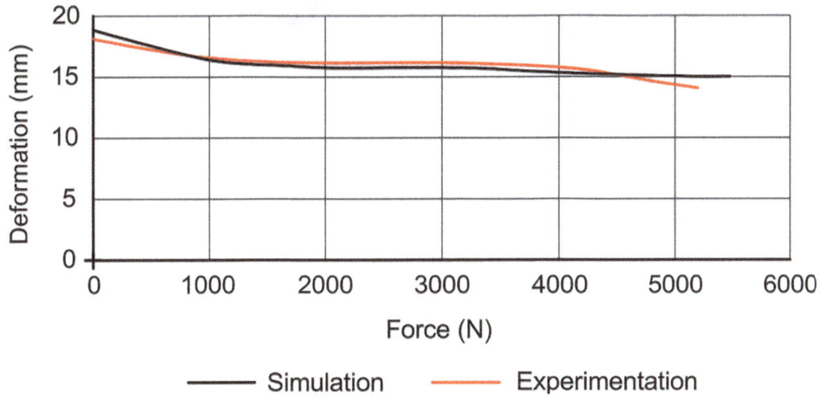

Fig. 11. Simulated vs. Experimental deformation.

As observed in Table 6, the shear stresses when deep drawing without a blank holder, and with a blank holder utilizing 2 or 4 springs, do not exceed the allowable shear stress of the bilayer material, which is 112 MPa. Therefore, the material does not exhibit shear failure.

However, during deep drawing with 8 blank holder springs—the force that prevents wrinkling in the flange a shear stress of 112.08 MPa is generated. This value is similar to the material's allowable shear stress (112 MPa), representing the ideal or admissible force for this study. When the calculated force is surpassed by implementing 10 springs in the die, a shear stress of 112.68 MPa is produced. This value exceeds the allowable shear stress (112 MPa), leading to a failure in the deep drawing process. Figure 16 indicates the locations of maximum shear stress for a deep-drawn part formed with 8 springs and without springs.

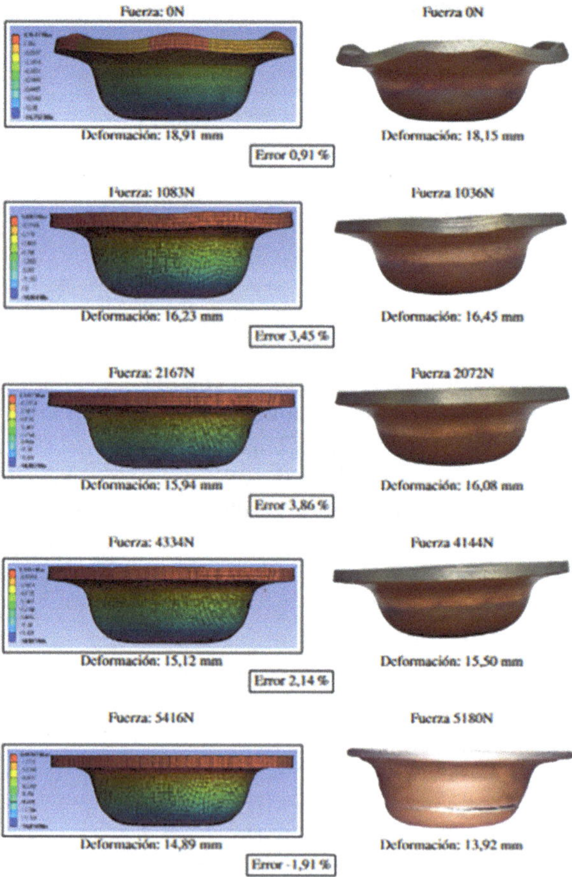

Fig. 12. Simulated vs. Experimental deformation.

Table 6. Simulated Shear Stresses.

N°	Calculated	Calculated	Calculated deep	Shear
compression	punch force	blank holder	drawing force	Stress
spring	(N)	force (N)	(N)	(MPa)
10	4207	5 416	9 623	112.68
8		4 334	8 541	112.08
4		2 167	6 374	111.16
2		1 083	5 290	110.64
0		0	4 207	109.17

Figure 13 illustrates the maximum shear stresses on the deep-drawn component, both with a blank holder actuated by 8 springs and without its action. The equivalent plastic strain indicates the part's maximum deformation value. Table 7 presents these values, resulting from the various forces applied during the process. The minimum simulated deformation at the bottom radius of the part is 0.56 mm when operating without a blank holder, and 0.61 mm when working with the calculated maximum admissible force of 4334 N on the blank holder.

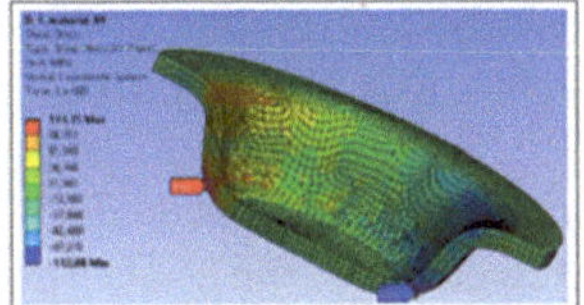

Fig. 13. Maximum Shear Stresses with and Without Springs.

When the blank holder force exceeds the calculated value, the simulation yields a plastic deformation of 0.67 mm, leading to failure. Figure 14 illustrates the equivalent plastic strain for a deep-drawn part using four springs in the blank holder, corresponding to a calculated force of 2167 N and a deformation of 0.57 mm. Furthermore, practical results show a force of 2072 N and a deformation of 0.52 mm.

Table 7. Simulated Shear Stresses.

N°	Calculated	Calculated	Calculated deep	Equivalent	Maximum
compression	punch force	blank holder	drawing force	plastic strain	strain
spring	(N)	force (N)	(N)	(mm/mm)	Percentage
10	4207	5 416	9 623	0.678	22.6
8		4 334	8 541	0.611	20.3
4		2 167	6 374	0.574	19.1

(*continued*)

Table 7. (*continued*)

N^{O}	Calculated	Calculated	Calculated deep	Equivalent	Maximum
2		1 083	5 290	0.576	19.2
0		0	4 207	0.563	18.7

An 8.77% error occurs relative to the initial thickness, which decreases to 4.03% if the initial error between the calculated and practical blank holder forces is compensated.

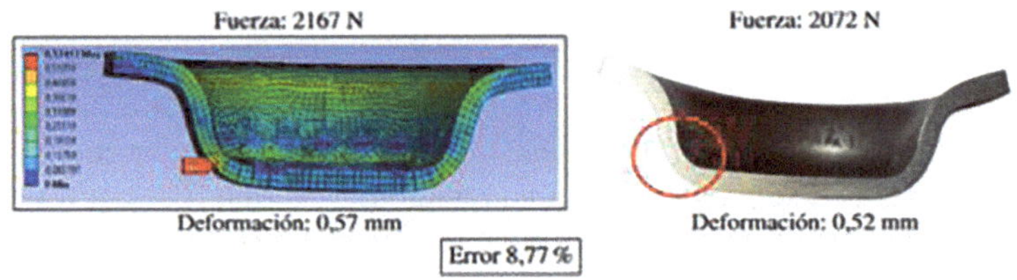

Fig. 14. Maximum Shear Stresses with and Without Springs.

4 Conclusions

Tensile tests performed on the bilayer material revealed that copper-clad aluminum exhibits a 14.94% improvement in elongation and a 26.35% increase in tensile strength compared to uncoated aluminum. According to the material producer, the interlayer bond strength exceeds 40 MPa. Tensile testing confirmed that the bimetallic sheet experiences tensile failure before delamination, leading us to consider the bilayer material as a single sheet for analysis.

Conducting experiments is a slow and costly process; therefore, it becomes necessary to utilize modified analytical models for single-material sheet metal deep drawing, which, combined with FEM, demonstrate a strong correlation between analytical and experimental results. Directional deformations, shear stresses, and equivalent plastic strain all show errors below 4% between the analytical and experimental studies. All results are satisfactory, except for the test employing a blank holder force greater than the calculated value. This is because explicit dynamic simulation does not depict material rupture; instead, this can be observed when the material's shear stress exceeds its allowable shear stress, which occurred in this specific test. For all other cases, the calculated blank holder force generates shear stresses equal to or less than the material's allowable shear stress, confirming that the material will not fail under forces below the calculated value. The anisotropy values obtained for the bimetallic material at 0°, 45°, and 90° exhibit a maximum deviation of merely 0.17. Consequently, no significant wrinkling was observed in the flange with forces below the admissible value, according to the chosen model.

References

1. Dehghani, F., Salimi, M.: Analytical and experimental analysis of the formability of cop-per stainless - steel 304L clad metal sheets in deep. Int. J. Adv. Manuf. Technol. **82**, 163–167 (2016)
2. Li, L., Nagai, K., Yin, F.: Progress in cold roll bonding of metals. Sci. Technol. Adv. Mater. **9**(2), 023001 (2008)
3. Shajari, Y., et al.: Formation of intermetallic compounds in Al–Cu interface via cold roll bonding. Surf. Eng. Appl. Electrochem. **58**(1), 41–50 (2022)
4. Uscinowicz, R.: The effect of rolling direction on the creep process of Al–Cu bimetallic sheet. Mater. Des. **49**, 693–700 (2013)
5. Shiran, M., Bakhtiari, H., Mousavi, S., Khalaj, G., Mirhashemi, S.: Effect of stand-off distance on the mechanical and metallurgical properties of explosively bonded 321 austenitic stainless steel - 1230 aluminum alloy tubes. Mater. Res. **20**, 291–302 (2017)
6. Sheng, L., Yang, F., Xi, T., Lai, C., Ye, H.: Influence of heat treatment on interface of Cu/Al bimetal composite fabricated by cold rolling. Compos. B Eng. **42**(6), 1468–1473 (2011)
7. Bykov, A.: Bimetal production and applications. Steel Transl. **41**(9), 778 (2011)
8. Chen, Y., Wang, A., Xie, J., Guo, Y.: Deformation mechanisms in Al/Al2Cu/Cu multilayer under compressive loading. J. Alloy. Compd. **885**, 160921 (2021)
9. Taali, S., Toroghinejad, M., Saeidi, N.: Architectured lightweight steel composite: evaluation of the effect of geometrical parameters and annealing treatments on deformation behavior. J. Mater. Res. Technol. **15**, 5414–5427 (2021)
10. Lesuer, D., Syn, C., Sherby, O., Wadsworth, J., Lewandowski, J., Hunt, W.: Mechanical behaviour of laminated metal composites. Int. Mater. Rev. **41**(5), 169–197 (1996)
11. Kumar, M., Choudhary, A.: Plastic wrinkling investigation of sheet metal product made by deep forming process: a FEM study. Int. J. Eng. Res. Technol. **3**(10), 186–191 (2014)
12. Demirci, H., Yacsar, M., Demiray, K., Karalí, M.: The theoretical and experimental investigation of blank holder forces plate effect in deep drawing process of AL 1050 material. Mater. Des. **29**(2), 526–532 (2008)
13. Yu, T., Johnson, W.: The buckling of annular plates in relation to the deep-drawing process. Int. J. Mech. Sci. **24**(3), 175–188 (1982)
14. Morovvati, M., Mollaei, B., Asadian, M.: A theoretical, numerical, and experimental investigation of plastic wrinkling of circular two-layer sheet metal in the deep drawing. J. Mater. Process. Technol. **210**(13), 1738–1747 (2010)
15. Anarestani, S.S., Morovvati, M.R., Vaghasloo, Y.A.: Influence of anisotropy and lubrication on wrinkling of circular plates using bifurcation theory. Int.J. Mater. Form. **8**(3), 439–454 (2014). https://doi.org/10.1007/s12289-014-1187-6
16. Darendeliler, H., Kaftanoglu, B.: Deformation analysis of deep-drawing by a finite element method. CIRP Ann. **40**(1), 281–284 (1991)
17. Flores, H.: Un algoritmo de concepto para el análisis explícito de procesos de embutición. Revista internacional de métodos numéricos **16**(4), 421–432 (2000)
18. Garrido, C., Celentano, D.; Castillo, J., Guerra, J.: Simulación del proceso de embutición de una tapa de embrague de una lavadora semiautomática. La Serena (2004)
19. Bernai, Y., Urama, R., Marty, J., Okoye, N.: Development of intelligent control of optimum parameters in deep drawing of sheet metal using genetic algorithm and finite element methods. Adv. Mater. Res. **690**, 2280–2290 (2013)

BY NC ND

Analysis of the Dynamic Behavior of the OPGW Cable–Spiral Damper System: An Experimental Approach

Damián Campos[1(✉)], Andrés Ajras[1], and Marcelo Piovan[2]

[1] Laboratorio de Ensayo de Conductores, Departamento de Mecánica Aplicada, Facultad de Ingeniería, Univ. Nacional del Comahue, Neuquen, Argentina
damian.campos@fain.uncoma.edu.ar

[2] Centro de Investigaciones en Mecánica Teórica y Aplicada, Univ. Tecnológica Nacional Facultad Regional Bahía Blanca - CONICET, Bahía Blanca, Argentina

Abstract. This study focuses on identifying nonlinear damping characteristics through the analysis of experimental data obtained from laboratory span tests conducted on an OPGW cable system equipped with a spiral-type damper. These dampers play a significant role in mitigating aeolian vibrations in overhead power transmission lines. The experimental data collected are used to assess the system's behavior under external excitation, enabling a deeper understanding of its dynamic response. Advanced signal processing techniques and numerical methods based on the KarhunenLoève transform are employed to analyze the data and extract relevant damping parameters. The results of this analysis provide valuable insights into the practical design and implementation of the damping system within the span of a transmission line.

Keywords: Spiral damper · cables · aeolian vibrations · KarhunenLoève

1 Introduction

Due to their extrinsic characteristics, medium- and high-voltage overhead power transmission lines are highly exposed to wind action. One of the consequences of this exposure is that the cables comprising these lines experience vibrations and oscillations due to various types of aerodynamic and aeroelastic instabilities.

The types of vibrations and oscillations mentioned above are closely related to the configuration of the line, particularly the arrangement of its conductors and cables. In the case of single-conductor-per-phase lines or the ground wire, the leading cause of vibration is vortex shedding, a phenomenon known as aeolian vibration. If the line consists of conductor bundles per phase, the primary cause of oscillation is associated with the wake effect generated by the windward conductor acting on the leeward conductor, a phenomenon referred to as sub-span

O. F. Farías Fuentes et al. (Eds.): CIBIM 2024, *Proceedings of the XVI Ibero-American Congress of Mechanical Engineering*, pp. 73–82, 2026.
https://doi.org/10.1007/978-3-032-22823-9_6

oscillation. Finally, galloping is a self-excited vibrational phenomenon characterized by low frequency and large amplitude, commonly occurring in conductors with ice accretion during winter conditions.

Among these, aeolian vibration is the phenomenon with the highest frequency. If not properly mitigated, it can lead to fatigue damage, including wear and/or breakage of cables, hardware, insulators, and clamps. In severe cases, it may even affect the structural behavior of the supporting towers. Specifically, in conductors, such damage is associated with the accumulation of bending fatigue cycles at the fixation points where the cables are secured to the hardware. The severity of the process depends on the level of stress/strain to which the individual wires of the conductor or cable are subjected, as well as the number of accumulated cycles [3,4].

To mitigate the harmful effects described above and prevent potential failures, it is necessary to reduce conductor and cable vibrations and oscillations to acceptable levels. This can be achieved by increasing the corresponding damping. There are two main ways to accomplish this: the first, more costly option, involves reducing the mechanical tension of the conductor to increase its internal damping. However, this leads to a greater sag and, consequently, taller support structures, with a direct impact on the overall cost of the line. The second, generally more economical alternative, is to equip conductors and ground wires with energy absorbers.

In this regard, the use of Stockbridge-type dampers is widespread for single conductors and ground wires. Although the original design, introduced in 1925, has been modified and optimized over time, it retains the same basic operating principle: two masses suspended at the ends of a stranded messenger cable clamped to the phase conductor or ground wire. The size and shape of the masses, as well as the entire geometry of the damper, affect its performance and the amount of energy it can dissipate at specific frequencies. For this reason, asymmetric designs are commonly used to optimize frequency response and energy dissipation across a broader range of frequencies.

When specifying these types of dampers, it is essential to consider the selected phase conductor and ground wire, as well as the expected oscillation frequencies and typical wind speeds in the region where the line is installed. Accordingly, the damping study should provide the optimal placement of the damper, which, in principle, is at the antinode of the standing wave. However, since multiple oscillation frequencies may be present, the damper design must ensure effective energy dissipation across the entire frequency range. This requirement aims to prevent the damper from being located at a node for any of the potential vibration modes [5].

In recent decades, the Spiral Vibration Damper (SVD) has become one of the most widely used impact dampers on the market. The SVD mitigates aeolian vibration by dissipating vibrational energy through impact with the cable. This type of damper is highly effective on small-diameter cables (such as OPGW cables), but its effectiveness decreases with larger conductors. This reduction is attributed to the mass-to-frequency ratio between the damper and the cable [6].

Given this context, the present study focuses on an OPGW cableSVD damper system. The damping efficiency of SVDs on these specific cables has been documented using field vibration measurements. In such analyses, field data are processed and evaluated to ensure that vibration levels remain below the fatigue endurance limit. However, these field measurements are highly costly and time-consuming. Therefore, it is recommended to conduct laboratory tests to assess the efficiency of the OPGW cableSVD damper system.

In the following discussion, the application of the laboratory testing technique recommended by the relevant standards is presented, along with the dynamic response characterization of the system using numerical methods based on the KarhunenLoève transform.

2 Methodology

2.1 Experimental Setup

Laboratory measurements of the energy dissipated by vibration dampers used in power transmission lines are employed by engineers to validate line designs and assess the suitability of a damper for a given application. Consequently, several methods have been developed to characterize the energy dissipation performance of various types of available dampers.

The IEEE Standard 664-1993 [7] describes four basic test methods: the power method, the Inverse Standing Wave Ratio (ISWR) method, the decay method, and the forced response method. These techniques require pure sinusoidal excitation to quantify energy dissipation accurately and have, therefore, been used almost exclusively for Stockbridge-type dampers. Impact-type dampers, such as the Spiral Vibration Damper (SVD), introduce an almost random component to the cable's motion during vibration, which complicates the application of the methods above. Therefore, the adopted testing procedure must be implemented meticulously, with particular attention to the data acquisition and processing systems.

In this study, the laboratory span was adapted to conduct the test using the ISWR method. Figure 1 shows the adopted setup, which includes a 30-m active cable span. The main components are the clamping devices at both ends, a system that maintains constant axial tension in the cable, and an electromechanical exciter.

The experimental test replicates the impact of aeolian vibrations on the cable. When the frequency of vortex-shedding-induced forces matches one of the natural frequencies of the cable span, aeolian vibration occurs. The wind speed (V) that triggers the vibration is related to the vibration frequency (f) through the Strouhal relationship (St), given by Eq. 1. This dimensionless number is approximately 0.185 for all typical transmission line cables [8]. The resonance frequencies are within the range of 0.18/D to 1.4/D, where D is the cable diameter in meters.

$$f = St\frac{V}{D} \tag{1}$$

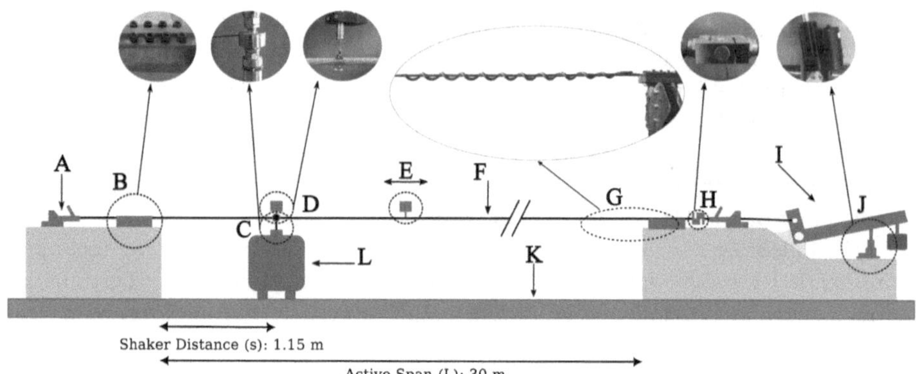

References: (A) Tension Clamp, (B) Rigid Clamp, (C) Load Cell (Excitation Force), (D) LVDT Sensor, (E) Accelerometer, (F) OPGW Cable, (G) SVD Damper, (H) Load Cell for Axial Tension, (I) Constant Tension Device, (J) Electric Linear Actuator, (K) Concrete Base, and (L) Shaker.

Fig. 1. Experimental setup.

Vibrations take the form of discrete standing waves with forced nodes at the support structures and intermediate nodes spaced at intervals along the span. The natural vibration frequencies (f_n) of a tensioned cable are analytically calculated using the following expression:

$$f_n = \frac{n}{2L}\sqrt{\frac{T}{m}} \tag{2}$$

where:

n: Number of standing wave loops in the span $(1, 2, 3, \ldots)$
L: Span length [m]
m: Mass per unit length of the cable [kg/m]
T: Axial tension in the cable [N]

To carry out the damping efficiency test, at least ten test points must be selected within the frequency range of interest. At each of these points, the wave on the cable must be stable, stationary, and exhibit at least three complete loops. The measurement points correspond to the first free node and antinode. At each frequency, the excitation amplitude must be adjusted so that the peak-to-peak velocity at the antinode reaches 200 mm/s. The following parameters must be measured and recorded: node amplitude, antinode amplitude, and loop length, as illustrated in the diagram shown in Fig. 2.

The power dissipated by the damper (P) and the damping efficiency of the system (e) are given by the following equations:

$$P = \sqrt{Tm}\frac{V_a^2}{2}\left(\frac{Y_n}{Y_0}\right) = ez\frac{V_a^2}{2} \tag{3}$$

$$e = \frac{Y_n}{Y_0} \tag{4}$$

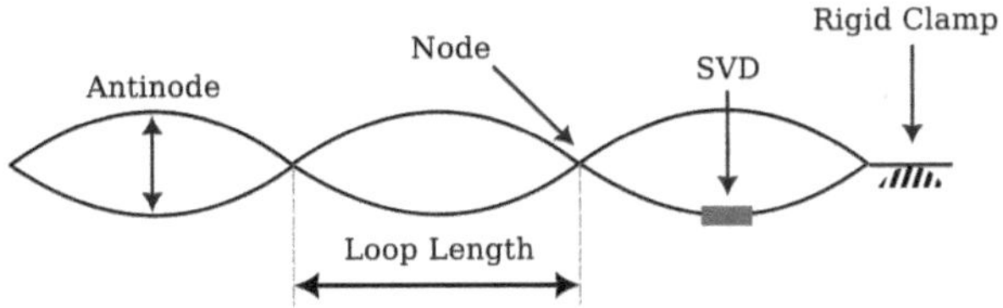

Fig. 2. Location of nodal and antinodal measurements.

where:

V_a: Velocity at the antinode
Y_n: Amplitude at the node
Y_0: Amplitude at the antinode
z: Characteristic impedance of the cable ($\sqrt{Tm}$)

A frequency inverter regulates the exciter, allowing resonance frequencies of the cable to be tuned within the range of interest. A load cell is used to measure the force of excitation. The test setup is sensitive to variations in axial tension. To control this load within a specific range, a closed-loop control system uses the load cell measurement to adjust a lever arm continuously via an electric linear actuator.

Since the system exhibits nonlinear behavior, it is not always possible to generate pure sinusoidal signals at resonance. Therefore, a Fourier Transform was applied to the signal to extract the amplitude and phase of the dominant vibration component.

2.2 Karhunen–Loève Transform

The Karhunen-Loève Transform (K-L), also known as Proper Orthogonal Decomposition (POD), is a key tool in multivariate data analysis, renowned for its ability to simplify and uncover patterns in complex datasets. Its application spans various disciplines such as engineering, physics, and economics, where multidimensional data is common [9–11].

The POD method begins by identifying the original variables and their relationships through the use of a covariance matrix. The decomposition of this matrix yields eigenvectors and eigenvalues that represent the principal directions and magnitudes of variation in the data. By projecting the data onto the subspace defined by the eigenvectors associated with the largest eigenvalues, POD reduces dimensionality without losing critical information [12].

Among its advantages, POD can reveal hidden structures in the data that are not readily apparent, thus facilitating pattern identification and the improvement of predictive models. Moreover, dimensionality reduction enhances computational efficiency and facilitates data analysis and visualization, allowing for faster calculations and clearer graphical representations [13].

However, although POD minimizes the mean square error between the original signal and its reduced linear representation, its main limitation lies in the fact that it only provides the best linear subspace approximation within the configuration space represented by the data. This may be a drawback when the data lies on a nonlinear subspace [14].

Given its widespread use in structural dynamics, it is relevant to provide an overview of POD as applied to the analysis of experimental acceleration data (a) obtained from the previously described setup.

The following presents the mathematical formulation of the proposed method. First, a series of snapshots is collected-these are essentially time-series data of the system response under various conditions. These instantaneous captures are organized into a matrix $Y \in \mathbb{R}^{N \times n}$, where N is the number of spatial dimensions, and n is the number of snapshots. The resulting matrix is then decomposed using the mathematical technique of Singular Value Decomposition:

$$Y = U \Sigma V^T \tag{5}$$

Here, U contains the left singular vectors (the POD modes), ΣV^T is a diagonal matrix of singular values, and V^T contains the right singular vectors. The POD basis is constructed from the left singular vectors associated with the largest singular values. This basis effectively captures the dominant features of the data, enabling a reduced representation of the system. The reduced-order model can be expressed as:

$$u_N(\mu) \approx \sum_{i=1}^{r} \xi_i(\mu) \chi_i(x) \tag{6}$$

where $u_N(\mu)$ is the full-order solution, $\xi_i(\mu)$ are the coefficients, and $\chi_i(x)$ are the spatial modes derived from POD. The proposed method was implemented by adapting the FDApy package, which is used for analyzing functional data in Python [15]. Since acceleration signals typically contain characteristic noise associated with this type of measurement, the applied algorithm smooths the curves in the time domain by estimating the probability density using the Epanechnikov kernel method [16].

3 Case Study and Results

The case study focuses on the installation project of SVD-type dampers on the OPGW cable of a 500 kV transmission line located in the southern region of Buenos Aires Province, Argentina. The proposal involves replacing the original damping system, which consists of Stockbridge-type dampers, as vibration records obtained from field measurements exceeded the recommended maximum levels. The tests were conducted under an applied tensile load of 7.5 kN. The main characteristics of both the OPGW cable and the SVD damper are summarized in Table 1.

The results obtained from the laboratory tests are presented below. Figure 3 illustrates the efficiency of the damping systems, comparing the performance of the existing Stockbridge-type dampers with that of the proposed SVD system. The latter demonstrated superior energy dissipation capacity across the entire range of tested frequencies. These results are consistent with previous comparative studies reported in the literature [17].

Table 1. System characteristics.

OPGW Cable	
Nominal diameter	16.5 mm
Nominal weight	549 kg/km
Cross-sectional area of armor	134.2 mm^2
Aluminum alloy strands	100.4 mm^2
Aluminum-clad steel strands	33.8 mm^2
Ultimate tensile strength	73.2 kN
Allowable tensile strength	40.3 kN
Elastic modulus	83 GPa
SVD Damper	
Diameter range	14.4–19.3 mm
Length	1.65 m
Weight	0.45 kg

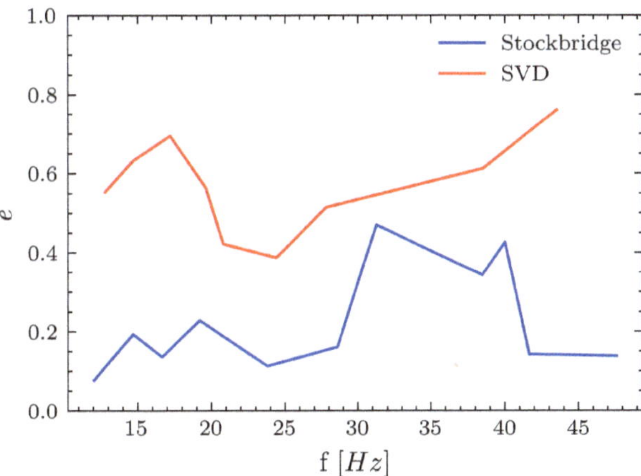

Fig. 3. Efficiency of the damping system.

It is important to mention that the tests conducted inherently involve uncertainty associated with the measurement procedures and the sensitivity of the instruments used-factors that are not addressed in the methodologies applied according to the reference standards.

Figure 4 shows the vibration records for different frequencies. In all cases, the measured acceleration patterns, the smoothed curve, and the reconstruction obtained through the application of Proper Orthogonal Decomposition (POD) are compared.

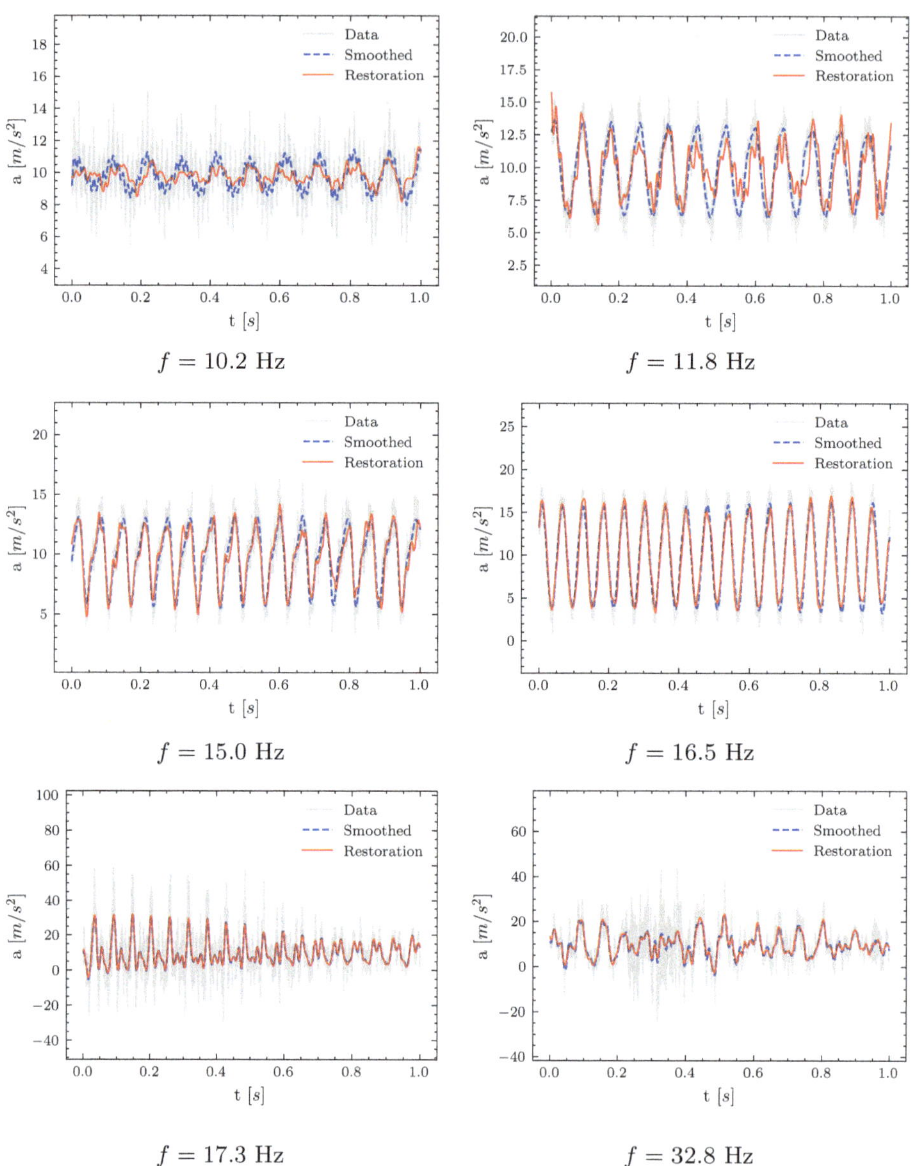

Fig. 4. Vibration records at different frequencies.

The results demonstrate an excellent fit to the actual behavior of the system, thanks to the use of the KarhunenLoève basis. This methodology enables a precise and efficient representation of the acceleration signal. It is observed that the principal components of the basis effectively capture the variability and the most relevant features of the system, resulting in a significant reduction of the mean squared error. This fit confirms the validity of the model and its ability to represent the underlying dynamics accurately.

Moreover, the computational cost is low, as the code requires only a few function calls to yield results within an approximate runtime of five minutes, utilizing the following hardware: an Intel i5-11400H processor and 16 GB of DDR4 RAM at 3200 MHz. This efficiency in terms of processing time and resource usage underscores the practicality of the method, particularly for applications that require a compact and reliable representation of complex data without incurring high system performance costs.

4 Conclusions

In this work, the methodology based on the KarhunenLoève (K-L) transform proved to be a powerful tool for the analysis and design of damping systems, offering several key advantages that significantly enhance dynamic analysis.

Firstly, the K-L transform enabled direct analysis of experimental data, which facilitated the modeling of the cableSVD damper system. Given the context in which the system equations are not fully defined or are difficult to derive, the K-L approach provided an effective approximation through model simplification and dimensionality reduction. This greatly improves the ability to understand and predict the system's behavior without requiring exhaustive knowledge of all parameters and governing equations.

Secondly, the K-L approach reduces computational requirements. By using a reduced model derived from the K-L transform, the computational complexity needed for simulating and analyzing the system is significantly lowered. This enables faster and more efficient simulations, making future studies involving uncertainty quantification feasible without a proportional increase in computational cost.

References

1. Chan, J.: Updating the EPRI transmission line reference book: wind-induced conductor motion. EPRI, vol. 3, no. 3 (2005)
2. Ghannoum, E., Chouteau, J.P., Miron, M., Yaacoub, S., Yoshida, K.: Optical ground wire for hydro-Quebec's telecommunication network. IEEE Trans. Power Deliv. **10**(4), 1724–1730 (1995)
3. Cosmai, U., Van Dyke, P., Mazzola, L., Lillien, J.L.: Conductor motions. In: Papailiou, K. (ed.) Overhead Lines. CIGRE Green Books, Springer, Cham (2017)
4. Papailiou, K.O.: Overhead Lines. Springer Handbook of Power Systems. Springer Handbooks (2021)
5. Wang, Z., Li, H.N., Song, G.: Aeolian vibration control of power transmission line using Stockbridge type dampers - a review. Int. J. Struct. Stab. Dyn. **21**(1), 1–29 (2021)

6. Sunkle, D.C., Olenik, J.J., Fullerman, M.D.: Determination of damping effectiveness of impact damper on ADSS cable. In: IEEE 9th International Conference on T&D Construction, Operation and Live-Line Maintenance, Montreal, QC, pp. 195–201 (2000)
7. IEEE Std 664-1993: IEEE Guide for Laboratory Measurement of the Power Dissipation Characteristics of Aeolian Vibration Dampers for Single Conductors, pp. 1–24 (1993)
8. Diana, G.: Modelling of Vibrations of Overhead Line Conductors. CIGRE Green Books. Springer (2018)
9. Kerschen, G., Golinval, J.C., Vakakis, A.F., Bergman, L.A.: The method of proper orthogonal decomposition for dynamical characterization and order reduction of mechanical systems: an overview. Nonlinear Dyn. **41**, 147–169 (2005)
10. Lin, W.Z., Lee, H.P., Lu, P., Lim, S.P., Liang, Y.C.: The relationship between eigenfunctions of Karhunen-Loève decomposition and the modes of distributed parameter vibration system. J. Sound Vib. **256**, 791–799 (2002)
11. Wolter, C., Trindade, M.A., Sampaio, R.: Obtaining mode shapes through the Karhunen-Loève expansion for distributed-parameter linear systems. Shock. Vib. **9**(4–5), 177–192 (2002)
12. Bellizzi, S., Sampaio, R.: POMs analysis of randomly vibrating systems obtained from Karhunen-Loève expansion. J. Sound Vib. **297**(3), 774–793 (2006)
13. Golovkine, S.: Statistical methods for multivariate functional data. Doctoral Thesis. ENSAI, Rennes (2021)
14. Feeny, B.F., Liang, Y.: Interpreting proper orthogonal modes in randomly excited vibration systems. J. Sound Vib. **265**(5), 953–966 (2003)
15. Golovkine, S.: FDApy: a Python package for functional data. arXiv:2101.11003 (2021)
16. Hastie, T., Tibshirani, R., Friedman, J.: The Elements of Statistical Learning: Data Mining, Inference, and Prediction. Springer Series in Statistic, 2nd edn. (2009)
17. Dulhunty, P.: Spiral vibrations dampers on AAC and AAAC conductors. In: 22nd International Conference on Electricity Distribution, Stockholm (2013)

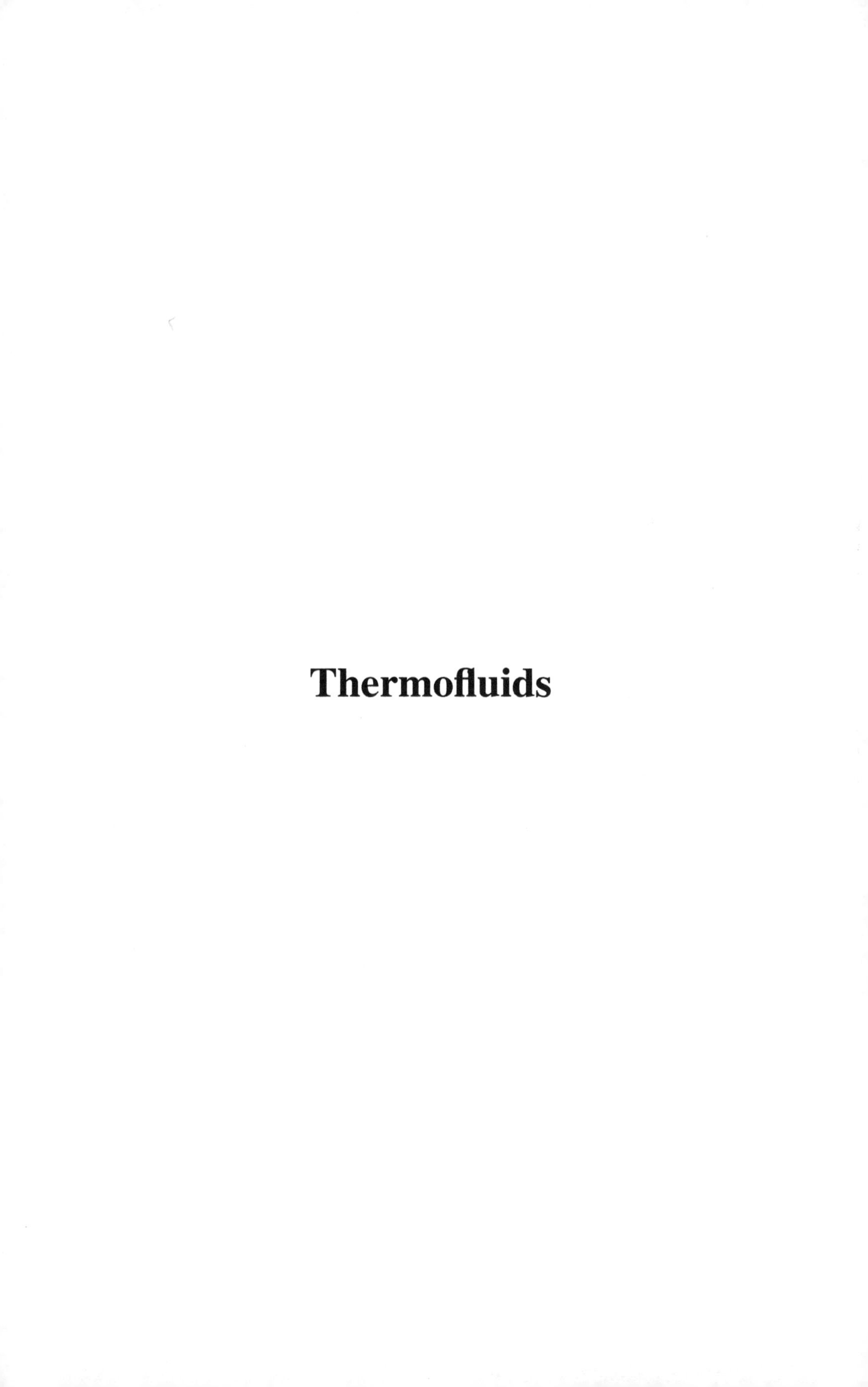

Thermofluids

Characterization of Ethanol Spray in a Commercial Injector Aimed its Application in a Bi-Fuel Ethanol/Biomethane Engine in Development for the Brazilian Vehicle Fleet

Gustavo Vieira Frez[1,2](✉), José Carlos de Andrade[3], Christian Jeremi Coronado Rodriguez[2], Túlio Augusto Zucareli de Souza[4], Jean Andrade Barbosa[3], Roberto Berlini Rodrigues da Costa[2], Luís Filipe de Almeida Roque[2], Luís Pedro Vieira Vidigal[2], Gabriel Marques Pinto[2], Nelly Vanessa Pérez Rangel[2], Davi José Souza Ferreira[2], and Vítor Brumano Andrade Cardinali[2]

[1] Mechanical Engineering Department, Federal Center for Technological Education, Celso Suckow da Fonseca UnED Angra dos Reis, Angra dos Reis, RJ, Brazil
gustavo.frez@cefet-rj.br

[2] Energy Conversion Technology Research Group (GETEC), Mechanical Engineering Institute, Federal University of Itajubá, Itajubá, MG, Brazil

[3] Associate Combustion and Propulsion Laboratory, National Space Research Institute, Cachoeira Paulista, Brazil

[4] Department of Mechanical Engineering, Federal University of Pampa, Alegrete, RS, Brazil

Abstract. To mitigate the environmental impacts from the automotive sector, the use of biofuels and innovative technologies becomes increasingly pressing. In this scenario, bi-fuel engines present themselves as a promising alternative. Considering the Brazilian energy matrix, engines of this type operating with ethanol and biomethane would meet commercial and logistical demands in a more comprehensive and effective way. The development of a technology of this size involves several steps, among which is the liquid fuel injection process, where it is desired to obtain the best characteristics that provide an ideal air-fuel mixture, which will influence the consumption, performance and emission levels of the engine. In this sense, this work brings an experimental study of the characterization of ethanol spray, using direct methods, laser diffraction and high-speed camera. The results obtained provide experimental data for modeling, input data and validation of computer simulations, contributing to the development of a bi-fuel ethanol/biomethane engine for light commercial vehicles in the national transport sector.

Keywords: bi-fuel engine · injection process · spray · ethanol · experimental techniques

O. F. Farías Fuentes et al. (Eds.): CIBIM 2024, *Proceedings of the XVI Ibero-American Congress of Mechanical Engineering*, pp. 85–102, 2026.
https://doi.org/10.1007/978-3-032-22823-9_7

1 Introduction

According to Associação Nacional dos Fabricantes de Veículos Automotores (ANFAVEA) data, the total accumulated number of licensed vehicles in Brazil increased by approximately 143.3% over the last 20 years, considering the period from June 2004 to June 2024 [1]. According to the Energy Research Company (EPE), a large portion of national road vehicles is powered by internal combustion engines that consume fossil fuels as their primary energy source, as reported in the 2023 National Energy Balance [2]. Specifically, 74.2% of the energy consumption corresponds to diesel and gasoline, while renewable fuels such as biodiesel and ethanol account for 23.8% [2].

The combustion of these conventional (fossil) fuels emits gases harmful to both the environment and human health. Consequently, there has been growing concern over sustainability issues in recent years, driving initiatives aimed at reducing these emissions. Regulatory standards such as EURO in the European Union, Tier in the United States, and Proconve in Brazil have become increasingly stringent with respect to emission limits. Currently, national pollutant emission control follows CONAMA resolutions L7 for light vehicles [3] and P8 for heavy vehicles [3], which establish maximum emission limits, expressed in g/km, for CO (carbon monoxide), NOx (nitrogen oxides), HC (hydrocarbons), and PM (particulate matter) [4].

This increased control over atmospheric pollutant emissions has driven two key developments: (i) the incorporation of fuels derived from renewable sources with the aim of reducing the consumption of fossil fuels, and (ii) the growing concern within the automotive industry to reinvent itself and seek the development of new technologies that align the maintenance of engine efficiency with the reduction of emission rates, thereby further contributing to the mitigation of pollutant emissions.

According to [2], the use of renewable fuels in the Brazilian energy matrix has been increasing over the last decades, especially in the transportation sector due to the mandatory blending of biodiesel with diesel and anhydrous ethanol with gasoline. In addition, programs such as Proálcool [5] stimulated the production of ethanol-powered automobiles, resulting in the insertion and consolidation of flex-fuel engines in the automotive sector [6], due to the solidity of the national sugar-alcohol sector and the existence of a robust and well-structured system of ethanol production and distribution, making it the most attractive option considering the environmental aspect [7].

In this sense, the development and insertion of new technologies, such as flex-fuel, dual-fuel and bi-fuel [8] for internal combustion engines that use parts or all non-fossil fuels, such as ethanol, is of great interest to the Brazilian automotive sector.

Among these more modern technologies, the bi-fuel, that considers the asynchronous use of two fuels that have different storage and injection systems, stands out in the European markets and in some Latin American countries [8, 9]. However, its current use considers CNG or LPG as the main fuel and gasoline as the fuel in recovery mode [9]. Adapting to the Brazilian reality, the development of a bi-fuel engine that uses ethanol and biomethane presents itself as a great potential for application as an alternative to achieve the goals of reducing emissions associated with mobility.

From this perspective, the design of the injection system is a crucial step, as it aims to achieve optimal air-fuel mixture characteristics, which directly influence flame propagation within the combustion chamber and, consequently, affect engine performance, pollutant emissions, and fuel consumption [10]. Therefore, spray characterization, which impacts air-fuel mixture formation [10], is an integral part of this design process.

Anand et al. (2010) [11] used experimental laser techniques to characterize the spray of 2- and 4-hole low-pressure port fuel injection (PFI) automotive injectors. Some quantitative spray characteristics were determined, including droplet size distribution, Sauter Mean Diameter (SMD), and spray penetration. A reduction in SMD values was observed when comparing the two-hole injector to the four-hole injector. The temporal evolution of the spray was also captured using high-speed imaging. It was noted that the breakup process is highly complex due to the interaction of the jets from the 2- or 4-hole injectors. Additionally, spray penetration over time was found to exhibit a linear profile.

Vásquez, Maia, and Costa (2011) [12] experimentally investigated the spray characteristics of a dual-pressure swirl injector using laser diffraction techniques with renewable fuels from the Brazilian energy matrix (ethanol and biodiesel). The injection pressure, p_{inj}, was varied for both fuels, and droplet size distributions along with their corresponding Sauter Mean Diameters (SMDs) were obtained and compared to theoretical formulations. For ethanol, an increase in p_{inj} resulted in a decrease in SMD, a trend corroborated by the shift of both the frequency and cumulative volume distributions to the left as p_{inj} increased, along with a reduction in the peak of the frequency curve. A similar behavior was observed for biodiesel. Additionally, it was found that the Radcliffe and Lefebvre equations provided the best fit for ethanol, while the Coutto-Carvalho equation yielded better agreement with the experimental data for biodiesel.

Nigra Júnior et al. (2015) [13] experimentally evaluated the macroscopic characteristics of ethanol, gasoline, and n-heptane sprays. A commercial injector with four equally spaced orifices was used, along with two measurement techniques: a patternator and a laser-based device. Measurements were conducted 100 mm below the injector outlet, under an injection pressure of 0.3 MPa and an injector pulse width of 5 ms. The spray cone angles obtained from both measurement techniques and among the different fluids were compared. The results showed good agreement between the two measurement methods, with only slight variations in spray angles. It was also observed that the ethanol spray exhibited a slightly larger cone angle compared to gasoline and n-heptane, which was attributed to ethanol's higher density and surface tension.

Dias et al. (2020) [14] experimentally analyzed the atomization behavior of ethanol using free-colliding liquid jets at atmospheric pressure. Shadowgraph imaging and laser diffraction techniques were employed to evaluate atomization characteristics as a function of injection pressure, jet velocity, and the Weber and Reynolds numbers for both ethanol jets. A decreasing trend in representative droplet diameters, specifically the Sauter Mean Diameter (SMD) and DV10, was observed with increasing collision angle between the jets. Additionally, the bimodal droplet size distributions tended to exhibit a separation between the two populations as jet velocity increased, which may be attributed to secondary atomization and droplet coalescence.

Padala et al. (2013) [15] investigated the spray characteristics of port fuel injection (PFI) ethanol sprays, considering variations in injection parameters and ambient airflow conditions. It was found that higher fuel injection rates led to longer actual injection durations, a greater number of droplets, and larger average droplet diameters. Conversely, shorter injection durations resulted in fewer droplets and smaller average diameters compared to longer injection durations. Additionally, the presence of crossflow air led to reductions in both the average droplet diameter and the total number of droplets. These reductions were attributed to the increased evaporation rate induced by the airflow.

Considering the importance of the injection process and fuel spray for automotive applications and aiming at the development of a bi-fuel ethanol/biomethane prototype engine for use in light commercial vehicles, this study presents an experimental investigation of ethanol spray characterization in order to obtain relevant data for modeling, input parameters, and validation of computational simulations within the context of the ongoing engine development project.

Thus, this article is part of a broader national initiative under Brazil's ROTA 2030 program, titled "High-efficiency bi-fuel ethanol and biomethane engine for application in light commercial vehicles: experimental testing, hybridization, dual-fuel operation with green hydrogen, and carbon footprint analysis," conducted at the Thermal Machinery Laboratory (LMT) of UNIFEI by researchers from GETEC.

2 Materials and Methods

The experimental measurements were conducted using a commercial six-hole electronic injector, which is currently employed in flex-fuel vehicles operating in Brazil.

2.1 Fuel Used in the Experiments

The fuel employed in the tests was hydrated ethanol derived from sugarcane, obtained from a fuel station accredited by the ANP (National Agency of Petroleum, Natural Gas, and Biofuels of Brazil). Its main physicochemical properties are presented in Table 1.

Table 1. Main physicochemical properties of hydrated ethanol.

Composition (% vol.)	95.1% C_2H_5OH 4.9% H_2O
Specific mass, at 20 °C	805.2 kg/m^3
MON Octane	91.8
Autoignition temperature	363 °C
Lower Calorific Value (PCI)	24.76 MJ/kg
Surface tension, at 26 °C	0.024 N/m
Kinematic viscosity, at 300 K	1.78 mm^2/s
Latent heat of vaporization	854.99 kJ/kg

2.2 Mass Flow Rates Measurement and Discharge Coefficient Calculation

Measurement of mass flow rates is important because it determines the amount of fuel to be injected during the intake process. Two types of flow measurements, commonly used in analyses of commercial injectors, were performed: static flow rate and dynamic flow rate.

Both static and dynamic flow rates were measured directly. For this purpose, a balance, stopwatch, and a graduated beaker were used, with the following precisions: ± 0.05 mL (beaker); ± 0.01 g (balance); and ± 0.01 s (stopwatch), for sample collection.

For the static flow rate, the mass of fuel passing through the continuously open injector was measured over a specific time interval, and the mass flow rate, $\dot{m}_{exp}$, in mg/s was obtained using the relation:

$$\dot{m}_{exp} = \frac{\Delta m}{\Delta t} \tag{1}$$

where Δm and Δt are, respectively, the mass and temporal variations at each measurement.

The theoretical mass flow rate, $\dot{m}_{theo}$, is obtained by the product of the injector orifices' total cross-sectional area, A_s, and the theoretical velocity at each orifice, given by Bernoulli's equation, as follows:

$$\dot{m}_{theo} = A_s\sqrt{2\rho\Delta p} \tag{2}$$

where ρ is the fluid density and Δp is the upstream pressure at the injector orifice.

Considering the experimentally measured real mass flow rate, defined in Eq. (1), and the theoretical mass flow rate, from Eq. (2), the discharge coefficient, C_D, of the injector is given by:

$$C_D = \frac{\dot{m}_{exp}}{\dot{m}_{teo}} \tag{3}$$

For the dynamic, or pulsed flow, rate, the mass was measured over a certain number of active injector pulses, determined based on simulated values corresponding to various engine speeds ranging from low to high RPM, resulting in a flow rate expressed in mg per pulse. The injector's inactive time (when the injector remains closed) was kept constant, and as the simulated engine speed increased, the injector's active time decreased accordingly. A representation of the square-wave electronic signal used to open and close the injector during pulsed measurements is shown in Fig. 1.

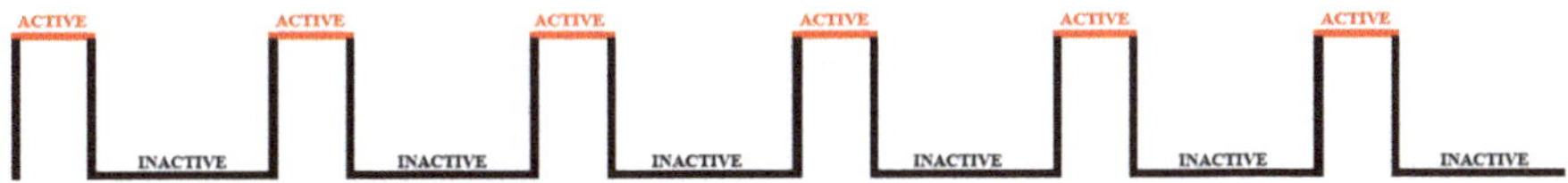

Fig. 1. Representation of a signal sent to the injector in a pulsed manner.

The flow measurements were repeated to obtain the mean values, $\overline{m}$, and the standard deviations, σ, calculated respectively by Eqs. (4) and (5).

$$\overline{m} = \frac{\sum(\dot{m}_j)}{n} \tag{4}$$

$$\sigma = \sqrt{\frac{\sum\left(\dot{m}_j - \overline{m}\right)^2}{n-1}} \tag{5}$$

where j represents each measurement of a total number n.

2.3 Droplet Diameter and Distribution Measurement

The droplet size in a spray plays a critical role in determining the efficiency of the combustion process and the level of emissions produced [17]. Therefore, accurate measurement of droplet size is important. Currently, optical techniques are considered the most advanced and precise methods for this type of measurement, with laser diffraction being a notable example.

In the present study, droplet size distributions of the spray were measured for pulsed jets using the Spraytec/Malvern® system (measurement range from 0.1 to 2000 μm with ± 1% accuracy), which employs the laser diffraction technique.

Measurements were conducted at a distance (*h*) of 75 mm from the spray nozzle outlet of the injector, as illustrated in Fig. 2, which provides a simplified schematic of the experimental setup, and the measurement system used.

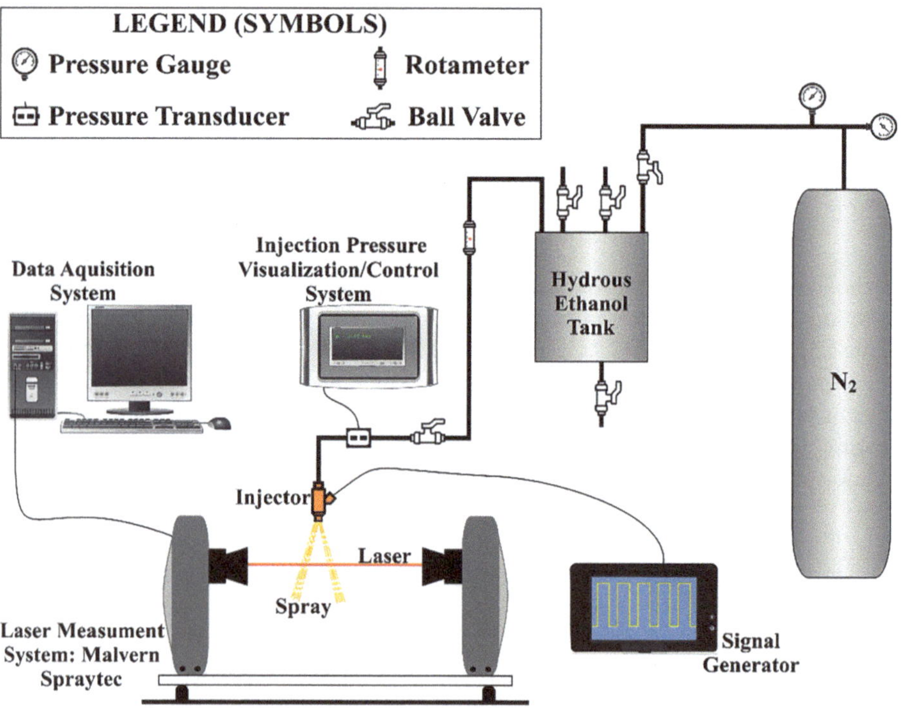

Fig. 2. Schematic representation of the atomization test bench.

In addition to the distribution of droplets, it was also obtained their representative diameters in relation to the total volume of the atomized liquid (DV10 and DV90) and the mean diameters of the atomized liquid, defined by Eq. (6), [18]:

$$D_{ab} = \left[\frac{\sum(N_i D_i^a)}{\sum(N_i D_i^b)} \right]^{\left(\frac{1}{a-b}\right)} \tag{6}$$

where a and b are constants corresponding to the investigated effect, and N_i and D_i are, respectively, the number of droplets and the mean diameter within droplet size range i.

The most commonly used mean diameter is D_{32}, also known as the Sauter Mean Diameter (SMD), which is particularly relevant to processes involving efficiency, mass transfer, and combustion.

2.4 Spray Angle Measurement and Liquid Penetration

Considering applications in internal combustion engines, spray angles have a direct impact on the direction and penetration of the spray toward the intake valve(s) (in the case of Port Fuel Injection – PFI systems). These angles can influence the amount of fuel retained in the intake port and, consequently, affect the air–fuel mixture admitted into the combustion chamber. Therefore, the characterization, definition, and determination of these angles are of foremost importance in the design of internal combustion engines.

Figure 3 illustrates the main spray angles of an injector intended for automotive applications, like the one considered in this study, as well as the configuration of the injector orifices used in the experiments.

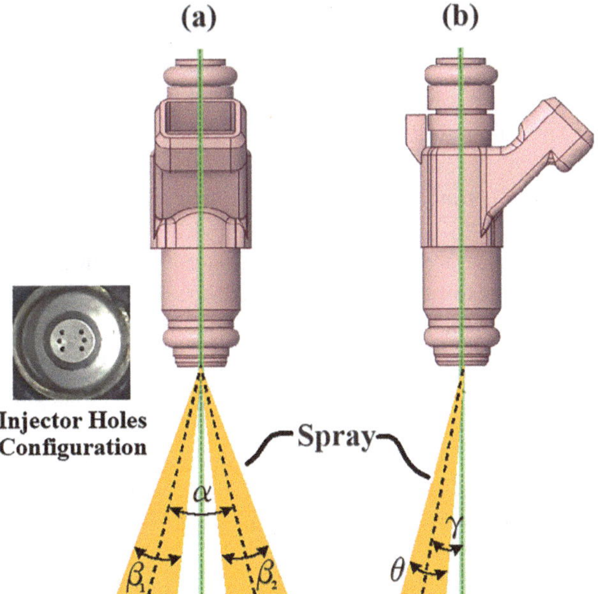

Fig. 3. Schematic illustration of the fuel spray angles in an injector designed for automotive applications, presented in (a) front view and (b) side view.

The measurements of the ethanol spray cone angles were carried out through the processing and analysis of images acquired using a high-speed camera, FASTEC model TS3–250.

To obtain these angles, the injector was positioned vertically so that the camera could capture images of the fuel spray from the front view, similar to that shown in Fig. 3(a). Subsequently, by rotating the injector while maintaining its vertical orientation, images of the spray were captured from the side view, corresponding to Fig. 3(b).

After capturing the images of the injector and spray analyzed in this study, they were processed and then loaded into a computational code developed in Matlab® with a GUI (Graphical User Interface) created by Vázquez (2011) [19].

Initially, the image must be calibrated in order to establish the scale between the pixels and the actual size of the object, thereby enabling accurate measurements. Following calibration, spray angles can be measured by selecting three relevant points for each angle, as well as distances between two defined points.

A sequential set of images was acquired using the high-speed camera and was used to estimate liquid spray penetration after the start of injection. To this end, the camera's frame rate was used to determine the time corresponding to each image, and the same computational code employed for measuring spray angles was used to measure the central spray penetration distances from the injector outlet, as indicated by the dashed lines in Fig. 3.

3 Results and Discussions

This section presents the results obtained in this study, following a sequence similar to that of Sect. 2, along with some discussions regarding these findings.

3.1 Mass Flow Rates and Discharge Coefficient

Based on the average of five measurements of mass and time, values of $\Delta m = 26.40$ mg, $\Delta t = 5.754$ s, and an average injection pressure of 3.68 bar were obtained.

Therefore, from Eqs. (1) and (4), the experimental average static mass flow rate.

$\overline{\dot{m}_{est}} = 4.59$ g/s. The standard deviation among the measurements, defined by Eq. (5), was $\sigma = 0.08$.

The static mass flow rate provided by the injector manufacturer is 4.85 g/s. Thus, the experimentally obtained result represents a percentage difference of -5.36% relative to the manufacturer's value, which may be related to the injector's usage time, given that it was already in service. Additionally, differences in temperature and pressure conditions, as well as variations in the fuels used, could have influenced the discrepancy between the results.

Furthermore, from Eqs. (2) and (3), considering that the upstream pressure at the injector orifice is given by the difference between the injection pressure and atmospheric pressure (101.325 kPa), the discharge coefficient $C_D = 0.60$ was obtained for this injector.

Typical C_D values for injectors vary according to their design, geometry, and operating conditions. However, for fuel injection applications in internal combustion engines at the intake port, generally the coefficient ranges are between 0.5 and 0.8. Therefore, the experimentally obtained value falls within this characteristic range for automotive fuel spray and atomization injectors.

To evaluate the effects of the injector active time, t_{act}, and, consequently, the simulated engine speed, ω, of a four-stroke engine on the dynamic mass flow rate, $\dot{m}_{dyn}$, of the fuel, were considered twelve active times corresponding to different engine speeds varying from 500 to 6000 rpm. For each test point, five measurements were taken and their averages and standard deviations calculated, as shown in Eqs. (4) and (5). Moreover, each measurement was taken considering 25 pulses and/or 25 injector active times, similar to the signal represented in Fig. 1.

Table 2 presents the mean values and standard deviations of the dynamic mass flow rates, as well as the injection pressure for each measurement point. Figure 4 illustrates a graph correlating the simulated engine speed with the injector's pulsed mass flow rate.

Table 2. Dynamic mass flow variations with injector active time.

ω [rpm]	t_{act} [ms]	p_{inj} [bar]	$\overline{\dot{m}_{dyn}}$ [mg/pulse]	σ [–]
500	30.0	3.67	118.668	0.274
1000	15.0	3.65	53.400	0.566
1500	10.0	3.66	33.900	0.500
2000	7.5	3.67	21.235	0.525
2500	6.0	3.67	18.032	1.100
3000	5.0	3.69	13.665	0.569
3500	4.286	3.68	10.091	0.726
4000	3.75	3.69	7.756	0.223
4500	3.333	3.69	7.177	0.205
5000	3.0	3.68	7.007	0.231
5500	2.727	3.72	3.335	0.202
6000	2.5	3.71	3.299	0.093

Based on the results presented in Table 2, it can be observed that the injection pressure, p_{inj}, presented a slight variation of only 0.04 bar, or approximately 1.1%, when the injector active time was modified. This variation did not significantly influence the flow rates measured in the tests conducted in this study.

Considering the relationship between the simulated engine speed, ω, and the average dynamic mass flow rate, $\overline{\dot{m}_{dyn}}$, , its behavior is best fitted by an exponential function with a good coefficient of determination (R^2), as shown in Fig. 4. It can also be observed that as engine speed increases, the mass flow rate decreases, since both the injector opening time and fuel flow are reduced. This indicates that engine operation at low speeds (idle conditions) requires a greater amount of fuel to be injected. However, this rate of decrease becomes less pronounced as the engine speed increases.

Similar conclusions can be observed from the behavior of $\dot{m}_{dyn} \times t_{at}$, which exhibits an excellent fit to a linear function, as shown in Fig. 5, varying in a directly proportional manner. In other words, a reduction in the injector active time results in a decrease in the dynamic mass flow rate.

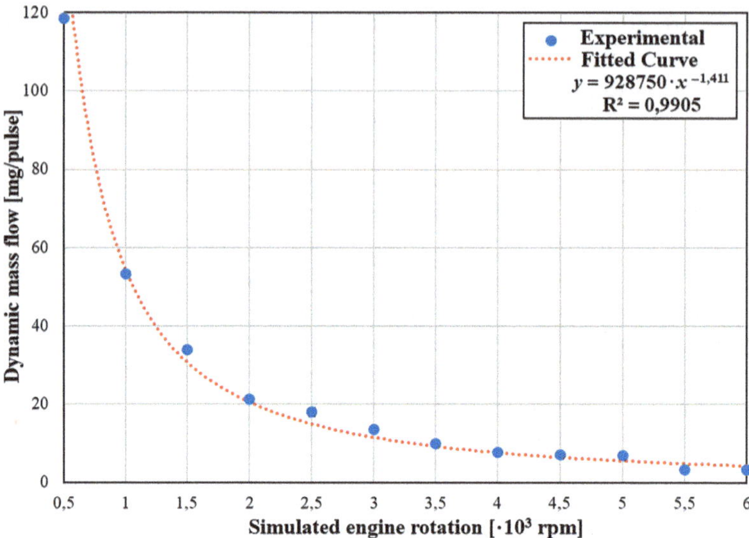

Fig. 4. Variation of Injector Dynamic mass flow as a function of simulated engine speed.

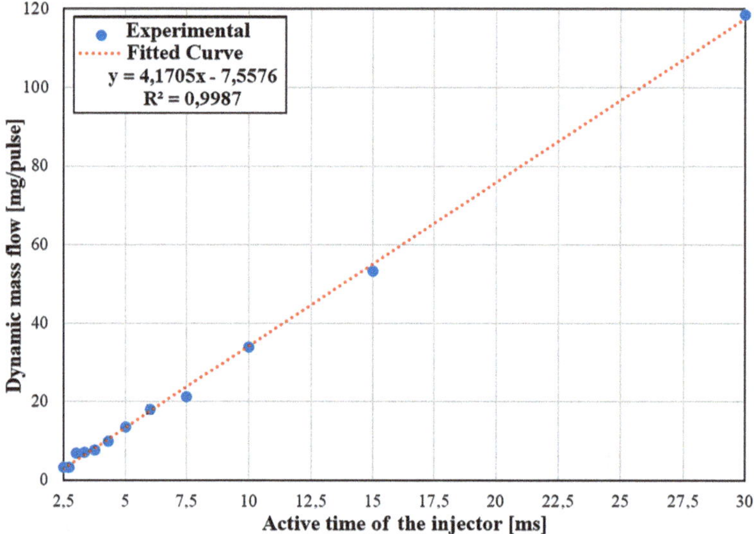

Fig. 5. Variation of injector dynamic mass flow as a function of its active time.

Finally, for all measured mass flow rates under the various respective conditions, the low standard deviations indicate a good consistency and homogeneity among the experimental data.

3.2 Droplet Diameter

The droplet size distributions, as well as their representative diameters, were obtained by considering six points from those used in the dynamic mass flow rate measurements (ranging from 2.5 to 15 ms, in 2.5 ms intervals).

Figure 6 presents the relationship between the main representative diameters and the active time of the injector. These results were obtained from measurements taken at a distance of $h = 75$ mm (see Fig. 2) and can be observed that the representative diameters vary directly with the injector active time.

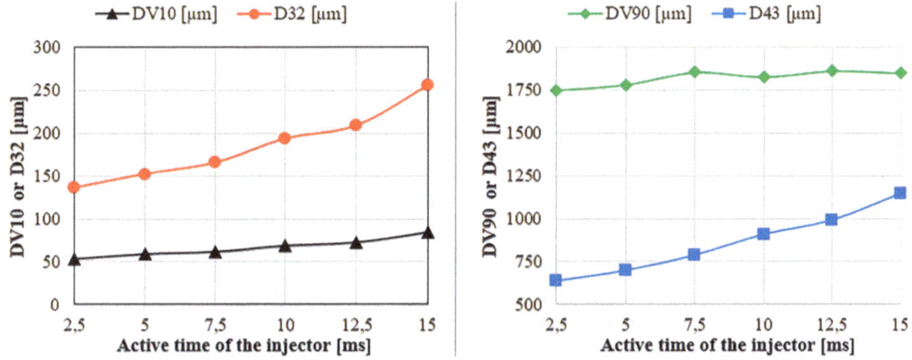

Fig. 6. Variation of the main representative diameters with injector active time.

The statistical distributions of droplet sizes for four injector active times, measured using the SprayTec laser system, are shown in Figs. 8 and 9.

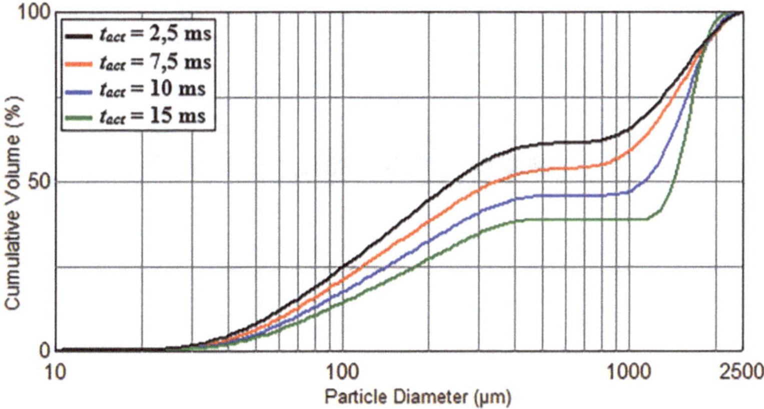

Fig. 7. Comparison of the cumulative droplet size distributions for different active times.

From Fig. 7 and Fig. 8, a bimodal distribution of droplet diameters can be observed, which was also reported by [14]. This bimodal characteristic becomes increasingly evident as the injector active time increases, resulting in a higher frequency of droplets with larger diameters. This bimodal nature is also reflected in the variations of the representative diameters, as shown in Fig. 6.

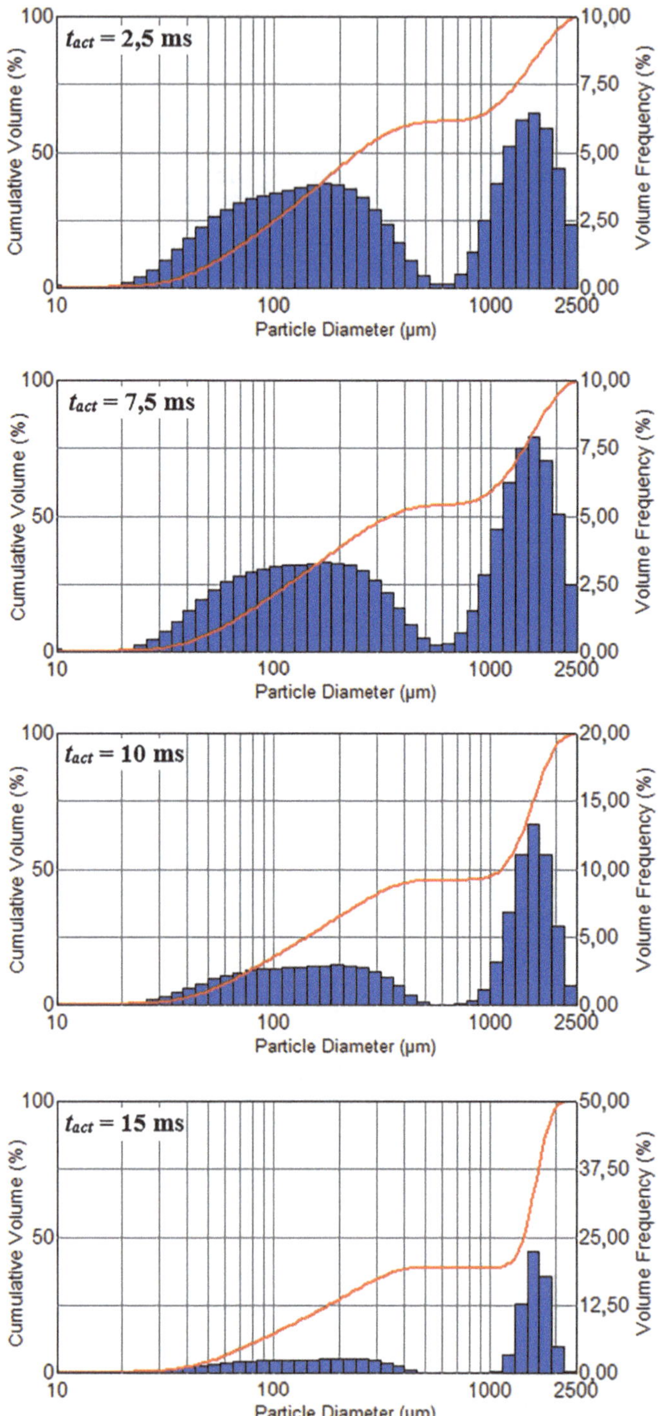

Fig. 8. Modal distribution of ethanol spray particle diameter for some injector active times.

The experimentally obtained droplet distribution curves are often fitted to mathematical models, such as the Rosin-Rammler or Log-Normal distributions, so that the spray atomization characteristics can be incorporated into computational simulations of internal combustion engines.

3.3 Spray Angles and Liquid Penetration

Figure 9 presents the front and side views obtained from the high-speed camera image processing. The positions y_1 and y_2 correspond to the regions where 99% and 60% of the ethanol spray penetration occur, respectively, measured at 3.3 ms after injector activation. The external measurement conditions were ambient pressure (1.01 atm) and 26 °C. Based on the methodology described in Sect. 2.4, the spray angles shown in Fig. 9 were obtained. It is possible to observe interactions between the spray jets, which is a typical feature of multi-hole injectors and results in larger droplet sizes [20].

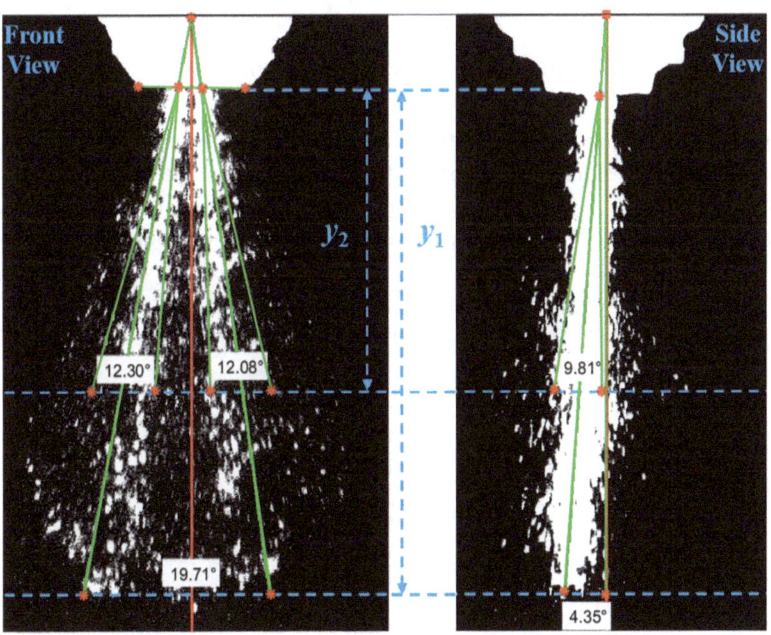

Fig. 9. Shadowgraph images of ethanol spray.

Thus, by comparison between Fig. 3 and Fig. 9, the following spray angles were obtained: $\alpha = 19.71°, \beta_1 = 12.30°, \beta_2 = 12.08°, \theta = 9.81°$, and $\gamma = 4.35°$.

The spray penetration was estimated using a sequence of images and the algorithm proposed by Vázquez [19], which was employed to measure the distances traveled by the fuel after the start of injection, measured along the central axis of the spray, as shown in Fig. 3 and Fig. 9, obtaining the graph in Fig. 10.

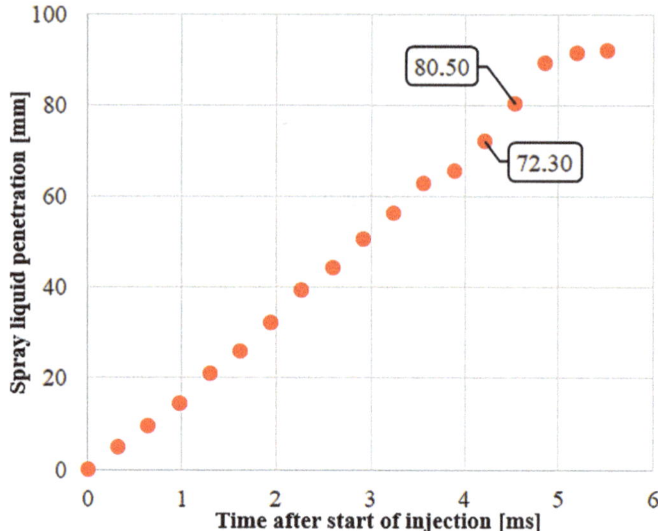

Fig. 10. Spray penetration vs. time.

It can be observed from Fig. 10 that the spray liquid penetration exhibits a linear profile over time after start of injection. Moreover, considering a penetration height (as illustrated by h in Fig. 2) equal to 75 mm, the time required to reach this height is approximately 4.32 ms under the experimental conditions considered and based on a linear fit of the data points highlighted in Fig. 10. Similar spray penetration profiles were reported by [11] for gasoline and by [15] for ethanol.

In Fig. 11 can be observed the spray evolution in the frontal view for a single injection pulse.

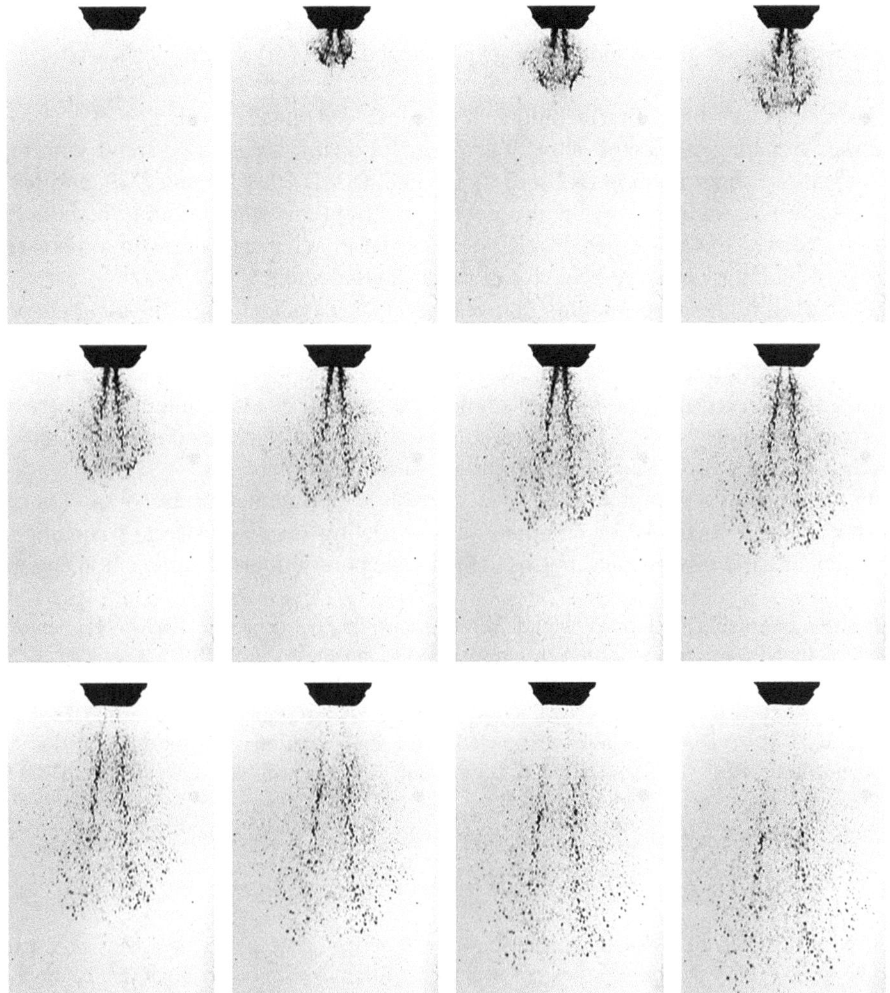

Fig. 11. Spray evolution during a pulse in frontal view.

4 Conclusions

This study presented the characterization of ethanol spray injections using a commercial six-hole injector commonly employed in Otto-cycle internal combustion engines with port fuel injection systems. Analyzing these characteristics is of significant importance, as they provide crucial information for achieving an optimal air-fuel mixture, which is essential for proper flame propagation, more efficient combustion, and reduced emissions.

The injector flow rates were measured directly, and good agreement was observed between the experimental measurements and the static flow rate specified by the manufacturer. Furthermore, the discharge coefficient obtained for the injector falls within the typical range for this type of application. It was also found that the dynamic flow rate decreases as the injector activation time is reduced or as the simulated engine speed increases. A linear fit applied to this experimental data ($\dot{m}_{dyn} \times t_{act}$) showed a high degree of accuracy.

Using the laser diffraction technique, it was observed that the mean droplet diameters increase with longer injector activation times, following a parabolic trend with high coefficients of determination (R2 > 0.98 for D32 (SMD), DV10, and D43, and R2 = 0.87 for DV90). Additionally, the droplet size distributions were found to be bimodal, with a tendency toward larger droplet sizes as the injector activation time increased, consistent with the observed behavior of the mean diameters.

Moreover, high-speed imaging allowed for clear visualization of the ethanol spray evolution over time, as well as for determining the spray penetration, which displayed an approximately linear profile. Finally, the characteristic spray angles were estimated, and it was observed that the injector's hole configuration leads to interaction between jets from different orifices. This interaction contributes to greater dispersion and the presence of larger droplets as the injector activation time increases.

In conclusion, a comprehensive and valuable experimental dataset was obtained, which can serve as input for computational modeling of ethanol sprays and for the validation of CFD simulations applied to the intake port of internal combustion engines.

Acknowledgments. The authors would like to thank the Fundação de Desenvolvimento da Pesquisa (FUNDEP) for supporting this research through the Rota 2030 Line V project (No. 2719262) and Intec No. 2719268 (FUNDEP 27192.01.INT01/2022.01-00), as well as FAPEMIG (Projects: APQ-01763-23 and RED-00090-21) and CNPq (Grants No. 308567/2023-4 and 442662/2023-8). The authors also acknowledge the collaboration with the Combustion and Propulsion Laboratory (LCP) of INPE/Cachoeira Paulista and the Biofuels Laboratory at UNIFEI.

References

1. ANFAVEA. Dados Estatísticos para Download. Séries históricas: Séries mensais, a partir de janeiro/1957, de autoveículos por segmento (automóveis, comerciais leves, caminhões, ônibus, total) de produção; licenciamento de nacionais, importados e total; exportações em unidades. https://anfavea.com.br/site/edicoes-em-excel/
2. Empresa de Pesquisa Energética (Brasil). Balanço Energético Nacional 2024: Ano base 2023. Empresa de Pesquisa Energética. EPE, Rio de Janeiro (2024). https://www.epe.gov.br/pt/publicacoes-dados-abertos/
3. IBAMA. Programa de controle de emissões veiculares (Proconve). https://www.gov.br/ibama/pt-br/assuntos/emissoes-e-residuos/emissoes/programa-de-controle-de-emissoes-veiculares-proconve
4. ANFAVEA. Conteúdos técnicos: Informações técnicas – Programas Automotivos – Emissões. https://anfavea.com.br/site/informacoes-tecnicas/
5. Grangeia, C., Santos, L., Lazaro, L.L.B.: The Brazilian biofuel policy (RenovaBio) and its uncertainties: an assessment of technical, socioeconomic and institutional aspects. Energy Conv. Manag. X **13**, 100156 (2022)
6. de Melo, T.C.C., et al.: Hydrous ethanol–gasoline blends – combustion and emission investigations on a Flex-Fuel engine. Fuel **97**, 796–804 (2012)

7. Souza, S.P., Seabra, J.E.A.: Integrated production of sugarcane ethanol and soybean biodiesel: environmental and economic implications of fossil diesel displacement. Energy Convers. Manag. **87**, 1170–1179 (2014)
8. Malaquias, A.C.T., et al.: A review of dual-fuel combustion mode in spark-ignition engines. J. Braz. Soc. Mech. Sci. Eng. **43**, 426 (2021)
9. Giechaskiel, B., et al.: Emissions of Euro 6 Mono- and Bi-fuel gas vehicles. Catalysts **12**(6), 651 (2022)
10. Heywood, J.B.: Internal Combustion Engines, 2nd edn. McGraw-Hill Education, New York (2018)
11. Anand, T.N.C., et al.: Laser-based spatio-temporal characterisation of port fuel injection (PFI) sprays. Int. J. Spray Comb. Dyn. **2**(2), 125–149 (2010)
12. Vásquez, R.A., Maia, F.F., Costa, F.S.: Droplet size distributions of biofuel sprays. In: Proceedings of the 21st Brazilian Congress of Mechanical Engineering (COBEM 2011), Natal, RN, Brazil, 24–28 October 2011 (2011)
13. Nigra Júnior, E.L.P., et al.: Comparative analysis of spray cone angles of ethanol, gasoline and n-heptane from a multi-hole port fuel injector. In: Proceedings of the IV Journeys in Multiphase Flows (JEM 2015), Campinas, SP, Brazil, 23–27 March 2015 (2015)
14. Dias, G.S. et al.: Hydrous ethanol atomization by impinging jets. In: Proceedings of the 18th Brazilian Congress of Thermal Sciences and Engineering (ENCIT 2020), 16–20 November 2020 (2020)
15. Padala, S., et al.: Imaging diagnostics of ethanol port fuel injection sprays for automobile engine applications. Appl. Therm. Eng. **52**(1), 24–37 (2013)
16. Da Costa, R.B.R., et al.: Experimental assessment of renewable diesel fuels (HVO/Farnesane) and bioethanol on dual-fuel mode. Energy Conv. Manag. **258**, 115554 (2022)
17. Azevedo, C.G.: Desenvolvimento de um sistema compacto de combustão sem chama visível utilizando um injetor blurry para queima de biocombustíveis. Thesis (D.Sc. in Engenharia e Tecnologia Espaciais/Combustão e Propulsão) – Instituto Nacional de Pesquisas Espaciais, São José dos Campos (2013)
18. Lefebvre, A.H.: Atomization and Sprays. Hemisphere Publishing Corporation, New York (1989)
19. Vázquez, R.A.: Desenvolvimento de um injetor centrífugo dual para biocombustíveis líquidos. Dissertation (M.Sc. in Propulsão e Combustão) – Instituto Nacional de Pesquisas Espaciais, São José dos Campos (2011)
20. Aoki, F., et al.: Spray Analysis of Port Fuel Injector. SAE Technical Paper Series, n° 2005-01-1154. From 2005 SAE World Congress Detroit, Michigan, 11–14 April 2005 (2005)

BY NC ND

Technical and Economic Assessment of an Integrated System for Drying and Gasifying Sewage Sludge Aiming at Electricity Generation

Ellen C. A. Cezarini[1], Osvaldo José Venturini[1](✉), Túlio Tito Godinho de Rezende[1], Zudivan Peterli[2], José Carlos Escobar Palacio[1], and York Castillo Santiago[3]

[1] Federal University of Itajubá, Av. BPS, 1303, Itajubá, MG 37.500-903, Brazil
osvaldo@unifei.edu.br
[2] Espírito Santo Federal University, Av. Fernando Ferrari, 514, Vitória, ES 29.075-910, Brazil
[3] Fluminense Federal University, Passo da Patria, 156, Niteroi, RJ 24210-240, Brazil

Abstract. The correct disposal of sewage sludge (SS) is the subject of studies worldwide. For Brazil, these studies have become crucial since 2014, when Law 12.305, which addresses the National Solid Waste Policy, reinforced that only residues that cannot be economically reused or recycled should be sent to landfills. In this scenario, using SS as feedstock to produce biofuels could be a viable alternative to its disposal in landfills. However, the sludge must undergo a thermal drying stage before being used as a biofuel feedstock, which may require large amounts of thermal energy. Solar thermal energy (heliothermal energy) can be considered an alternative to complement the energy consumed in SS thermal drying. Considering this scenario, this work presents a technical and economic assessment of an integrated system designed to dry and gasify SS to produce syngas, which is then burned in an internal combustion engine (ICE) to generate electricity. The sludge is dried by combining energy recovered from the ICE with solar thermal energy harvested with parabolic trough collectors. The results show that using the ICE exhaust gases to supplement the thermal demand of the sludge dryer reduced the solar field area to 33.2% of the amount needed if the dryer's thermal demands were supplied exclusively with heliothermal energy. The levelized cost of electricity generated (LCOE) was 107.85 US$/MWh, a value close to the selling price of electricity generated from municipal solid waste (MSW), as established in auctions run by the Brazilian National Agency of Electric Energy (ANEEL).

Keywords: sewage sludge · drying and gasification · electricity generation

O. F. Farías Fuentes et al. (Eds.): CIBIM 2024, *Proceedings of the XVI Ibero-American Congress of Mechanical Engineering*, pp. 103–120, 2026.
https://doi.org/10.1007/978-3-032-22823-9_8

1 Introduction

Increased globalization and the growth of the global population have led to a rise in energy consumption over the years, with fossil fuels remaining the dominant energy source [1]. In addition, concerns about climate change and sustainable development encourage increasing the share of renewable energies in the world's energy mix [2].

Brazil has one of the most renewable electricity matrices in the world [3], with 83.79% of the total electricity coming from renewable sources. However, the Brazilian energy matrix still has considerable room to replace fossil fuels with renewable energy since the latter accounts for only 47.9% of Brazil's total energy supply [4].

On the other hand, according to the World Health Organization [5], more than 1.5 billion people worldwide lack access to essential sanitation services. In Brazil, Law No. 14,026 (Legal Framework for Basic Sanitation) came into force in 2020, aiming to universalize essential sanitation services by 2033. However, by 2023, 100 million Brazilians still lacked access to sewage collection, and 35 million to treated water [6].

According to the Law of the Legal Framework for Basic Sanitation, an increase in the generation of sewage sludge is anticipated, as one of the objectives of this Framework is to achieve 90% coverage of sewage collection and treatment for the entire population. At the same time, since 2014, Law 12.305, which establishes the National Solid Waste Policy, has reinforced the restriction on sending materials to landfills that are only ultimate waste and not economically viable for reuse or recycling [7].

Sewage sludge is a waste that is inevitably generated during the sewage treatment stages [8]. Sludge can contain numerous toxic substances, including pathogens, metals, and organic contaminants, which can lead to severe environmental pollution [9]. Several procedures are employed to safely treat and dispose of sanitary sludge, including anaerobic digestion, dehydration, drying, thermal sanitization, and landfilling [10]. However, other technologies, such as pyrolysis and gasification, can produce fuels from SS and other wastes with a low environmental impact [11]. An example of a facility using SS to generate energy in Brazil is the Franca Wastewater Treatment Plant (WWTP), which in 2018 initiated a project to produce biomethane from the sludge generated at the WWTP for use in vehicles [12].

Singh et al. [13] investigated the potential for energy recovery from SS by incineration and anaerobic digestion, obtaining, respectively, values in the range of 555 to 1068 kWh/ton and 315 to 608 kWh/ton (dry basis). The authors concluded that these technologies can help reduce the volume of sludge by 60 to 70% and the associated costs of transporting waste to a landfill. Lee et al. [14] investigated the conversion of SS into synthesis gas using a laboratory-scale steam gasification system, obtaining promising results regarding gasification efficiency and gas yield.

Capodaglio and Callegari [15] state that thermochemical processes, such as pyrolysis and gasification, have the advantage of processing sludge quickly and generating versatile final products. However, the high moisture content of raw sludge makes it unsuitable for use in thermochemical processes without a preliminary drying stage, which requires large amounts of energy, thereby reducing the overall efficiency of the thermochemical route and increasing specific treatment costs. These authors suggest that thermochemical processes can be an ideal complement to anaerobic digestion, provided that the initial costs are reduced and the energy demands, particularly those associated with the sludge drying stage, are minimized.

The thermal drying process promotes the reduction of sludge moisture through water evaporation and stands out from other dewatering methods since, according to Svensson [16], it promotes a significant reduction in sludge volume and results in its proper sanitization, making it possible to reduce costs and risks during transport and handling, as well as making the material suitable for energy production through combustion, pyrolysis, and gasification processes. However, thermal drying is energy-intensive, with consumption ranging from 0.79 to 1.20 kWh per liter of water removed, aiming to raise the total solids content from 25% to 90% [17]. According to Svensson [16], implementing thermal sludge drying using solar energy combined with biogas burning could be a sustainable alternative for the future of SS management.

Considering this scenario, this work presents a technical and economic assessment of generating electricity using ICE powered by syngas produced from the gasification of dried SS. It is considered that the drying of previously dewatered sludge with a total solids (TS) content of 25% can be achieved using heliothermal energy in conjunction with heat recovery from the exhaust gases of the internal combustion engine. The relevance of this work is justified by the fact that the energy use of SS is aligned with the UN Sustainable Development Goal (SDG) number 6 (Drinking Water and Sanitation) since an unwanted by-product is transformed into a renewable energy source, contributing to the sustainable management of water resources, reducing water pollution and promoting efficiency in wastewater treatment. SDG 7 (Clean and Affordable Energy) is also contemplated since the proposed system generates energy from the biomass produced by the treatment of sewage SS, as well as SDG 12 (Responsible Consumption and Production) since utilizing energy from SS aligns with the goal of reducing waste generation and fostering sustainable consumption and production practices.

2 Materials and Methods

The objectives of this work were achieved by performing an energy balance of the integrated system, which incorporates the processes of sludge drying, gasification, and electricity cogeneration. The integrated system considered in this work is shown in Fig. 1.

The results obtained by Peterli [18], regarding the characteristics of the SS and the solar field design, were used as inputs for the gasification process and to determine the thermal energy required by the dryer. In addition, a technical evaluation of the gasification of dry SS using air as the gasification agent was conducted, considering the chemical equilibrium model presented by Zainal et al. [19], which enabled the evaluation of syngas composition under various gasification conditions. Next, the cogeneration system was modeled using GateCycle™ to assess the potential for electricity generation and heat recovery from the ICE exhaust gases.

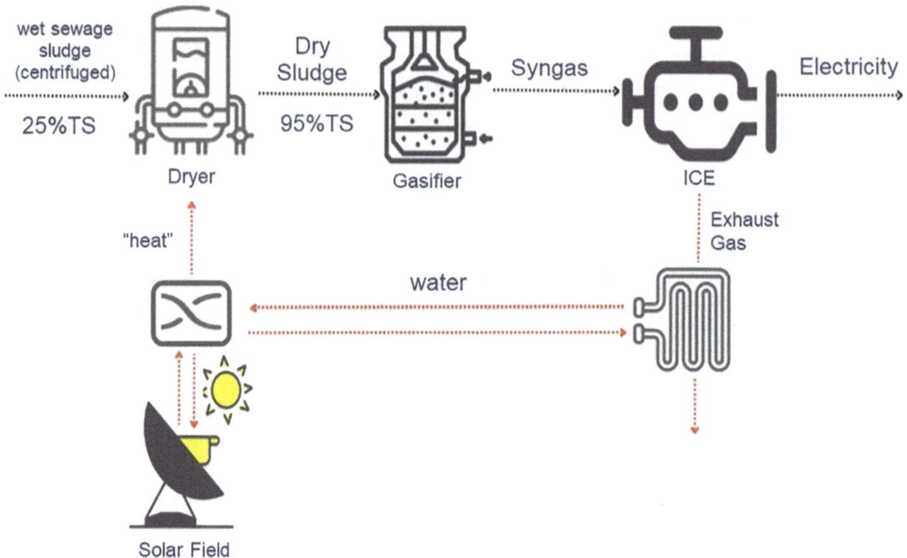

Fig. 1. Schematic of the proposed integrated system: sludge drying, gasification, and electricity cogeneration. Source: the authors.

2.1 Sewage Sludge Characterization

This study analyzed samples of centrifuged SS produced at Mulembá I WWTP, located in Vitória (ES), Brazil. The centrifuged sludge was collected and treated on a hot plate and in an oven. The first treatment stage aimed to produce dry sludge that would be mixed with the centrifuged sludge to generate samples with varying total solids (TS) concentrations. These samples were then subjected to thermal drying experiments to characterize the energy demand of the process and the drying rates.

The proximate analysis of the SS was performed according to the ASTM D3172 standard, using a Thermogravimetric Analyzer (TGA) model 701 from LECO, with a precision of 0.3%. For the ultimate analysis, the sludge samples were evaluated in a PerkinElmer CHNS/O 2400 Series II elemental analyzer (0.2% precision), according to ASTM D5373 and D4239 standards, allowing the carbon (C), hydrogen (H), nitrogen (N), and sulfur (S) concentrations to be determined, with the oxygen (O) content estimated by difference. The sludge heating value was determined using an IKA calorimetric pump, model C2000 (precision $< 0.4\%$). All the experiments to characterize the sewage sludge were conducted in the laboratories of the Excellence Group in Thermal Power and Distributed Generation (NEST) of the Federal University of Itajubá (UNIFEI).

2.2 Gasification Process Modeling

To simulate the dry sludge gasification process, a computer code was developed based on the work presented by Zainal et al. [19], who proposed a model to simulate downdraft gasifiers based on the principle of chemical equilibrium. The model enables determining the composition of syngas by knowing the elemental composition of the fuel (sludge) and establishing the gasification temperature, assuming all considered reactions are in thermodynamic equilibrium. The overall gasification reaction considered is given by Eq. (1).

$$CH_yO_z + wH_2O + 3,7mN_2 = x_1H_2 + x_2CO + x_3CO_2 + x_4H_2O + x_5CH_4 + 3.7mN_2 \tag{1}$$

where w and m correspond to the stoichiometric coefficients of H_2O and O_2 in the reactants; x_1, x_2, x_3, x_4, and x_5 are the stoichiometric coefficients of the products; y and z are the amount of hydrogen and oxygen per mole of carbon in the fuel. In addition, the other two main gasification reactions considered are (Eq. 2 and 3):

$$C + 2H_2 = CH_4 \tag{2}$$

$$CO + H_2O = CO_2 + H_2 \tag{3}$$

The equilibrium constants of the methanation reaction (Eq. 2) and water displacement (Eq. 3) are given by:

$$K_1 = \frac{x_5}{x_1^2} \tag{4}$$

$$K_2 = \frac{x_1x_3}{x_2x_3} \tag{5}$$

Using the First Law of Thermodynamics and the principle of minimizing Gibbs' free energy (the equilibrium condition), a set of nonlinear equations is obtained, which requires a numerical solution [19]. A computational code was developed to calculate the necessary thermodynamic properties directly from an integrated property library, as described in a previous work of Rezende et al. [20]. The model was validated by comparing its results with data from the literature [19], using the same biomass, and the maximum relative deviation in gas composition was inferior to 4.8%.

2.3 Heliothermal System

The heliothermal system is modeled based on Eqs. (6), (7), and (8) and considering the energy balance between the solar field (production) and the dryer's energy demand (heating the drying air).

$$A_{CSP} = \frac{\Delta H_{dr}}{DNI\ \varepsilon} \tag{6}$$

$$\Delta H_{sec} = \dot{m}_{Lu}\bar{c}_{p\,Lu}(T_{Lf} - T_{Li}) + \dot{m}_{w,ev}h_{lv} + \text{losses} \tag{7}$$

$$\bar{c}_{p\,Lu} = \frac{W}{W+1}\bar{c}_{p\,ag} + \frac{1}{W+1}(1,434 + 3,29\ T_L) \tag{8}$$

where ΔH_{dr} is the thermal demand; $\dot{m}_{Lu}$ is the wet sludge flowrate; $\dot{m}_{w,\,ev}$ is the flowrate of water removed in the dryer; T_{Li} and T_{Lf} are the initial and final temperature of the sludge, respectively; h_{lv} is the latent heat of water evaporation; A_{CSP} is the mirrored area of the solar field; ε is the efficiency of the solar field (parabolic trough); and DNI is the hourly normal solar irradiation.

2.4 Cogeneration System

The modeling of the cogeneration system was carried out using GateCycle™ software, version 6.1.2. The system comprises a turbocharged internal combustion engine that burns the syngas produced by the gasification of dried sludge (see Fig. 2).

The engine's exhaust gases flow through a heat exchanger (HX1), where water is heated to 120 °C at 500 kPa, recovering energy from the exhaust gases. This hot water flow mixes with water from the heliothermal system (CSP1), and the resulting flow is used to supply the thermal demand of the sludge dryer, i.e., it heats the air used for drying the sludge. The "DEM. SEC" component, used in GateCycle to impose heat demands on fluid streams, represents the dryer's thermal demands. Upon leaving this component, the water flow at a temperature of 75 °C splits, with part returning to the heliothermal system (CSP2) and the remainder to the engine heat recovery unit (HX1). The performance curves supplied by the ICE manufacturer were inserted in a GateCycle component to simulate combustion engines. The ICE used the CATG132 turbocharged engine from Caterpillar Inc., which has a rated power of 400 kW. The temperature of the ICE exhaust gases leaving the heat exchanger (HX1) was limited to 135 °C, given the possibility of sulfur compounds in the exhaust gases.

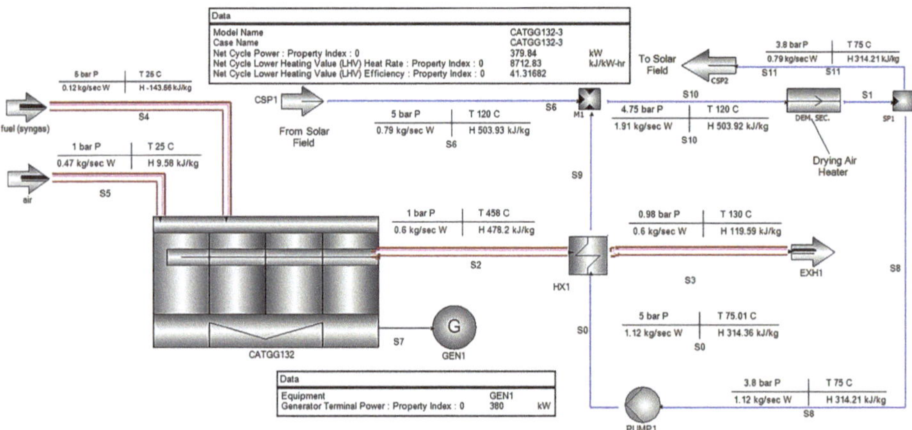

Fig. 2. GateCycle™ model of the cogeneration system. Source: prepared by the authors.

2.5 Economic Analysis

In this work, to assess the economic viability of the proposed system, the following financial indicators were considered: i) NPV (Net Present Value); ii) IRR (Internal Rate of Return); iii) Payback Time; and iv) LCOE (Levelized Cost of Electricity), which provides information on the cost of the electricity generated, taking into account all the energy generated and costs over the lifetime of the project, and is determined by Eq. (9).

$$\mathrm{LCOE} = \sum_{t=1}^{n} \frac{\frac{I_t + O\&M_t + F_t}{(1+i)^t}}{\frac{EE_t}{(1+i)^t}} \tag{9}$$

where i is the interest rate; n is the project's life; t is the year considered; I_t is the total investment (CAPEX) annualized; $O\&M_t$ are the maintenance (fixed + variable) costs at year t; F_t is the fuel costs at year t; and EE_t is the electricity generated at year t.

The parameters considered for the economic analysis are presented in Table 1. The minimum attractive rate of return (MARR) used is based on the SELIC rate (Brazilian federal funds rate) of July 2024 [21].

Table 1. Parameters used in economic analysis (source [22][1], [23][2], [24][3], [21][4])

Parameters	Value	Units
Degradation of gasification system efficiency	0.35[1]	% a.a
System availability	95.0[1]	%
project's life	25[1]	year
Maintenance and Operation Costs	2.0[2]	% a.a
Administration Costs	0.5[2]	% a.a
Interest Rate	10.0[3]	% a.a
Minimum attractive rate of return (MARR)	11.75[4]	% a.a

3 Results

As mentioned, the SS used in this experiment comes from the Mulembá I WWTP, located in Vitória (ES), which has a nominal capacity of treating 17,626 m3/day of sewage. As a reference scenario for the analyses presented (base scenario), the continuous processing of 0.171 kg/s of sludge with 25% TS was considered to obtain sludge with 95% TS, which implies vaporizing 0.126 kg/s of water in the dryer. Under these conditions, using Eq. (7) and assuming that the sludge is heated from 25 °C to 85 °C [18], the thermal demand for drying results in 0.59 kWh/kg of sludge (25% TS), or 0.81 kWh/kg of water removed. This value shows a relative deviation of only 1.8% from the lower end of the range reported by [17], which is 0.79 kWh/kg of water removed.

The ultimate and proximate composition of the sludge is shown in Table 3. The elemental composition obtained from the samples analyzed in this work agrees with the data presented by [25], with a maximum relative deviation of 10.1% for oxygen. The lower heating value obtained was 17.4 MJ/kg (0.06%w) (Table 2).

Table 2. Ultimate and proximate composition of activated sludge (0.06% moisture content).

Elemental composition [%]			Proximate composition [%]	
Elements	This Work	Collard et al. [25]		
C	40.0 ± 0,3	36.2	Fixed carbon	13.02 ± 0.19
H	5.8 ± 0,1	5.5	Volatile	67.81 ± 0.45

(continued)

Table 2. (*continued*)

Elemental composition [%]			Proximate composition [%]	
Elements	This Work	Collard et al. [25]		
O	47.2 ± 0,4	52.0	Moisture	0.06 ± 0.01
N	7.0 ± 0,1	6.3	Ash	19.11 ± 0.21
S	n.a	n.a		

In Vitória (ES), the atmospheric air has a daily average temperature of 23.2 ± 3.6 °C and RH of 82 ± 12% based on the monitoring of INMET's (Brazilian National Institute of Meteorology) A-612 weather station, values close to those reported by Marchioro [26], who describes averages of 24.9 °C temperature and 78% RH for a more extended monitoring period. The DNI data obtained from 612 is shown in Fig. 3.

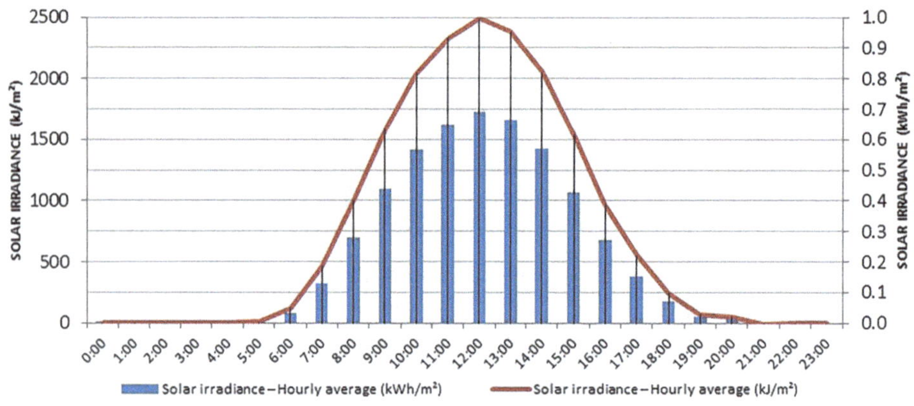

Fig. 3. DNI profile for Vitória/ES (source: [18])

Based on Fig. 3, the average daily DNI obtained is 4.96 kWh/m2, which falls within the range of 4.6 to 5.1 kWh/m2 for the Metropolitan Region of Vitória [27].

Table 3 summarizes the values obtained for the main characteristics of the solar field and dryer, assuming that the thermal energy demanded is supplied exclusively by the solar field (CSP), which consists of parabolic trough collectors.

Table 3. The main results for the solar field (CSP) and thermal dryer.

SOLAR FIELD			
Solar Collector Area	3,640.0	m^2	
Thermal Energy Production	8,410.0	kWh/day	
THERMAL DRYER			
INPUT	Energy Demand	8,258.0	kWh/day

(*continued*)

Table 3. (*continued*)

	Sludge flow (25% TS)	14,768.0	kg/day
OUTPUT	Total flow (sludge + water)	0.171	kg/s
	Dry sludge flow	0.043	kg/s
	Removed water(evaporated)	0.126	kg/s

Regarding the thermochemical conversion process, using the equilibrium model proposed by [18], considering air as the gasification agent, and based on the composition of the sludge presented in Table 1, the gasification temperature and the lower heating value (LHV) of the syngas were determined, for different equivalence ratios (ER), as shown in Fig. 4. Given the low proportion of fuel to oxidizer, using ERs lower than 0.20, from a practical point of view, can lead to instabilities and low efficiencies in the gasification [28]. Hence, the analysis considered ERs ranging from 0.25 to 0.40.

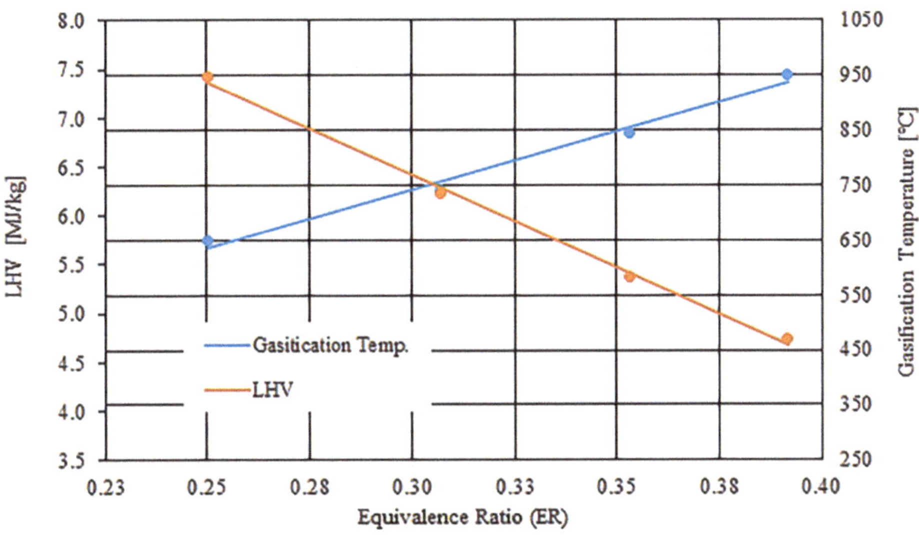

Fig. 4. LHV and gasification temperature as a function of the ER

The decrease in the LHV of the syngas produced with the increase in ER (Fig. 4) occurs because higher ERs imply more oxygen availability, promoting the oxidation of the fuel species formed [29]. Hence, the higher release of energy (heat) increases the gasification temperature. The syngas compositions for the same values of ERs are shown in Fig. 5.

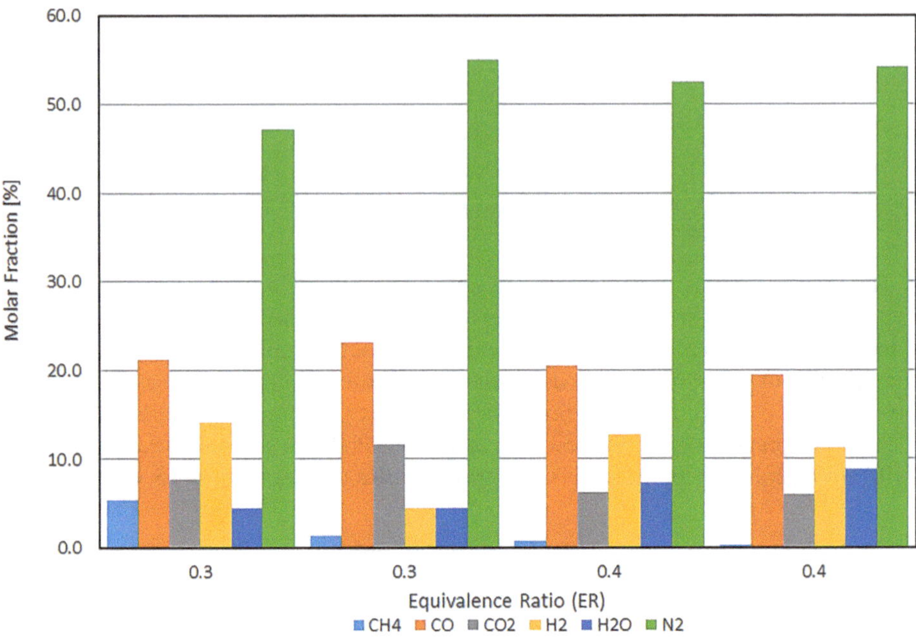

Fig. 5. Syngas composition for different ERs.

The syngas flow rate produced during gasification was determined considering the dry sludge flow rate presented in Table 3. The ICE model developed in GateCycle™ was then used to estimate the potential for electricity generation and energy recovery (heat) from exhaust gases through water heating. The results obtained for the cogeneration system with different scenarios regarding the ER are presented in Table 4.

Table 4. Results obtained for the cogeneration system

Scenario	A	B	C	D
ER	**0,25**	**0,31**	**0,35**	**0,40**
Syngas flow[kg/s]	0.13	0.15	0.16	0.17
Syngas LHV [MJ/kg]	7.42	6.22	5.37	4.74
ICE Net Power [kW]	379.2	373.2	351.6	331.4
ICE Efficiency [%]	41.3	41.1	40.9	40.5
Exhaust gas flow [kg/s]	0.597	0,587	0.556	0.526
Exhaust gas temperature [°C]	458.2	459.7	465.1	470.1
Hot water flow (cogen system) [kg/s]	1.13	1.11	1.07	1.04
Available thermal power [kW] (cogen system)	213.3	210.9	202.8	197.8

Considering the flow and temperature of the water exiting the cogeneration system (HX1), along with the thermal demand of the dryer (8,258 kWh/day) as outlined in

Table 4, it was then determined how much of this demand can be supplied by the cogeneration system. Additionally, the new required area of the solar field was determined for the various ERs evaluated (scenarios A, B, C, and D). The results obtained are presented in Table 5. It is observed that scenario A presents the best performance regarding ICE exhaust energy recovery since 5,120 kWh/day of thermal energy can be supplied to the sludge dryer, reducing the area of the solar field to 1,209 m^2, which corresponds to 33.2% of the area required in the base case (3,640 m^2).

Table 5. Influence of cogeneration thermal energy in the solar field area (mirrored area).

Scenario	A	B	C	D
Electricity generation [MWh/year]	3154	3104	2925	2760
Thermal energy recovered (cogen) [kWh/day]	5120	5063	4868	4747
Energy demand from solar field [kWh/day]	3138	3347	3541	3662
Required solar collector area [m^2]	1209	1230	1308	1386
Percentage of original (base case) area [%]	33.2	33.7	35.9	38.1

Figure 6 shows the annual electricity generation over the project's life, considering a yearly efficiency loss of the power generation systems due to gasification system degradation of 0.35% p.a. [22]. In the first year, the energy generated is 3,154,391 kWh/year, and at the end of its useful life, generation drops to 2,899,818 kWh/year.

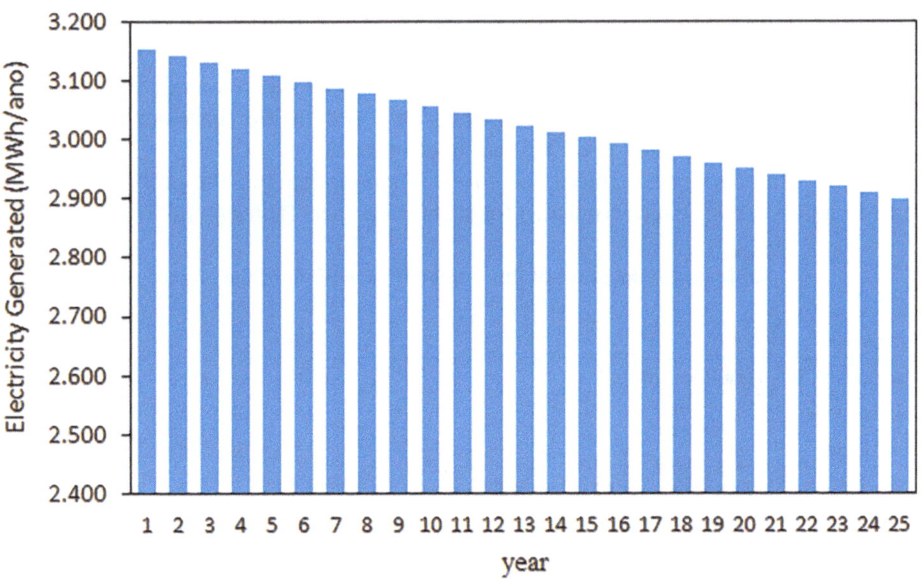

Fig. 6. Annual electricity generation over the project's life.

From the results obtained, scenario A was selected for economic evaluation, as it presents the highest potential for electricity generation and results in the smallest area of the solar field.

The cost of the proposed integrated system components was estimated based on the literature. According to Lourinho et al. [30], the specific cost of cogeneration systems, consisting of "biomass gasifier + ICE + recovery boiler", varies from 3,900.00 U$/kWe (850 kWe plant) to 5,450.00 U$/kWe (450 kWe plant). Thus, this study considered value of 5,000.00 U$/kWe for the cogeneration system.

The cost of the solar field was calculated based on the work of Schuknecht et al. [31], who indicated a range of values from 100.00 to 150 U$/m^2. The technical literature reveals a wide range of costs for sludge thermal dryers. Based on the work of Cerqueira [32], the specific costs of thermal dryers are approximately 296.8 US$/kg.day (dry base). Youssef and Kahil [33] reported specific costs of 486.9 US$/kg.day (dry base). Therefore, in this work, since the dryer was designed to have a low manufacturing cost, a specific cost of 390.0 US$/kg.day (dry base) was applied, corresponding to an intermediate value in the indicated range. It should be noted that the specific costs indicated were updated, taking into account the Chemical Engineering Plant Cost Index (CEPCI) for the year 2023 ($CEPCI_{2023} = 794,3$) and the corresponding CEPCIs for the year of publication of the works cited ($CEPCI_{2016} = 541.7$ for [33] and $CEPCI_{2019} = 607,5$ 2019 for [32]).

For the sake of establishing a base for comparison, it was also considered that when the sludge is used to generate electricity, the dewatered sludge (25% TS) is no longer sent for final disposal in a landfill, thus avoiding the costs of transportation and the fees charged by the landfill operator. Foladori et al. [34] reported an average cost of wet sludge (20% TS) disposal in Europe of 129.5 US$/ton, including transportation. A case study for Brazil [35] indicates a cost of 48.5 US$/ton for transporting and disposing of sewage sludge (25% TS) in a landfill in Paulinia city (SP). Therefore, in this study, the avoided cost of transporting and final disposing of the dewatered sludge (25% TS) is considered revenue in the cash flow of the proposed system, and, based on the reported values, a conservative value of 40.0 US$/ton was used.

Table 6 summarizes the initial costs for the components of the proposed system, considering the conditions established for scenario A (see Tables 4 and 5).

Table 6. Components of the initial cost (CAPEX) of the project

Component	Cost
Gasifier + ICE + Heat Recovery Unit	US$ 1,895,000.00
Thermal Drier	US$ 1,448,928.00
Solar Field	US$ 175,305.00
BOP (balance of plant): 7.5% of other components	US$ 263,942.00
Sub-total	**US$ 3,783,147.00**
Administration (construction phase) (5.0%)	US$ 189,158.00
Engineering (2.5%)	US$ 94,579.00
Total	**US$ 4.066.912,00**

Regarding the tariffs for commercializing the electricity generated, a range of values was used, considering prices practiced in ANEEL auctions [36] for energy produced from different biomasses, ranging from 67.3 US$/MWh (sugarcane) to 125.1 US$/MWh (MSW). Thus, the lower and upper limits of the range are formed by these values, and an intermediate value (96.2 US$/MWh) was also considered.

Based on the values described above, the project's cash flows were elaborated, and the economic indicators (NPV, IRR, and payback) were calculated and presented in Table 7. The project was not viable at electricity tariffs of 67.3 and 96.2 US$/MWh. The best economic indicators are obtained when the electricity tariff corresponds to 125.1 US$/MWh, presenting a positive NPV, an IRR higher than the MARR, and a payback time within the project's useful life.

Table 7. Results of the Economic Analysis

Electricity Tariff US$/MWh	NPV US$	IRR %	Payback years
67.3	−909,697.86	6.8	–
96.2	−104,415.52	9.6	–
125.1	700,866.83	12.2	14.16

The LCOE obtained, considering the conditions presented for scenario A, was 107.85 US$/MWh, a value within the range of prices for the sale of energy from other biomasses in ANEEL auctions, approaching the price proposed for energy produced from MSW [36].

Finally, the influence of the sale price of electricity and CAPEX on the economic indicators NPV and IRR of the proposed system was evaluated. The range of values previously used for electricity was considered, with the upper limit being increased by 20%, which results in 155 US$/MWh. The sensitivity analysis also considered a variation of 20% of CAPEX about the reference value calculated for scenario A (Table 6). The results of this sensitivity analysis are presented in Fig. 7 for the NPV and in Fig. 8 for the IRR.

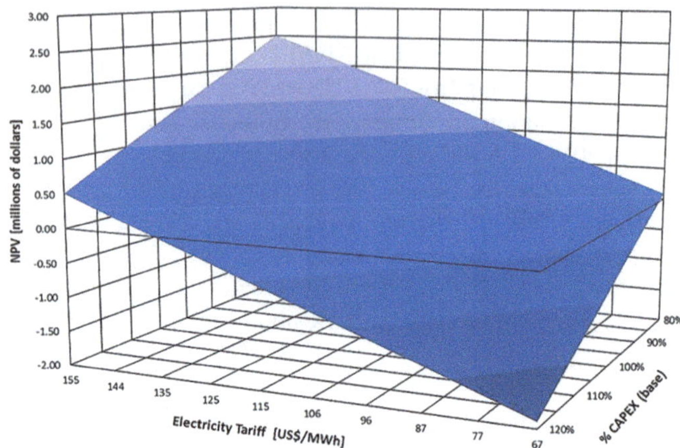

Fig. 7. NPV as a function of the electricity tariff and the percentage of the reference CAPEX.

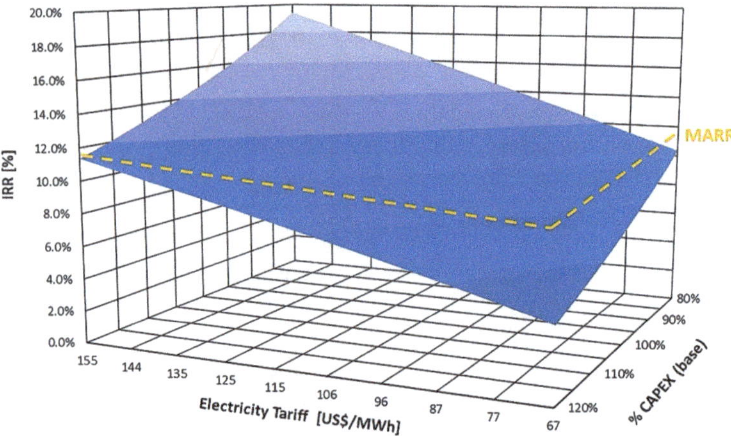

Fig. 8. IRR as a function of the electricity tariff and the percentage of the reference CAPEX.

The analysis of Figs. 7 and 8 shows that if CAPEX increases by 20%, the system is not economically viable, regardless of the electricity tariff considered. For a 10% CAPEX increase, the system begins to present positive economic indicators (NPV > 0 and IRR > MARR) only when electricity is traded at a price above 139.5 US$/MWh, a value higher than that used for electricity generated from MSW, according to ANEEL auctions [36]. On the other hand, if CAPEX is reduced by 20%, economic indicators become favorable when commercialized electricity prices are just above 85.0 US$/MWh, a value close to the lower limit of the electricity tariff range applied.

4 Conclusions

The present work evaluated the technical and economic feasibility of an integrated system for drying and gasifying sewage sludge, aiming at electricity generation through an internal combustion engine. The thermal demand of the sludge dryer was supplied by combining the use of solar thermal energy and the heat recovered from the exhaust gases of the ICE.

The composition of the sewage sludge used in the experiments was close to the values reported by other authors. Submitting the dry sludge to a gasification process, using air as the gasification agent, a syngas with LHV of 7.42 MJ/kg was obtained when the ER is 0.25. This LHV value enables the use of the syngas to generate electricity by burning it in internal combustion engines.

By burning the syngas in an internal combustion engine, an electrical power of 379.2 kW was obtained. Additionally, it was possible to recover 55,120 kWh/day of thermal energy from the engine's exhaust gas. It was observed that the energy recovered from the exhaust gases was insufficient to meet the thermal demands of the sludge dryer. However, considering the hybrid system proposed, in which solar energy was harvested in parabolic troughs through solar collectors, the thermal demand of the sludge dryer can be adequately met. The use of the energy from the ICE exhaust gases allowed for a reduction in the mirrored area of the solar field to 33.2% of the required value when the thermal demands of the dryer were met exclusively with heliothermal energy (base case). It should be noted that when drying is carried out only with heliothermal energy, without using sludge energy, a significant revenue would be neglected: the electricity generated.

The economic analysis verified that using the sludge for electricity generation in a cogeneration system, recovering energy from the ICE exhaust gases to partially meet the thermal demands of the sludge dryer, can be a viable option from an economic perspective. For the conditions evaluated and considering the sale price of electricity at 125.1 US$/MWh, the system presented a positive NPV (US$700,866.83), an IRR of 12.2% higher than the adopted MARR (11.75%), and a payback time of 14.16 years (within the project's lifespan). In addition, the LCOE obtained for this scenario was 107.85 US$/MWh, close to the value established by ANEEL for the sale of electricity generated from MSW (125.10 US$/MWh). Although these economic indicators may seem modest, the sensitivity analysis showed that it is possible to obtain better financial performance. One of the possibilities would be to provide tariff incentives for the commercialization of electricity generated from this waste (sludge) or for the acquisition and installation of the proposed system (solar collectors, cogeneration system, etc.), as well as incentives for the development of the national industry, aiming at reducing production costs of the components of this system.

Finally, it is emphasized that the proposed system allows for the energetic and economic valorization of waste, whose production volume tends to increase in the coming years and whose traditional disposal methods are costly, not always well accepted by society, and, in the case of sewage sludge landfilling, lead to significant environmental issues.

Acknowledgments. The authors would like to thank the National Council for Scientific and Technological Development - CNPq (process nº 317527/2021-5), the Research Support Foundation of the State of Minas Gerais – FAPEMIG (process nº APQ-02413-17), and the Carlos Chagas Filho Foundation for Research Support in the State of Rio de Janeiro – FAPERJ (SEI processes E- 26/200.387/2025 and E- 26/200.388/2025) for their collaboration and financial support in the development of this work.

References

1. Chen, R., Sheng, Q., Dai, X.: Upgrading of sewage sludge by low-temperature pyrolysis: fuel properties and behavior. Fuel **300**, 121007 (2021)
2. IEA: World Energy Outlook 2023. International Energy Agency (2023)
3. ANEEL. Sistema de Informação da Geração da ANEEL. https://dadosabertos.aneel.gov.br/dataset/siga. Accessed 15 May 2024
4. EPE. Balanço Energético Nacional – 2023. https://www.epe.gov.br/pt/publicacoes/balanco-energetico-nacional-2023. Accessed 03 May 2024
5. WHO Homepage. https://www.who.int/news-room/fact-sheets/detail/sanitation. Accessed 21 Sept 2024
6. Agência Senado. Situação do Saneamento Básico Ainda é "Catastrófica" no Brasil. https://www12.senado.leg.br/noticias/materias/2023/1/16/situacao-do-saneamento-basico. Accessed 12 July 2024
7. Costa, I.M., Dia, M.F., Robaina, M.: Evaluation of the efficiency of urban solid waste management in Brazil by data envelopment analysis and possible variables of influence. Waste Disp. Sustain. Energy **6**, 283–295 (2024)
8. Cezarini, E.C.A., Peterli, Z., Santiago, Y.C., Venturin, O.J.: Sewage sludge energy potential assessment: characterization approach. Chem. Eng. Trans. **109**, 247–252 (2024)
9. Gomes, L.A.: Aproveitamento do lodo gerado em estações de tratamento de esgoto e a relação com o meio ambiente. Master's thesis (in poetuguese), Univ. Federal de Minas Gerais, Brasil (2019)
10. Ding, A., et al.: Life cycle assessment of sewage sludge treatment and disposal based on nutrient and energy recovery. Sci. Total Environ. **769**, 144451 (2021)
11. Santiago, Y.C., et al.: Techno-economic assessment of producer gas from heavy oil and biomass co-gasification aiming electricity generation in rankine cycle. Processes **10**, 2358 (2022)
12. Miki, R. E.: Biometano produzido a partir de biogás de ETEs e seu uso combustível veicular. Revista DAE (in portuguese) **66**, 209 (2018)
13. Singh, V., Phuleria, H.C., Chandel, M.K.: Estimation of energy recovery potential of sewage sludge in India. J. Clean. Prod. **276**, 122538 (2020)
14. Lee, U., Dong, J., Chung, N.J.: Experimental investigation of sewage sludge solid waste conversion to syngas using high temperature gasification. Energy Conv. Manag. **158**, 430–436 (2018)
15. Capodaglio, A.G., Callegari, A.: Energy and resources recovery from excess sewage sludge. Res. Conserv. Recycl. Adv. **19**, 200184 (2023)
16. Svenson, I.: Using solar thermal energy in the sludge drying process of a wastewater treatment plant - a case study of Kattastrand. Master's thesis, Chalmers University of Technology (2021)
17. Baresel, C., Lüdtke, M.: Slamtorkning som en del av slamhantering vid Syvab Himmerfjärdsverket. Results from R&D cooperation Syvab-IVL, report B2276.2017, Swedish Environmental Research Institute (2017)

18. Peterli, Z.: Avaliação energética e experimental da secagem em baixa temperatura de lodo sanitário com o uso de energia heliotérmica. Master's thesis (in portuguese), Federal University of Itajubá – UNIFEI (2019)
19. Zainal, Z.A., Ali, R., Lean, C.H., Seetharamu, K.N.: Prediction of performance of a downdraft gasifier using equilibrium modeling for different biomass materials. Energy Convers. Manag. **42**(12), 1499–1515 (2001)
20. Rezende, T.T.G., et al.: Technical and economic potential for hydrogen production from biomass residue gasification in the state of Minas Gerais in Brazil. Int. J. Hydrogen Energy **101**, 358–378 (2025)
21. BACEN. Taxas de Juros Básicas. Banco Central Brasileiro. https://www.bcb.gov.br/controleinflacao/historicotaxasjuros. Accessed 02 Sept 2024
22. Luz, F.C., et al.: Techno-economic analysis of municipal solid waste gasification for electricity generation in Brazil. Energy Convers. Manag. **103**, 321–337 (2015)
23. IRENA - Renewable Energy Technologies: Cost Analysis Series - Biomass for Power Generation. www.irena.org/publications. Accessed 10 Sept 2024
24. Marsiglia, M.C.M., Gonzalez, C.A.D., Lora, E.E.S., Andrade, R.V., Maya, D.M.Y.: Feasibility study of a 1 MWe RDF gasification plant for sustainable power generation in Brazil: Insights from pilot plant data. Biomass Bioenerg. **199**, 107893 (2025)
25. Collard, M., Teychené, B., Lemée, L.: Comparison of three different wastewater sludge and their respective drying processes. J. Environ. Manag. **203**, 760–767 (2017)
26. Marchioro, E.: A Incidência de Frentes Frias no Município de Vitória (ES). Revista ACTA Geografia, 49–60 (2012). (in portuguese)
27. Lima, A. C., Delpupo, A. M., Silva, B. F., Scarpatti, M. P., Almeida, P. V. D.: A energia solar no Espírito Santo. ASPE, Vitória – ES (2013)
28. Sidek, F. N., Samad, N. A.: Review on effects of gasifying agents, temperature and equivalence ratio in biomass gasification process. In: IOP Conference Series – Materials and Science Engineering, Kuantan (2020)
29. Henao, N.C., et al.: Energy and economic assessment of a system integrated by a biomass downdraft gasifier and a gas microturbine. Processes **10**, 2377 (2022)
30. Lourinho, G., Alves, O., Garcia, B., Rijo, B., Brito, P., Nobre, C.: Costs of gasification technologies for energy and fuel production: overview, analysis, and numerical estimation. Recycling **8**, 49 (2023)
31. Schuknecht, N., McDaniel, J., Filas, H.: Achievement of the $100/m^2 parabolic trough. In: AIP Conference Proceedings, vol. 2033, no. 1 (2018)
32. Cerqueira, P. L. W.: Custos de Desaguamento e Higienização de Lodo em ETEs com Reatores UASB Seguidos de Pós-Tratamento Aeróbio: Subsídios para Estudos de Concepção. Master's thesis (in portuguese), Universidade Federal do Paraná- UFP (2019)
33. Youssef, S., Kahil, M. A.: Solar sludge drying for medina Al-Munawarah sewage treatment plant in the Kingdom of Saudi Arabia. J. Environ. Eng. **142**(12) (2016)
34. Foladori, P., Andreottola, G., Ziglio, G.: Sludge Reduction Technologies in Wastewater Treatment Plants, 1st edn. IWA Publishing Ltd., London (2010)
35. Martins, S.F., Esperancini, M.S.T., Quintana, N.R.G., Barbosa, F.S.: Análise Econômica da Produção de Lodo de Esgoto Compostado para Fins Agrícolas na Estação de Tratamento de Esgoto de Botucatu - SP. Energia na Agricultura **36**(2), 218–229 (2021)
36. ANEEL. Edital do Leilão n° 4/2022. https://www2.aneel.gov.br/cedoc/. Accessed 29 Aug 2014

BY NC ND

Theoretical-Experimental Analysis of Water Injection in Diesel Cycle Engines for Emission Research

Mateus Evangelista Dos Santos(✉) and Márcio Andrade Rocha

Instituto Federal da Bahia - Campus Jequié, Jequié, BA 45201-570, Brazil
prosel.jqe@ifba.edu.br

Abstract. Vehicles have always had a significant impact on society, serving as agents of change. They are the focus of intensive research aimed at improving performance and sustainability. This study examines the introduction of water mist into the combustion chamber of diesel engines to reduce emissions through an academic experiment. The research involves collecting and analyzing data before and during water mist injection, with a focus on mitigating harmful gases emitted by vehicle engines, such as nitrogen oxides (NO_X), sulfur dioxide (SO_2), and nitrogen monoxide (NO)compounds known to intensify acid rain. Additionally, these pollutants contribute to respiratory illnesses, impaired cardiovascular function, and vascular disorders. The study aims to reduce the volume of environmentally and health-damaging gases, potentially encouraging future research and advancements in the field while supporting the ongoing development of Diesel-cycle engines. The experiment was conducted at the Thermal Machines Laboratory of the Federal Institute of Bahia Jequié campus, using a stationary Buffalo 5.0 Diesel-cycle engine and S-10 diesel fuel as the test fuel.

Keywords: Water mist · experiment · diesel engines · emissions

1 Introduction

In 1892, Rudolf Christian Karl Diesel patented the internal combustion engine. Machines that use internal combustion convert the thermal energy generated by fuel combustion into mechanical energy [1].

With the increasing vehicle fleet and the need for mitigating increasingly frequent climatic phenomena, combustion engines have been the focus of research to improve environmental performance. In this context, engines using electrical sources have become a more common proposal by major manufacturers, as they offer sustainable long-term plans for the automotive industry. However, the major issue lies in the circulating vehicles that use combustion engines, which remain significant polluting agents and exacerbate greenhouse gas (GHG) emissions and other gases harmful to nature and humans. Reducing the quantity of gases harmful to the environment and human health can stimulate

O. F. Farías Fuentes et al. (Eds.): CIBIM 2024, *Proceedings of the XVI Ibero-American Congress of Mechanical Engineering*, pp. 121–137, 2026.
https://doi.org/10.1007/978-3-032-22823-9_9

future studies and advancements in the field, aiding in the continuous development of Diesel-cycle engines.

Thus, seeking techniques to mitigate harmful gas emissions, this study aims to use water in the form of mist in diesel engines to evaluate potential benefits in emission control. According to [2], the temperature in this region is directly related to pollutant emissions, making it important to experiment with combustion chamber cooling techniques to improve emission reduction in engines. As highlighted by [3], water injection in combustion engines is widely adopted due to its proven effectiveness in controlling knock and reducing NOx (nitrogen oxide) emissions.

The study involves the collection and analysis of data before and during water mist injection, focusing on reducing emissions and other harmful substances. In this scenario, the research will address gaps by analyzing the impact of water addition on emission reduction, promoting greener automotive technologies.

The water mist method has proven promising, which may encourage future research and innovations in the field. Additionally, the experimental system is simple, economical, and easy to adapt, requiring no major engine modifications.

The main objective is to analyze the water mist technique in the combustion chamber, investigating the mitigation of pollutant emissions. Specific objectives were established, including testing the water mist injection technique, using easily acquired components, and comparing pollutant mitigation at different water proportions and engine speeds.

2 Theoretical Framework

2.1 Pollutant Gases in Engine Exhaust

In 1966, exhaust gas regulations began in the United States, and the automotive industry started investing in electronic control systems. Additionally, the 1973 oil crisis led the U.S. government to enact regulations that promoted fuel conservation. These rules were based on the average fuel consumption of vehicles produced by each manufacturer [4].

The significant increase in gas emissions, primarily due to the growing number of vehicles, was evidenced by a 9.5% rise in CO_2 (carbon dioxide) emissions in Brazil in 2020, according to a new environmental study by the Climate Observatory Center, a partnership of [5]. China remains the world's largest CO_2 emitter, with 10.06 billion metric tons in 2018 [6].

According to [7], toxic components expelled from the engine - the entire exhaust gas composition - vary depending on the engine's generational status. The primary pollutant gases are NO_x (nitrogen oxides), CO (carbon monoxide), and SO_2 (sulfur dioxide).

Nitrogen Oxides (NO_x): [8] These compounds form in the combustion chamber through the reaction between fuel and atmospheric air. NO_2 in contact with sunlight forms O_3 (ozone), a potent photochemical oxidant. They can cause respiratory infections and asthma. Another photochemical reaction converts NO into nitric acid and nitrates, both potential contributors to acid rain, which may lead to forest destruction and pollution of lakes and rivers. NO_x forms due to the high temperature of burned gases behind the flame front, through reactions between nitrogen and oxygen in the air under high-temperature chemical equilibrium. The hotter the combustion and the more oxygen available, the greater the quantity of these compounds formed [9].

Carbon Monoxide (CO): This colorless, odorless gas is difficult to detect. Prolonged exposure to small doses can reduce cardiac function, cause vascular problems, anemia, and slow fetal development. CO persists in the atmosphere for about a month, gradually oxidizing to CO_2. Its production is primarily controlled by the air-fuel mixture ratio [9].

Besides being slightly flammable and produced by incomplete fuel combustion, CO can cause fatal asphyxiation when hemoglobin (responsible for oxygen transport in the body) binds to carbon monoxide instead, making it a toxic substance [8].

Carbon Dioxide (CO_2): Road transport significantly contributes to CO_2 emissions, making it the most impactful anthropogenic greenhouse gas [5].

Sulfur Dioxide (SO_2): This forms when burning fossil fuels containing sulfur, which oxidizes into sulfur dioxide (SO_2) and sulfur trioxide (SO_3), collectively known as sulfur oxides (SO_x).

Sulfur oxides are considered both primary and secondary pollutants. Secondary pollutants are particularly harmful to the environment, responsible for acid rain formation [10].

2.2 Combustion Chamber Temperature

Combustion is a chemical process in which the carbon and hydrogen of a fuel combine with oxygen from the air to form CO_2 and H_2O (complete case) and CO, CO_2, NOx, HC, and H_2O (incomplete case) as a reaction. The most important role of combustion is to be an exothermic reaction that releases energy to heat the gases and facilitate engine operation, [11]. Results obtained by [2] indicate that NOx formation demonstrates an increase proportional to temperature, especially above approximately 1800 K, where this trend intensifies rapidly. Reducing peak temperatures proves to be an effective strategy to mitigate NOx emissions, as it implies lower energy availability for breaking the N-N triple bond, crucial in the initial stage of NOx formation.

According to [12], in a literature review, water injection is proven to reduce internal cylinder and exhaust gas temperatures, alleviate combustion knock, improve combustion phasing, and reduce NOx emissions. This technology has attracted significant interest. In recent years, efforts have been made to further improve fuel efficiency and comply with exhaust gas regulations.

[13], in agreement with [14], states that autoignition promotes an increase in temperature and pressure inside the combustion chamber. This uncontrolled rise reduces engine power and increases the production of harmful gases. The study suggests controlling combustion temperature through cooling via water spray, exhaust gas recirculation (EGR) control, and reducing intake and exhaust valve opening times.

2.3 Water Injection

The concept of water injection is closely related to the development of energy conversion processes through combustion. Bertram Hopkinson emphasized in 1913 that “the introduction of water into internal combustion engines is not new.“ The use of water for explosion suppression and internal cooling of gas engines proved so successful that Hopkinson began designing power plants that relied exclusively on water as a cooling system [15, 16].

According to the authors in [17], recent studies demonstrate that water injection can increase engine power, optimize energy efficiency, and reduce fuel consumption while contributing to the reduction of pollutant emissions. [17] indicates that water addition can enhance engine power, improve energy efficiency, and lower fuel consumption, in addition to reducing harmful emissions. [18] complements this perspective, stating that water injection plays a crucial role in lowering the high temperature of the engine charge, mitigating the likelihood of knock. Furthermore, it eliminates the need for large volumes of fuel, offering a beneficial alternative to conventional charge cooling. The results obtained by [18] show performance improvements when introducing water injection at a 10% ratio. These efficiency benefits not only enable emission reductions but are also significant for decreasing fuel consumption.

In the study by [19], a computational simulation of a diesel-cycle engine operating with diesel/biodiesel/water vapor mixtures was conducted using a zero-dimensional model to investigate fuel flow rates and exhaust gas emissions (NOx, CO, and CO_2). The experimental results showed that water injection reduced NOx emissions by approximately half without significantly affecting fuel consumption. Moreover, adding water to the intake system has a limited effect on pressure, varying depending on operating conditions, as stated by the author in [20]. Using a theoretical framework methodology, it can be affirmed that the analysis of water injection into the air-fuel mixture aims to reduce greenhouse gas (GHG) emissions during combustion while simultaneously enhancing engine performance in terms of power, torque, and volumetric efficiency. This improvement is achieved through the cooling of engine combustion chambers and the optimization of fuel combustion efficiency.

Water injection, with its effective cooling effect on the cylinder combustion process, has attracted extensive attention in recent years due to its potential for knock mitigation and NOx reduction, as noted in [12].

The author [21] conducted a study focusing on the implementation of electronic management systems in the engines used and systems that allow for water injection control. The experiment confirmed that the water injection management system was successfully introduced. This enables precise definition of the water injection start timing and adjustment of the water injection timing to match the fuel injection timing. This implementation has proven promising for application in academic experimental research.

The addition of water into the combustion chamber is highly beneficial for reducing combustion temperature and suppressing knock occurrence, among other advantages. However, there is a scarcity of studies on direct water injection in SI engines with hydrogen-enriched natural gas [22].

To control combustion properly, one of the most important aspects of dual-fuel operation is injection, which can vary in terms of location (in the intake or directly into the combustion chamber), timing, and quantity [23].

2.4 Thermal and Volumetric Efficiency with Water Mist Injection

Today, despite the advent of electric motors, gasoline and diesel remain the primary fuels for powering cars and heavy-duty vehicles. Therefore, various strategies to optimize engine operation are used as tools to maximize overall efficiency and reduce pollutant emissions [24].

As reinforced by [25], the implementation of water injection can result in fuel savings ranging from 3% to 8%. [16] show that water addition can improve engine performance, optimize energy efficiency, reduce fuel consumption, and even contribute to lowering pollutant emissions.

[26] The water injection technique is not yet fully developed for Otto-cycle engines that use homogeneous charge compression ignition (HCCI). HCCI is an Otto-cycle engine that operates with auto-ignition, meaning ignition occurs through compression. Thus, the results showed that water injection effectively controlled combustion phasing and heat release rate, tripling the maximum operational load. Water injection played a crucial role in extending HCCI operation under higher loads, acting as a combustion phasing control mechanism.

[24] In the study on internal combustion engines: a systematic review using bibliometric analysis, the objective was to assess the influence of water injection on engine performance parameters. In this case, engine tests were conducted with E27 fuel, using two injection systems—PFI and DI—with a compression ratio of 15.0:1 and indirect water injection at 800,000 pascals of IMEP. The research results indicate that with water injection, it was possible to observe faster mixture combustion, and the water acted as an anti-knocking agent, increasing the octane number of the mixture.

3 Methodology

3.1 Methods and Approach

The study is classified as experimental. As highlighted by [27], in general, experiments represent the best example of scientific research. Essentially, the experimental design consists of determining a study object, selecting the variables that could influence it, and defining the methods for controlling and observing the effects that the variable produces on the object.

The adopted approach will be quantitative to obtain a comprehensive understanding of the phenomena under study through instrumentation, data analysis, and interpretation. Relevant literature on this topic was selected using Google Scholar, organized into subtopics with initial studies on the subject dating back to 1913 and current studies within the analysis range of recent articles from 2018 to 2024.

The research environment will be the Thermal Machines Laboratory at the Federal Institute of Bahia, located on the Jequié campus in the state of Bahia.

3.2 Development of the Experiment

The primary objective of water spraying in the form of mist is to cool the combustion chamber and reduce emissions. The proposed adaptation consists of the indirect integration of water injection systems into diesel engines, as shown in Fig. 1 below:

Figure 1 highlights the development of the system, which will include the following components and steps: the ultrasonic humidifier will generate the mist, which will be introduced into the intake duct, taking advantage of the engine's natural aspiration through adjustments and pipe adaptations. This will include leak verification and correction, as an ideally sealed system is required. Next, the water mist injection system

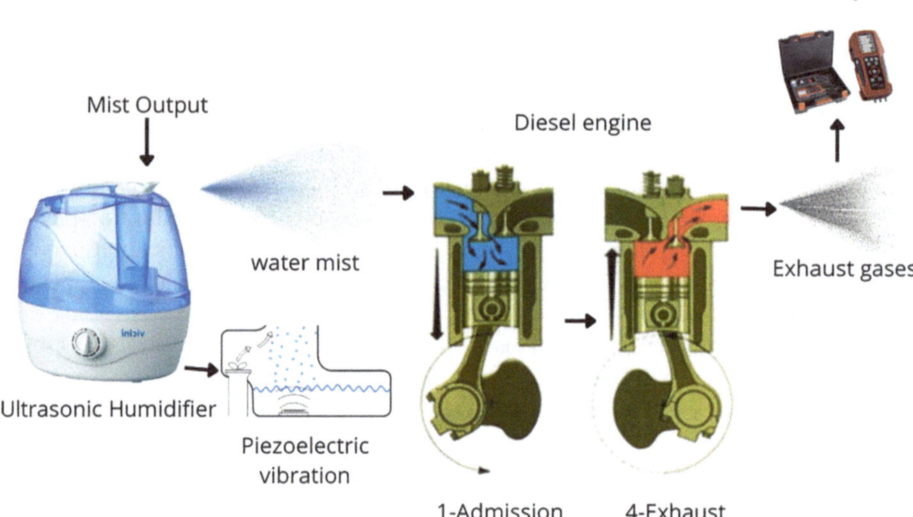

Fig. 1. Operation of the proposed water mist system. Source: Author 2024.

will be configured, initially considering the determination of the amount to be injected; this process will allow for precise system calibration. The probe will be coupled to the engine exhaust, where it will capture and analyze exhaust gases through experimental apparatuses, and their measurements will be saved for each condition.

3.3 Determination of Diesel Engine Revolutions Per Minute (RPM)

To obtain varied results and evaluate performance under different conditions, three distinct rotation speeds were selected: 2/4 of the maximum rotation (1800 RPM), 3/4 of the maximum rotation (2700 RPM), and 4/4. A geometric method was used to measure the engine rotations. The setup was based on reference marks on the metal rod.

3.4 Water Consumption Determination for Ultrasonic Humidifiers

The Vicini humidifier was operated at different power settings (low, medium, high, and maximum) for 10-min periods. The initial and final mass of water in the humidifier's reservoir was measured using a high-precision scale for each power setting. The density of water was assumed to be 1 g/ml to simplify consumption calculations (Table 1).

Table 1. Humidifier Specific Water Consumption. Source: Author 2024.

Power (Water Mist)	Consumption (g/h)	Consumption (ml/h)
P1	18	18
P2	30	30

(continued)

Table 1. *(continued)*

Power (Water Mist)	Consumption (g/h)	Consumption (ml/h)
P3	72	72
P4	270	270

3.5 Data Collection and Interpretation

The analysis of emissions will be based on the targeted parameters. For the analysis of NOx and other exhaust gases, samples were collected directly from the engine exhaust while the engine was running, following a 10-min warm-up period to simulate real operating conditions. Five consecutive 30-s analyses were performed for each scenario: without water, 18 mL/h, 30 mL/h, 72 mL/h, and 270 mL/h. To analyze the exhaust gases, a device was used—the OPTIMA 7 gas analyzer, Fig. 2 shows the equipment.

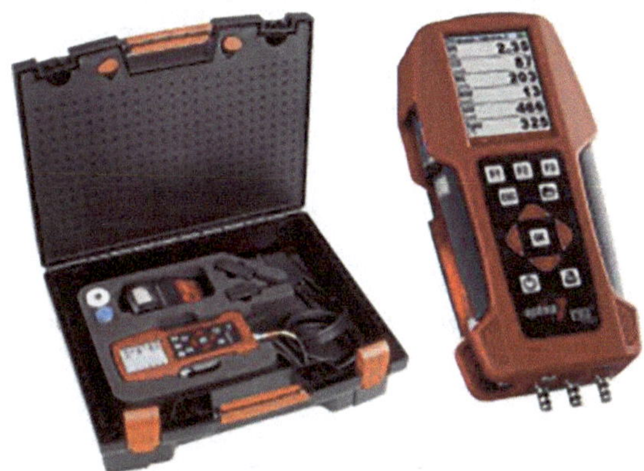

Fig. 2. Equipment and accessories. Source: mru.instruments.com

By following these steps detailed in Fig. 3, you will ensure the correct use of the OPTIMA 7 gas analyzer, obtaining accurate and reliable results. The device's intuitive interface facilitates navigation and adjustment of settings, making the measurement process efficient and accessible to any user.

Fig. 3. Measurement steps interface on the equipment. Source: mru.instruments.com

3.6 Materials Used and Their Characteristics

Several specific equipment and materials were used. The Buffalo 5.0 Diesel engine was chosen due to its quick accessibility and availability in the laboratory, in addition to having the desired technical characteristics for the experiment. This engine is single-cylinder, with a displacement of 219 cc, natural aspiration intake, oil lubrication with a pump, maximum torque of 10 N, compression ratio of 20:1, and a piston stroke of 57 mm. For the water injection system, we used a Vicini VCL-200 ultrasonic humidifier, with a power of 35W, energy consumption of 0.035 kWh, maximum spray capacity of 250 ml/h, and a water reservoir capacity of 2.2 L, operating on automatic bivolt voltage (127 V–220 V). This was selected for its adaptability and ease of handling. For data collection, the OPTIMA 7 gas analyzer, model 420076US1, was used, capable of measuring oxygen (O_2), carbon dioxide (CO_2), sulfur dioxide (SO_2), hydrogen sulfide (H_2S), carbon monoxide (CO), nitric oxide (NO), nitrogen dioxide (NO_2), and pressure difference. This equipment was chosen for its ability to measure gases relevant to the study. The pipes used were flexible, as were the rubber hoses, all selected for their compatibility with the necessary adaptations for the experiment. S-10 Diesel fuel: 2 L, was chosen for being commonly available at fuel stations. Support bench: Made of steel and wood, for equipment support. Metal clamp: For securing pipes and hoses. Precision scale: For high-accuracy mass measurements. Stopwatch: For time control during experiments. Graduated cylinder: For precise volumetric measurements of the liquids used.

4 Results and Discussions

The analyses were all performed with S-10 diesel in the engine without the system and with the system at different water mist proportions as a strategy to mitigate harmful gas emissions.

4.1 Nitrogen Oxides (NOx)

Nitrogen oxide, in the S-10 diesel at 1800 revolutions per minute, obtained a reduction as shown in Fig. 4, as well as with the results in Fig. 5, with the engine operating at 2700 revolutions per minute, obtained mitigation in relation to the engine without the water injection system.

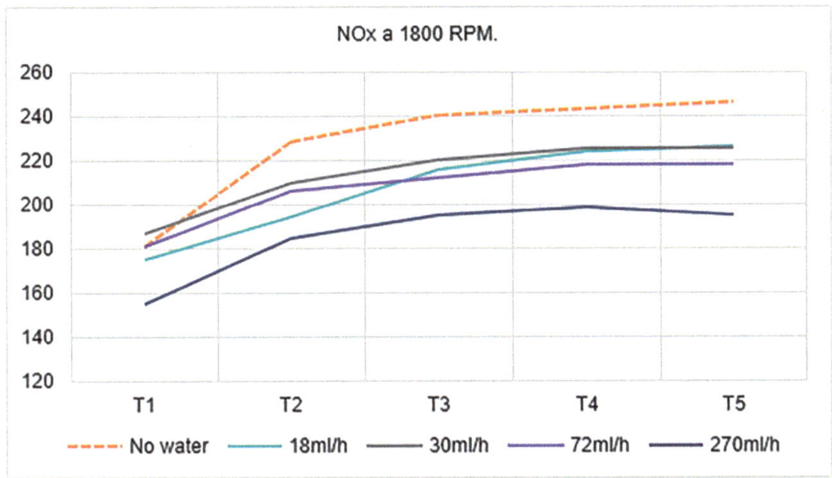

Fig. 4. NO_X emissions at 1800 RPM. Source: author 2024.

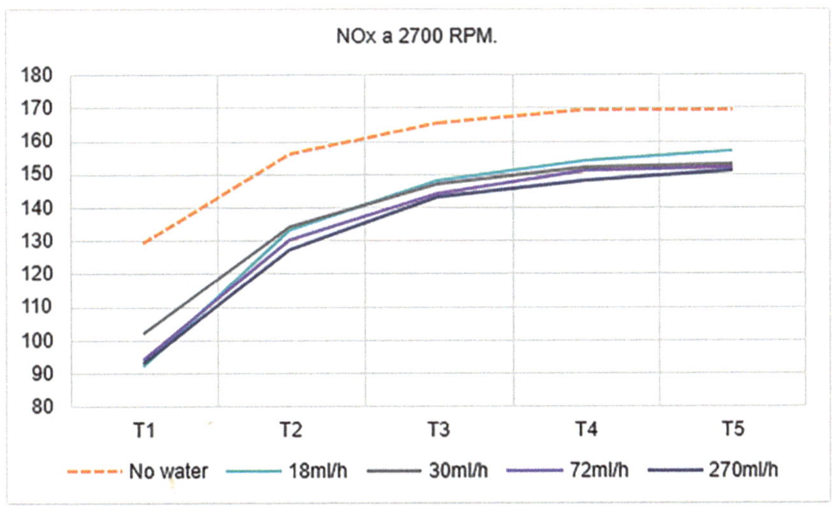

Fig. 5. NOX emissions at 2700 RPM. Source: author 2024.

Figures 1 and 2 indicate that all experimental water proportions yielded better results in reducing NOx emissions. Furthermore, the rate of 270 ml/h performed more consistently, ensuring a substantial reduction in NOx. The graph above shows that all experimental water proportions led to improved NOx emission reduction. Additionally, the 270 ml/h rate exhibited the most consistent behavior, providing a significant decrease in NOx.

Consistent with the findings of [28], water was identified as an effective method for NOx management. The authors state that it increases the thermal capacity of the working fluid, resulting in lower combustion temperatures. This reduction in emissions can be explained by the decrease in peak combustion temperature due to water cooling, since NOx is primarily produced at high temperatures.

4.2 Nitrogen Monoxide (nO)

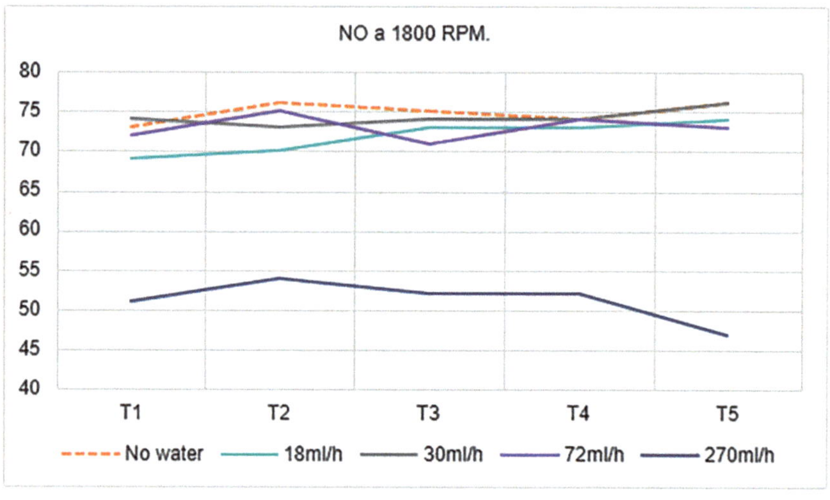

Fig. 6. NO emissions at 1800 RPM. Source: author 2024.

The result indicates that NO at 1800 rpm achieved reductions of 18 ml/h, 72 ml/h, with the highest reduction observed at 270 ml/h, which is the highest percentage of water used. However, the 30 ml/h percentage showed a value similar to the system without water, thus presenting no variation (Fig. 6).

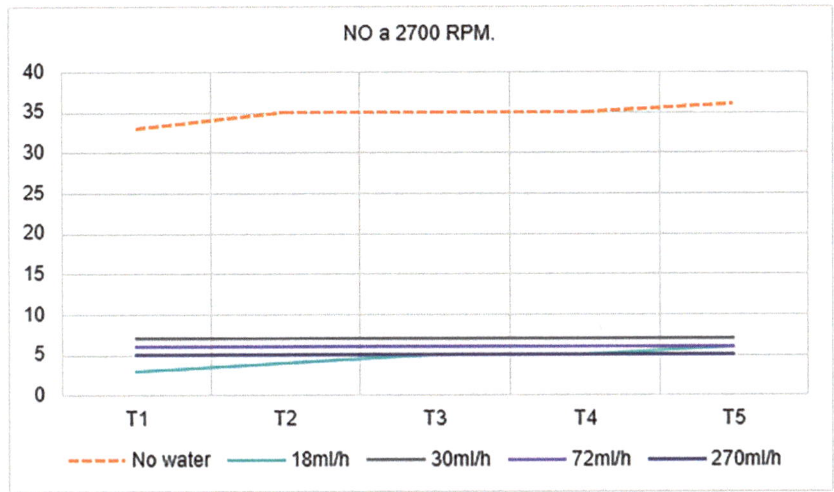

Fig. 7. NO emissions at 2700 RPM. Source: author 2024.

The NO result at a rotation of 2700 rpm showed a mitigation of emissions across all water percentages applied to the engine, with the 270 ml/h proportion standing out (Fig. 7).

4.3 Carbon Monoxides (CO)

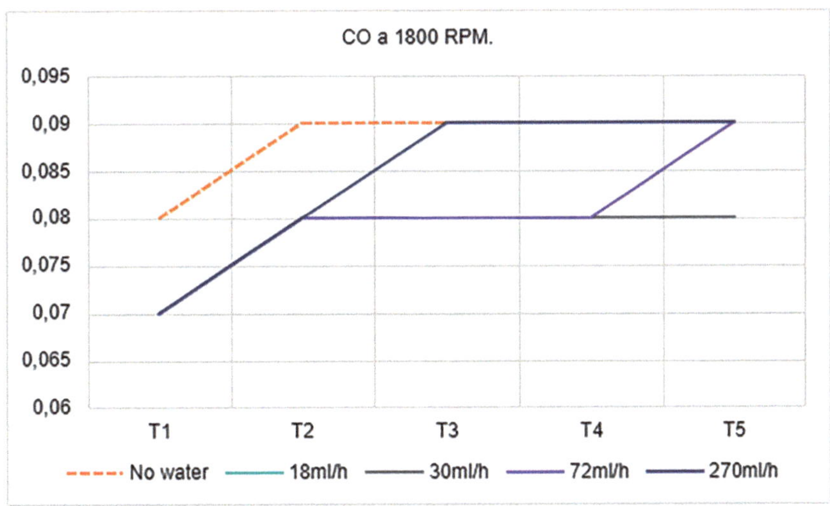

Fig. 8. CO emissions at 1800 RPM. Source: author 2024.

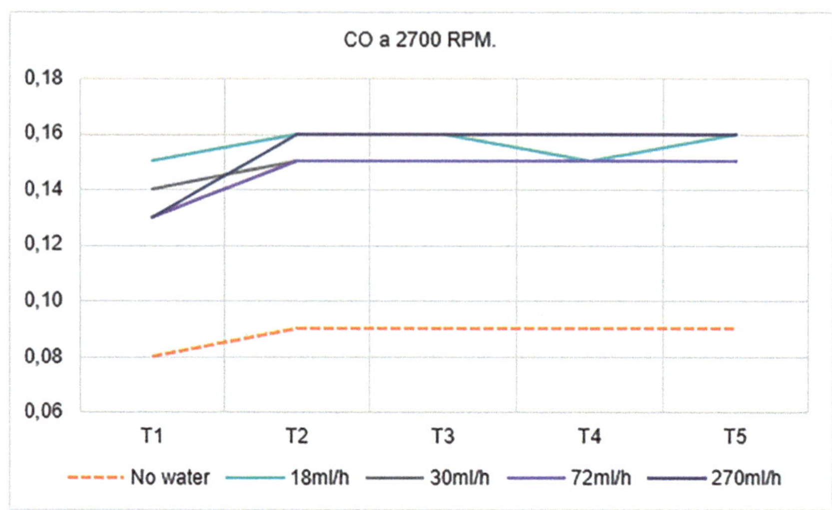

Fig. 9. CO emissions at 2700 RPM. Source: author 2024.

CO emissions are predominantly influenced by the air/fuel ratio. The graph shows that, at 1800 RPM, emissions are low. This occurs due to the nature of the mixtures used in diesel engines, which are already inherently lean. With the introduction of water injection, an additional reduction in emissions is observed, except at the point where the injection rate reaches 270 ml/h, resulting in a decrease in the air/fuel ratio (Figs. 8 and 9).

On the other hand, at 2700 RPM, emissions are higher. This is due to the reduction in the air/fuel ratio associated with higher engine speed, leading to an even leaner mixture for fuel combustion.

4.4 Carbon Dioxide (CO_2)

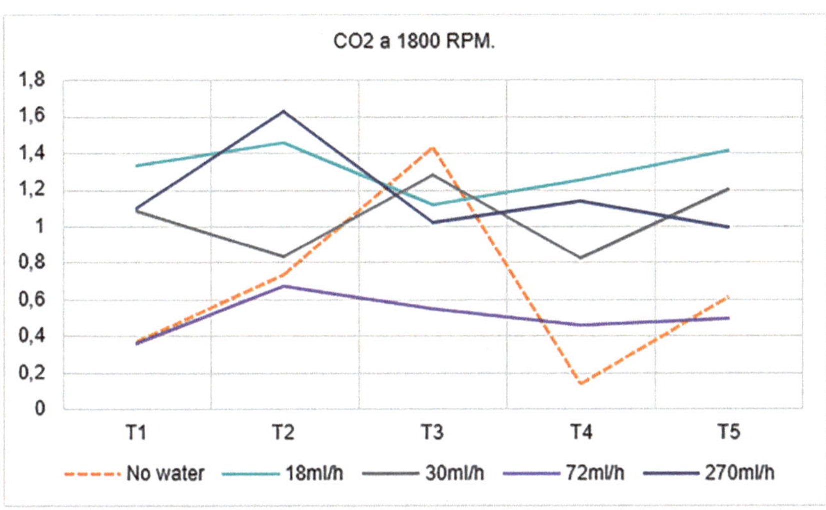

Fig. 10. CO_2 emissions at 1800 RPM. Source: author 2024.

Based on the analysis of the carbon dioxide production graph at 1800 rpm rotation, a reduction only occurred at the 72 ml/h water percentage. For the other values, there was an increase compared to the system without water, with the highest increase occurring at the 18 ml/h proportion (Fig. 10).

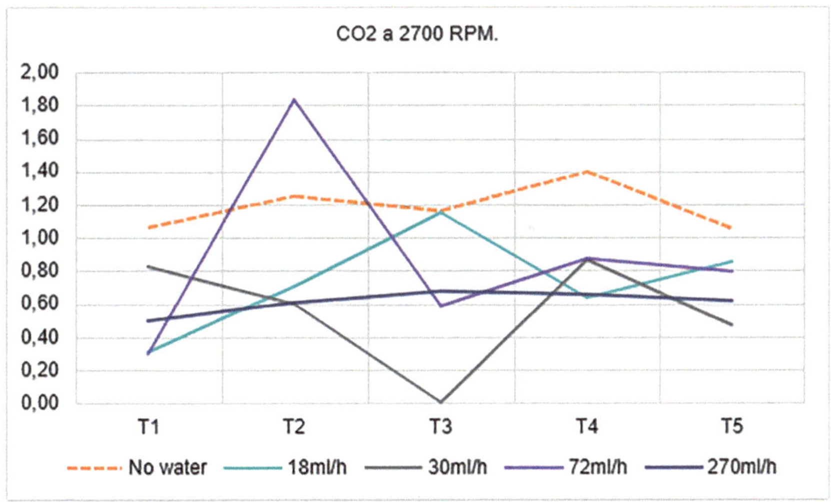

Fig. 11. CO_2 emissions at 2700 RPM. Source: author 2024.

The values obtained at the 2700 rpm rotation show a reduction in emissions across all water percentages, with the greatest decrease observed at the 30 ml/h ratio (Fig. 11).

4.5 Sulfur Dioxide (sO_2)

Based on the analyses of the sulfur oxide (SO_2) emission graphs, a significant reduction in emissions is observed at 1800 RPM when water is injected. All values at different water ratios showed a sharp decrease in SO_2 emissions; however, the 72 ml/h ratio had the highest reduction. Thus, it can be stated that water injection has an effective potential for reducing SO_2 emissions at 1800 RPM. In contrast, operation at 2700 RPM shows an increase in emissions across all tested ratios (Figs. 12 and 13).

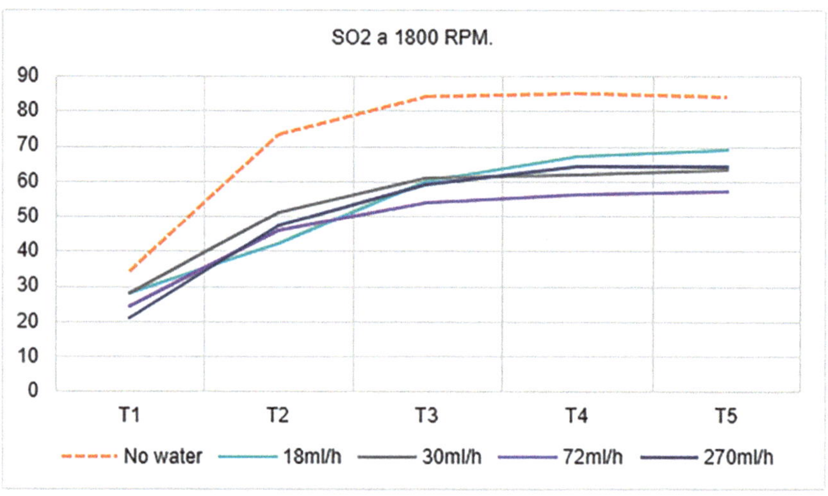

Fig. 12. SO_2 emissions at 1800 RPM. Source: author 2024.

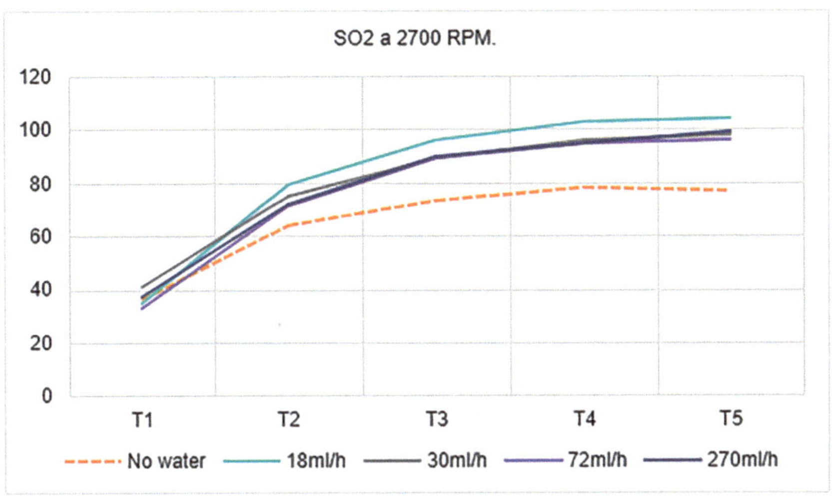

Fig. 13. SO_2 emissions at 2700 RPM. Source: author 2024.

5 Final Considerations

Based on the obtained results, the proposed methodology was efficient for quantifying emissions of NOx, CO_2, CO, and NO from the diesel engine, showing a tendency toward lower pollutant emissions.

Based on the water input data from Sect. 4, it was possible to identify the percentage variation in emissions. For NOx, all water injection rates reduced emissions, with the 270 ml/h ratio standing out for its greater consistency at both engine speeds. CO_2 showed its best results at 2700 and 3600 rpm with the 30 ml/h rate. In the case of CO, the 30 ml/h rate was more effective at 1800 rpm, while at 2700 rpm, no ratio was efficient due to this gas's relationship with the mixture's richness. Regarding NO, all ratios showed significant reductions, with 270 ml/h achieving the greatest reduction.

Aligned with the low cost of the system used in the experiment, this makes the process easier to replicate for academic research in the field. The use of household and easily accessible materials increases the availability for testing the proposed technique. Thus, this study, employing alternative equipment for this technique, proves promising as it achieves emission reduction characteristics similar to experiments conducted with specialized equipment.

It is important to note that this study did not consider the potential oxidation effects that water may cause to the engine. Additionally, no engine performance assessment was conducted to determine whether water interferes with it.

For future work: A comparative analysis between S-10 diesel and S-500 diesel using water mist is intended; A study of engine power when exposed to different water mist percentages; An investigation into potential engine damage that the water mist technique may cause.

References

1. Tillmann, C.A.C.: Motores de combustão interna e seus sistemas. Instituto Federal de Educação, Ciência e Tecnologia, Pelotas (2013)
2. Wilson, J.P.: Effects of water injection and increased compression ratio in a gasoline spark ignition engine. Ph.D. Thesis. University of Idaho (2011)
3. Bernal, J.L.L.: Study of the operation of an internal combustion engine, fueled by indirect injection of ethanol and direct injection of water into the combustion chamber. Ph.D. Thesis. [s.n.] (2019)
4. Ribbens, W.B.: Understanding Automotive Electronics, 5th edn. Butterworth-Heinemann, Woburn (1998)
5. Potenza, R.F. et al.: Análise das emissões brasileiras de gases de efeito estufa e suas implicações para as metas climáticas do Brasil 1970–2020. SEEG 54 (2021). https://energiaeambiente.org.br/wp-content/uploads/2023/04/SEEG-10-anos-v5.pdf. Accessed 25 July 2024
6. Ricardo, J.: Economia e Negócios: Os 5 países que mais produzem dióxido de carbono (CO2) (2021)
7. Carvalho, A.: Desempenho e emissões de gases de um MCI-diesel utilizando óleo diesel e mistura de biocombustível. M.Sc. Thesis. Universidade de São João Del Rei (2014)
8. COMPET.: A saúde da população e a Poluição atmosférica. Programa Economizar (2006)
9. Martins, J.: Motores de combustão interna. 4th edn. [s.l.: s.n.] (2006)
10. Vernghanini Filho, R.: Emissão de óxidos de enxofre (SOx) na combustão industrial. Revista IPT: Tecnologia e Inovação **4**(14) (2020). https://revista.ipt.br/index.php/revistaIPT/article/view/119. Accessed 23 Sept 2024
11. Giacosa, D.: Endotermic Engines. 3rd edn. Editorial Dossat (1988)
12. Zhu, S., et al.: A review of water injection applied on the internal combustion engine. Energy Conver. Manag. **184**, 139–158 (2019). https://doi.org/10.1016/j.enconman.2019.01.042
13. Akal, D., et al.: A análise de hidrogênio uso em interno combustão motores (gasolina-GLP-diesel) de combustão desempenho aspecto. Int. J. Hydrogen Energy **45**, 35257–35268 (2020)
14. Faizal, M., et al.: Review of hydrogen fuel for internal combustion engines. J. Mech. Eng. Res. Dev. **42**(3), 35–46 (2019)
15. Hopkinson, B.: A new method of cooling gas-engines. Proc. Inst. Mech. Engineers **85**(1), 679–715 (1913). https://doi.org/10.1243/PIME_PROC_1913_085_008_02
16. Dryer, F.L.: Water addition to practical combustion systems-concepts and applications. In: Symposium (International) on Combustion, pp. 279–295. Elsevier (1977)
17. Mingrui, W., et al.: Water injection for higher engine performance and lower emissions. J. Energy Inst. **90**(2), 285–299 (2017). https://doi.org/10.1016/j.joei.2015.12.003
18. Da Rocha, D.D.: Estudo da injeção de água e alteração da razão volumétrica de compressão para mitigação da detonação e melhoria do desempenho de um motor de combustão interna. M.Sc. Thesis. Universidade Federal de Minas Gerais (2018)
19. Novaes, T.L.C.C.: Modelo zero dimensional e estudo paramétrico de um motor ciclo Diesel operando com misturas diesel/biodiesel/vapor de água. Ph.D. Thesis. Universidade Federal de Pernambuco (2018)
20. Corcetti, L.L.A.: Sustentabilidade Automobilística: comparação entre sistemas de acionamentos dos automóveis atuais utilizando o método AHP. M.Sc. Thesis. Instituto Politécnico do Porto (2019). https://www.proquest.com/openview/b19e76bac2e20e0b3162f7fede4ca0c8/1?pq-origsite=gscholar&cbl=2026366&diss=y. Accessed 10 Aug 2024
21. Resende, R.M.O.: Desenvolvimento e programação de um sistema de injeção e ignição de um motor de combustão interna. M.Sc. Thesis. Universidade do Minho (2019)
22. Wang, J., et al.: Numerical investigation of water injection quantity and water injection timing on the thermodynamics, combustion and emissions in a hydrogen enriched lean-burn natural

gas SI engine. Int. J. Hydrogen Energy **45**(35), 17935–17952 (2020). https://doi.org/10.1016/j.ijhydene.2020.04.146
23. Hall, C., Kassa, M.: Advances in combustion control for natural gas-diesel dual fuel compression ignition engines in automotive applications: a review. Renew. Sustain. Energy Rev. **148**, 111291 (2021). https://doi.org/10.1016/j.rser.2021.111291
24. Amaral, L.V.; Ferreira, A.G.: Combustíveis, combustão e motores de combustão interna: Uma revisão sistêmica. 1st edn. Poisson, Belo Horizonte (2024)
25. Boretti, A.; Scalzo, J.: Influence of water injection on performance and emissions of a direct injection jet ignition engine. J. Environ. Sci. Eng. Technol. **3**(1), 29–35 (2015). https://savvysciencepublisher.com/index.php/jeset/article/view/360/337. Accessed 24 Oct 2024
26. Telli, G.D., et al.: An experimental study of performance, combustion and emissions characteristics of an ethanol HCCI engine using water injection. Appl. Therm. Eng. **204**, 118003 (2022). https://doi.org/10.1016/j.applthermaleng.2021.118003
27. Gil, A.C.: Métodos e técnicas de pesquisa social, 6th edn. Atlas (2008)
28. Mazlan, N.A., et al.: Effects of different water percentages in non-surfactant emulsion fuel on performance and exhaust emissions of a light-duty truck. J. Clean. Prod. **179**, 559–566 (2018). https://doi.org/10.1016/j.jclepro.2018.01.143

Development and Optimization of Intelligent Models for Predicting the Efficiency of a Small-Scale Refrigeration System Through Machine Learning

Paulo Rafael Costa Silva[1], Jullyene Stephanie Santos Barbosa[1], Stiven Gutemberg Figueira Rolim[1], Alvaro Antonio Villa Ochoa[1,2](✉), Sérgio da Silva Franco[1,2], Kilvio Alessandro Ferraz[1], José Ângelo Peixoto da Costa[1,2], Gustavo de Novaes Pires Leite[1,2], and Noelle D.'Emery Gomes Silva[1,2]

[1] Federal Institute of Pernambuco, Recife, PE 50740-545, Brazil
ochoaalvaro@recife.ifpe.edu.br
[2] Federal University of Pernambuco, Recife, PE 50670-901, Brazil

Abstract. This work involves the development and optimization of intelligent models to predict the energy efficiency of an industrial refrigeration system using machine learning techniques. The system studied is a small-scale mechanical compression refrigeration prototype with a nominal capacity of approximately 1 kW that uses R404A as the refrigerant. The implemented methodology includes the use of the Python Sklearn library to develop regression models that predict energy efficiency, namely, the coefficient of performance (COP) and the energy efficiency ratio (EER), using easily accessible measurable variables (electrical, thermal, and hydraulic) as inputs. The main contribution of this study lies in introducing predictive models capable of optimizing the energy performance of refrigeration systems. The intelligent regression models developed proved effective in predicting the COP and EER of the refrigeration prototype, achieving excellent metrics during the final testing phase (R2 above 70–80%, MAE and RMSE below 0.5), indicating a low average difference between actual and predicted values and thus confirming the high accuracy of the models. The results demonstrate the possibility of developing control strategies based on easily measurable physical variables, significantly improving the energy efficiency of the refrigeration system.

Keywords: Intelligent methods · Refrigeration · Machine learning · EER · COP

1 Introduction

The high energy consumption of mechanical compression refrigeration systems is a significant issue due to inefficient operation and control, resulting in increased energy usage, lower efficiency, and intensified mechanical wear of the compression system. Several studies highlight the need to improve control and operational systems to maximize

O. F. Farías Fuentes et al. (Eds.): CIBIM 2024, *Proceedings of the XVI Ibero-American Congress of Mechanical Engineering*, pp. 138–155, 2026.
https://doi.org/10.1007/978-3-032-22823-9_10

cooling capacity and reduce power consumption, thereby achieving higher performance coefficients (COP) [1–4].

Machine learning techniques have been used in several areas of engineering for detecting and diagnosing faults [5–7], verifying the energy behavior of absorption refrigeration systems [8, 9], classifying vibration impacts on humans [10], as well as determining a maintenance policy for mechanical compression refrigeration devices through reinforcement learning [11, 12].

The literature underscores the extensive investigation into enhancing the control and operation of refrigeration systems [13, 14], aiming to increase cooling capacity, reduce energy consumption, and thus achieve better performance coefficients (COP) [1]. Artificial intelligence and machine learning techniques [15, 16] have shown promising results in various sectors by identifying and predicting behavioral patterns in processes based on historical data. This enhances system control, allowing more accurate adaptation to real-world conditions and improving overall efficiency. On the other hand, having verified refrigerant gas detection in refrigeration equipment [17], they proposed a method for detecting refrigerant leaks in industrial refrigeration systems using machine learning. With real data and calculated performance indicators (superheating, subcooling), models such as Extremely Randomized Trees (EXT) were compared, standing out for their accuracy (average error of ~ 4.1 mm). Tested in 19 installations, it detected 14 leaks (34.7% gradual loss and 57.4% rapid), issuing early alerts. The method demonstrated adaptability to seasonal variations, validating its practical application in complex industrial environments.

These machine-learning techniques have also been used to predict the energy behavior of refrigeration systems, whether by absorption or vapor compression [1, 18]. In this context, de Sousa et al. [18] presented intelligent regression models to simulate the energy behavior of an absorption chiller using temperature and flow rate parameters, demonstrating a promising alternative for analyzing such equipment. Focusing on the energy efficiency of refrigeration systems, Franco et al. [1] applied parameter identification methods to determine the coefficients of a PID controller's transfer function, aiming to increase cooling capacity and overall system efficiency through control of superheating and electronic expansion valve opening. The developed models improved the refrigeration system's EER by 21% to 32%. Aiming to expand the exergy evaluation of chillers using models to optimize energy efficiency [19], researchers carried out a comparison of four models (fitted equation, physical parameters, refrigeration cycle, and neural network) to predict energy consumption through the application of the Optimization Potential Index (OPI) on real data. The artificial neural network (ANN) model showed the lowest error (~2.8%), enabling accurate classification (adequate/acceptable/inadequate) via the OPI in a real case study.

Meanwhile, Silva et al. [20] investigated improving energy efficiency using intelligent models and machine learning to evaluate the operational behavior of a refrigeration system. The developed models accurately estimated the system's dynamic behavior under full and partial loads. Other applications of these intelligent techniques have also been used to improve building energy efficiency. Ahmad et al. [21] presented a comparative study that used decision trees and neural networks to predict electricity consumption for heating, ventilation, and air conditioning (HVAC) in a hotel. Their methodology offered a viable approach and expanded sustainable energy use by analyzing regional climate histories. Xing et al. [22] proposed using Input Convex Neural Networks (ICNN) to optimize chiller loading in multi-chiller refrigeration systems. The study compared ICNN with classical models (MLP, RF, AdaBoost), demonstrating superior accuracy (R2 ~ 0.98) and better generalization to unseen data. Using a real system with three chillers, the ICNN approach combined with the Broyden-Fletcher-Goldfarb-Shanno (BFGS) algorithm achieved about 6% energy savings, outperformed traditional and meta-heuristic strategies and converged 100 times faster. This method ensures global optima and operational efficiency, validating its practical application in air conditioning systems.

Hence, this work presents the development of intelligent models to predict the energy behavior of refrigeration systems using machine learning techniques, such as K-nearest neighbors, Random Forest, and Extra Trees, considering easily measurable inputs such as temperatures, flow rates, and currents. The main contributions of this study are directed toward identifying more efficient control and operation strategies for refrigeration systems, such as:

- *Application of Machine Learning Models (Random Forest, SVM, and Artificial Neural Networks) to estimate the efficiency of a small refrigeration system using experimental data*
- *Proposal of an alternative method for evaluating refrigeration systems' energy efficiency through intelligent prediction, which is faster and more efficient than traditional measurement methods.*

2 Experimental Methodology

This section presents the experimental setup, the evaluated test procedures, and the relevant operational variables through a commercial refrigeration prototype.

2.1 Experimental Setup

The small-scale refrigeration prototype uses R-404A as the refrigerant fluid and has a nominal cooling power of approximately 1.00 kW, as shown in Fig. 1. This system features a secondary circuit that uses a 50% ethylene glycol solution as the secondary fluid, simulating the refrigeration system's thermal load.

The main components of the prototype are: (1) HP 20 monitor, (2) chilled water tank, (3) Tube-in-Tube heat exchanger, (4) condenser, (5) Arduino Uno system, (6) Full Gauge control panel, (7) electronic expansion device, (8) hermetic compressor, (9) 34 W electric pump, and (10) electric resistors.

2.2 Tests Performed

The experiments were conducted by varying the superheating levels (5 °C and 8 °C) and the electronic valve opening at 60%, 80%, and 100%. Each test lasted 3 h: one hour for prototype stabilization and two hours for data collection of physical quantities such as suction, liquid line, discharge temperatures, low and high pressures, secondary fluid

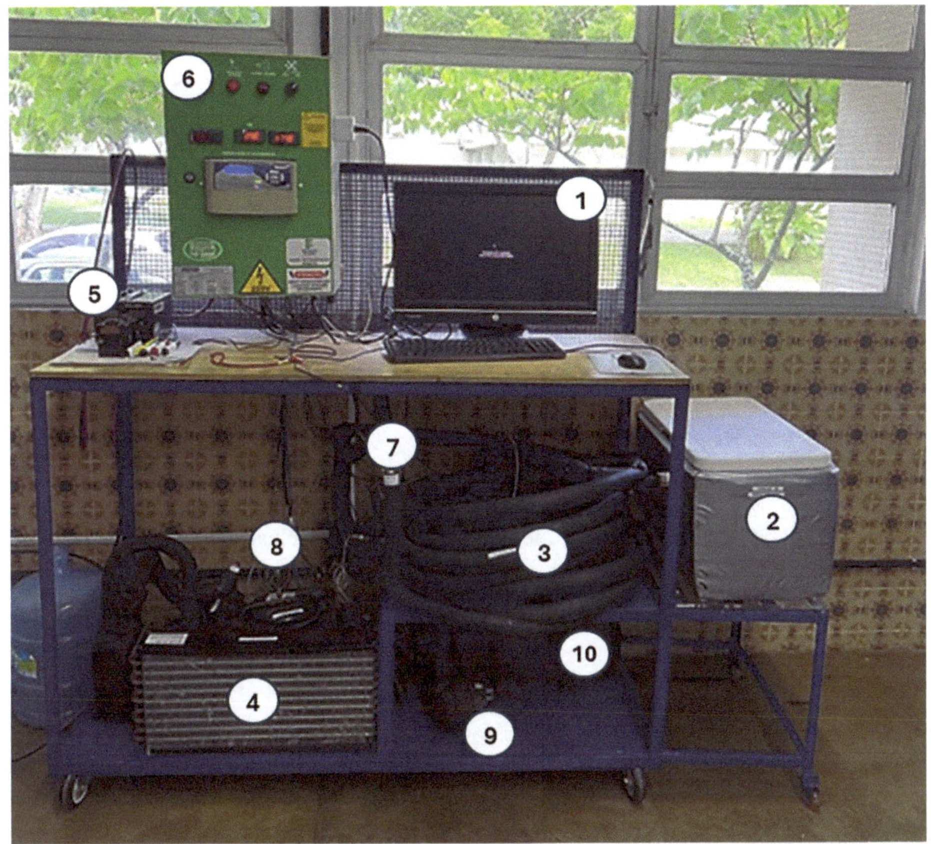

Fig. 1. Refrigeration prototype.

flow rate, power, electric current, and electrical consumption, among others. Data were collected every 10 s. The system's temperature setpoint was configured to −3 °C with a hysteresis of 5 °C.

2.3 Collected Data – Relevant Variables

Among the quantities collected, the most representative variables were selected based on the calculation of COP and EER. These included some temperature variables (suction, liquid line, discharge, glycol solution), volumetric flow rate of the glycol solution, suction and discharge pressures, active energy (consumption), electric current, and active power. The COP and EER were calculated using the following equations:

$$Q_{ev} = \dot{m}_1 \cdot \Delta H \tag{1}$$

$$COP_{elet} = \frac{Q_{ev}}{P_{useful}} \tag{2}$$

$$\text{EER} = \frac{Q_{ev} \cdot time \cdot FT}{E_{useful}} \tag{3}$$

where:
FT:: Conversion Factor (3.413)
$\dot{m}_1$**:** Refrigerant Mass Flow Rate
Coefficient of Performance
$\boldsymbol{Q_{ev}}$: Evaporator Heat Rate
P_{useful}: Useful power
E_{useful}: Useful energy
ΔH: Enthalpy variation

3 Intelligent Regression Methods

In the development of intelligent prediction models for the COP and EER of the refrigeration prototype, three regression methods were used: K-Nearest Neighbors (KNN), Random Forest (RF), and Extra Trees (ET), implemented using Python's Scikit-Learn library [23].

3.1 KNN (Regression)

The K-Nearest Neighbors (KNN) regression method uses the K most similar (nearest) data points in the training dataset to estimate the value of a new observation. This method is commonly used in applications such as price forecasting, weather pattern prediction, and wind power estimation [24]. It determines the distance between the new observation and all observations in the training data, with the Euclidean distance being the most commonly used metric.

3.2 Random Forest (Regression)

The Random Forest method is a machine learning algorithm based on the decision tree paradigm [25], widely used for regression analyses and assessing various variables' impact. This method uses the CART (Classification and Regression Trees) algorithm, a supervised machine learning approach, employing the bagging concept in which only a subset of variables is randomly selected as candidate variables.

3.3 Extra Trees (Regression)

The Extra Trees (ET) method presents a significant improvement in ensemble learning derived from the Random Forest (RF) method [26]. It uses a set of unpruned regression trees generated individually using a traditional top-down methodology. The key difference from RF is that ET employs a deterministic splitting method when developing individual trees. This way, it randomly selects a split point for each feature and chooses the best split among those options.

3.4 Evaluation Metrics (Regression)

The models' performance was evaluated using the following metrics: coefficient of determination (R2), mean squared error (MSE), root mean squared error (RMSE), and mean absolute error (MAE).

- Coefficient of Determination (R2): Evaluates the model's fit, ranging from $-\infty$ to 1. The closer to 1, the better the model fits the predicted values. Values near 0 suggest that the model only predicts the meaning of the test dataset. Negative values indicate a poor model fit.
- Mean Squared Error (MSE): Measures prediction accuracy by quantifying the average squared difference between actual and predicted results.
- Root Mean Squared Error (RMSE): Calculates the average difference between predicted and actual values, like MAE, but squares the difference and takes the square root. It's less sensitive to variance than MAE.
- Mean Absolute Error (MAE): Measures the average difference between actual and predicted values. It ranges from 0 to $+\infty$, with values closer to 0 indicating less error.

3.5 Computational Implementation

Figure 2 shows the computational methodology flowchart used to develop the intelligent prediction models for the COP and EER of the refrigeration prototype. The methodology has three stages: preprocessing, processing, and postprocessing.

- Preprocessing involves data filtering, removing outliers, consolidating experimentally collected data across various operating conditions (superheating, valve opening), and selecting relevant variables for predicting COP and EER.
- Processing includes selecting regression methods (KNN, RF, ET) and splitting the dataset into three portions: 50% for training, 25% for validation and hyperparameter tuning, and 25% for final testing. Hyperparameter optimization is performed using 5-fold cross-validation and evaluated with selected metrics (R2, MSE, RMSE, MAE).
- After model optimization and performance metric verification, postprocessing focuses on the final testing phase. Finally, COP and EER prediction models are presented via plots of these metrics over time.

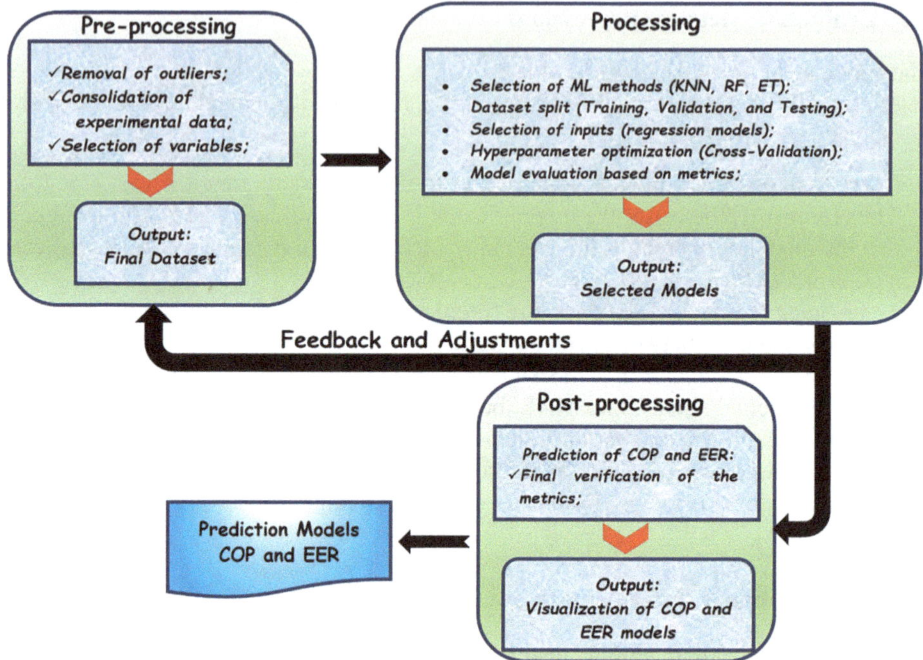

Fig. 2. Flowchart of the computational methodology.

4 Analysis and Discussion of Results

This section presents the results obtained from the implemented methodology using the collected experimental data, structured as follows:

- Selected Models – Input variables
- Intelligent Model Results – Training Phase
- Intelligent Model Results – Validation Phase
- Intelligent Model Results – Test Phase
- Final Performance of COP and EER Models

4.1 Select Models - Input

Four sets of directly measurable physical quantities (inputs) were selected to describe the operation of the refrigeration prototype and predict COP and EER behavior. The following equations represent the models:

Model 1 (M1):

$$\{COP, EER\} = f(T_{suc}, T_{sol}, T_{desc}) \tag{4}$$

Model 2 (M2):

$$\{COP, EER\} = f\left(T_{suc}, T_{linha\_liq}, T_{saq}\right) \tag{5}$$

Model 3 (M3):

$$\{COP, EER\} = f(V_{sol}, Aber_{val}, P_{desc}) \tag{6}$$

Model 4 (M4):

$$\{COP, EER\} = f(Cor, P_{desc}, T_{suc}, V_{sol}) \tag{7}$$

Model 1 (M1) presents a set of temperature variables of the refrigeration circuit, suction (T_{suc}) and discharge (T_{desc}) temperatures and the temperature of the ethylene glycol solution ($Tsol$) of the secondary circuit;

Model 2 (M2) presents a set of temperature variables exclusive to the refrigeration circuit, suction (T_{suc}) and discharge (T_{desc}) temperatures and superheat (T_{saq});

Model 3 (M3) presents a set of variables related to the solution flow, volumetric flow rate of the solution (V_{sol}), the opening of the electronic expansion device ($Aber_{val}$) and discharge pressure (P_{desc});

Model 4 (M4) presents a set of electrical variables, current (Cor), volumetric flow rate of the solution (V_{sol}), and discharge pressure (P_{desc});

4.2 Results of the Intelligent Models – Training Phase

Model training was carried out using 50% of the dataset, as described in Sect. 3.5. The results, shown in Figs. 3 and 4, demonstrate the metrics of Models 1 to 4 for predicting COP and EER.

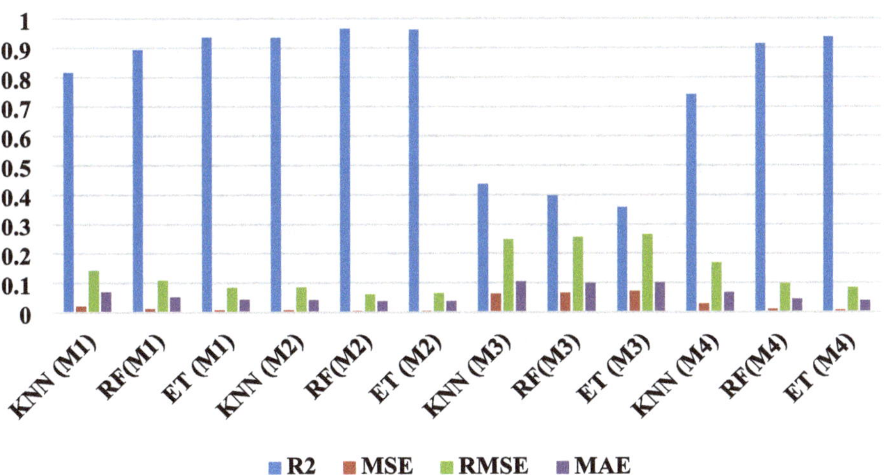

Fig. 3. Metrics of the models developed to estimate COP during the training phase.

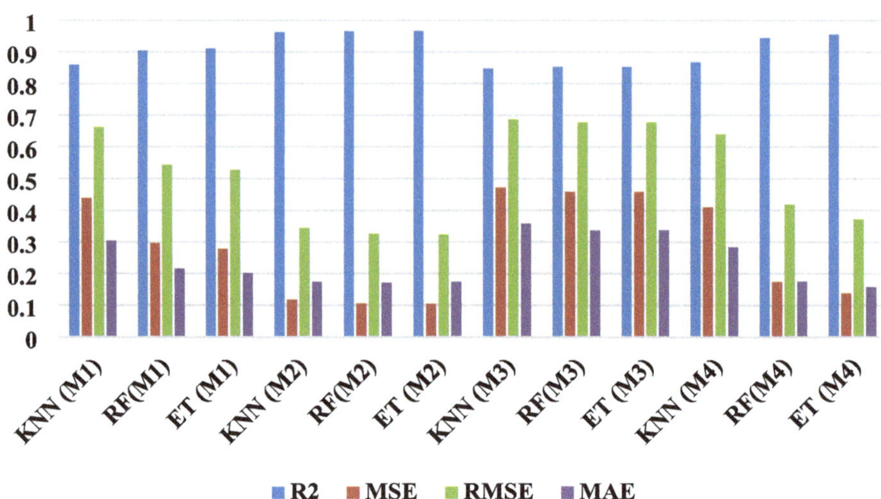

Fig. 4. Metrics of the models developed in estimating EER during the training phase.

The metrics show a good fit for both COP and EER prediction. The RF and ET methods in Models 1 and 2 achieved the best performances. The KNN method yielded lower metric values compared to the others.

4.3 Results of the Intelligent Models – Validation Phase

Model validation was carried out through hyperparameter optimization of the three methods (KNN, RF, ET) using 25% of the dataset. The hyperparameters used in the optimization were as follows: KNN = number of neighbors: [3, 5, 7, 9, 11], weights: ['uniform,' 'distance'], p: [1, 2], for the RF = number of trees: [100, 200, 300], minimum number of samples: [1, 2], maximum number of features: ['auto', 'sqrt', 'log2']}, and for the ET = number of trees: [100, 200, 300], maximum depth of the tree: [None, 10, 20, 30], minimum number of samples split: [2, 5, 10], minimum number of samples: [1, 2, 4], maximum number of features: ['auto', 'sqrt', 'log2']}. The optimized hyperparameters for each model and method are summarized in Table 1.

Table 1. Optimized hyperparameters

Model	Method	Hyperparameters
M1	KNN	number of neighbors = 3; p = 1; weights = uniform
	RF	number of trees = 100; minimum number of samples = 1; maximum number of features = auto
	ET	maximum depth of the tree = None; maximum number of features = auto; minimum number of samples = 1; minimum number of samples split = 2; minimum number of samples = 1; number of trees = 100
M2	KNN	number of neighbors = 3; p = 1; weights = uniform

(continued)

Table 1. (*continued*)

Model	Method	Hyperparameters
	RF	number of trees = 100; minimum number of samples = 1; maximum number of features = auto
	ET	maximum depth of the tree = None; maximum number of features = auto; minimum number of samples = 1; minimum number of samples split = 2; minimum number of samples = 1; number of trees = 100
M3	KNN	number of neighbors = 3; p = 1; weights = uniform
	RF	number of trees = 100; minimum number of samples = 1; maximum number of features = auto
	ET	maximum depth of the tree = None; maximum number of features = auto; minimum number of samples = 1; minimum number of samples split = 2; minimum number of samples = 1; number of trees = 100
M4	KNN	number of neighbors = 3; p = 1; weights = uniform
	RF	number of trees = 100; minimum number of samples = 1; maximum number of features = auto
	ET	maximum depth of the tree = None; maximum number of features = auto; minimum number of samples = 1; minimum number of samples split = 2; minimum number of samples = 1; number of trees = 100

Figures 5 and 6 present the metric results for Models 1 to 4 validation phase for COP and EER prediction.

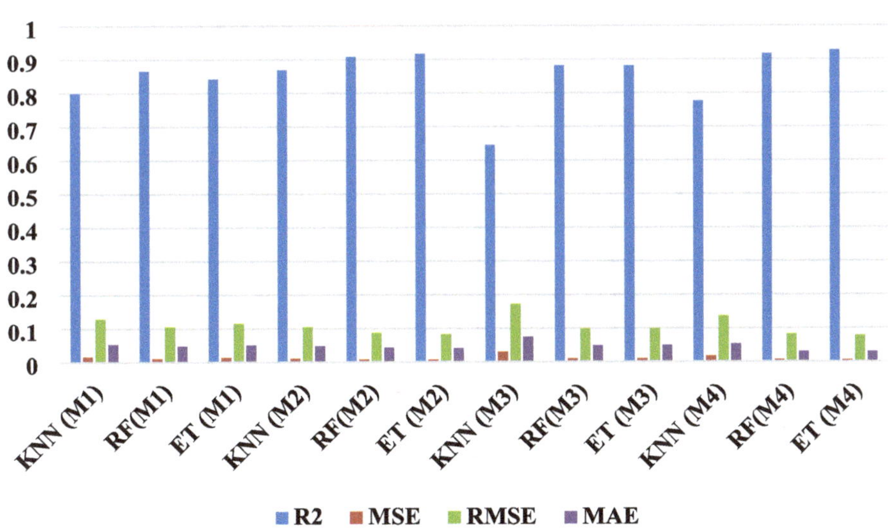

Fig. 5. Metrics of the models developed to estimate COP during the validation phase.

During this validation step, metrics improved due to hyperparameter optimization, and all models and methods showed enhanced performance compared to the training phase. R2 values exceeded 80% for most models, except for KNN in Models 3 and 4.

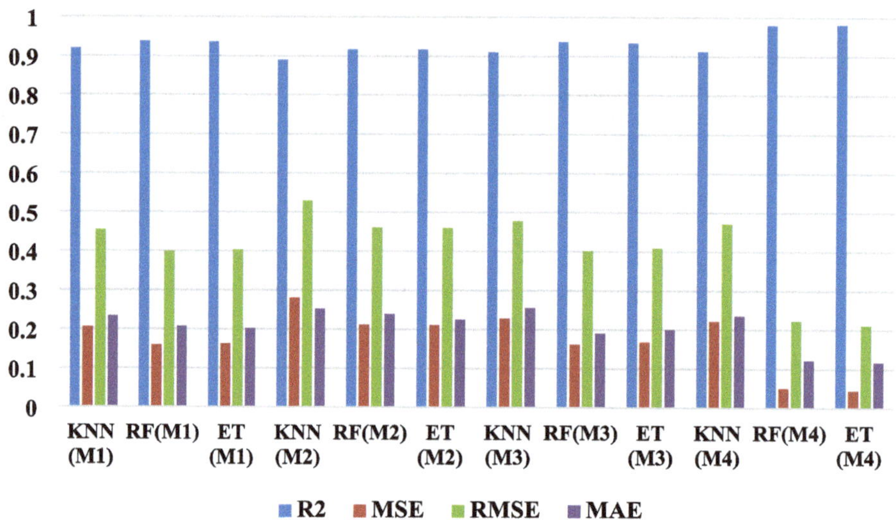

Fig. 6. Metrics of the models developed in estimating EER during the validation phase.

4.4 Results of the Intelligent Models – Testing Phase

In the final phase, the last 25% of the dataset was used to test the optimized models. Figures 7 and 8 present the metric results for the test phase of Models 1 to 4.

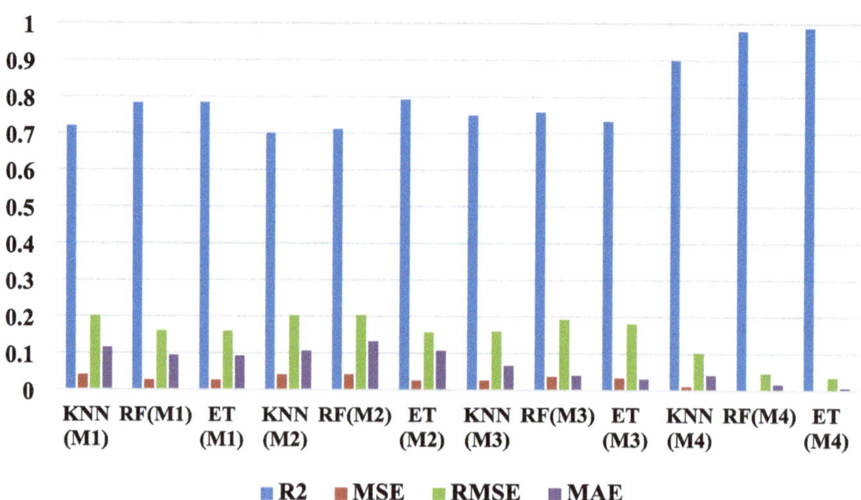

Fig. 7. Metrics of the models developed to estimate COP during the testing phase.

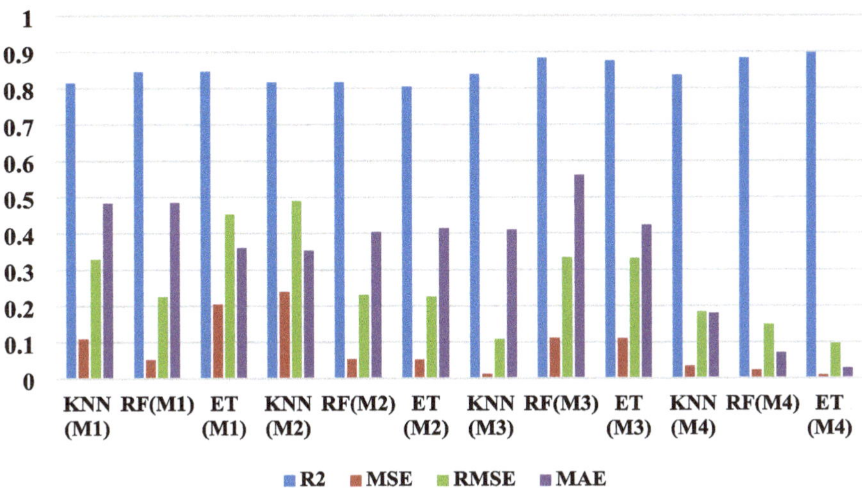

Fig. 8. Metrics of the models developed in estimating EER during the testing phase.

The R2 values were above 70% for all models and methods in COP prediction and above 80% for EER. MAE values were below 0.5, indicating high model accuracy, and RMSE was below 0.4 for almost all models. These results validate the high performance of the models in predicting COP and EER with different input combinations.

4.5 Results of the Intelligent Models – Testing Phase

The transient behavior of the prototype's efficiency was analyzed through COP and EER using the final 25% of the dataset. Figure 9 shows the COP behavior using Model 1 and three methods (KNN, RF, ET), all showing good agreement with actual data over time. RF and ET demonstrated the best fit. Similar results were obtained for Models 2, 3, and 4 (not shown due to space limitations).

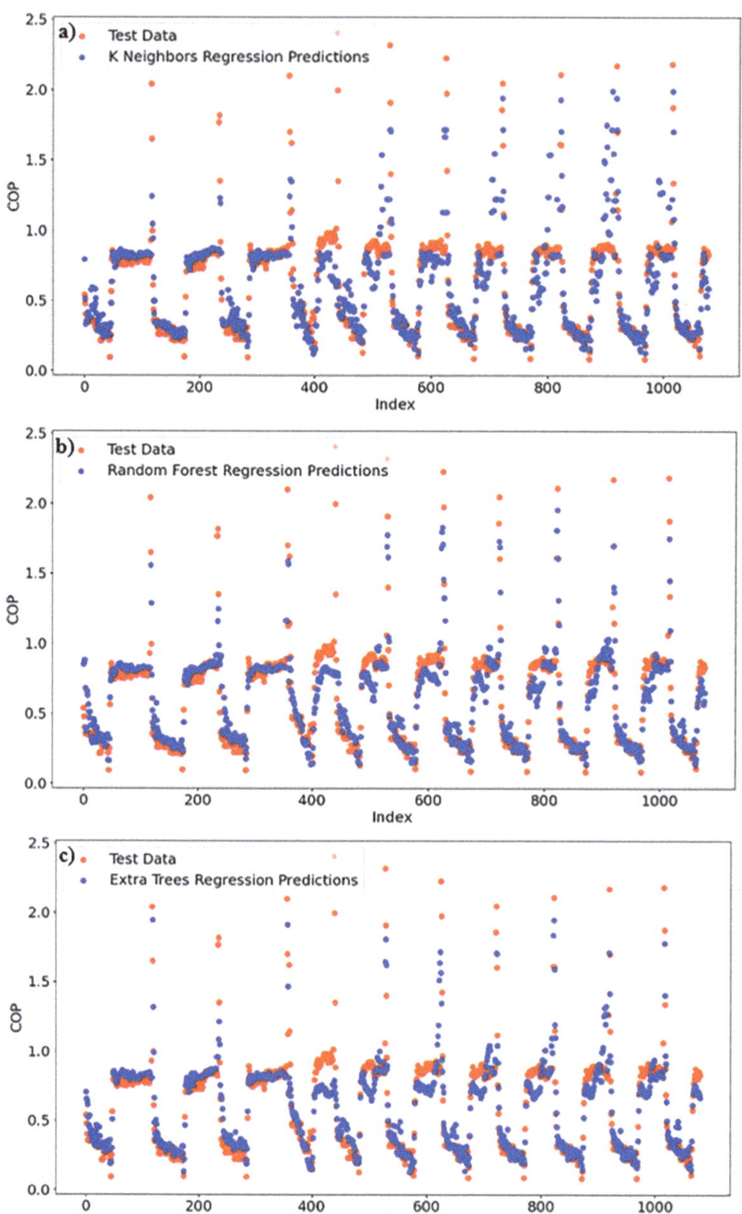

Fig. 9. COP profile through model 1 and values calculated with real data. a) KNN. b) RF. c) ET.

Figure 10 presents the EER behavior for Model 1, following a similar trend as the COP results, with RF and ET again yielding the most consistent adjustments. Despite minor deviations during brief system spikes, the models proved effective for simulating and optimizing the operation of refrigeration systems.

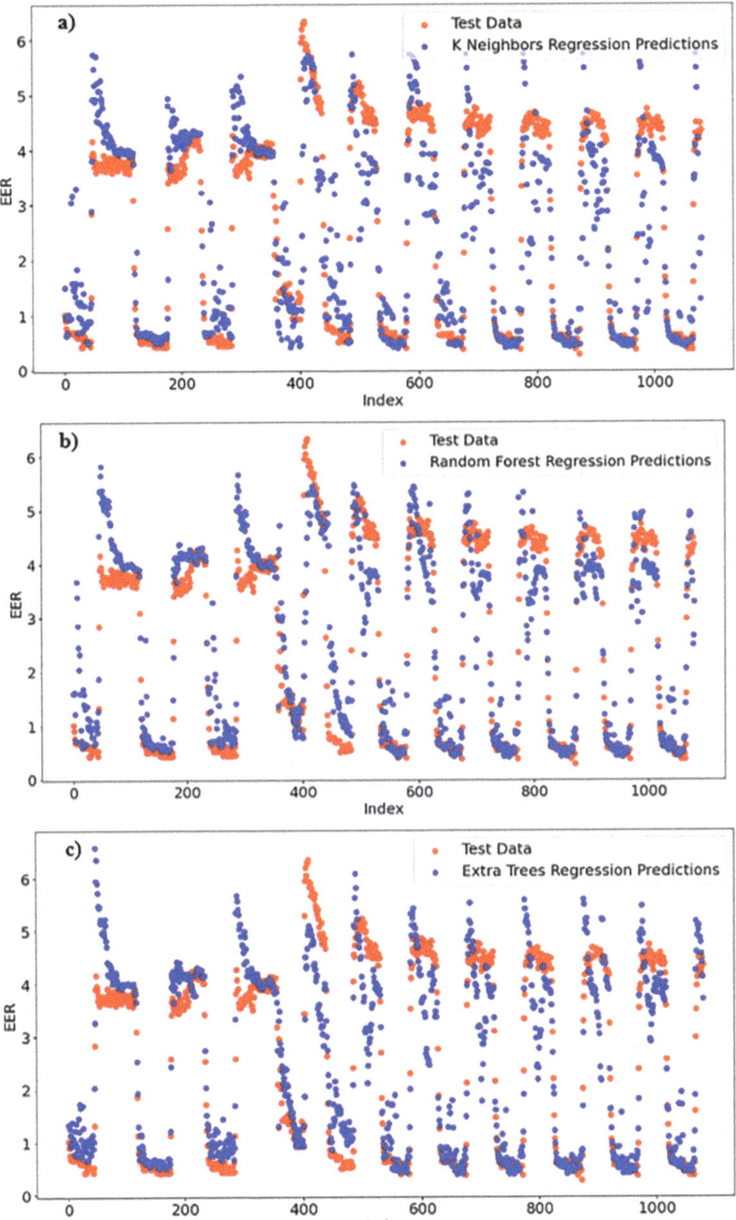

Fig. 10. EER profile through model 1 and values calculated with real data. a) KNN. b) RF. c) ET.

4.6 Critical Discussion of the Results and Challenges

The intelligent regression models developed in this study using K-Nearest Neighbors (KNN), Random Forest (RF), and Extra Trees (ET) demonstrated high predictive accuracy in estimating the coefficient of performance (COP) and energy efficiency ratio

(EER) of a small refrigeration prototype, reaching values above 70–80% in R2 and relatively low values of MAE and RMSE (< 0.5). These results are consistent with the trends reported in the recent literature on the application of machine learning (ML) techniques in thermal systems [11, 27]. As shown in Adelekan et al. [15] highlighted the strong potential of AI methods, including Random Forest and neural networks, to improve the performance and efficiency of refrigeration, air conditioning, and heat pump systems, achieving comparable levels of predictive performance when the datasets are appropriately organized and the models are adequately adjusted. This was also proven in the work developed by Panahizadeh et al. [16], where ML models were used to predict the behavior of absorption chillers, demonstrating their applicability. Compared to the literature presented [22, 28, 29], this study offers a significant contribution by developing intelligent and accurate models based only on easily measurable physical parameters (such as temperatures, pressures, flow rates, and electrical current). This approach greatly simplifies implementation in real-world scenarios, reducing the need for expensive or complex instrumentation and making real-time optimization more feasible.

Although promising results were achieved under controlled laboratory conditions through a refrigeration prototype, implementing machine learning models in refrigeration systems still needs further exploration and intensification [29, 30]. Thus, several challenges must be faced to overcome the limitations of these models. Therefore, future work should focus on i) The construction of standardized and high-quality data from several refrigeration systems for training more robust and general models, ii) conduct extensive testing with other equipment and varying operating conditions to validate model performance, and finally, iii) explore hybrid approaches that combine physical modeling and machine learning to improve predictive accuracy and reliability.

5 Conclusion

It presented the development of intelligent regression models for predicting the energy performance of a small-scale refrigeration prototype using only physically measured quantities from the system. The most relevant conclusions are:

- The intelligent regression models developed proved effective in predicting the Coefficient of Performance (COP) and the Energy Efficiency Ratio (EER) of the refrigeration prototype, achieving excellent performance metrics (R2 above 70% and 80%, and MAE and RMSE below 0.5). This suggests a low average difference between actual and predicted values, confirming the high accuracy of the models.
- The Random Forest (RF) and Extra Trees (ET) methods showed better adjustment and coherence with the actual values over the system's operation period for both COP and EER when compared with the K-Nearest Neighbors (KNN) method.
- The implemented methodology—dividing the dataset into training, validation, and testing stages, along with cross-validation and hyperparameter optimization—contributed to developing precise and robust models for predicting the energy efficiency of the refrigeration prototype.
- This methodology can be extended and adapted to other refrigeration systems and energy-related equipment, expanding its reach and impact on energy optimization.

Acknowledgments. The first and second authors would like to thank CNPq for the PIBIC scholarships. The fourth and fifth authors also thank CNPq for their Productivity Scholarships (grant numbers 3303417/2022-6 and 303200/2023–5). All authors express their gratitude to the company Fullgauge and IFPE for their support in developing this project. The authors also acknowledge the financial support provided by CNPq for the research project – Universal Call No. 402323/2016-5.

References

1. Franco, S.D.S., Henríquez, J.R., Ochoa, A.A.V., Da Costa, J.A.P., Ferraz, K.A.: Thermal analysis and development of PID control for electronic expansion device of vapor compression refrigeration systems. Appl. Thermal Eng. **206**, 118130 (2022)
2. Carreño-Zagarra, J.J., Villamizar, R., Moreno, J.C., Guzmán, J.L.: Active disturbance rejection and PID control of a one-stage refrigeration cycle. IFAC-PapersOnLine. **51**(4), 444–449 (2018)
3. Luyben, W.L.: Control of compression refrigeration processes with superheat or saturated boiling. Chem. Eng. Process. - Process Intensif. **138**(1), 97–110 (2019)
4. González, J.M., Ochoa, A.A.V., Cardemil, J.M., Godoy, F., Zapata, M.Z.: Improving the off-design modeling of a commercial absorption chiller. Energy Conv. Manag. **326**, 11947 (2025)
5. Leite, G.N.P., et al.: Alternative fault detection and diagnostic using information theory quantifiers based on vibration time-waveforms from condition monitoring systems: application to operational wind turbines. Renew. Energy **164**(1), 1183–1194 (2021)
6. Velásquez, R.M.A., Tataje, F.A.O., Ancaya-Martínez, M.D.C.E.: Early detection of faults and stall effects associated to wind farms. Sustain. Energy Technol. Assess. **47**, 101441 (2021)
7. de Lima Munguba, C.F., et al.: Fault detection framework in wind turbine pitch systems using machine learning: development, validation, and results. Eng. Appl. Artif. Intell. **138**, 109307 (2024)
8. Naik, M.M.M., Murthy, V.S.S., Durga Prasad, B.: An intelligent energy-efficient vapor absorption refrigeration system for inlet air cooling of CCPPS. J. Braz. Soc. Mech. Sci. Eng. **44**(10), 489 (2022)
9. Alaskaree, E.H.: Design and implementation of a solar - powered absorption cooling system for the Baghdad Pearl gas station : a heat transfer analysis. Heat Mass Transf. **61**(43), 1–19 (2025)
10. da Silvam, R.L.A., et al.: Operational and intelligent analysis under the ergonomics approach of the prevalence of musculoskeletal disorders in container operators. Int. J. Ind. Eng. Theory Appl. Pract. **30**(3), 763–780 (2023)
11. Bala, A., et al.: Artificial intelligence and edge computing for machine maintenance-review. Artif. Intell. Rev. **57**(5), 1–12 (2024)
12. de Lima Munguba, C.F., Leite, G.D.N.P., Ochoa, A.A.V., Droguett, E.L.: Condition-based maintenance with reinforcement learning for refrigeration systems with selected monitored features. Eng. Appl. Artif. Intell. **122**, 106067 (2023)
13. Dong, X., Ma, Y., Wang, Y., Chen, Q., Liu, Z., Jia, X.: An improved power flow calculation method based on linear regression for multi-area networks with information barriers. Int. J. Electr. Power Energy Syst. **142**(1), 108385 (2022)
14. Silva, S.F., et al.: Análise experimental do desempenho de um protótipo de refrigeração por compressão mecânica para conservação de alimentos através do uso de dispositivos de expansão Termostática e eletrônica. In: Conbrava 2019, ABRAVA (2019)
15. Adelekan, D.S., Ohunakin, O.S., Paul, B.S.: Artificial intelligence models for refrigeration, air conditioning and heat pump systems. Energy Rep. **8**(1), 8451–8466 (2022)

16. Panahizadeh, F., Hamzehei, M., Farzaneh-Gord, M., Villa, A.A.O.: Evaluation of machine learning-based applications in forecasting the performance of single effect absorption chiller network. Therm. Sci. Eng. Prog. **26**(1), 101087 (2021)
17. Mtibaa, A., Sessa, V., Guerassimoff, G., Alajarin, S.: Refrigerant leak detection in industrial vapor compression refrigeration systems using machine learning. Int. J. Refrig. **161**(1), 51–61 (2024)
18. Alcântara, S.C.S., et al.: Development of a method for predicting the transient behavior of an absorption chiller using artificial intelligence methods. Appl. Therm. Eng. **231**(1), 120978 (2023)
19. Brenner, L., Tillenkamp, F., Ghiaus, C.: Refrigeration machine modeling for exergy-based performance and optimization potential evaluation of chillers in real field plants. Int. J. Refrig. **131**(1), 775–785 (2021)
20. Silva, P.R.C., et al.: Intelligent regression modeling for performance prediction of a vapor compression refrigeration prototype using machine learning techniques. In: Papers Book, COBEM 2023, pp. 1–10 (2023)
21. Ahmad, M.W., Mourshed, M., Rezgui, Y.: Trees vs Neurons: comparison between random forest and ANN for high-resolution prediction of building energy consumption. Energy Build. **147**(1), 77–89 (2017)
22. Xing, S., Zhang, J., Mu, S.: An optimization-oriented modeling approach using input convex neural networks and its application on optimal chiller loading. Build. Simul. **17**(4), 639–655 (2024)
23. Pedregosa, F., et al.: Scikit-learn: machine learning in python. J. Mach. Learn. Res. **12**, 2825–2830 (2011)
24. Naghibi, S.A., Dashtpagerdi, M.M.: Evaluation of four supervised learning methods for groundwater spring potential mapping in Khalkhal region (Iran) using GIS-based features. Hydrogeol. J. **25**(1), 169–189 (2017)
25. Breiman, L.: Random forests. Mach. Learn. **45**(1), 5–32 (2001)
26. Geurts, P., Ernst, D., Wehenkel, L.: Extremely randomized trees. Mach. Learn. **63**(1), 3–42 (2006)
27. Brenner, L., Tillenkamp, F., Ghiaus, C.: Refrigeration machine modeling for exergy-based performance and optimization potential evaluation of chillers in real field plants. Int. J. Refrig. **131**(1), 775–785 (2021)
28. Panjapornpon, C., Bardeeniz, S., Hussain, M.A., Vongvirat, K., Chuayock, C.: Energy efficiency and savings analysis with multirate sampling for petrochemical process using convolutional neural network-based transfer learning. Energy AI **14**(1), 100258 (2023)
29. Miller, C., et al.: The ASHRAE great energy predictor iii competition: overview and results. Sci. Technol. Built Environ. **26**(10), 1427–1447 (2020)
30. Ramanipriya, M., Anitha, S.: An imperative need for machine learning algorithms in heat transfer application: a review. J. Therm. Anal. Calorim. **150**(1), 49–75 (2024)

Modelling of an Atmospheric Water Harvesting System Under Climate Conditions of Ibero-American Cities with Water Scarcity

Cristian Cuevas(✉), Aitor Cendoya, Matías Pezo, and Daniel Sacasas

Departamento de Ingeniería Mecánica, Facultad de Ingeniería, Universidad de Concepción, Casilla 160-C, Concepción, Chile
crcuevas@udec.cl

Abstract. In Chile and the world, freshwater scarcity is a reality that has been exacerbated in recent years due to climate change and global population growth. Among the alternatives that are being explored to exploit new water sources, atmospheric water harvesting systems based on refrigeration systems seems to be a good alternative for isolated communities located in the countryside. To evaluate the potential and performance of this type of water generator, a system capable of producing 100 L of water per day at an ambient temperature of 20°C and a relative humidity of 60% has been designed and subsequently evaluated in four cities in countries with water scarcity: Durango (México), Lucena (Spain), Ovalle (Chile) and Petorca (Chile). Among the studied cities, Ovalle presented the highest volumes of collected water, with 44,202 L per year and an average Specific Energy Consumption of 0.65 kWh/L, then Durango with 28,177 L and a SEC of 0.71 kWh/L, after Lucena with 24,708 L and an SEC of 0.58 kWh/L and finally Petorca with 23,992 L and an SEC of 0.62 kWh/L.

Keywords: Drought · Atmospheric Water Generator · Refrigeration System

1 Introduction

Freshwater sources for human consumption includes mainly rivers, lakes, and groundwater, which in recent years have been affected by overexploitation and climate change [1]. The main sectors competing for freshwater are agriculture, municipalities, and industry. Globally, for countries with high water consumption, these are divided into 69%, 12%, and 19%, respectively. In the case of the American continent (North, Central, and South America), the average water extraction is divided in 51% for agriculture, 15% for municipal consumption, and 34% for industrial consumption. It is estimated that water consumption for the three sectors grows by around 0.5% per year (baseline taken from 2000 to 2010) [2].

Drought has affected 55 million people worldwide, causing 1,100 deaths per year and causing economic damages accounting near to USD 5.4 billion [3]. According to the UN [3], access to freshwater is a human right, crucial for the dignity of all people.

O. F. Farías Fuentes et al. (Eds.): CIBIM 2024, *Proceedings of the XVI Ibero-American Congress of Mechanical Engineering*, pp. 156–167, 2026.
https://doi.org/10.1007/978-3-032-22823-9_11

Every government and public service company must guarantee that all members of the population have access to this basic service and its sanitation.

The indicator used to define water scarcity is the amount of renewable water per capita [4], which is summarized in Table 1. In addition, this table presents the level of water stress, which corresponds to the symptoms that appear under a water scarcity scenario, such as frequent and serious restrictions on water use, increasing conflicts between users and competition for water, decreasing standards of reliability and services, crop failures and food insecurity.

Table 1. Conventional definitions of water stress levels.

Water availability (per capita in m^3/year)	Level of water stress
Less than 500	Absolute water scarcity
500–1.000	Chronic water scarcity
1.000–1.700	Regular water stress
More than 1.700	Localized or occasional water stress

According to the IEP [5], in Ibero-America, Chile, Mexico, Spain and Portugal will be the countries with the greatest vulnerability to water stress in 2040 under a business-as-usual scenario, as shown in Fig. 1.

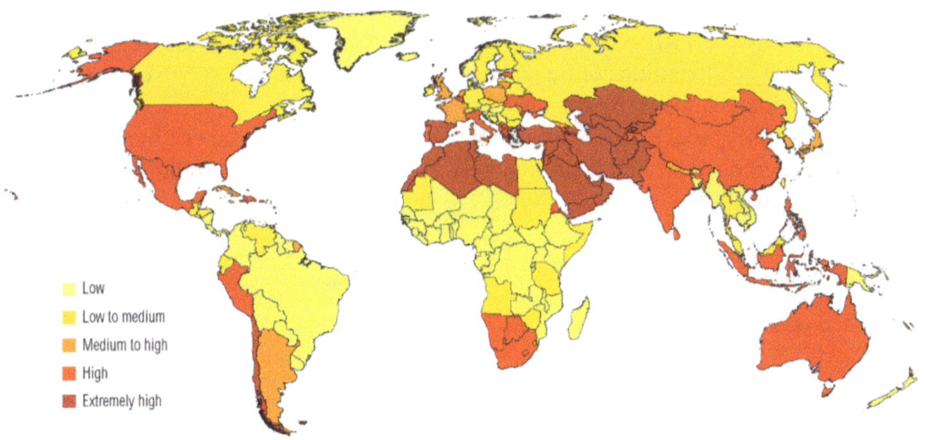

Fig. 1. Country-Level Water Stress in 2040 under the Business-As-Usual Scenario [5].

Alternative sources to rivers, lakes and groundwater for providing freshwater to people are: 1) seawater, which can be converted into freshwater through different systems, such as reverse osmosis desalination and thermal desalination, 2) water present in fog in the form of water droplets, and 3) water present in the ambient air, which can be trapped through different methods, being the most known systems those using adsorption materials, those able to cool the air below its dew point temperature, such as active refrigeration systems, and radiative methods using radiative heat exchange with the sky.

Specifically, this study is focused on Atmospheric Water Generators (AWGs) using vapour compression refrigeration systems. Numerous studies have been conducted to evaluate these kind of systems experimentally and/or numerically [6–9]. The main indicator used to evaluate them is the Specific Energy Consumption (SEC), which quantifies the amount of energy consumed by the system per litre of harvested water. This indicator varies within very wide ranges, from 0.4 kWh/litre to values of 6 kWh/litre. It depends on many factors, including ambient temperature, available air humidity, and the characteristics of the water harvesting system (components and control).

According to the previous discussion, it is interesting to evaluate these systems considering the climate conditions of cities having problems of water scarcity, especially cities and small towns located far from the coast and that are supplied with freshwater using water tanker trucks.

2 Methodology

Four cities with high level of water scarcity have been taken into account for this analysis: Ovalle (Chile), Petorca (Chile), Durango (Mexico), and Lucena (Spain).

2.1 Meteorological Data for the Cities Considered

The meteorological data for the cities considered in this study are obtained from the Meteonorm software [10], which considers a Typical Meteorological Year. Figure 2 shows the hourly distributions of the dry bulb temperature and the relative humidity for the cities considered, sorted by frequency of occurrence.

Of the four cities analysed, Ovalle exhibits a different behaviour, with most of its temperatures concentrated around approximately 12 °C and with higher relative humidities, with values above 40%.

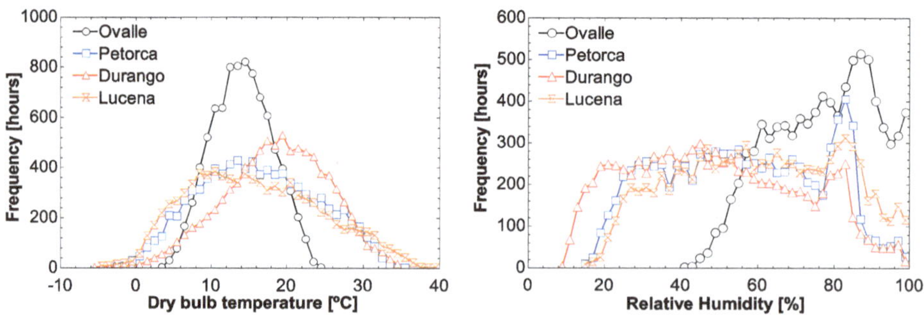

Fig. 2. Hourly distribution of the a) Dry bulb temperature, and b) Relative humidity.

2.2 Water Harvesting System

The water harvesting system evaluated in this study is based on a vapour compression refrigeration system, as shown in Fig. 3. It consists of a variable-speed scroll compressor, a tube-and-fin condenser, an electronic expansion valve, a tube-and-fin evaporator, and a variable-speed axial fan. The system modulates the fan speed and the compressor speed to maximize the collected water. For the control of the system, it is considered that the air exhausting the evaporator has a constant temperature of 6 °C, to maximize the condensed water and to avoid frosting formation on the evaporator surface.

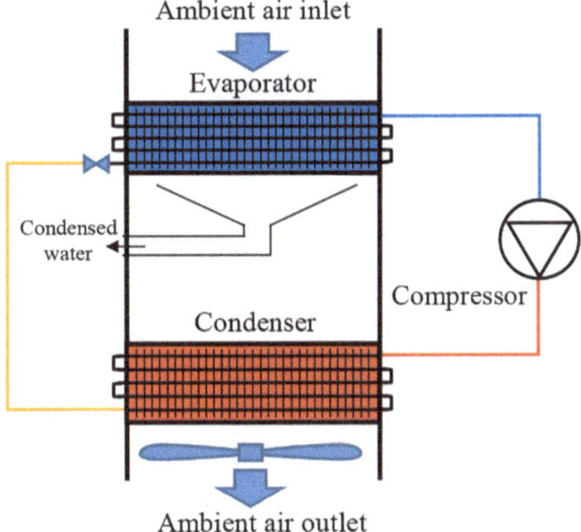

Fig. 3. Atmospheric water generator.

This last assumption limits the use of this system to climate conditions with dew point temperatures above 6 °C. According to this, and comparing this value with the meteorological data of the cities considered, this system will be unable to operate for 43 (0.5% of the total hours of the year) hours in Ovalle, 1,734 (19.8% of the total hours of the year) hours in Petorca, 760 (8.7% of the total hours of the year) hours in Durango, and 1,802 (20.6% of the total hours of the year) hours in Lucena due to this temperature restriction.

The refrigerant used in the atmospheric water generator is R410A.

2.3 Mathematical Model

The water harvesting system is modelled using a modular approach, with each model having its own equations, inputs, parameters, and outputs. This modularity allows a better connection between the different models, resulting in a better way to define the overall model.

Compressor Model

The compressor is modelled according to the model proposed by Winandy et al. [11], which is shown in Fig. 4.

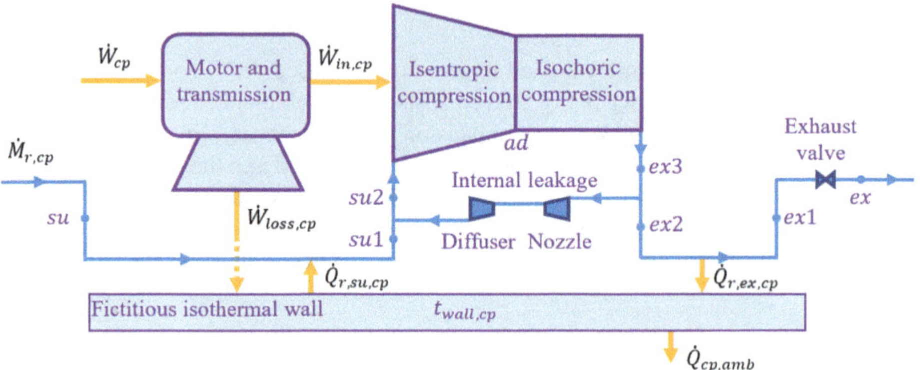

Fig. 4. Compressor principle modelling.

This model predicts the refrigerant mass flow rate, the compressor's electrical input power, and the refrigerant outlet temperature. For the refrigerant mass flow rate, the model considers the internal leakages in the compressor, which become more significant when it operates at low speeds. Thus, the compressor's intake flow is determined according to Eq. (1).

$$\dot{M}_{r,in,cp} = \dot{M}_{r,su1,cp} + \dot{M}_{r,leak,cp} \tag{1}$$

This mass flow rate is related to the compressor swept volume through Eq. (2).

$$\dot{M}_{r,in,cp} = \frac{V_{s,cp} \cdot N_{cp}}{v_{r,su2,cp}} \tag{2}$$

The compressor input power is determined with a semi-empirical model, which considers the internal power delivered to the refrigerant and the compressor electromechanical loss, as indicated in Eq. (3).

$$\dot{W}_{cp} = \dot{W}_{in,cp} + \dot{W}_{loss,cp} \tag{3}$$

The internal power is decomposed into an isentropic compression power and an isochoric compression power, as presented in Eq. (4).

$$\dot{W}_{in,cp} = \dot{M}_{r,in,cp} \cdot \left[\left(h_{r,ad,cp} - h_{r,su2,cp} \right) + v_{r,ad,cp} \cdot \left(P_{r,ex3,cp} - P_{r,ad,cp} \right) \right] \tag{4}$$

The electromechanical loss are determined with the empirical correlation shown in Eq. (5), which presented better agreement with catalogue data during the validation of the compressor model.

$$\dot{W}_{loss,cp} = \dot{W}_{loss0,cp} + \dot{W}_{loss2,cp} \cdot \left(\frac{N_{cp}}{N_{ref,cp}} \right)^2 \tag{5}$$

Finally, the refrigerant exhaust temperature is determined from the compressor energy balance, which is given by Eq. (6).

$$\dot{W}_{cp} = \dot{M}_{r,su1,cp} \cdot \left(h_{r,ex,cp} - h_{r,su,cp} \right) + \dot{Q}_{cp,amb} \tag{6}$$

Condenser Model

The condenser is discretized into three zones, as proposed by Cuevas et al. [12], defined by the refrigerant states: superheated vapour, two-phase, and subcooled liquid zone, as shown in Fig. 5. Each one of these zones is modelled using the ε-NTU method, which is more stable for numerically solving the proposed models.

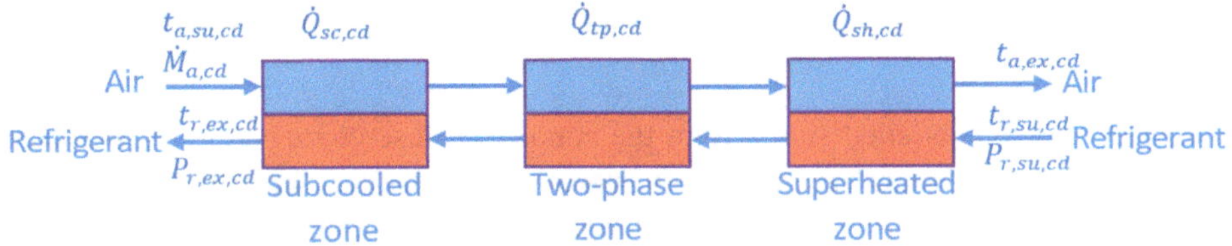

Fig. 5. Condenser principle modelling.

Each zone is described through three equations, which are here presented only for the superheated zone through Eqs. (7), (8) and (9).

$$\dot{Q}_{r,sh,cd} = \dot{M}_{r,cd} \cdot \left(h_{r,su,sh,cd} - h_{r,ex,sh,cd}\right) \tag{7}$$

$$\dot{Q}_{a,sh,cd} = \dot{M}_{a,cd} \cdot \left(h_{a,ex,sh,cd} - h_{a,su,sh,cd}\right) \tag{8}$$

$$\dot{Q}_{sh,cd} = \varepsilon_{sh,cd} \cdot \dot{C}_{min,sh,cd} \cdot \left(t_{r,su,sh,cd} - t_{a,su,sh,cd}\right) \tag{9}$$

where the effectiveness of the heat exchanger is determined by considering a counterflow configuration, according to Eq. (10).

$$\varepsilon_{sh,cd} = \frac{1 - \exp\left(-NTU_{sh,cd} \cdot \left(1 - \omega_{sh,cd}\right)\right)}{1 - \omega_{sh,cd} \cdot \exp\left(-NTU_{sh,cd} \cdot \left(1 - \omega_{sh,cd}\right)\right)} \tag{10}$$

The subcooled zone is modelled in the same way, unlike the two-phase zone, where the effectiveness is determined through Eq. (11).

$$\varepsilon_{tp,cd} = 1 - \exp\left(-NTU_{tp,cd}\right) \tag{11}$$

On the refrigerant side, the convective heat transfer coefficients are determined with the correlations proposed by Baehr and Stephan [13] for laminar regime, by Gnielinski [13] for turbulent regime in single phase, and by Thome et al. [14] for condensation. On the air side, the convective heat transfer coefficient is determined with an empirical law fitted from a figure presented in Shah and London [15] for a fin-and-tube heat exchanger.

In the case of the pressure drop, on the refrigerant side it is determined as indicated in Eq. (12).

$$\Delta P_{r,zi,cd} = \Delta P_{f,r,zi,cd} + \Delta P_{m,r,zi,cd} + \Delta P_{s,r,zi,cd} \tag{12}$$

It considers three contributions: the pressure drop due to the friction ($\Delta P_{f,r,zi,cd}$), the pressure variation due to the density change (acceleration) ($\Delta P_{m,r,zi,cd}$) and the singular pressure drop due to the U-bend fittings of the heat exchanger ($\Delta P_{s,r,zi,cd}$).

On the air side, it is determined as indicated in Shah and London, as indicated in Eq. (13).

$$\Delta P_{a,cd} = \Delta P_{a,c,cd} + \Delta P_{a,f,cd} + \Delta P_{a,m,cd} + \Delta P_{a,e,cd} \tag{13}$$

It considers four contributions: the pressure drop during the contraction $\Delta P_{a,c,cd}$, the pressure drop due to the friction $\Delta P_{a,f,cd}$, the pressure variation due to the density change $\Delta P_{a,m,cd}$ and the pressure drop during the expansion $\Delta P_{a,e,cd}$.

Evaporator Model

The evaporator is modelled similarly to the condenser, discretizing it into two zones: two-phase and superheated zone. Since this exchanger is designed to condense moisture from the ambient air, it must be modelled considering both heat and mass transfer. It is also possible that moisture from the air may not condense, which is why dry and wet conditions could coexist in this heat exchanger. In particular, in this case this heat exchanger is designed to permanently work in wet regime, thus, in the modelling it is assumed that the evaporator is working under fully wet regime. The modelling principle used for this heat exchanger is presented in Fig. 6.

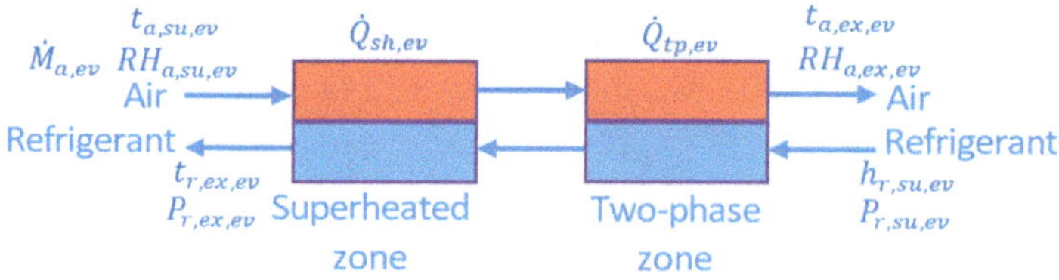

Fig. 6. Evaporator principle modelling.

The energy balance on the air side is defined though Eqs. (14) and (15).

$$\dot{Q}_{a,wet,coil} = \dot{M}_{a,coil} \cdot \left(h_{a,su,wet,coil} - h_{a,ex,wet,coil}\right) - \dot{M}_{a,coil} \cdot \left(W_{a,su,wet,coil} - W_{a,ex,wet,coil}\right) \cdot h_{wl,coil} \tag{14}$$

$$\dot{Q}_{a,wet,coil} = \dot{C}_{fa,wet,coil} \cdot \left(t_{wb,su,wet,coil} - t_{wb,ex,wet,coil}\right) \tag{15}$$

It is assumed that the driving potential defining the heat flow is the air wet bulb temperature, which is a good approach to model the air side heat transfer.

The air conditions at the evaporator exhaust are determined using the assumption proposed in ASHRAE [16], which assumes that the air conditions are the result of the mixing of the air entering the evaporator and the saturated air in contact with the heat evaporator surface, which is at the surface temperature (contact temperature) of the evaporator. It is summarized through Eqs. (16), (17), (18) and (19).

$$h_{a,su,wet,coil} - h_{a,ex,wet,coil} = \varepsilon_{c,wet,coil} \cdot \left(h_{a,su,wet,coil} - h_{c,wet,coil}\right) \tag{16}$$

$$W_{a,su,wet,coil} - W_{a,ex,wet,coil} = \varepsilon_{c,wet,coil} \cdot \left(W_{a,su,wet,coil} - W_{c,wet,coil}\right) \tag{17}$$

$$\varepsilon_{c,wet,coil} = 1 - exp\left(-NTU_{c,wet,coil}\right) \tag{18}$$

$$NTU_{c,wet,coil} = \frac{1}{R_{a,coil} \cdot \dot{C}_{a,dry,coil}} \tag{19}$$

Similar to the condenser, on the refrigerant side, the convective heat transfer coefficients are determined with the same correlations in single phase, but in two-phase with the Shah correlation [17].

Axial fan Model

The fan is modelled using an empirical model depending on 3 dimensionless factors: flow factor, pressure factor and power factor, which are presented in Eqs. (20), (21) and (22), respectively.

$$\phi_{fan} = \frac{\dot{V}_{a,fan}}{A_{fan} \cdot U_{fan}} \tag{20}$$

$$\psi_{fan} = \frac{\Delta P_{total,fan}}{P_{dyn,periph,fan}} \tag{21}$$

$$\lambda_{fan} = \frac{\phi_{fan} \cdot \psi_{fan}}{\varepsilon_{s,fan}} \tag{22}$$

These factors are correlated among them though polynomial laws, where one of them is defined as an independent variable and the other two depending on this independent variable. The polynomials laws are indicated in Eqs. (23) and (24). In this modelling, the pressure factor is considered as independent variable.

$$\lambda_{fan} = a_0 + a_1 \cdot \psi_{fan} + a_2 \cdot \psi_{fan}^2 \tag{23}$$

$$\phi_{fan} = b_0 + b_1 \cdot \psi_{fan} + b_2 \cdot \psi_{fan}^2 \tag{24}$$

The constants of these polynomial laws are determined from the axial fan catalogue.

System Model

The system model is obtained by interconnecting each one of the models presented previously to create the overall model presented in Fig. 7. In this figure, the variables coloured in black represent the model inputs, the variables coloured in red must be verified to avoid any working condition outside the equipment domain and the other variables are used for analysis and to connect the different sub-models.

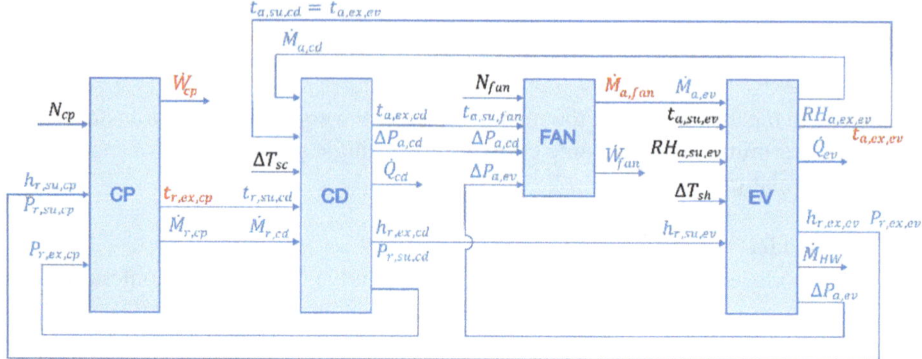

Fig. 7. System model.

As previously mentioned, the system will operate with a constant evaporator discharge air temperature of 6 °C. The model results will allow to determine the monthly water production and the monthly specific energy consumption of the system, according to Eqs. (25) and (26).

$$V_{HW} = \int_{\tau_{initial}}^{\tau_{final}} \dot{V}_{HW} \cdot d\tau \tag{25}$$

$$SEC = \int_{\tau_{initial}}^{\tau_{final}} \frac{\left(\dot{W}_{cp} + \dot{W}_{fan}\right)}{\dot{V}_{HW}} \cdot d\tau \tag{26}$$

2.4 Simulation Considerations

The system is simulated over a Typical Meteorological Year. Monthly data are classified using the BIN method, with wet-bulb temperature as a reference.

The simulations consider the following system constraints: compressor speed between 15 Hz and 120 Hz, maximum compressor input power of 3.6 kW, and air mass flow rate between 0.14 kg/s and 0.7 kg/s. When the operating conditions are outside these limits, the system is forced to operate within these safe conditions by modifying the compressor speed or the evaporator exhaust air temperature.

For the simulations, the system model is first run at a compressor speed of 120 Hz and an evaporator outlet air temperature of 6 °C. The model determines the compressor input power and the required air mass flow rate. If the maximum input power is exceeded, it is set to the maximum value, and the model determines the speed at which the compressor should operate or if the air mass flow rate exceeds the maximum, it is set to the maximum value.

3 Results and Discussion

Figure 8 presents the monthly water production of the four cities considered. A marked seasonal trend is observed in the water production, with higher production during summer and lower production during winter. Ovalle is the city with the highest annual water production, with 44,202 L, followed by Durango with 28,177 L, then Lucena with 24,708 L, and finally Petorca with 23,992 L.

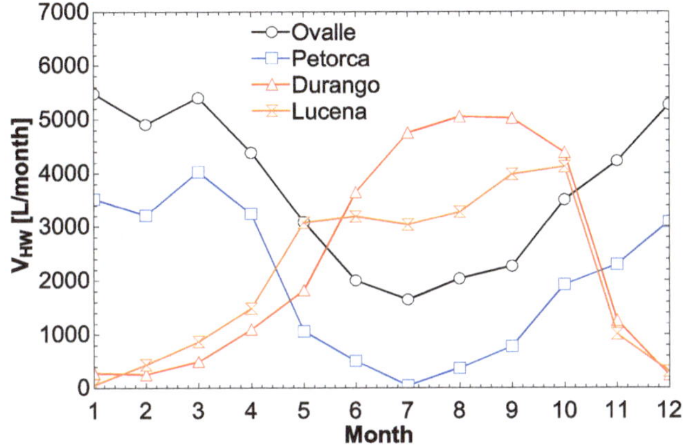

Fig. 8. Monthly water production for the considered cities.

Regarding the SEC, presented in Fig. 9, it varies between 0.35 and 1.15 kWh/L, with Lucena being the city with the lowest average SEC value of the 4 cities considered.

Ordered from better to the worse SEC, Lucena is the first with an average SEC of 0.58 kWh/L, then Petorca with an SEC of 0.62 kWh/L, followed by Ovalle with 0.65 kWh/L and finally Durango with 0.71 kWh/L.

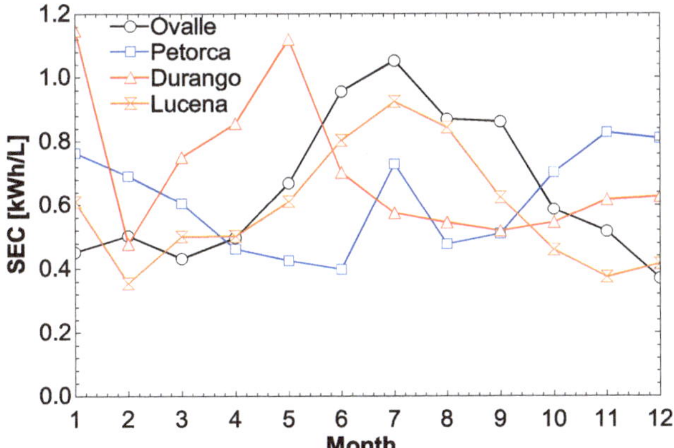

Fig. 9. Specific Energy Consumption of the system.

4 Conclusions

An atmospheric water harvesting system is modelled based on a vapour-compression refrigeration system using R410A as the refrigerant. The system is designed to collect 100 L of water per day at an ambient condition of 20 °C and 60% relative humidity.

The performance of this type of system is closely related to the climate conditions and to the season or month of the year of the location where this system is operating. A higher water production is observed during summer, due to the more favourable conditions during this period: higher water potential and higher temperature. Of the four cities analysed, Ovalle shows the highest water harvesting rate.

Regarding Specific Energy Consumption, it varies between 0.35 and 1.15 kW/L, with the city of Lucena having the lowest SEC.

Acknowledgments. This work was funded by the ANID FONDEF ID22I10051.

References

1. FAO: Coping with water scarcity – An action framework for agriculture and food security. https://www.fao.org/4/i3015e/i3015e.pdf. Accessed 26 May 2025
2. FAO: AQUASTAT - FAO's Global Information System on Water and Agriculture, https://www.fao.org/aquastat/en/overview/methodology/water-use. Accessed 26 May 2025
3. UNESCO World Water Assessment Programme: Leaving No One Behind, https://unesdoc.unesco.org/ark:/48223/pf0000367306. Accessed 26 May 2025
4. Falkenmark, M., Widstrand, C: Population and Water Resources: a delicate balance. Population Bulletin. Population Reference Bureau, Washington, USA (1992)
5. Institute for Economics & Peace. Ecological threat register 2020: understanding eco-logical threats, resilience and peace, Sydney (2020)
6. Bagheri, F.: Performance investigation of atmospheric water harvesting systems. Water Resour. Ind. **20**, 23–28 (2018)
7. Patel, J., Patel, K. Mudgal, A., Panchal, H., Kumar Sadasivuni, K.: Experimental investigations of atmospheric water extraction device under different climatic conditions. Sustain. Energy Technol. Assessments, **38**, 100677 (2020)
8. Ravesh, G., Goyal, R., Tyagi, S.K.: Advances in atmospheric water generation technologies. Energy Convers. Manage. **239**, 114226 (2021)
9. Wang, X., Xu, B., Liu, Q., Yang, Y., Chen, Z.: Enhancement of vapor condensation heat transfer on the micro- and nano-structured superhydrophobic surfaces. Int. J. Heat Mass Transf. **177**, 121526 (2021)
10. Meteonorm Version 8 - Meteonorm (en). https://mn8.meteonorm.com/en/meteonorm-version-8. Accessed 26 May 2025
11. Winandy, E., Saavedra, C., Lebrun, J.: Experimental analysis and simplified modelling of a hermetic scroll refrigeration compressor. Appl. Therm. Eng. **22**(2), 107–120 (2002)
12. Cuevas, C., Lebrun, J., Lemort, V., Ngendakumana, P.: Development and validation of a condenser three zones model. Appl. Therm. Eng. **29**, 3542–3551 (2009)
13. Incropera, F., Bergman, T., Lavine, A., DeWitt, D.: Fundamentals of Heat and Mass Transfer. 7th Edn. Wiley, Hoboken (2011)
14. Thome, J.R., El Hajal, J., Cavallini, A.: Condensation in horizontal tubes, part 2: New heat transfer model based on flow regimes. Int. J. Heat Mass Transf. **46**, 3365–3387 (2003)

15. Kays, W.M., London, A.L.: Compact Heat Exchangers. McGraw-Hill, New York (1955)
16. ASHRAE Handbook: HVAC Systems and Equipment. American Society of Heating, Refrigerating and Air-Conditioning Engineers, Atlanta (2016)
17. Shah, M.M.: A new correlation for heat transfer during boiling flow through pipes. ASHRAE Trans. **82**, 66–86 (1976)

Development of an Algorithm for the Aerodynamic Evaluation of Airfoils Aimed at the Design of Horizontal Axis Wind Turbines

Felipe Eduardo de Farias(✉), Angie Lizeth Espinosa Sarmiento, and Diego Mauricio Yepes Maya

Federal University of Itajubá, Itajubá, MG 37500-903, Brazil
{d2021013069,angieespinosa,diegoyepes}@unifei.edu.br

Abstract. In the aerodynamics of Horizontal Axis Wind Turbines (HAWT), the airfoil plays a crucial role in the sizing and performance of these turbines. Thus, this work addresses the creation of a Python® algorithm for the aerodynamic evaluation of airfoils, using XFOIL® to calculate lift and drag coefficients. Additionally, to overcome the limitations of XFOIL® in values beyond the pre-stall, we used the Montgomerie extrapolation technique, which extends the aerodynamic polar curves to a wider range of angles of attack. This approach allows for a more accurate and comprehensive analysis of HAWT performance under various operational conditions. Finally, the work was compared with QBlade® to validate its effectiveness, highlighting its contribution to the advancement of wind energy turbine designs.

Keywords: XFOIL® · Montgomerie · Lift Coefficient · Horizontal Axis Wind Turbines – HAWT

1 Introduction

The airfoil is a crucial component in the aerodynamic design and overall efficiency of Horizontal Axis Wind Turbines (HAWTs), as its aerodynamic properties directly influence wind-to-electric energy conversion. A thorough understanding of airfoil behavior enables the optimization of blade geometry and pitch angle, thus enhancing turbine performance[1]. The studies conducted by [2] in 2011 on aerodynamic profiles marked a significant advancement in wind turbine efficiency. In their work, [2] thoroughly investigated the characteristics of airfoil profiles, focusing on optimizing lift and reducing drag—key factors for efficient turbine performance. However, the sharp increase in oil prices following the 1973 energy crisis triggered a significant wave of investments and research initiatives, resulting in numerous projects aimed at maximizing power generation from alternative energy sources [3]. In the context of wind turbines, these efforts have primarily focused on refining aerodynamic profiles to enhance aerodynamic performance and, consequently, improve energy conversion efficiency.

O. F. Farías Fuentes et al. (Eds.): CIBIM 2024, *Proceedings of the XVI Ibero-American Congress of Mechanical Engineering*, pp. 168–178, 2026.
https://doi.org/10.1007/978-3-032-22823-9_12

The study conducted by [4] in 2013 investigated the sensitivity of the aerodynamic performance of airfoils used in microgeneration wind turbines. The methodology employed computational tools, integrating XFLR5® and EasyCFD® software alongside experimental data to determine the lift, C_L, and drag, C_D coefficients across a range of angles of attack (α). Subsequently, the AirfoilPre spreadsheet was utilized to extend these coefficients to higher angles of attack. The article also analyzed stall delay models, providing a comprehensive understanding of airfoil behavior under various operational conditions.

In the study conducted by [5] in 2011, a comprehensive numerical investigation was carried out on various aerodynamic profiles used in wind turbines. Two- and three-dimensional Computational Fluid Dynamics (CFD) simulations were employed to analyze aerodynamic behavior at low Reynolds numbers, ranging from 55,000 to 100,000. The numerical results were compared with XFOIL® predictions and experimental wind tunnel data. Based on instantaneous flow visualizations, the study discussed key aspects of transitional flow regimes and the accuracy of two-dimensional computations.

Chapter four of the book by [6] focuses specifically on the aerodynamic profiles of wind turbine blades. The chapter discusses the use of tools such as XFOIL®, RFOIL®, and CFD techniques for predicting the aerodynamic characteristics of airfoils. It also examines methods for representing airfoil behavior at high angles of attack (α), as well as the proper application of correction approaches to account for the effects of three-dimensional flow on the blades. In this context, the chapter establishes a relevant dataset for aerodynamic blade design and concludes with a forward-looking perspective on future research directions in the field of aerodynamic profiles for wind turbines [6].

This study presents the development of a Python® based algorithm for the aerodynamic evaluation of airfoils, with a focus on the design of a Horizontal Axis Wind Turbine (HAWT). The adopted methodology involved the implementation of XFOIL® to compute aerodynamic coefficients. However, to overcome XFOIL® limitations under post-stall conditions, the Montgomerie extrapolation technique was applied, enabling the estimation of lift and drag coefficients, C_L, C_D, over an extended range of angles of attack, α. This approach allows for a more accurate and comprehensive assessment of HAWT performance under various operational conditions. Furthermore, the results were benchmarked against QBLADE®, a well-established tool in wind turbine analysis, to validate the proposed methodology.

2 Methodology

The methodology was fully implemented in Python®, with an emphasis on flexibility, reproducibility, and modular integration. It was structured in two stages: the first involved obtaining the aerodynamic polar curves (C_L, C_D) of the airfoils using XFOIL®, and the second consisted of applying Montgomerie analytical extrapolation over a 360° range of angle of attack, α, thereby eliminating the need for experimental data. Developed by Mark Drela in 1986, XFOIL® is a tool composed of multiple subroutines that provide an interactive environment for a range of operations, including: (i) direct analysis, (ii) inverse and mixed design, and (iii) manipulation of two-dimensional airfoil geometries in subsonic flows, both viscous and inviscid, under low Reynolds number conditions.

XFOIL adopts a fully implicit approach for coupling the viscous and inviscid flow components, using a panel method with linear vortex distributions to model the inviscid region and an integral boundary layer method to represent the viscous layers [7].

The theoretical structure underlying XFOIL® is illustrated in Fig. 1, which shows the hierarchical derivation of the flow equations. On the left, the inviscid flow is simplified progressively from the Navier–Stokes equations to Euler and potential flow, culminating in the Laplace equation under steady, incompressible conditions. On the right, the Reynolds-averaged equations and empirical models provide the basis for the viscous boundary layer analysis. This combined formulation allows XFOIL® to produce two-dimensional aerodynamic results with sufficient accuracy for pre-design stages of wind turbine blades [7].

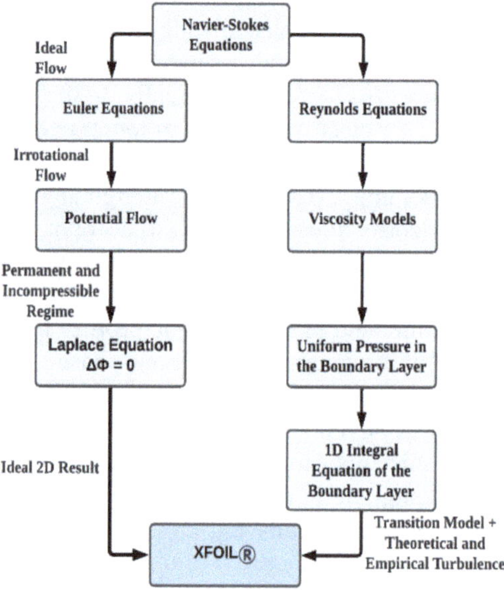

Fig. 1. Flowchart representing the operation of XFOIL®, which was implemented in Python for this study.

Furthermore, based on Fig. 1, which illustrates the XFOIL® workflow, the code integrates ideal potential flow models with simplified turbulence and transition approaches to include viscous effects. The coupling between inviscid potential flow and viscous boundary layer is solved iteratively: first, the two-dimensional inviscid potential flow is computed to determine the velocity field around the airfoil, which then serves as input for the boundary layer calculation. The boundary layer modifies the potential flow by adding displacement thickness, requiring updates to the potential flow solution. This iterative process is known as the Interactive Boundary Layer (IBL). Several numerical schemes have been developed to solve this coupling efficiently [8].

Additionally, XFOIL® solves the Interactive Boundary Layer (IBL) equations and incorporates the One-Dimensional Integral Method. In this approach, the boundary layer equations are integrated across its thickness, making the boundary layer properties dependent on the streamwise position along the airfoil surface. [9].

According to the theoretical framework of XFOIL® and as illustrated in Fig. 2, the software was implemented using a methodological approach in which XFOIL® is invoked through an input variable within the computational environment. This variable enables the insertion of airfoil profile data through two distinct methods:

- External Method: In this case, the user provides the full path to the file in which the airfoil coordinate data is stored on the local device. This approach allows the software to directly access the specified location to retrieve the aerodynamic profile data.
- Internal Method: Alternatively, the user can input only the name of the airfoil. In this method, the software queries an internal database managed by the AeroSandbox library. This library is responsible for storing and retrieving airfoil data, thereby simplifying both the insertion and management of aerodynamic profiles.

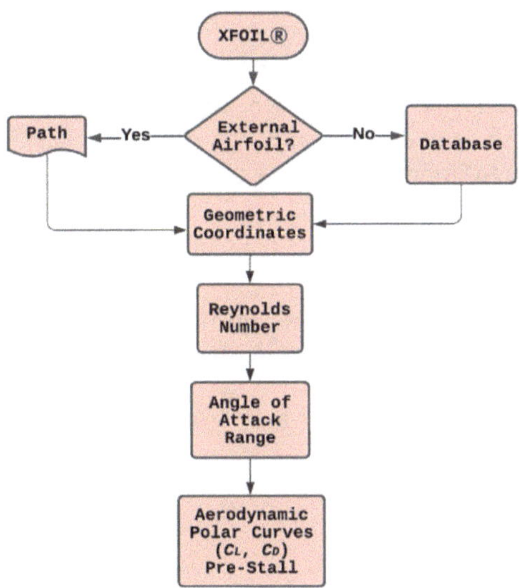

Fig. 2. Flowchart representing the implemented XFOIL methodology.

Once the airfoil coordinates are entered and processed, the code proceeds to generate a graphical plot of the airfoil contour. Subsequently, based on the specified Reynolds numbers and the defined range of angles of attack, α, XFOIL® performs the calculations required to determine the lift, and drag, coefficients, C_L, C_D, of the profile. These calculations are essential for evaluating aerodynamic performance pre-stall. Therefore, the methodology implemented using XFOIL® enables efficient integration of geometric airfoil data with the analytical process, providing detailed and accurate results that are critical for both academic studies and practical applications in the design of horizontal axis wind turbines.

2.1 Post-stall Aerodynamic Polaris

Aerodynamic polar curves describe the variation of lift, C_L, and drag, C_D, coefficients as a function of the angle of attack, α. However, such data are typically limited to the pre-stall region, as is the case with results generated by XFOIL®. In wind turbines, on the other hand, stall can be intentionally exploited as a means of limiting power output under high wind conditions. To expand the applicability of aerodynamic tables in these scenarios, the Montgomerie method extends the angle of attack, α, values up to 360°, which is essential for aerodynamic codes used in aeroelastic simulations. These extended data are also critical in methods used to analyze C_p vs. (TSR) Tip-Speed Ratio, which account for a wide angular variation along the blade span. Since these curves serve as input parameters for (BEM) Blade Element Momentum models, high angles of attack, α, may occur in certain regions—particularly near the blade root—making continuous modeling along the entire blade span essential.

2.2 Montgomerie Extrapolation

Montgomerie extrapolation is used to extend the aerodynamic polar curves over the full 360° range of angle of attack, α. The procedure begins with the pre-stall curves generated using XFOIL®, which serve as input parameters. The extrapolation of the lift coefficient, C_L, is performed through interpolation between two functions: $t_{(\alpha)}$, which represents potential flow —i.e., an ideal regime without flow separation—and $s_{(\alpha)}$, which models the limiting behavior of the airfoil as a flat plate under full separation. The function $t_{(\alpha)}$, given by Eq. (1), describes an ideal circulatory flow in which vorticity effects around the body are neglected. His simplification allows a more direct characterization of the velocity field, with streamlines aligning with the airfoil contour [10].

$$t_{(\alpha)} = C_{L_0} + C_{L\alpha} \cdot \alpha \tag{1}$$

where C_{L_0} is the lift coefficient at $\alpha = 10°$, and $C_{L\alpha}$ is the slope of the linear lift curve in the pre-stall regime, as shown in Eq. (2).

$$C_{L\alpha} = \frac{C_{L_{\alpha 10}} - C_{L_{\alpha 1}}}{\alpha_{10} - \alpha_1} \tag{2}$$

where $C_{L_{\alpha 10}}$ and $C_{L_{\alpha 1}}$ represent the lift coefficients corresponding to the angles of attack $\alpha = 10°$ and $\alpha = 1°$, respectively.

The function $s_{(\alpha)}$, corresponds to the condition of full flow separation over the airfoil, which leads to aerodynamic performance loss and causes the airfoil to behave, in a simplified manner, like a flat plate [11]. This function is obtained using Eq. (3).

$$s_{(\alpha)} = 1 + \frac{C_{L\alpha_0}}{\sin(45°)} \cdot \sin(\alpha) \cdot C_{D_{\alpha 90}} \cdot \sin(\beta) \cdot \cos(\beta) \tag{3}$$

where α is the angle of attack, $C_{D_{\alpha 90}}$ is the drag coefficient at $\alpha = 90°$, and β is an angular correction factor that accounts for the effects of the rounded leading edge and curvature of the airfoil. The value of β is computed using Eq. (4), considering that $C_{L90} = 0.08$ [11].

$$\beta = \alpha - 57.6 \cdot C_{L_{90}} \cdot \sin(\alpha) - \alpha_0 \cdot \cos(\alpha) \tag{4}$$

To perform the interpolation between $t_{(\alpha)}$ and $s_{(\alpha)}$, a transition function $\mathcal{F}(\alpha)$ is defined, as shown in Eq. (5), which quantifies the degree of similarity between (potential flow) behavior and the separated flow regime characteristic of the post-stall condition:

$$\begin{aligned} F(\alpha) = \{1, & \textit{ linear regime(pre} - \textit{stall)} \\ & 0, \textit{full separation regime(pos} - \textit{stall)}\} \end{aligned} \tag{5}$$

The transition between the attached and separated flow regimes is handled through the transformation function $\mathcal{F}(\alpha)$, which weights the contributions of the $t_{(\alpha)}$ and $s_{(\alpha)}$ curves in estimating the lift coefficient $C_L(\alpha)$, as expressed in Eq. (6), [10].

$$C_L(\alpha) = \mathcal{F}(\alpha) \cdot t(\alpha) + [1 - \mathcal{F}(\alpha)] \cdot s(\alpha) \tag{6}$$

where $t_{(\alpha)}$ represents the potential flow, $s(\alpha)$ represents the separated flow, and the function $\mathcal{F}(\alpha)$ depends on a differential angle $\Delta\alpha$, defined by Eq. (7).

$$\Delta\alpha = \alpha_M - \alpha \tag{7}$$

where α_M is the anchor point of the transition—typically close to the angle of attack α associated with—$C_L ma'x$ at which the $t_{(\alpha)}$ curve deviates from the ideal potential flow behavior. Moreover, the simplest and most effective functional form to represent $\mathcal{F}(\alpha)$, as suggested by Montgomerie through numerical fitting, is given by Eq. (8).

$$\mathcal{F}(\alpha) = \frac{1}{1 + k \cdot \Delta\alpha^4} \tag{8}$$

where, k, is a parameter that controls the width of the transition. Higher values of k, result in a sharper transition between flow regimes, while lower values lead to smoother and more gradual transitions.

The constant k, can be computed, along with α_M, , from two points (α_1,f_1) and (α_2, f_2) located within the transition region of the lift curve C_L, using Eq. (9).

$$k = \left(\frac{1}{f_2} - 1\right) \cdot \frac{1}{(\alpha_2 - \alpha_M)^4} \tag{9}$$

where α_M, is defined by Eq. (10), which provides the central point of the transition between the attached and separated flow regimes.

$$\alpha_M = \frac{\alpha_1 - G \cdot \alpha_2}{1 - G} \tag{10}$$

where G, is defined by Eq. (11), which establishes its relationship with the transition parameters used in modeling the function $\mathcal{F}(\alpha)$.

$$G = \sqrt[4]{\frac{\frac{1}{f_1} - 1}{\frac{1}{f_2} - 1}} \tag{11}$$

This method ensures that the lift curve $C_L(\alpha)$ is smoothly interpolated between the physical-ideal (potential flow) and fully separated flow limits, maintaining continuity and consistency in the solution. Furthermore, the drag coefficient $C_D(\alpha)$ extrapolation is performed based on the lift curve $C_L(\alpha)$. Initially, the difference between the ideal lift predicted by potential flow, $t_{(\alpha)}$, and the actual extrapolated lift curve $C_L(\alpha)$, is computed. This difference, denoted as ΔC_L, is defined by Eq. (12), [12].

$$\Delta C_L = t(\alpha) - C_L(\alpha) \tag{12}$$

The drag variation associated with flow separation, ΔC_D, is then obtained based on an empirical relation identified by [10], which is expressed in Eq. (13).

$$\Delta C_D = 0{,}13 \cdot \Delta C_L \tag{13}$$

Accordingly, the ideal drag coefficient $C_{D_I(\alpha)}$, which represents the combined contribution of separation drag and skin friction drag, is determined using Eq. (14).

$$C_{D_I(\alpha)} = \Delta C_D + C_{D_f(\alpha)} \tag{14}$$

where ΔC_D is the drag variation associated with flow separation, and $C_{D_f(\alpha)}$ is the frictional drag component, Eq. (15), estimated based on the geometry of the airfoil.

$$C_{D_f(\alpha)} = 1{,}25 \cdot \left(\frac{t_m}{c}\right)^2 \tag{15}$$

where t_m represents the maximum thickness of the airfoil, and, c, is the chord length of the profile. To represent the maximum drag under full stall conditions—when the airfoil behaves like a flat plate—Eq. (16) is used.

$$C_{D_P(\alpha)} = C_{D90} \cdot \sin^2(\alpha) \tag{16}$$

where C_{D90} is the drag coefficient at $\alpha = 90°$, commonly assumed to be 2.0 for thin airfoils.

The final interpolation of the total drag coefficient is performed using the transition function $\mathcal{F}(\alpha)$, previously defined in the lift coefficient extrapolation process, resulting in the final expression, Eq. (17).

$$C_D = \mathcal{F}(\alpha) \cdot C_{D_I(\alpha)} + [1 - \mathcal{F}(\alpha)] \cdot C_{D_P(\alpha)} \tag{17}$$

where $\mathcal{F}(\alpha)$ is the transition function, $C_{D_I(\alpha)}$ is the ideal drag coefficient, and $C_{D_P(\alpha)}$ is the maximum drag coefficient under full stall conditions.

3 Results and Discussion

The methodology validation began with the NACA 23112 airfoil at a Reynolds number of 1,200,000, as illustrated in Fig. 3, where the geometric properties of the profile were input into the system. These coordinates were read and plotted, as shown in Fig. 4, ensuring the profile geometry was correctly represented before proceeding with the aerodynamic analysis.

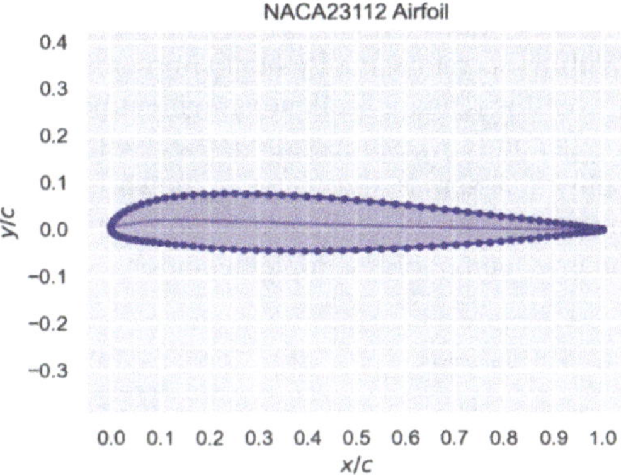

Fig. 3. – Representation of the NACA 23112 airfoil in Python.

The validation of the lift, C_L and drag, C_D comparing results from the XFOIL® implementation with those generated by QBLADE®, demonstrated a high degree of agreement, as shown in Fig. 4. The results were found to be extremely similar across 38 different angles of *attack*, α, values. For the lift coefficient, C_L, the percentage difference exceeded 2% only at higher angles of attack, specifically between $\alpha = 22^o$ and 25°, while for the remaining 36 points the difference remained below 2%. Regarding the drag coefficient, C_D, five angles of attack exhibited errors greater than 4%, with all other points showing smaller differences.

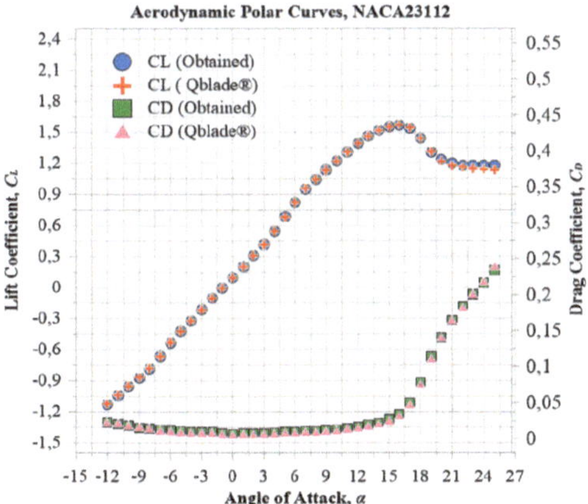

Fig. 4. Comparison between pre-stall aerodynamic values and the QBLADE software.

The validation of the post-stall aerodynamic polar curves was conducted by comparing the results with data provided by the QBLADE® software. The analysis used the NACA 23112 airfoil. The results, presented in Fig. 4, showed strong agreement between the proposed Montgomerie extrapolation and those generated by QBLADE®. The main discrepancies in the lift coefficient (C_L), as illustrated in Fig. 5 (a), were observed within the angle of attack (α) range of 30o to 50º, with a maximum error of approximately 17%. Within this range, the Python® implementation tended to underestimate the results. Outside this range, excellent agreement between data was observed. Regarding the drag coefficient (C_D) extrapolation, the results demonstrated high accuracy with variations below 5% across all points, as shown in Fig. 5 (b). Therefore, the aerodynamic polar curves exhibited excellent overlap, reinforcing the robustness and fidelity of the Montgomerie methodology implementation.

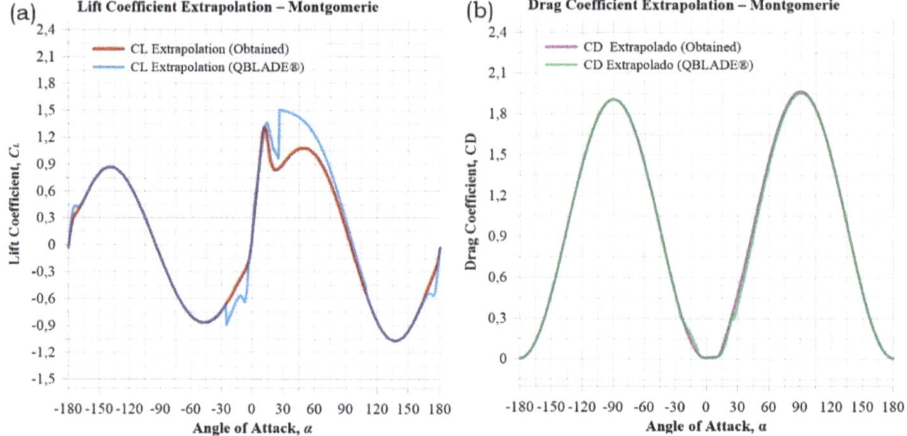

Fig. 5. Comparison between the Montgomerie methodology and results obtained from QBLADE® software, (a) Lift coefficient C_L (b) Drag coefficient C_D.

4 Conclusions

This work presented the development of a Python® based algorithm for the aerodynamic evaluation of airfoils, focusing on the design of HAWTs. The adopted methodology involved implementing XFOIL® to compute aerodynamic polar curves in the pre-stall region and applying Montgomerie extrapolation to estimate lift and drag coefficients, C_L, C_D, over a wide range of angles of attack, α. The results were compared with those obtained from the QBLADE® software. The discrepancies observed between the Python-based methodology and QBLADE® results can be attributed to differences in numerical implementation and data processing. However, the overall strong similarity between the curves validates the methodology for both pre-stall and post-stall analyses. Thus, this work provides a solid foundation for future research in aerodynamic evaluation of wind turbines, contributing to the advancement of design and performance analysis techniques for HAWTs.

Acknowledgments. The authors would like to acknowledge the Fundação de Amparo à Pesquisa do Estado de Minas Gerais (FAPEMIG) for the financial support provided through grant APQ-00653–22, under call 001/2022, related to the project "Numerical and Experimental Analysis of Small Wind Turbines for Applications in Remote Regions of Brazil," registered at DPI UNIFEI under record number PVDI297–2022. We also thank the Institutional Scientific Initiation Scholarship Program (PIBIC) for funding our research. Our gratitude is extended to the Federal University of Itajubá (UNIFEI) for providing an environment conducive to personal and intellectual development, as well as to the TIES and FDT research groups – UNIFEI, for their continued support and encouragement of scientific inquiry.

References

1. Grasso, F.: Usage of numerical optimization in wind turbine airfoil design. J. Aircr. **48**(1), 248–255 (2011). https://doi.org/10.2514/1.C031089;PAGE:STRING:ARTICLE/CHAPTER
2. Anderson, J.D.: Fundamentals of Aerodynamics. McGraw-Hill (2011). Accessed 23 May 2025
3. Hau, E.: Wind Turbines : Fundamentals, Technologies, Application, Economics. Springer, Cham (2013)
4. J. Antonio et al., "Sensitivity Analysis of Aerodynamic Performance of Airfoils Used in Small Wind Turbines"
5. Gross, A., Fasel, H.: Numerical Investigation of Different Wind Turbine Airfoils (2011). https://doi.org/10.2514/6.2011-557
6. Timmer, W.A., Bak, C.: Aerodynamic characteristics of wind turbine blade airfoils. Adv. Wind Turbine Bl. Des. Mater. 129–167 (2023). https://doi.org/10.1016/B978-0-08-103007-3.00011-2
7. de Souza, B.S.: Projeto de perfis aerodinâmicos utilizando técnicas de otimização mono e multiobjetivos. Universidade Federal de Itajubá (2008)
8. de Paula, L.B.: Estudo da melhoria de performance aerodinâmica de um perfil aplicado a veículos aéreos não tripulados. Pontifícia Universidade Católica do Rio de Janeiro (2019)
9. Drela, M.: XFOIL: an analysis and design system for low Reynolds number airfoils. LOW REYNOLDS NUMBER Aerodyn. PROC. CONF., NOTRE DAME, U.S.A., JUNE 5–7, 1989 }EDITED BY T.J. MUELLER]. (LECTURE NOTES, no. 54), Berlin, Germany, Springer-Verlag, 1989, pp. 1–12 (1989). https://doi.org/10.1007/978-3-642-84010-4_1
10. Montgomerie, B.: Methods for Root Effects, Tip Effects and Extending the Angle of Attack Range to ±180°, with Application to Aerodynamics for Blades on Wind Turbines and Propellers (2004)
11. Rolón Ortiz, H.A., Villamizar González, Y., Acevedo Peñaloza, C.H.: Metodología para el cálculo de coeficientes de sustentación y arrastre en perfiles aerodinámicos simétricos aplicado a turbinas Darrieus. Redes Ing, vol. 8, no. 2 SE-Reporte de caso, pp. 92–100 (2017). https://doi.org/10.14483/2248762X.12502
12. Mejía De Alba, M., García Fernández, L., Gutiérrez Almonacid, M.: Metodología de obtención de los coeficientes de sustentación y arrastre para un rango amplio de números de Reynolds y ángulos de ataque para aplicaciones en turbinas eólicas. Av. Investig. en Ing., **13**(1), 53 (2010). https://doi.org/10.18041/1794-4953/AVANCES.1.357

BY NC ND

Simulation Study of the Hydrodynamic Effect of the Number of Impulse Valves on the Operating Characteristics of a Hydraulic Ram Pump

Fran Reinoso-Avecillas[1](✉), Nelson Jara-Cobos[2], Olena Leonidivna Naidiuk[2], and Jorge Brito-Tigre[3]

[1] Research and Development Group in Simulation, Optimization and Decision Making, Master in Mechanical Engineering, Universidad Politécnica Salesiana, Cuenca, Ecuador
freinoso@ups.edu.ec

[2] Research and Development Group in Simulation, Optimization and Decision Making, Mechanical Engineering Program, Universidad Politécnica Salesiana, Cuenca, Ecuador
{njara,oneira}@ups.edu.ec

[3] Graduate of the Mechanical Engineering Program, Universidad Politécnica Salesiana, Cuenca, Ecuador
jbritot@est.ups.edu.ec

Abstract. The purpose of this research is to study through simulation the hydrodynamic effect of the number of impulse valves on the operating characteristics of a hydraulic ram pump. For this, a case study is identified and characterized, the physical model is built, the computational domain is defined, and the boundary conditions are established to develop a mesh. Using Ansys Fluent software, a two-dimensional study simulates the base case and several scenarios by increasing the number of impulse valves. The simulation results show that by increasing the number of impulse valves to three, there is an improvement of efficiency of the hydraulic ram pump that reaches a value of 24.11%, compared to the 4.3% efficiency of a single-valve hydraulic ram pump.

Keywords: Hydraulic ram · Impulse valve · Fluent

1 Introduction

Water supply for human consumption and irrigation in remote rural areas worldwide faces multiple challenges that compromise the well-being and development of these communities [1–3]. Although water sources are often abundant in these areas, they are usually located at great distances and frequently contaminated, increasing health risks. Additionally, water infrastructure tends to be deficient or non-existent, making permanent and safe access to water difficult [2, 3].

Moreover, reliance on conventional energy sources for water pumping proves costly, especially in regions with limited access to electricity. In this context, technologies such

O. F. Farías Fuentes et al. (Eds.): CIBIM 2024, *Proceedings of the XVI Ibero-American Congress of Mechanical Engineering*, pp. 179–194, 2026.
https://doi.org/10.1007/978-3-032-22823-9_13

as the hydraulic ram pump—which harnesses hydraulic energy without the need for external energy—is presented as a viable and sustainable solution, thus, maximizing its efficiency and promoting its use is essential to improve rural living conditions and support sustainable agricultural development, mainly in Latin America [3, 4].

Hydraulic ram pumping is a technology that dates back to the 18th century. Its mechanical simplicity and ability to operate without an external power source have maintained its relevance in contemporary applications. However, the optimization of its components and a thorough understanding of its hydrodynamic behavior still represent significant technical challenges [5, 6].

In a hydraulic ram pump, the number and arrangement of impulse valves are critical elements directly influencing its operation and performance. Impulse valves regulate water flow and create pressure waves within the system, affecting the pump cycle and, ultimately, its capacity to lift water to desired heights [6, 7].

This study focuses on computational analysis and simulation of the hydrodynamic effect of the number of impulse valves on the operating characteristics of a hydraulic ram pump. Using Computational Fluid Dynamics (CFD), we aim to model and simulate different valve configurations to understand how these changes affect key parameters such as overall efficiency, energy loss, and flow stability.

Through detailed simulations, this study aims to provide a theoretical and practical framework for the optimization of hydraulic ram design. It is expected that the results obtained will not only improve the understanding of the internal behavior of these systems, but also provide specific recommendations for the improvement of their performance, thus contributing to the development of more efficient and sustainable water pumping technologies.

The methodology includes creating detailed geometric models of the pump, configuring realistic boundary conditions based on experimental data, and utilizing advanced CFD software for simulation. Results will be analyzed in terms of mass and volumetric flow patterns and pressure distributions for each impulse valve configuration.

In conclusion, this study aims to establish a clear relationship between the number of impulse valves and the performance of a hydraulic ram pump, offering technical guidance for its design and application under various operating conditions.

2 Materials and Methods

2.1 Principle of Operation of the Hydraulic Ram Pump

The hydraulic ram pump is a hydraulic engine that uses the energy of a liquid volume (usually water) located at a higher elevation (e.g., a river drop, dam, or ditch) to raise part of that liquid to a greater height, using the physical phenomenon known as water hammer [8].

This pump (see Fig. 1) does not require any external energy source—be it mechanical or hydraulic. It uses the momentum of a descending water stream to induce a controlled water hammer, whose overpressure is retained in an accumulator (c). The pump includes waste (*vd)* and check (*vr*) valves, whose opening and closing dynamics, regulation, and calibration are related to pressure and flow rate [9, 10].

The water hammer is triggered when a flowing water stream reaches a preset value, regulated by the closure of the waste valve. The resulting overpressure opens the check valve, allowing fluid into the air chamber. When the depression phase begins, the check valve closes, isolating the derived flow, and once at rest, the waste valve reopens by gravity or a spring, starting a new cycle. Repetition of this process causes the accumulator tank's outlet pressure to become pulsating, requiring measurement tanks to collect the fluid [10, 11].

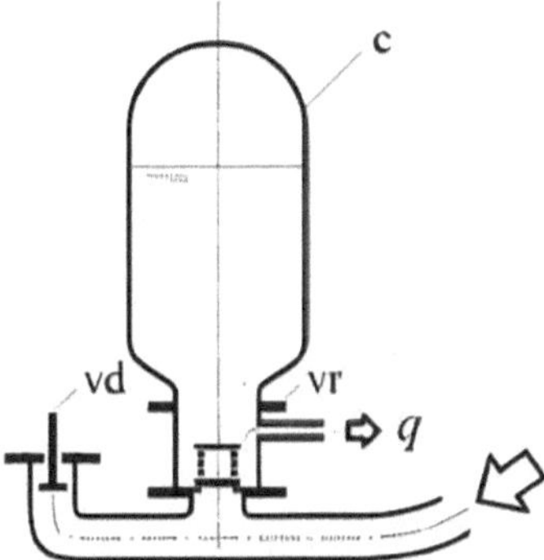

Fig. 1. Schematic diagram of hydraulic ram pump. Source [3, 12].

Several design variants have been developed to improve operating conditions, such as variable flow multi-valve and multipulse hydraulic ram pumps [11].

2.2 Multipulse Hydraulic Ram

The multipulse hydraulic ram pump involves installing impulse valves in series in the main delivery pipe, taking advantage of available flow and residual energy normally lost in a single-valve system, thereby improving performance [13].

The operating principle is similar to the conventional ram. Volumetric flow is distributed among the valves under the condition [14]:

$$Qv_1 = Qv_2 = Qv_3 \cdots Qv_n \tag{1}$$

where Qv is the flow through the valves. Placing the valves in series provides energy gains by lifting multiple smaller masses at staggered times, increasing hydrodynamic pressure for each cycle. Without the additional valves, residual energy would be wasted and could lead to faster inlet pipe wear [14].

Adding valves addresses major operational issues of conventional pumps, such as excessive weight, high water volumes for operation, and low efficiency [14, 15].

The efficiency is defined by Eq. (2):

$$\eta = n_u / n_b \tag{2}$$

where n_u is the useful output power and n_b the absorbed input power. The output power n_u is calculated as:

$$n_u = Q_b \times P_s \times g \times \rho \quad (3)$$

where Q_i is the pumped flow (m^3/s), P_i is the outlet pressure, g is gravity (9.81 m/s2), and ρ is water density (1000 kg/m^3). Similarly, absorbed power n_b is:

$$n_b = Q * P * g * \rho \quad (4)$$

where Q_i is the inlet flow and P_i the inlet pressure.

2.3 Governing Equations

Water flow in the pump delivery pipe is considered viscous, including dissipative transport phenomena like friction and thermal/mass diffusion, which increase flow entropy. To analyze and model the wind field, we solve conservation equations of mass, energy, and momentum (Navier-Stokes equations) [16, 17]:

$$\frac{D\rho}{Dt} + \rho(\nabla \cdot V) = 0 \quad (5)$$

$$\rho c \frac{DT}{Dt} = -\mathrm{p}\nabla \cdot \mathrm{V} + \nabla \cdot (\mathrm{k}\nabla\mathrm{T}) + \phi_{\mathrm{v}} \quad (6)$$

$$\rho(\mathrm{D}V/\mathrm{D}t) = -\nabla\mathrm{p} + \nabla \cdot \overline{\overline{\tau'}} + \rho f \quad (7)$$

where ρ is density, V is velocity, c is specific heat at constant volume, T is temperature, p is pressure, k is thermal conductivity, ϕ_v is work by viscous forces, $\overline{\overline{\tau'}}$ is viscous stress tensor, and f is body force per unit volume.

2.4 Computational Modeling of the Hydraulic Ram

This section describes the numerical simulation process of the multipulse hydraulic ram prototype using Computational Fluid Dynamics (CFD). The numerical study uses data and the configuration of the multipulse hydraulic ram presented in [12], which serves as the case study for this research.

2.5 Computational Tools

Autodesk Inventor® 2020 3D CAD software was used to create the physical model of the multipulse hydraulic ram. For CFD modeling of the water flow in the ram pump, ANSYS FLUENT® 2020 R2 software was used.

2.6 Physical Model

Figure 2 presents the CAD design of the physical model of the multipulse hydraulic ram pump used in this case study.

In Table 1, the general dimensions of the multipulse hydraulic ram pump are presented.

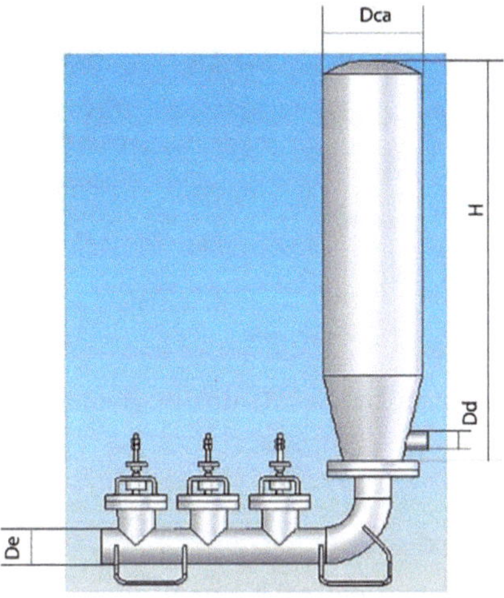

Fig. 2. Physical model of this case study. Source: Adapted from [12].

Table 1. Main Dimensions of the Hydraulic Ram. Source: Adapted from [12].

Description	Dimensions [inches]
Inlet diameter (*De*)	2
Check valve diameter (*Dvc*)	1
Air chamber diameter (*Dca*)	6
Air chamber height (*H*)	20
Outlet diameter (*Dd*)	1

2.7 Computational Domain

The computational domain is defined based on the CAD design. To reduce computational cost, a 2D model is created using ANSYS Workbench's Design Modeler (see Fig. 3), which is sufficient to simulate the water flow in the hydraulic ram.

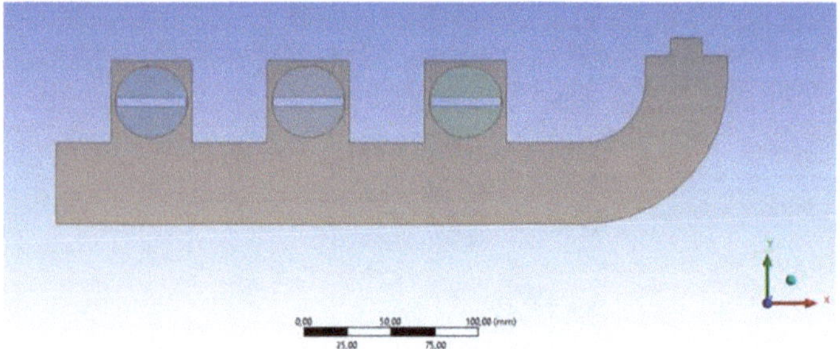

Fig. 3. 2D Computational Domain of the Multipulse Hydraulic Ram Pump. Source: Own Authorship.

2.8 Mesh Generation

A 2D mesh is generated with quadrilateral elements, based on the finite volume method (FVM). Figure 4 shows the meshing, with elements that vary in shape depending on the region during optimization.

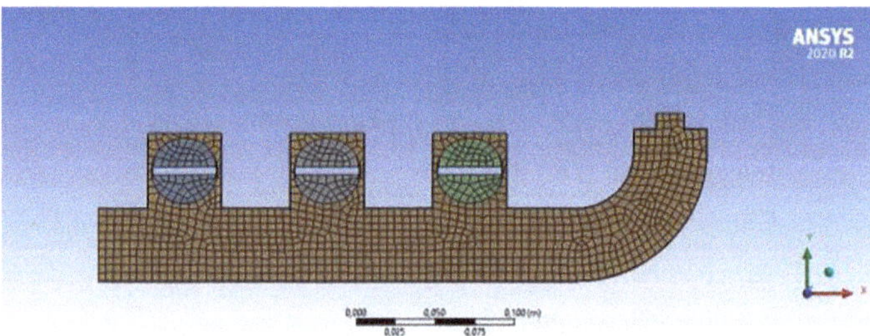

Fig. 4. Computational Domain Mesh of the Hydraulic Ram Pump. Source: Own Work.

The mesh is unstructured, with an optimal total of 685782 elements. Each valve domain is independently sized to allow future manipulation of each part. To simulate the moving mesh effect, contact regions are created between the valve and its stationary seat. This setup enables the transfer of information from one body to another, requiring that the mesh elements in these two regions be similar. The characteristics of the moving mesh are shown. (see Fig. 5).

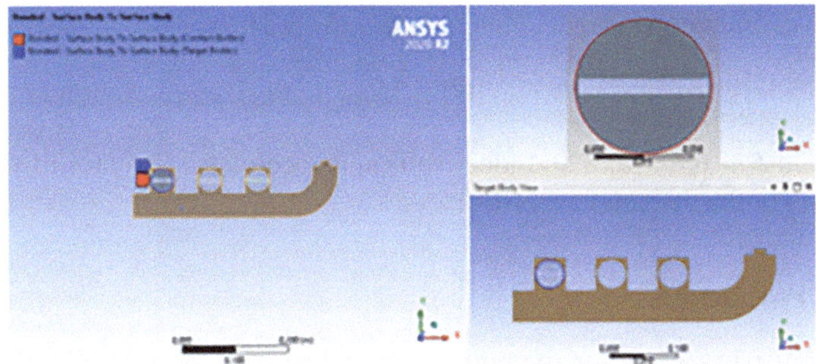

Fig. 5. Characteristics of moving mesh in the impulse valve 1 contact region. Source: Own elaboration.

In the mesh convergence or independence study, the discretization or refinement of elements is performed, using the discharge pressure of the hydraulic ram pump as the variable. Table 2 presents the results of the discretization process.

Table 2. Discretization errors in the computational domain meshing of the Multipulse Hydralic Ram Pump. Source: Based on [12].

Case	Number of Elements	Element Size [m]	Pressure [Pa]	Error [%]
1	2.172	5.0E-03	17.49	29.69
2	49.997	8.0E-04	24.65	10.59
3	119.078	5.0E-04	10.45	53.10
4	312.331	3.0E-04	23.22	4.17
5	685.782	2.0E-04	21.90	1.76
6	1152.280	1.5E-04	22.29	–

2.9 Boundary Conditions

The boundary conditions or physical limits of the computational domain are presented in Table 3.

Table 3. Boundary conditions of the computational domain for the Multipulse Hydraulic Ram Pump. Source: Own elaboration.

Region / Variable	Boundary Condition	Unit
Pump body	Symmetry	–
Flow inlet	Mass Flow Inlet (5% turbulence)	kg/s
Flow outlet	Mass Flow Outlet	kg/s
Impulse valve outlet	Outlet_v1 – Mass Flow Outlet	kg/s
Impulse valve domain	Domain valve	–

The error estimation is carried out using Eq. (8) [12]:

$$Error[\%] = 100 \times (Nua - Nui)/Nui \tag{8}$$

where *Nua* and *Nui* the actual and ideal values, respectively. The mesh from case 5 is chosen, as it exhibits the smallest error.

2.10 Input Parameters

The main input parameters for the numerical analysis of the hydraulic ram pump are presented in Table 4.

Table 4. Operating data of the Multipulse Hydraulic Ram Pump. Source: Taken from [12].

Parameter	Unit	Value
Flow velocity (v)	m/s	2.697
Angular velocity (ω)	rad/s	6.28
Max volumetric flow rate (Q)	m^3/s	5.46e-3
Mass flow rate ($\dot{m}$)	kg/s	5.46
Cycle time (Tc)	S	1
Pumped volume per cycle (Q_b)	m^3	1.28e-4
Dynamic pressure (P_D)	Pa	1201.25
Valve closing time (T_v)	S	0.00313
Max pressure (P_{max})	MPa	2
Min pressure (P_{min})	MPa	1.96

In Fig. 6, the boundary conditions for the different parts of the computational domain of the Multipulse Hydraulic Ram Pump are identified.

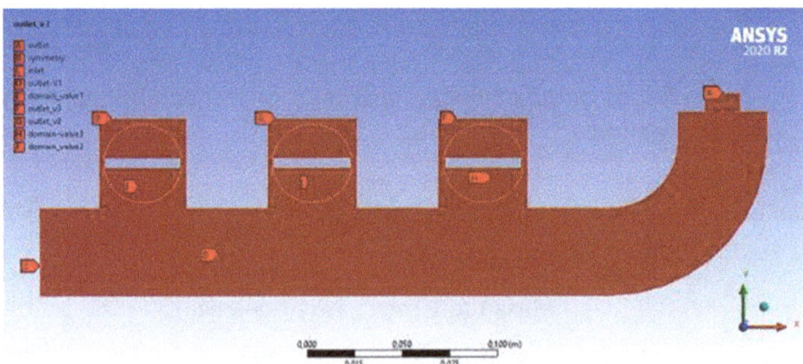

Fig. 6. Identification of the boundary conditions of the computational domain for the Multipulse Hydraulic Ram Pump. Source: Own elaboration.

For this type of CFD modeling, the two-equation turbulence model k -ϵ *Realizable* is included taking into account that the occurrence of circular trajectories will not be on a large scale, the solver will be pressure based, considering that water is an incompressible fluid, the quality of the residuals will be set as 1×10^{-4}. The parameters and boundary conditions established above will complement the solution.

2.11 Simulation Scenarios

To numerically simulate the water flow in the Multipulse Hydraulic Ram Pump, three scenarios are defined (see Table 5):

- E1 – one impulse valve
- E2 – two impulse valves
- E3 – three impulse valves

A value of zero (0) in the table indicates that the corresponding valve is blocked.

Table 5. Scenario Setup. Source: Own elaboration.

Scenario	Valve 1	Valve 2	Valve 3	Output Variable
E1	0	0	1	Flow rate
E2	0	1	1	Flow rate
E3	1	1	1	Flow rate

Procedure Validation

The validation of the procedure is carried out by comparing the numerical results obtained from the different simulations with the experimental data on the performance of the hydraulic ram pump reported in [18]. This reference presents inlet mass flow values of 5.46 kg/s and outlet flow of 1 kg/s, with an efficiency of 18.3%. These results were obtained using a dead weight of 1400 g and a stroke of 3 mm, with three impulse valves. Under these conditions, the error in overall performance is 0.23% compared to the numerical values, which are around 18.53%.

Simulation of Scenario E1

Figure 7 shows the velocity contours with the impulse valve in a closed position, where the flow progression is observed after 1 s of integration under the aforementioned conditions.

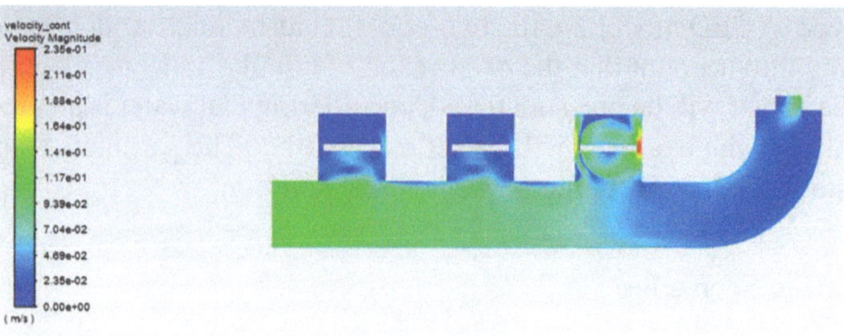

Fig. 7. Velocity contour – E1, impulse valve closed position. Source: Own elaboration

In Fig. 8, the flow progression is observed after 2.5 S of simulation, with the impulse valve in the open position. A reduction in flow at the outlet area is noticeable, indicating that part of the fluid is wasted at the moment the impulse valve opens.

Fig. 8. Velocity contour – E1, impulse valve open position. Source: Own elaboration.

In Fig. 9, the pressure contours are shown with the impulse valve in the closed position. Under this condition, the outlet pressure of the hydraulic ram pump is higher than when the impulse valve is open, as can be observed in Fig. 10. The analysis is carried out at simulation times of 1 s and 2.5 s, respectively.

Fig. 9. Pressure contours – E1, impulse valve closed position. Source: Own elaboration

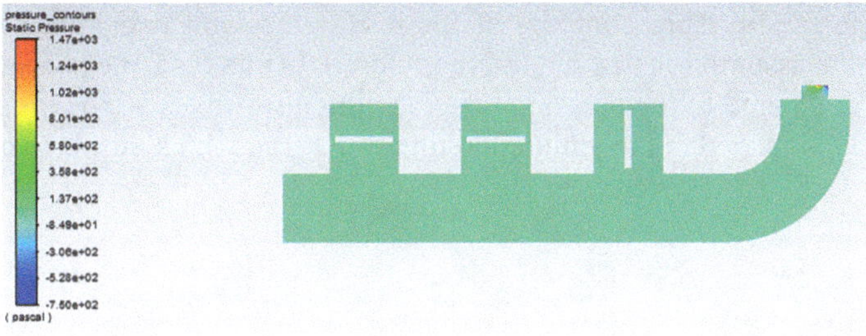

Fig. 10. Pressure contours – E1, impulse valve open position. Source: Own elaboration

Simulation of Scenario E2

As in scenario E1, a numerical simulation is performed with two impulse valves, considering valve 1 as a static element, as shown in Fig. 11.

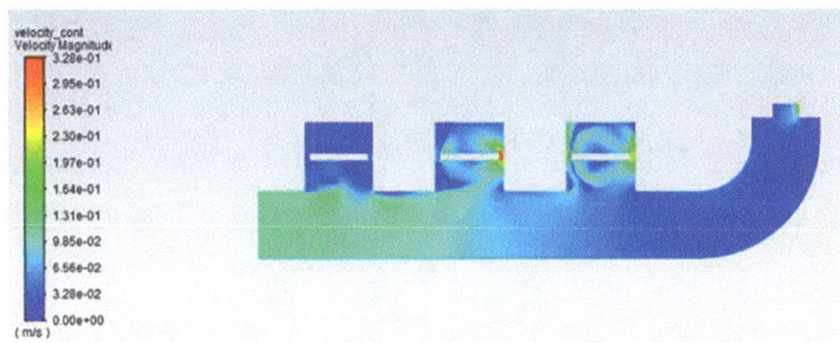

Fig. 11. Velocity contours – E2, impulse valves closed position. Source: Own elaboration.

In Fig. 12, the velocity contours are shown at a simulation time of 2.5 s, where a variation in the outlet velocity at the analysis point is observed compared to the condition when the valve is closed.

Fig. 12. Velocity contours – E2, impulse valves open position. Source: Own elaboration

In Fig. 13, the pressure contours of the ram pump with two impulse valves are shown under the assumption that one valve is closed. In this case, the pressure increases significantly compared to the pressure when the impulse valves are open (see Fig. 14). The evaluation is carried out at simulation times of 1 s and 2.5 s, respectively.

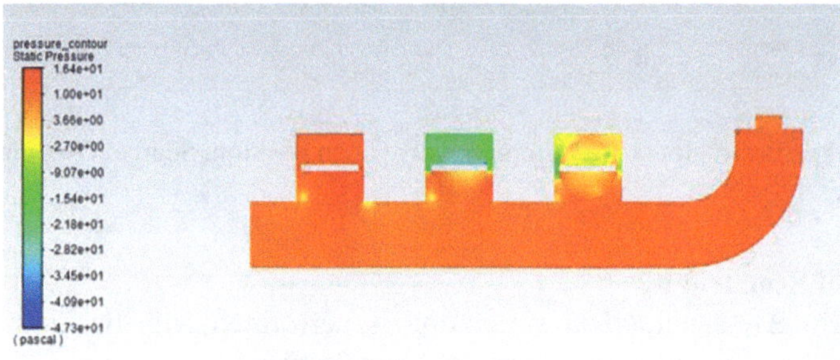

Fig. 13. Pressure contours – E2, impulse valves closed position. Source: Own elaboration

Fig. 14. Pressure contours – E2, impulse valves open position. Source: Own elaboration

Simulation of Scenario E3

Simulation of Scenario E3 In Fig. 15, the flow velocity is shown while the impulse valves are assumed to be in the closed position, after 1 s of numerical simulation. In Fig. 16, the velocity contour is presented after 2.5 s of simulation with the valve in the open position.

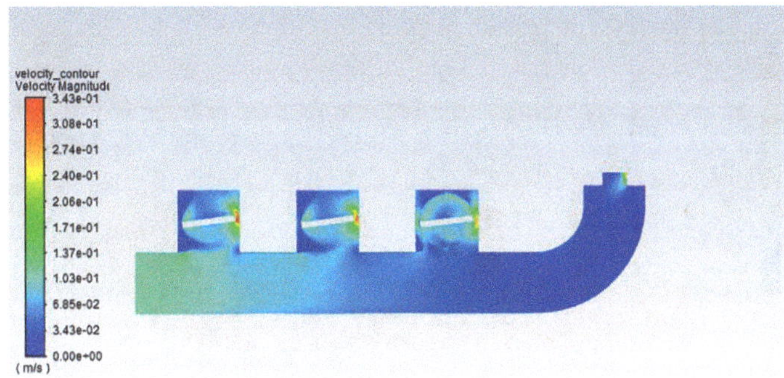

Fig. 15. Velocity contours – E3, impulse valves closed position. Source: Own elaboration

Fig. 16. Velocity contours – E3, impulse valves open position. Source: Own elaboration

Figures 17 and 18 show the static pressure contours—first with the impulse valves in the closed position, and in the following figure with the valves in the open position—at simulation times of 1 s and 2.5 s, respectively.

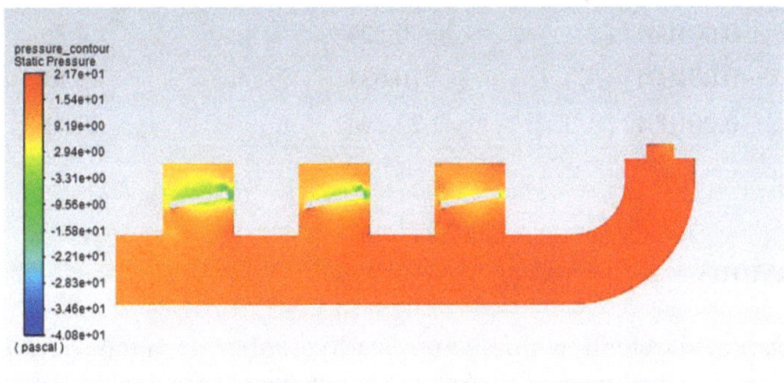

Fig. 17. Pressure contours - E3 closed valve positions. Source: Own elaboration.

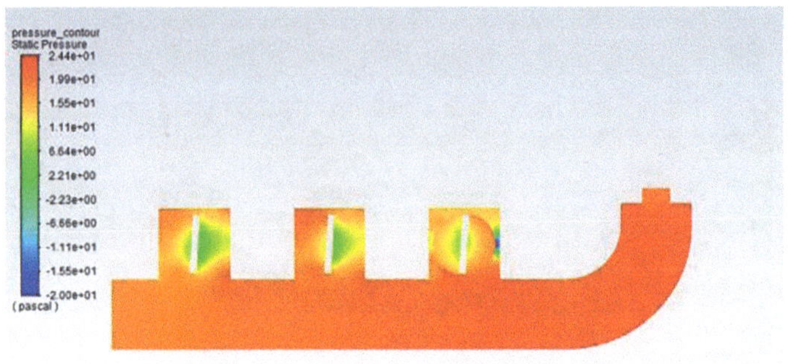

Fig. 18. Pressure contours - E3 open valve positions. Source: Own elaboration.

3 Results Analysis

The three simulation scenarios are compared, focusing on flow velocity, static pressure, mass flow, and volumetric flow. As expected, mass and volumetric flow remain constant at 0.99 kg/s and 0.001001 m^3/s, respectively. However, pressure increases from 0.327 Pa to 22.194 Pa as more impulse valves are added (see Table 6) (Table 7).

Table 6. Simulation Results

Scenario	Mass Flow [kg/s]	Vol. Flow [m^3/s]	Static Pressure [Pa]
E1	1	0.001001	0.327
E2	1	0.001001	10.631
E3	0.999	0.001001	22.194

Table 7. Efficiency Calculations

Scenario	Output Flow [m^3/s]	Outlet Pressure [Pa]	Efficiency [%]
E1	0.001001	0.327	4.36
E2	0.001001	10.631	11.55
E3	0.001001	22.194	24.11

4 Conclusions

This study conducted water flow simulations in a hydraulic ram pump with added impulse valves, the models developed with Ansys Fluent were validated against experimental results, showing good agreement in flow, pressure fields, and performance. Main conclusions:

- The simulation procedure validated using known data for a multipulse hydraulic ram allows referencing the error margin (approx. 0.23%) and defining model constraints and boundary conditions.
- Mesh independence study reduced computational cost, achieving a precision error of 1.75% with high-quality meshing.
- The k–ε turbulence model effectively represented the water hammer effect, producing coherent simulation results.
- Static discharge pressure increases with more impulse valves. With three valves, it reaches 22.19 Pa.
- Both the mass flow rate and the volumetric flow rate show no significant variation as the number of impulse valves increase during the operation of the multipulse hydraulic ram pump used in this case study.
- The overall efficiency of the hydraulic ram pump increases significantly as more impulse valves come into operation, reaching its maximum with three valves. The increase ranges from 4.35% to 24.11%.

References

1. Lentini, E.: Hacia una agenda de seguridad hídrica para América Latina y el Caribe 2030 (2022)
2. Bretas, F., et al.: Agua para el futuro: estrategia de seguridad hídrica para América Latina y el Caribe (2020)
3. Urquiza, A., Billi, M.: Seguridad hídrica y energética en América Latina y el Caribe: definición y aproximación territorial para el análisis de brechas y riesgos de la población. Documentos de Proyectos (2020)
4. Cercado Damiany, W.A.: Gestión gubernamental del uso del agua con fines de riego en Ecuador. (BABAHOYO: UTB, 2022) (2022)
5. Ávila-González, E., Ríos-Hernández, A., Morejón-Mesa, Y., Campos-Cuní, B.: Evolución histórica de las fuentes energéticas empleadas en el abasto de agua y riego agrícolas. Revista Ingeniería Agrícola **11**, 47–57 (2021)
6. Li, J., et al.: Mathematical model of hydraulic ram pump system. Proc. Inst. Mech. Eng. Part A: J. Power Energy **236**, 1097–1108 (2022)
7. Zeidan, M., Ostfeld, A.: Hydraulic ram pump integration into water distribution systems for energy recovery application. Water **14**, 21 (2021)
8. Rajaonison, A., Rakotondramiarana, H.T.: Experimental validation of a mathematical model of the operation of a hydraulic ram pump with a Springs system. Am. J. Appl. Sci. **17**, 135–140 (2020)
9. Ahmad, R., Sarip, S., Hashim, K.A.B., Suhot, M.A.: Design and development of a hydraulic ram pump for the rural communities. In: 2020 IEEE 8th R10 Humanitarian Technology Conference (R10-HTC), pp. 1–5. IEEE (2020)
10. Sarip, S., et al.: Design, analysis and fabrication of UTM hydraulic ram pump for water supply in remote areas. Indonesian J. Electr. Eng. Comput. Sci. **17**, 213–221 (2020)
11. De las Heras Jiménez, S.: Fluidos, Bombas e Instalaciones Hidráulicas. (Universitat Politecnica de Catalunya. Iniciativa Digital Politecnica) (2019)
12. Reinoso, F., Bustamante, A., Quezada, M.: Diseño de un prototipo de bomba de ariete hidráulico multipulsor de abastecimiento de agua para irrigación (2009)
13. Maatki, C.: Enhancing the efficiency and stability of hydraulic ram pump design. Phys. Fluids **37**, 023131 (2025)

14. A Manual On The Hydraulic Ram For Pumping Water S. B. Watt
15. Vega Frias, K.: Comparación de rendimiento de dos bombas de ariete hidráulico multipulsoras en paralelo para el abastecimiento de agua (2024)
16. Reinoso Avecillas, F.Z., Jara Cobos, N.G., Gómez Del Pino, P.J., Nieto Londoño, C. CARACTERIZACIÓN DEL FLUJO DE AIRE EN COLINAS PARA EL EMPLAZAMIENTO DE PARQUES EÓLICOS. ings 17 (2016). https://doi.org/10.17163/ings.n15.2016.02
17. Versteeg, H.K.: An Introduction to Computational Fluid Dynamics the Finite Volume Method, 2/E. (Pearson Education India) (2007)

Proposal to Improve the Chilean Building Thermal Regulations by Implementing the Spanish Building Thermal Regulations: First Steps in Residential Buildings

Luis María López-Ochoa(✉), Jesús Las-Heras-Casas, Pablo Olasolo-Alonso, and César García-Lozano

TENECO Research Group, Department of Mechanical Engineering, University of La Rioja, Calle San José de Calasanz, 31, 26004 Logroño, La Rioja, Spain
luis-maria.lopezo@unirioja.es

Abstract. This work studies and analyzes how the Chilean building thermal regulations can be improved by implementing the Spanish building thermal regulations. Initially, the thermal envelopes of a multi-family building in Santiago de Chile that meet the different building thermal regulations are defined, and later, the corresponding energy demands for heating are evaluated through energy simulation. Applying the Spanish building thermal regulations, a reduction of more than 70% is achieved in the energy demand for heating. This work shows progress towards achieving sustainable buildings with nearly zero energy consumption to obtain a highly energy-efficient and decarbonized Chilean residential building stock.

Keywords: Chilean Building Thermal Regulations · Spanish Building Thermal Regulations · Sustainable Building · Nearly Zero-Energy Building · Residential Sector · Santiago de Chile

1 Introduction

According to the study Uses of Energy in Homes – Chile 2018 [1], in 2018, the final energy consumption was 50,763 GWh and, with a residential building stock of 6,280,475 dwellings, the average final energy consumption per dwelling was 8083 kWh/year. With respect to the energy carrier, 39.6% of the final energy consumption corresponded to firewood, 31.4% corresponded to gas (liquefied petroleum gas and natural gas), 25.7% corresponded to electricity, 2.6% corresponded to paraffin and 0.8% corresponded to pellets. In addition, with respect to the use, 53% of the final energy consumption was allocated to heating and air conditioning, 20% to domestic hot water and 4% to lighting.

In Chile, Article 4.1.10 of the General Ordinance of Urbanism and Constructions (OGUC) [2] contains the Thermal Regulation, which establishes the thermal conditioning requirements that buildings must meet, depending on the thermal zone where they are located, and which came into force in 2007. These requirements focus mainly on the

O. F. Farías Fuentes et al. (Eds.): CIBIM 2024, *Proceedings of the XVI Ibero-American Congress of Mechanical Engineering*, pp. 195–207, 2026.
https://doi.org/10.1007/978-3-032-22823-9_14

maximum values of thermal transmittance that must be met by the roofs, walls, ventilated floors and windows. A total of 19.7% of Chilean dwellings were built from 2008 onwards [1], so approximately 20% of the residential building stock was built according to the OGUC [2]. Furthermore, in the residential building stock, while 60.8% of the dwellings has thermal insulation in the roofs, only 26.8% has thermal insulation in the walls [1]. In 2018, through the Sustainable Construction Standards (ECS) [3], energy efficiency standards were subsequently established for the design and construction of buildings, along with energy performance goals for their operation. The ECS [3] establishes the requirements for buildings to operate efficiently, generating the least possible environmental impact and proposing the incorporation of efficient air conditioning, water heating and lighting systems, as well as support systems based on renewable energy. All this with the aim of contributing to the reduction of both energy demand and energy consumption in the residential sector by promoting passive solar design and the use of energy-efficient equipment, renewable energy and efficient energy use habits in the different stages of the project.

Of the research works carried out in Chile, the following stand out: Larrea-Sáez et al. [4] evaluated the energy and environmental impacts of the thermal insulation of residential buildings in different thermal zones and discovered that the current building thermal regulations must be improved to reduce energy consumption in cities in the south of the country. Furthermore, Larrea-Sáez et al. [5] emphasized the need to prioritize the improvement of the thermal envelopes of buildings, followed by the improvement of the heating systems, in cities in the center-south of the country. Finally, Flamant et al. [6] studied the possibilities of renovation of residential buildings in Antofagasta, Santiago de Chile, Concepción and Punta Arenas, and reported that improving the thermal envelopes of buildings can achieve energy savings of up to 97%.

For residential buildings in three Latin American countries (Argentina, Brazil and Chile) and three European countries (Spain, Portugal and France), whose thermal envelopes meet the requirements of their corresponding building thermal regulations depending on the climate zone where they are located, Bienvenido-Huertas et al. [7] found that the energy demands for heating and cooling in the three European countries are lower than in the three Latin American countries and that buildings located in similar climates, but in different countries, have very different energy performances.

On the one hand, the European Energy Performance of Buildings Directive 2010 [8] defines nearly zero-energy building as that building that has a very high energy performance, for which the nearly zero or very low amount of energy required should come predominantly from renewable sources, including energy from renewable sources produced on-site or nearby. On the other hand, the European Energy Performance of Buildings Directive 2018 [9] seeks to achieve a highly energy-efficient and decarbonized building stock, as well as to guarantee that with long-term renovation strategies, it is possible to transform existing buildings into nearly zero-energy buildings. The different updates of the European Energy Performance of Buildings Directive [8, 9] were progressively transposed into Spanish national legislation through various updates of the Basic Document on Energy Saving of the Technical Building Code (CTE-DB-HE) [10]. The evolution of the implementation of the European Energy Performance of Buildings Directive [8, 9] in Spain was studied and analyzed in [11] and [12]. Given the basic requirements of the

CTE-DB-HE [10], this work focuses on the basic requirement related to the conditions for the control of energy demand. The implementation of the basic requirement related to the minimum contribution of renewable energy to meet the demand for domestic hot water of the CTE-DB-HE [10] in Chile, as well as the comparison with the corresponding Chilean regulations, were addressed in [13].

The aim of this work is to evaluate the possibility of improving the requirements of the OGUC [2] and the recommendations of the ECS [3], implementing the basic requirement related to the conditions for the control of energy demand of the CTE-DB-HE [10]. The achievement of sustainable buildings with nearly zero energy consumption is key to obtaining a highly energy-efficient and decarbonized Chilean residential building stock. The main novelty in this work compared with other works is the proposal to improve the Chilean building thermal regulations by implementing the Spanish building thermal regulations and carrying out energy simulations of a study residential building. Being a very ambitious work, this work focuses on the first steps carried out for residential buildings in Santiago de Chile with respect to the thermal envelope of the building and the energy demand for heating.

2 Methodology

The methodology followed in this work is as follows:

a. Selection of the study residential building.
b. Determination of the equivalences between the Spanish climate zone according to the CTE-DB-HE [10] and the Chilean thermal zones according to the OGUC [2] and the ECS [3].
c. Definition of the different case studies.
d. Energy simulation of the different case studies.
e. Evaluation of the possibility of improving both the OGUC [2] and the ECS [3] by implementing the CTE-DB-HE [10].

2.1 Study Residential Building

The study residential building is a multi-family building (Fig. 1) composed of a ground floor and five upper floors, with a square area of 484.00 m^2, a height of each floor of 3.00 m and a hipped roof with a height of 2.00 m. On the ground floor is the entrance and the garage, and floors 1 to 5 have four types of dwellings each floor, with a total habitable area of 2216.57 m^2. The thermal envelope is delimited by floors 1 to 5 and the space under roof. Furthermore, the study residential building was used to study and analyze the evolution of the Spanish building thermal regulations through the implementation of the European Energy Performance of Buildings Directive [8, 9] in [11, 12] and [14].

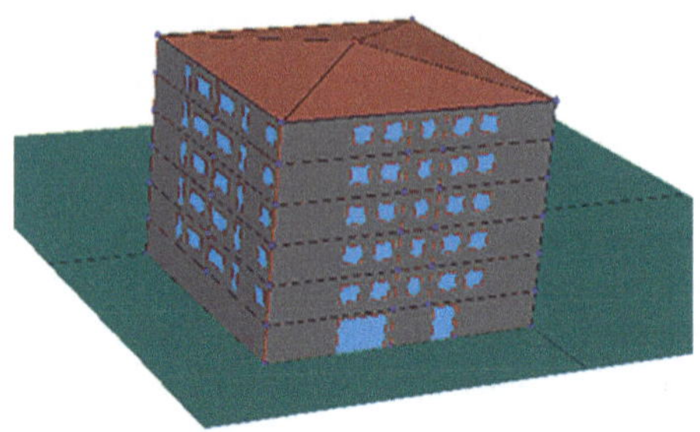

Fig. 1. 3D model of the study residential building.

2.2 Chilean Thermal Zones and Spanish Climate Zone Studied

This work was carried out in the city of Santiago de Chile, which corresponds to thermal zone 3 according to the OGUC [2] and thermal zone D according to the ECS [3]. The annual heating degree-days with a base temperature of 15 °C are between 751 °C·day and 1000 °C·day, both included [15]. According to the standard of the housing climate zoning for Chile and recommendations for architectural design [16], the thermal zone of Santiago de Chile is considered a zone with a Mediterranean climate with mild temperatures, winters of 4 to 5 months, normal vegetation, rainfall and frosts increasing towards the south, intense insolation in summer, especially towards the northeast, moderate daily oscillation of temperature, increasing towards the east, and winds from the southwest. In addition, the Köppen-Geiger climate of Santiago de Chile is Csb [17], a Mediterranean climate with winter rain.

In Spain, Cuenca is the most comparable city to Santiago de Chile, according to climatic and geographical parameters (Köppen-Geiger classification, altitude, distance to the sea, temperature, heating degree-days, rainfall, latitude and radiation) according to Sanhueza-Durán et al. [18]. Therefore, given that Cuenca is located in climate zone D2 according to the CTE-DB-HE [10], it can be assimilated that the climate zone corresponding to Santiago de Chile according to the CTE-DB-HE [10] is climate zone D2.

2.3 Case Studies

In this work, three cases corresponding to each of the building thermal regulations are studied and analyzed for Santiago de Chile:

a. A residential building that meets the OGUC [2].
b. A residential building that meets the ECS [3].
c. A residential building that meets the CTE-DB-HE [10].

Each of the building thermal regulations establishes the limit thermal transmittances of the different elements of the thermal envelope of the buildings that cannot be exceeded, depending on the thermal or climate zone where the buildings are located [2, 3, 10]. Those limit thermal transmittances, according to the different building thermal regulations [2, 3, 10], are presented in Tables 1 and 2. In addition, the CTE-DB-HE [10] establishes some recommended thermal transmittances for the different elements of the thermal envelope of buildings, which are presented in Table 2, with which to meet other requirements established by the basic requirement related to the conditions for the control of energy demand of the CTE-DB-HE [10]. Finally, the compositions and main characteristics of the roof, walls and first floor framework of the study residential building are presented in Tables 3–5 and are based on those used in [12] and [14].

Table 1. Limit thermal transmittances, in $W/m^2{\cdot}K$, for the different elements of the thermal envelope of the study residential building according to the OGUC [2] and the ECS [3].

Element	OGUC	ECS
Roof	0.47	0.38
Walls	1.90	0.80
First floor framework[(a)]	0.70	0.70
Windows	3.60[(b)]	3.60[(c)]

Note: (a) The first floor framework is equivalent to a ventilated floor; (b) the windows use hermetic double glazing and meet the established percentages of the glazed surface by the OGUC [2]; and (c) the windows meet the percentages of the glazed surface established by the ECS [3]

Table 2. Limit and recommended thermal transmittances (U_{lim} and U_{rec}), in $W/m^2{\cdot}K$, for the different elements of the thermal envelope of the study residential building according to the CTE-DB-HE [10].

Element	U_{lim}	U_{rec}
Roof	0.35	0.22
Walls	0.41	0.27
First floor framework*	0.65	0.48
Windows	1.80	1.60

Note: * The first floor framework is a floor in contact with a non-habitable space according to the CTE-DB-HE [10].

Table 3. Composition and main characteristics of the roof of the study residential building, based on [12] and [14].

Material	Thickness (m)	Conductivity (W/m·K)	Density (kg/m^3)	Specific heat (J/kg·K)	Thermal resistance ($m^2{\cdot}K/W$)
Wafer or ceramic tile	0.015	1.000	2000	800	

(continued)

Table 3. (*continued*)

Material	Thickness (m)	Conductivity (W/m·K)	Density (kg/m^3)	Specific heat (J/kg·K)	Thermal resistance (m^2·K/W)
Cement or lime mortar for masonry and for rendering/plastering 500 < d < 750	0.040	0.300	625	1000	
Sublayer felt	0.001	0.050	120	1300	
XPS Expanded with CO_2 (0.034 W/m·K)	*	0.034	38	1000	
Sublayer felt	0.001	0.050	120	1300	
Bitumen felt or sheet	0.003	0.230	1100	1000	
Sublayer felt	0.001	0.050	120	1300	
Cement or lime mortar for masonry and for rendering/plastering 500 < d < 750	0.020	0.300	625	1000	
Expanded clay (loose aggregate)	0.100	0.148	538	1000	
Unidirectional forged concrete infill, 300 mm span	0.300	1.422	1240	1000	
Unventilated 10 cm horizontal air chamber	–	–	–	–	0.18
Plasterboard (PYL) 750 < d < 900	0.015	0.250	825	1000	

Note: * indicates that the thickness of the thermal insulation is variable

Table 4. Composition and main characteristics of the walls of the study residential building, based on [12] and [14].

Material	Thickness (m)	Conductivity (W/m·K)	Density (kg/m^3)	Specific heat (J/kg·K)	Thermal resistance (m^2·K/W)
Cement or lime mortar for masonry and for rendering/plastering 500 < d < 750	0.015	0.300	625	1000	
Perforated metric or Catalan brick of ½ foot 40 mm < G < 60 mm	0.115	0.667	1140	1000	

(*continued*)

Table 4. (*continued*)

Material	Thickness (m)	Conductivity (W/m·K)	Density (kg/m^3)	Specific heat (J/kg·K)	Thermal resistance (m^2·K/W)
Cement mortar with high resistance to filtration	0.015	1.800	625	800	
MW Mineral wool (0.031 W/m·K)	*	0.031	40	1000	
Slightly ventilated 2 cm vertical air chamber	–	–	–	–	0.085
MW Mineral wool (0.031 W/m·K)	*	0.031	40	1000	
Low density polyethylene (LDPE)	0.001	0.330	920	2200	
Plasterboard (PYL) 750 < d < 900	0.015	0.250	825	1000	

Note: * indicates that the thickness of the thermal insulation is variable

Table 5. Composition and main characteristics of the first floor framework of the study residential building, based on [12] and [14].

Material	Thickness (m)	Conductivity (W/m·K)	Density (kg/m^3)	Specific heat (J/kg·K)
Wafer or ceramic tile	0.015	1.000	2000	800
MW Mineral wool (0.031 W/m·K)	*	0.031	40	1000
Unidirectional forged concrete infill, 300 mm span	0.300	1.422	1240	1000
MW Mineral wool (0.031 W/m·K)	*	0.031	40	1000
Plasterboard (PYL) 750 < d < 900	0.015	0.250	825	1000

Note: * indicates that the thickness of the thermal insulation is variable

2.4 Energy Simulation of the Case Studies

LIDER-CALENER Unified Tool (HULC) [19] was used to carry out energy simulations of the different case studies and to evaluate the energy demand for heating. HULC [19] has been widely used in research on Spanish building thermal regulations, as in [12] and [14]. In addition, the following conditions were considered:

- The location of the study residential building is the city of Cuenca, the most comparable Spanish city to Santiago de Chile [18].
- The ventilation flow is 660.00 l/s, meeting the basic requirement related to the indoor air quality of the Basic Document on Health of the Technical Building Code [20].
- Thermal bridges are those considered by default by HULC [19].

The method followed for the energy simulation using HULC [19] carried out in this work and obtaining the energy demands for heating is based on the method followed in [18] and [21]. On the one hand, González-Avilés et al. [21] assimilated the Köppen-Geiger Cfb climate of Pessac (France) to the climate zone C1 according to the CTE-DB-HE [10], in a similar way as was done in [18].

Furthermore, González-Avilés et al. [21] carried out energy simulations of a building located in a non-Spanish city with HULC [19], assimilating its climate to a Spanish climate zone according to the CTE-DB-HE [10], in a similar way to the method used in this work. On the other hand, Sanhueza-Durán et al. [18] reported that similar values of energy demand for heating were obtained for the different dwellings of a multi-family building, both in Santiago de Chile and in Cuenca, using the Chilean and Spanish procedures for the energy performance certification of dwellings and buildings. Therefore, the energy demands for heating obtained by HULC [19] are validated.

3 Results

In Sect. 3.1, the thermal transmittances used for each of the case studies are presented, compared and analyzed, while, in Sect. 3.2, the energy demands for heating obtained for each of the case studies are presented, compared and analyzed.

3.1 Thermal Transmittances

Table 6 presents the thermal transmittances used, as well as the required thicknesses of thermal insulation, to meet each of the building thermal regulations studied [2, 3, 10] (Tables 1 and 2).

Table 6. Thermal transmittance (U), in W/m^2·K, and the required thickness of thermal insulation (t), in m, to meet each of the building thermal regulations studied [2, 3, 10].

Element	OGUC		ECS		CTE-DB-HE	
	U	T	U	t	U	t
Roof	0.47	0.020	0.38	0.036	0.22	0.100
Walls	1.89	0.002*	0.79	0.022	0.27	0.098
First floor framework	0.70	0.030	0.70	0.030	0.48	0.048
Windows	3.51	–	3.51	–	1.66	–

Note: * The air chamber disappears and a single layer of thermal insulation is used.

With respect to applying the OGUC [2], the thermal transmittance of the roof is reduced by 19.15% by applying the ECS [3], which requires an additional thickness of thermal insulation of 16 mm, and by 53.19% by applying the CTE-DB-HE [10], which requires an additional thickness of thermal insulation of 80 mm (Table 6). With respect to applying the ECS [3], the thermal transmittance of the roof is reduced by 42.11% by applying the CTE-DB-HE [10], which requires an additional thickness of thermal insulation of 64 mm (Table 6).

With respect to applying the OGUC [2], the thermal transmittance of walls is reduced by 58.20% by applying the ECS [3], which requires an additional thickness of thermal insulation of 20 mm, and by 85.71% by applying the CTE-DB-HE [10], which requires an additional thickness of thermal insulation of 96 mm (Table 6). While applying the OGUC [2], only a single layer of thermal insulation is used, applying both the ECS [3] and the CTE-DB-HE [10], two layers of thermal insulation are used, with the same thickness, separated by an air chamber, as in [12] and [14] (Tables 4 and 6). With respect to applying the ECS [3], the thermal transmittance of walls is reduced by 65.82% by applying the CTE-DB-HE [10], which requires an additional thickness of thermal insulation of 76 mm (Table 6).

The thermal transmittance of the first floor framework is the same applying both the OGUC [2] and the ECS [3] (Table 6). With respect to applying the OGUC [2] or the ECS [3], the thermal transmittance of the first floor framework is reduced by 31.43% by applying the CTE-DB-HE [10], which requires an additional thickness of thermal insulation of 18 mm (Table 6).

The thermal transmittance of the windows is the same applying both the OGUC [2] and the ECS [3]. With respect to applying the OGUC [2] or the ECS [3], the thermal transmittance of the windows is reduced by 52.71% by applying the CTE-DB-HE [10] (Table 6). Applying the OGUC [2] and the ECS [3], the composition of the different windows of the study residential building is double glass, with a thermal transmittance of 3.30 W/m^2·K and a solar factor (g-value) of 0.750, and a metal frame with a thermal bridge break between 4 and 12 mm, with a thermal transmittance of 4.00 W/m^2·K and an absorptivity of 0.70. In addition, 30% of the window is covered by the frame and the air permeability is 27.00 m^3/h·m^2. While applying the CTE-DB-HE [10], the composition of the different windows of the study residential building is low-emissive double glass, with a thermal transmittance of 1.60 W/m^2·K and a solar factor (g-value) of 0.700, and a

3-chamber polyvinyl chloride frame, with a thermal transmittance of 1.80 $W/m^2 \cdot K$ and an absorptivity of 0.70. In addition, 30% of the window is covered by the frame and the air permeability is 9.00 $m^3/h \cdot m^2$.

3.2 Energy Demands for Heating

Figure 2 presents the energy demands for heating obtained by HULC [19] for the study residential building, which meets each of the building thermal regulations studied [2, 3, 10] (Table 6).

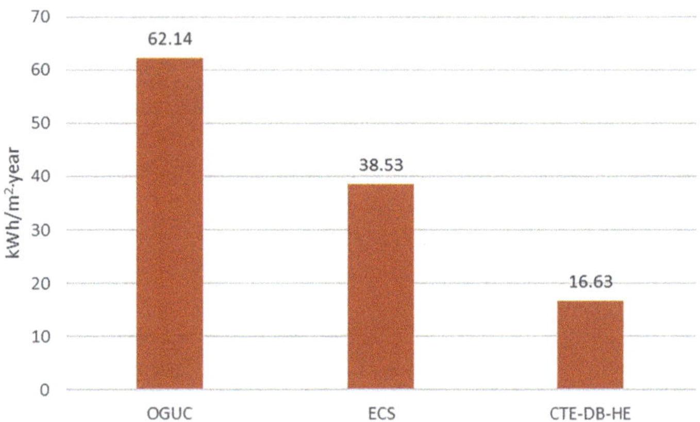

Fig. 2. Energy demands for heating obtained, in $kWh/m^2 \cdot year$, for the study residential building, which meets each of the building thermal regulations studied [2, 3, 10].

Applying the OGUC [2], the energy demand for heating is 26.89% lower than that corresponding to dwellings in multi-family buildings in the Chilean residential building stock (85 $kWh/m^2 \cdot year$ [22]) (Fig. 2).

The energy demand for heating is reduced by 37.99% by applying the ECS [3] and by 73.24% by applying the CTE-DB-HE [10], with respect to applying the OGUC [2], while the energy demand for heating is reduced by 56.84% by applying the CTE-DB-HE [10], with respect to applying the ECS [3] (Fig. 2).

Analyzing only the energy demand for heating obtained and taking into account the level of energy efficiency provided by the Chilean energy rating scale for homes [22] and the savings in energy demand for heating obtained (Fig. 2), an E could be assigned to the residential building that meets the OGUC [2] (current standard), a D to the residential building that meets the ECS [3] (efficiency of good standard with moderate increase in investment) and an A to the residential building that meets the CTE-DB-HE [10] (efficiency of excellent standard with noticeable increase in investment).

4 Conclusions

Based on the results obtained in this work, the Chilean building thermal regulations could be improved by reducing the thermal transmittances of the different elements of the thermal envelope, thereby achieving a considerable reduction in the energy demand for heating of the buildings. By applying the CTE-DB-HE [10], the energy demand for heating of the study residential building can be reduced by more than 50%, with respect to applying the ECS [3], and more than 70%, with respect to applying the OGUC [2]. In this way, by implementing the Spanish thermal building regulations, it is possible to advance towards the achievement of sustainable Chilean buildings with nearly zero energy consumption to obtain a highly energy-efficient and decarbonized residential building stock.

Finally, the following could be addressed in future works: (a) How the basic requirement related to the conditions for the control of energy demand of the CTE-DB-HE [10] could be implemented for both multi-family buildings and single-family houses in different cities in all the Chilean thermal zones; and (b) how other basic requirements related to the limitation of energy consumption or the minimum generation of electrical energy from renewable sources of the CTE-DB-HE [10] could be implemented in Chile.

References

1. Chilean Ministry of Energy, In-Data, Technological Development Corporation (CDT): Uses of Energy in Homes – Chile 2018 (Usos de la Energía de los Hogares – Chile 2018) (2019). https://www.energia.gob.cl/sites/default/files/documentos/informe_final_caracterizacion_residencial_2018.pdf. Accessed 01 Mar 2024
2. Chilean Ministry of Housing and Urbanism: Article 4.1.10 of the General Ordinance of Urbanism and Constructions (Artículo 4.1.10 de la Ordenanza General de Urbanismo y Construcciones) (2005). https://nuevo.leychile.cl/servicios/Consulta/Exportar?radioExportar=Normas&exportar_formato=pdf&nombrearchivo=DTO-192_04-ENE-2006&exportar_con_notas_bcn=False&exportar_con_notas_originales=False&exportar_con_notas_al_pie=False&hddResultadoExportar=245882.2006-01-04.0.0%23. Accessed 01 Mar 2024
3. Chilean Ministry of Housing and Urbanism: Sustainable Construction Standards for Homes in Chile, Volume II: Energy (Estándares de Construcción Sustentable para Viviendas de Chile, Tomo II: Energía) (2018). https://csustentable.minvu.gob.cl/wp-content/uploads/2018/03/EST%C3%81NDARES-DE-CONSTRUCCI%C3%93N-SUSTENTABLE-PARA-VIVIENDAS-DE-CHILE-TOMO-II-ENERGIA.pdf. Accessed 01 Mar 2024
4. Larrea-Sáez, L., Cuevas, C., Casas-Ledón, Y.: Energy and environmental assessment of the chilean social housing: Effect of insulation materials and climates. J. Clean. Prod. **392**, 136234 (2023). https://doi.org/10.1016/j.jclepro.2023.136234
5. Larrea-Sáez, L., Muñoz, E., Cuevas, C., Casas-Ledón, Y.: Optimizing insulation and heating systems for social housing in Chile: Insights for sustainable energy policies. Energy, **290**, 130024 (2024). https://doi.org/10.1016/j.energy.2023.130024
6. Flamant, G., Bustamante, W., Schmitt, C., Bunster, V., Osorio, C.: Thermal and environmental evaluation of mid-rise social housing retrofit under different climate conditions. J. Build. Eng. **46**, no. 103724 (2022). https://doi.org/10.1016/j.jobe.2021.103724
7. Bienvenido-Huertas, D., Oliveira, M., Rubio-Bellido, C., Marín, D.: A comparative analysis of the international regulation of thermal properties in building envelope. Sustainability (Switzerland), **11**(20), 5574 (2019). https://doi.org/10.3390/su11205574
8. European Union: Directive 2010/31/EU of the European Parliament and of the Council of 19 May 2010 on the energy performance of buildings (recast) (2010). https://eur-lex.europa.eu/legal-content/EN/TXT/PDF/?uri=CELEX:32010L0031. Accessed 01 Mar 2024
9. European Union: Directive (EU) 2018/844 of the European Parliament and of the Council of 30 May 2018 amending Directive 2010/31/EU on the energy performance of buildings and

Directive 2012/27/EU on energy efficiency (Text with EEA relevance) (2018). https://eur-lex.europa.eu/legal-content/EN/TXT/PDF/?uri=CELEX:32018L0844. Accessed 01 Mar 2024
10. Spanish Ministry of Housing and Urban Agenda: Basic Document on Energy Saving of the Technical Building Code (Documento Básico de Ahorro de Energías del Código Técnico de la Edificación) (2022). https://www.codigotecnico.org/pdf/Documentos/HE/DBHE.pdf. Accessed 01 Mar 2024
11. López-Ochoa, L.M., Las-Heras-Casas, J., Olasolo-Alonso, P., López-González, L.M.: Towards nearly zero-energy buildings in Mediterranean countries: fifteen years of implementing the Energy Performance of Buildings Directive in Spain (2006–2020). J. Build. Eng. **44**, 102962 (2021). https://doi.org/10.1016/j.jobe.2021.102962
12. López-Ochoa, L.M., Las-Heras-Casas, J., González-Caballín, J.M., Carpio, M.: Towards nearly zero-energy residential buildings in Mediterranean countries: the implementation of the energy performance of buildings directive 2018 in Spain. Energy, **276**, 127539 (2023). https://doi.org/10.1016/j.energy.2023.127539
13. López-Ochoa, L.M., Verichev, K., Las-Heras-Casas, J., Carpio, M.: Solar domestic hot water regulation in the Latin American residential sector with the implementation of the energy performance of buildings directive: the case of Chile. Energy, **188**, 115985 (2019). https://doi.org/10.1016/j.energy.2019.115985
14. López-Ochoa, L.M., Las-Heras-Casas, J., López-González, L.M., García-Lozano, C.: Environmental and energy impact of the EPBD in residential buildings in cold Mediterranean zones: the case of Spain. Energy Build. **150**, 567–582 (2017). https://doi.org/10.1016/j.enbuild.2017.06.023
15. Chilean Ministry of Housing and Urbanism: Thermal Regulation Application Manual: Thermal Zoning Plans, zoning of heating degree-days at the communal level (Manual de Aplicación de la Reglamentación Térmica: Planos de Zonificación Térmica, zonificación de grados día de calefacción a nivel comunal) (2006). https://www.curriculumnacional.cl/614/articles-230010_recurso_6.pdf. Accessed 01 Mar 2024
16. Chilean National Institute of Standardization: Official Chilean Standard NCh1079:2008, Architecture and Construction - Housing climate zoning for Chile and recommendations for architectural design (Norma Chilena Oficial NCh1079:2008, Arquitectura y Construcción - Zonificación climático habitacional para Chile y recomendaciones para el diseño arquitectónico) (2008). https://tipbook.iapp.cl/ak/a871a9028e2d17e03ecbc8b26cef7a6b0b0b13b5/embed/view/minvu-nch01079-2008-049. Accessed 01 Mar 2024
17. Chilean Ministry of National Assets, Geospatial Data Infrastructure (IDE) Chile: IDE Chile map viewer (Visor de mapas IDE Chile). https://www.geoportal.cl/geoportal/map/4. Accessed 01 Mar 2024
18. Sanhueza-Durán, F., Gómez-Soberón, J.M., Valderrama-Ulloa, C., Ossio, F.: A Comparison of energy efficiency certification in housing: a study of the chilean and spanish cases. Sustainability (Switzerland), **11**(17), 4771 (2019). https://doi.org/10.1016/10.3390/su11174771
19. HULC: LIDER-CALENER Unified Tool, version 2.0.2412.1173 (date of update 11 May 2023) (Herramienta Unificada LIDER-CALENER, versión 2.0.2412.1173 (fecha de actualización 11 de mayo de 2023)). https://www.codigotecnico.org/Programas/HerramientaUnificadaLIDERCALENER.html. Accessed 01 Mar 2024
20. Spanish Ministry of Housing and Urban Agenda: Basic Document on Health of the Technical Building Code (Documento Básico de Salubridad del Código Técnico de la Edificación) (2022). https://www.codigotecnico.org/pdf/Documentos/HS/DBHS.pdf. Accessed 01 Mar 2024
21. González-Avilés, Á.B., Pérez-Carramiñana, C., Galiano-Garrigós, A., Ibarra-Coves, F., Lozano-Romero, C.: Analysis of the energy efficiency of Le Corbusier's dwellings: the

Cité Frugès, an opportunity to reuse garden cities designed for healthy and working life. Sustainability (Switzerland), **14** (8), 4537 (2022). https://doi.org/10.3390/su14084537

22. Chilean Ministry of Housing and Urbanism: Energy Rating of Homes (Calificación Energética de Viviendas). https://www.calificacionenergetica.cl. Accessed 01 Mar 2024

Dynamic Thermal Simulation of a Double-PCM Roofing in a Semi-arid Climate

J. A. Contreras-Aguilar[1], M. Gijón-Rivera[2](✉), and C. I. Rivera-Solorio[1]

[1] Escuela de Ingeniería y Ciencias, Tecnológico de Monterrey, 64700 Monterrey, N.L., Mexico
[2] Escuela de Ingeniería y Ciencias, Tecnológico de Monterrey, 72453 Monterrey, PUE , Mexico
miguel.gijon@tec.mx

Abstract. An annual energy assessment of a building prototype with a double-layered phase change material (PCM) in the roof system is presented. The building was thermally modeled using the energy simulation software DesignBuilder. The annual energy consumption of the building with a double-layered PCM roof was assessed under the semiarid climatic conditions of Monterrey City in Mexico. The results compare the base case with the PCM roof configuration. The results showed that the energy saved was 1,858.44 kWh per year, which represents a reduction of 7.8%. The return on investment could be up to 6.7 years, and the internal rate of return could be up to14.4%. The emissions savings were 808.42 kgCO_2eq per year.

Keywords: Buildings · phase change material (PCM) · double layer · semiarid climate · energy analysis

1 Introduction

The building sector exhibits significant energy demand, accounting for approximately 30% of global energy consumption and contributing 27% to global greenhouse gas emissions [1]. Furthermore, energy consumption attributed to air conditioning, heating, and ventilation systems in buildings globally accounts for approximately 47% of total energy consumption [2]. In Mexico, the energy demand of the building sector constitutes 18% of the energy consumption, according to various sources [3]. Conversely, building occupants spend approximately 80–90% of their time indoors, necessitating significant requirements for indoor thermal comfort to enhance their quality of life and productivity [4]. In Mexico, 30% of the energy consumed by buildings is attributable to indoor thermal comfort applications [5], with increased demands for air conditioning systems in the northern states of the country [6].

Hence, various passive design strategies have been implemented to increase the energy efficiency of buildings and reduce the energy demand related to thermal comfort [7, 8]. These construction design techniques utilize solar or wind resources to improve indoor comfort conditions without the need for additional electrical energy or mechanical/electromechanical climate control systems. Particularly in warm climates, these

O. F. Farías Fuentes et al. (Eds.): CIBIM 2024, *Proceedings of the XVI Ibero-American Congress of Mechanical Engineering*, pp. 208–217, 2026.
https://doi.org/10.1007/978-3-032-22823-9_15

passive techniques can focus on the building envelopes through proper solar orientation, passive ventilation techniques, insulating materials, solar protection systems, reflective coatings, evaporative systems, and latent storage systems. In this regard, the roof is the envelope element that receives the highest solar radiation during the day, making it crucial to improve the thermal performance of the element to reduce heat gains inside the building. Some of the passive design strategies applied to the roof include reducing solar absorptivity through the application of reflective materials, enhancing thermal resistance with insulating materials, and increasing thermal mass to reduce indoor thermal fluctuations by using phase change materials (PCM) [9].

Phase change materials (PCM) possess both sensible (temperature change) and latent (phase change) thermal energy storage capabilities, with the latter having a higher impact because it absorbs or releases more energy during phase transition compared to any other type of material [10–13]. PCMs can be classified according to their composition (organic, inorganic, or eutectic) and their method of incorporation (microencapsulation, macroencapsulation, shape stabilization, or a combination) [12–17]. Various brands are commercially available on the market; however, paraffin and hydrated salts are the most commonly used for building envelopes [14–16]. PCM melts by absorbing heat from the external environment, allowing the interior temperature to stabilize; when the external temperature decreases, PCM releases the previously stored energy, preventing excessive drops in interior air temperatures [15]. Therefore, PCM helps to reduce energy consumption in HVAC systems, provides interior temperature control, and delays the peak energy demand for heating and cooling [18].

Literature reports indicate that a single layer of phase change material (PCM) in walls or roofs has not been the most effective option for annual evaluations. This is due to the incompatibility between the fusion temperature of the material and the seasonal effects in climates with significant variability throughout the year. For instance, Khadir et al. [19] reported a seasonal dependency in the behavior of single-layer PCM, with reductions in heat transfer between 25% and 47% using a single layer, but with large PCM thicknesses between 10 and 30 cm. In studies applying multiple layers of PCM, Panchal et al. [20] concluded that a double-layer PCM configuration could mitigate seasonal effects and improve the efficiency of PCM applications in building envelopes. Similarly, Pasupathy and Velraj [21] evaluated the effect of a double-layer PCM in a building roof for an annual evaluation and found that this configuration achieved a more thermally comfortable space. Additionally, Rehman et al. [22] proposed a double-layer configuration for bricks in both warm and cold climates, finding a good thermal response in summer and winter for reducing the interior temperature of a building in Islamabad. Finally, Gao et al. [23] also proposed a PCM configuration to improve envelope conditions and thermally adapt it throughout an annual evaluation. They found that double PCM was capable of meeting energy demands in summer and winter; however, it did not show optimal performance in several months. Double PCM increased the thermal delay by an average of 4.6 h in summer and winter. Based on the state-of-the-art discussion, this study proposes an annual energy evaluation of a Mexican residential building with a double-layer PCM roof located in a climate with abrupt changes between winter and summer in Monterrey, Mexico. A comparison is proposed between a conventional concrete roof configuration and the roof with a double-layer, inorganic, commercially available PCM to observe its advantages in annual thermal adaptation, energy savings, and indoor thermal comfort.

2 Methodology

2.1 Climate Data

Monterrey City is the capital of the State of Nuevo León in Mexico. It was selected due to its significance in the country, its energy consumption for air conditioning, and its extreme climatic conditions. The state ranked third in terms of gross domestic product in 2022 and second in terms of annual energy consumption for cooling purposes, representing a great case study for the application of PCM in building roofs. Monterrey is located in the northwest of the country at an altitude of 534 m above sea level (AMSL) (25.65°N, – 100.29°W); the prevailing climate is semi-arid (Köppen BS). Climatic data were obtained from the DesignBuilder database (EPW files), including solar radiation, ambient temperature, relative humidity, and wind speed and direction.

2.2 Physical Model

The geometry of the building to be modeled consists of a single floor with the main façade oriented to the west. It has floorplan dimensions of 6.0 × 6.0 m with asymmetry and a height of 3.0 m from the ground. Figures 1, 2, and 3 represent the floor plan, west façade, and east façade, respectively, with the dimensions and location of each envelope element.

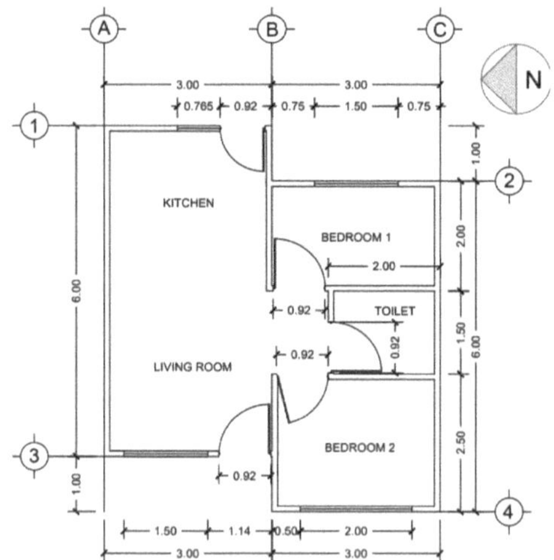

Fig. 1. Building floor plan (Dimensions in meters).

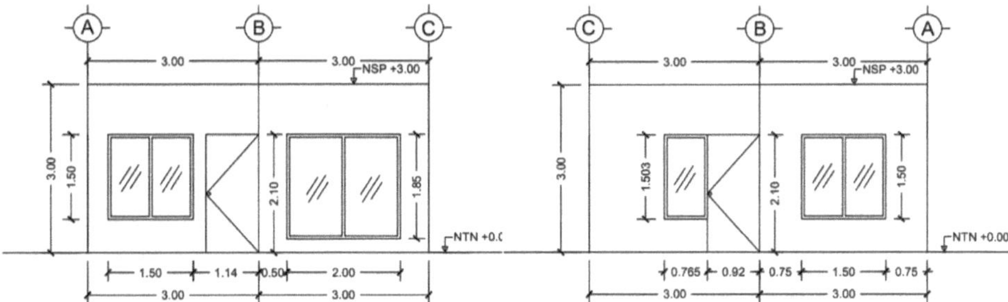

Fig. 2. West facade view.

Fig. 3. East facade view.

The materials applied to the building envelope are the same as those used in previous thermal comfort research in buildings in Mexico [9]. The thermal properties of the materials are presented in Table 1, indicating the material layer, location, and their thermophysical properties. The analysis consists of performing a transient numerical simulation considering a basic HVAC system in the building. The general characteristics and system details are presented in Table 2.

Table 1. Thermophysical properties of construction materials.

Component	Material	Thickness [m]	Thermal absortance [-]	Solar absortance [-]	Conductivity [W/m-K]
Roof without PCM					
External	Concrete	0.200	0.90	0.60	1.13
Internal	Plastering	0.013	0.90	0.50	0.50
Roof with PCM					
External	Concrete	0.200	0.90	0.60	1.13
Layer 2	PCM 25	0.006	-	-	0.54 (liquid) 1.09 (solid)
Layer 3	PCM 29	0.012	-	-	0.54 (liquid) 1.09 (solid)
Internal	Plastering	0.013	0.90	0.50	0.50
Floor					
External	Cast Concrete	0.200	0.90	0.60	1.13
Layer 2	Floor/Roof Screed	0.070	0.90	0.73	0.41
Internal	Ceramic Tiles Dry	0.030	0.90	0.60	1.20
External walls					
External	Plaster (Lightweight)	0.013	0.90	0.50	0.16

(continued)

Table 1. (*continued*)

Component	Material	Thickness [m]	Thermal absortance [-]	Solar absortance [-]	Conductivity [W/m-K]
Layer 2	Brickwork Outer	0.100	0.90	0.70	0.84
Internal	Plaster (Lightweight)	0.013	0.90	0.50	0.16
Internal walls					
External	Plaster (Lightweight)	0.013	0.90	0.50	0.16
Layer 2	Brickwork Outer	0.100	0.90	0.70	0.84
Internal	Plaster (Lightweight)	0.013	0.90	0.50	0.16
Window					
Frame	Aluminum (no break)	0.040	0.30	0.30	160.00
Glass	Sgl Blue 6mm	0.006	-	-	-

Table 2. Overall building characteristics.

Characteristics	Description
Absorptance	0.6 for roof/0.5 for wall
Occupation	3 people
Infiltration	0.5 ACH
HVAC system	DX package
Setpoint	24 °C
COP	2.5
HVAC operation	weekdays: 00:00 a 06:00, 0.5; 06:00 a 18:00, 1.0; 18:00 a 24:00, 0.5. Weekend: 00:00 a 24:00, 1.0

2.3 Building Modeling

DesignBuilder software was used for building energy analysis, and the geometry was modeled with the integrated module. The numerical model is solved using the dynamic building energy simulation tool Energy Plus v8.1. The simulation is a transient hourly analysis performed from January 1 to December 31. To model and solve PCM melting and solidification processes, the Phase Change Material module was activated, considering the Hysteresis mode and the Finite Difference solution algorithm.

The main simulation assumptions are: 1) The properties of the concrete remain constant; 2) The properties of the PCM are constant, except for the thermal conductivity coefficient, which differs only for the liquid and solid phases; 3) The HVAC system is a DX packaged unit; 4) An occupancy of 3 people and a comfort temperature of 24 °C are assumed; 5) The building was selected based on previous studies of Mexican housing classification [15]; 6) The double layer of PCM is 12 mm with a melting temperature of 29 °C and 6 mm with 25 °C, based on a previous CFD analysis.

3 Results

3.1 Validation

The ASHRAE-140 BESTEST was used to verify the accuracy of the simulation procedure, given that this standard is used to evaluate building energy analysis programs. The Benchmark 600 FF case from ANSI/ASHRAE 140-2021 was replicated using DesignBuilder to compare and report the error percentage for the maximum, minimum, and average indoor air temperature. The model is considered as a single thermal zone, and an average annual temperature of 26.1 °C was obtained, representing an error of less than 1%. Similarly, the maximum and minimum temperatures obtained for the year have an error of less than 1% when compared to the Benchmark Solution.

3.2 Results and Discussion

Figure 4 represents the energy consumed by the HVAC system to heat the building and achieve an indoor temperature of 24 °C with and without PCM added to the roof. It can be observed that energy consumption for heating the building is reduced in all months when using PCM. The months with the highest energy consumption for heating coincide with the winter season; conversely, the months with the lowest energy consumption for heating correspond to the summer period.

January was the month that shows the greatest reduction in energy consumption for heating by applying the PCM configuration in the roof. The energy was reduced from 2535.9 to 2403.9 kWh, resulting in a difference of 132 kWh, which represents a reduction of 5.2%.

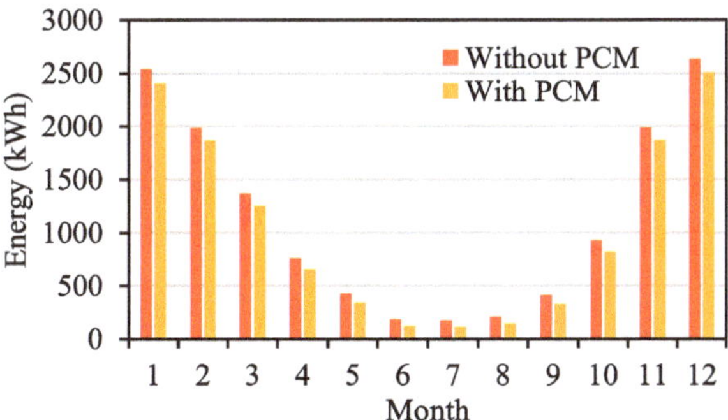

Fig. 4. Monthly energy consumption for heating.

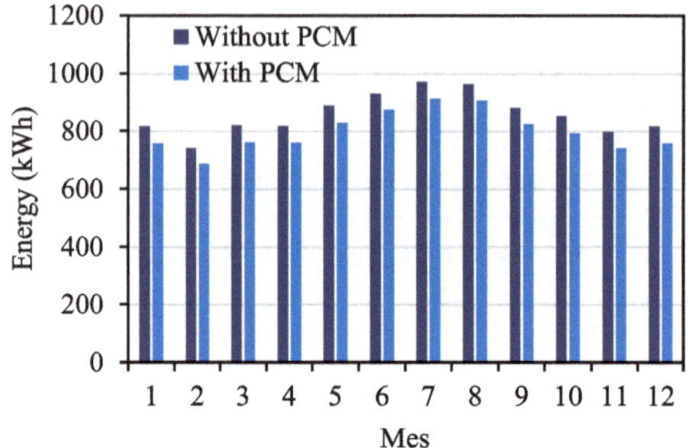

Fig. 5. Monthly energy consumption for cooling.

Figure 5 shows the energy consumption for cooling the building, reaching a setpoint temperature of 24 °C with and without the integrated PCM in the roof. It can be seen that, similarly, energy consumption is reduced in all months when applying the PCM. Additionally, the months with the highest energy consumption for cooling are reversed, with summer being the period of highest energy consumption and winter the lowest.

Figure 6 compares the annual energy consumption for cooling and heating with and without the application of the PCM configuration. The greatest annual energy reduction is due to heating, with a reduction of 8.7%. The energy reduction due to cooling is equivalent to 6.6%. Finally, the HVAC system consumes a total of 23,864.59 kWh without the PCM and 22,006.15 kWh with the PCM applied to the roof. This represents a total energy reduction due to the HVAC system of 1,858.44 kWh, equivalent to 7.8% (Fig. 7).

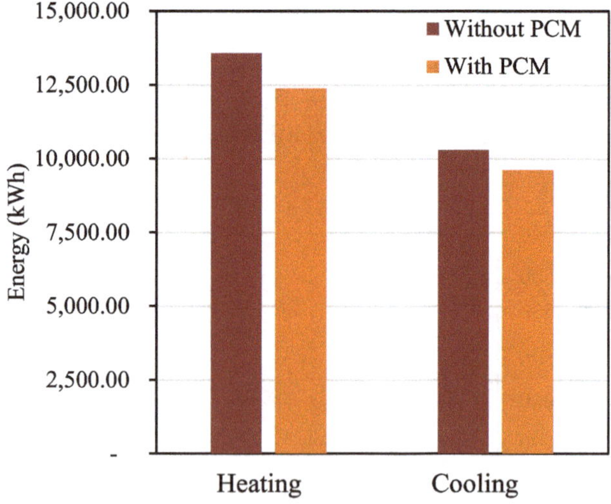

Fig. 6. Annual energy consumption for heating and cooling.

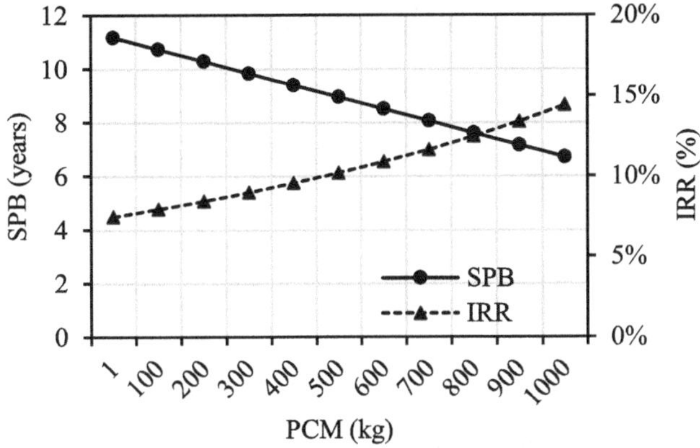

Fig. 7. SPP e IRR per amount of PCM.

According to the Federal Electricity Commission (CFE), for every kilowatt-hour of electricity consumed, 0.435 kg of CO_2 is emitted. Consequently, the CO_2 emissions mitigated by the PCM configuration on the roof amount to 808.42 $kgCO_2e$ for this type of building located in Monterrey, Mexico.

4 Conclusions

From the analysis of the results obtained from the dynamic simulation of utilizing a double layer of PCM applied to the roof of a building in Monterrey, the following conclusions can be made regarding energy consumption, economic analysis, and environmental impact:

- The energy consumed for heating was higher during the months of November to February. Conversely, the energy consumed for cooling was greater during the months of June to August. These months correspond to the coldest and warmest months of the year, respectively.
- The amount of energy required for heating, when integrating the double layer of PCM in the roof, was reduced by 8.7%, while for cooling it was reduced by 6.6%. The total energy consumption saved in the HVAC system was 1,858.44 kWh, representing a reduction of 7.8% over the year.
- If the PCM is purchased at retail, the investment in the material had a payback period of 11.18 years due to energy savings, with a return rate of 7.5%. However, if the material is purchased wholesale, the investment was retrieved in 6.71 years, with a return rate of 14.4%, making this PCM configuration economically attractive.
- Emissions are reduced by 808.42 $kgCO_2e$ yearly. If a greater number of buildings include PCM in their roofs, the reduction in emissions released into the atmosphere could significantly impact on the pollution levels Monterrey in Monterrey City.

References

1. IEA, https://www.iea.org/reports/buildings, last accessed 2023,11/10

2. Kreider, J., Curtis, P.: Heating and cooling of buildings: Design for efficiency, 2nd edn. Taylor & Francis, USA (2010)
3. SENER. https://www.cenace.gob.mx/Paginas/Info/MarcoRegulatorio.aspx. Accessed 10 Nov 2023
4. Liu, F., Yan, L., Meng, X.: A review on indoor green plants employed to improve indoor environment. J. Build. Eng. **53**, 104542 (2022)
5. SENER. https://www.conuee.gob.mx/transparencia/boletines/Cuadernos/cuadernoNo.10.pdf. Accessed 10 Nov 2023
6. USAID. https://mexico-cooling.lbl.gov/wp-content/uploads/sites/38/2019/09/mexico-cooling-factbook.pdf. Accessed 10 Nov 2023
7. Rodriguez, C., et al.: Passive design strategies and performance of net energy plus houses. Energy Build. **83**, 10–22 (2014)
8. IEA. https://iea.blob.core.windows.net/assets/deebef5d-0c34-4539-9d0c-10b13d840027/NetZeroby2050-ARoadmapfortheGlobalEnergySector_CORR.pdf. Accessed 10 Nov 2023
9. Mousavi, S., Gijón-Rivera, M., Rivera-Solorio, C., Godoy, C.: Energy, comfort, and environmental assessment of passive techniques integrated into low-energy residential buildings in semi-arid climate. Energy Build. **263**, 112053 (2022)
10. Oró, E., de Gracia, A., Castell, A., Farid, M., Cabeza, L.: Review on phase change materials (PCMs) for cold thermal energy storage applications. Appl. Energy **99**, 513–533 (2012)
11. Hasan, Z., et al.: Challenges of the application of PCMs to achieve zero energy buildings under hot weather conditions: a review. J. Energy Storage **64**, 107156 (2023)
12. Faisal, M., et al.: Inorganic phase change materials in thermal energy storage: a review on perspectives and technological advances in building applications. Energy Build. **252**, 111443 (2021)
13. Alothman, A., Nandini, S., Krishnaraj, L.: A phase change material for building applications - a critical review. Mater. Today: Proc. **56**, 1858–1864 (2022)
14. Madessa, H.: A review of the performance of buildings integrated with phase change material: opportunities for application in cold climate. Energy Procedia **62**, 318–328 (2014)
15. Kosny, J.: PCM-Enhanced Building Components: An Application of Phase Change Materials in Building Envelopes and Internal Structures. Springer, United Kingdom (2015)
16. Lamrani, B., Johannes, K., Kuznik, F.: Phase change materials integrated into building walls: An updated review. Renew. Sustain. Energy Rev. **140**, 110751 (2021)
17. Wang, X., Li, W., Luo, Z., Wang, K., Shah, S.: A critical review on phase change materials (PCM) for sustainable and energy efficient building: design, characteristic, performance and application. Energy Build. **260**, 111923 (2022)
18. Cabeza, L., Castell, A., Barreneche, C., de Gracia, A., Fernández, A.A.: Materials used as PCM in thermal energy storage in buildings: a review. Renew. Sustain. Energy Rev. **15**, 1675–1695 (2011)
19. Khdair, A., Rumman, G., Muhammad, B.: Developing building enhanced with PCM to reduce energy consumption. J. Build. Eng. **48**, 103923 (2022)
20. Panchal, J., Modi, K., Patel, V.: Development in multiple-phase change materials cascaded low-grade thermal energy storage applications: a review. Clean. Eng. Technol. **8**, 100465 (2022)
21. Pasupathy, A., Velraj, R.: Effect of double layer phase change material in building roof for year-round thermal management. Energy and Buildings **40**, 193–203 (2008)
22. Rehman, S., Sheikn, S., Kausar, Z., McCormack, S.: Numerical simulation of a novel dual layered phase change material brick Wall for human comfort in hot and cold climatic conditions. Energies **14**, 4032 (2021)
23. Gao, Y., Liu, Z., Gao, Y., Mao, W., Zuo, Y., Li, G.: Employing the double-PCM (phase change material) layer to improve the seasonal adaptation of building walls: a comparative studies. J. Energy Storage **66**, 107404 (2023)

24. Valencia, E., Gijón-Rivera, M., Rivera-Solorio, C.: Energy, economic, and environmental assessment of the integration of phase change materials and hybrid concentrated photovoltaic thermal collectors for reduced energy consumption of a school sports center. Energy Build. **293**, 113198 (2023)

Manufacturing Processes

Role of Polarizability in the Deformation and Fracture of Graphene Sheets

Enrique Wagemann Herrera[1(✉)], Pablo Caniu Villablanca[1], Elton Oyarzúa Vargas[2], and Cristian Alberto Cuevas Barraza[1]

[1] Departamento de Ingeniería Mecánica, Universidad de Concepción, Concepción, Chile
ewagemann@udec.cl

[2] Department of Chemistry, Faculty of Natural Sciences (Nat), Aarhus University, Aarhus, Denmark

Abstract. In this study, molecular dynamics simulations were used to investigate the impact of polarizability on the mechanical properties of monolayer and multilayer graphene. Tensile tests were simulated on graphene with 1, 2, and 3 layers, comparing models that consider and do not consider polarizability. From the stress-strain curves (σ vs. ϵ), the elastic modulus was determined, with values ranging from 923.1 to 957.5 GPa for non-polarizable graphene and from 899.6 to 932.1 GPa for polarizable graphene. The results indicate that polarizability introduces a slight flexibility to the material, although its effect in the elastic region is minimal. However, polarizability significantly reduces the tensile strength of graphene, resulting in earlier fractures due to the destabilization of C-C bonds by the presence of non-permanent dipoles in carbon atoms. These findings help strengthen existing models for simulating the mechanical properties of graphene.

Keywords: Graphene · Molecular Dynamics · Deformation · Mechanical properties

1 Introduction

Graphene [1] (GE) is a two-dimensional material composed of carbon (C) atoms arranged in a hexagonal structure similar to a honeycomb. The molecular structure of graphene endows it with a series of extraordinary physical properties, such as high thermal conductivity [2], high electrical conductivity [3], exceptional mechanical strength [4], and flexibility [5]. These properties make graphene a promising material for various technological applications, including the fabrication of electronic devices [6], sensors [5], composite materials [7], and energy storage systems [8].

Despite the remarkable properties of graphene, its behavior under deformation and the conditions leading to its fracture are research areas that still require a deeper understanding. The deformation and fracture of graphene sheets are influenced by the atomic nature of the fundamental structures that compose this material. In particular, the nature of its intramolecular (sp2-type) bonds plays a fundamental role in determining the

O. F. Farías Fuentes et al. (Eds.): CIBIM 2024, *Proceedings of the XVI Ibero-American Congress of Mechanical Engineering*, pp. 221–229, 2026.
https://doi.org/10.1007/978-3-032-22823-9_16

mechanical properties of graphene. Thus, one factor that could have a significant impact is the polarizability of carbon atoms.

Polarizability, a measure of the electronic deformability of an atom in response to an electric field, can significantly affect interatomic interactions and, therefore, the mechanical properties of the material. However, the specific relationship between polarizability and the properties of deformed graphene is still not fully understood. Understanding how polarizability influences the mechanical response of graphene during deformation and eventual fracture is crucial for developing more robust and optimized applications of this versatile material.

In this study, molecular dynamics (MD) simulations are employed to investigate how polarizability influences the deformation process and eventual fracture of monolayer and multilayer graphene sheets. MD simulations enable the modeling of the behavior of monolayer and multilayer graphene sheets under different tensile conditions, providing a detailed atomistic view of the mechanisms involved in the deformation and fracture of graphene. Two graphene models are compared: a first model, where polarizability is not considered, and a second model, which adds polarizability to the first model using Drude oscillators.

2 Methodology

All molecular dynamics (MD) simulations were performed using the Large-scale Atomic/Molecular Massively Parallel Simulator (LAMMPS) software package [9]. The visualization of simulations was carried out using the Open Visualization Tool (OVITO) [10]. The simulations mimic uniaxial tensile tests by progressively extending the simulation box along one of the principal directions of the Graphene sheet.

The simulated system consists of one or multiple stacked graphene sheets. Figure 1 shows a snapshot of the simulated system. The system is periodic in the x and y directions, ensuring the graphene sheet links with itself across the simulation box boundaries. In the z direction, non-periodic boundary conditions are applied to prevent the system from interacting with itself.

In all simulations, a timestep of 1 fs was used. The intramolecular interactions of the graphene sheet were modeled using a Tersoff-type potential, calibrated to reproduce the mechanical properties of graphene [11]. In cases considering more than one graphene layer, interlayer van der Waals interactions were modeled using a Lennard-Jones (12–6) potential with parameters [12]: $\sigma = 3.414$ Å y $\epsilon = 0.2312$ kJ/mol, reproducing an interlayer distance of 3.4 Å.

The polarizability of graphene was modeled using Drude oscillators [13]. In this model, polarization is simulated through two point charges with opposite signs: the Drude Particle (DP) and the Drude Core (DC). A DC is assigned to each C atom composing the graphene sheet, with a DP bonded to each DC via a Hookean potential. The Drude Particles have a mass of 0.4 g/mol, subtracted from the corresponding carbon atom to maintain the total mass of the system. The parameters used are taken from [14], calibrated to reproduce interfacial properties in water-graphene systems via quantum simulations.

The system was equilibrated in the $NP_{xy}T$ ensemble (constant Number of particles, Pressure, and Temperature) at 300 K and 1 bar for 500 ps. During this initial stage, a

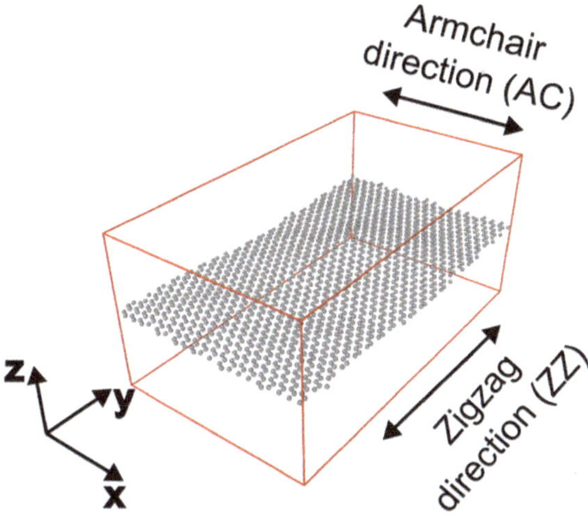

Fig. 1. Schematic representation of the simulated system. A graphene sheet, periodic in x and y directions, is deformed by progressively increasing the simulation box in the armchair (AC) or zigzag (ZZ) directions. Red lines represent the simulation box boundaries. Grey spheres represent the carbon atoms that form the graphene sheet. Source: Own elaboration.

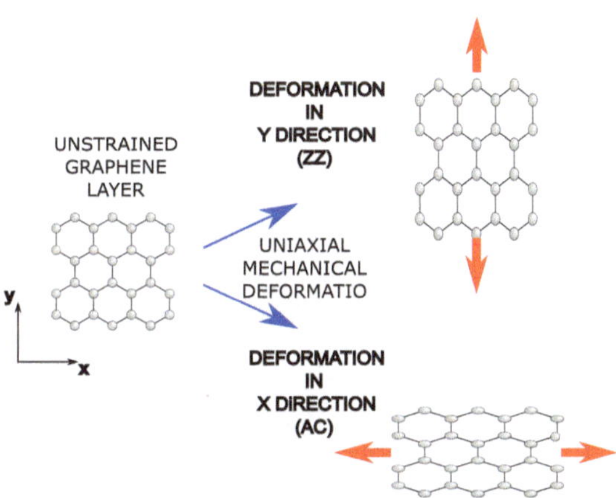

Fig. 2. In-plane mechanical deformation of a graphene monolayer. Two types of uniaxial deformations are shown: in the y direction (ZZ) and in the x direction (AC). Source: Own elaboration.

Nosé-Hoover thermostat with a coupling constant of 100 fs was applied. Pressure was controlled using a Nosé-Hoover barostat in the x and y directions (no barostat was applied in the z direction, normal to the graphene plane) with a coupling constant of 1000 fs.

After achieving equilibrium, the system was subjected to one of two independent uniaxial deformations: in the x direction (AC) or in the y direction (ZZ), as illustrated in Fig. 2. The simulation mimics a tensile test by progressively increasing the size of the simulation box along one direction, leveraging the periodicity of the material to maintain atomic connectivity. The deformation rate was 1 Å/fs. These tensile simulations were conducted in the NP_iT ensemble, where i is the direction normal to the applied deformation, at 300 K and 1 bar. The barostat was applied only along the direction perpendicular to deformation (e.g., if deformation was along x, the barostat acted along y) normal al plano de la hoja de grafeno) con una constante de acoplamiento de 1000 fs.

2.1 Results

Figure 3 illustrates the evolution of one of the simulations performed (graphene monolayer, deformation in the y direction). Three phases are shown: an initial equilibrium phase ($\epsilon = 0$), a second phase where the graphene is already deformed ($\epsilon = 0.1$), and a third phase where, as a result of the applied deformation, the graphene reaches its critical point and fractures ($\epsilon = 0.26$). It is important to note that in the simulations, the rupture of graphene occurs almost instantaneously, making it impossible to identify the mechanisms involved within the simulated timeframes. Similar fracture morphologies were observed for deformation in the armchair (AC) direction.

The resistance to deformation exerted by the graphene sheet is characterized through the stress tensor, particularly its component σ_{ii}, where i is the direction.

of the applied deformation. The stress tensor is directly extracted from the simulations. The strain ϵ is calculated using Eq. 1:

$$\epsilon = \frac{l_i - l_{i,0}}{l_{i,0}} \tag{1}$$

where l_i represents the length of the graphene sheet at the evaluated instant and $l_{i,0}$ the equilibrium length. These values are calculated based on the size of the simulation box along the deformed direction. The equilibrium length is estimated from the average box size during the last 250 ps of $NP_{xy}T$ ensemble equilibration.

Figure 4a presents the stress-strain curve for non-polarizable graphene with 1, 2, and 3 layers under uniaxial deformation in the zigzag (ZZ) direction. A clear elastic region is observed up to approximately 0.06 strain, characterized by linear behavior of the σ-ϵ curve (zoom shown in Fig. 4b). Beyond this strain, the loss of linearity suggests the onset of plastic deformation until rupture, occurring around 0.16 strain for three-layer graphene.

During rupture, benzene rings at the fracture zone are broken, and filaments ("flagella") form connecting the fracture edges. These filaments are expected to eventually break as well under greater strains, although such phenomena were not observed within the simulated timeframes.

It should be noted that not all simulated cases achieved fracture; for instance, in Fig. 4a, only the three-layer graphene showed complete fracture.

The elastic region is characterized by the elastic modulus (E), calculated as:

$$E = \frac{\partial \sigma_{ii}}{\partial \epsilon} \tag{2}$$

The slope was estimated using a linear moving fit across the elastic region. Table 1 summarizes the calculated elastic moduli for non-polarizable graphene with 1 to 3 layers in both AC and ZZ directions.

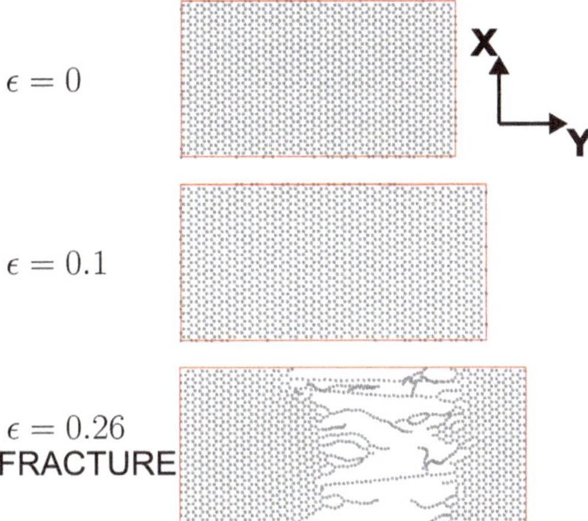

Fig. 3. Development of mechanical deformation in a graphene monolayer under uniaxial tension in the y (ZZ) direction. Phases shown: equilibrium, deformation, and fracture. Source: Own elaboration.

As shown, increasing the number of layers slightly increases the elastic modulus. Differences between AC and ZZ directions are minimal, suggesting that graphene's elastic properties are relatively isotropic.

Regarding the effect of polarizability on elastic properties, Table 2 presents the results for polarizable graphene systems. It is observed that polarizability slightly reduces the elastic modulus, indicating a small increase in the material elasticity. This may be associated with the destabilization of C-C bonds due to the presence of non-permanent dipoles at each carbon atom.

Although a slight increase in the material elasticity is noted, the difference is around 2%, suggesting that polarizability has a minimal influence on the elastic properties in the elastic region.

In contrast, polarizability significantly affects the fracture behavior. Polarizable systems exhibit earlier fractures compared to non-polarizable systems, as shown in Fig. 5, which compares the stress-strain curves of polarizable and non-polarizable graphene in the ZZ direction. In polarizable systems, fractures occurred at lower strain values across all simulated cases, at ca. 105 GPa, while in non-polarizable systems, fracture was observed at higher strain values (ca. 120 GPa). This behavior is attributed to the destabilization of C-C bonds induced by the presence of non-permanent dipoles associated with polarizability.

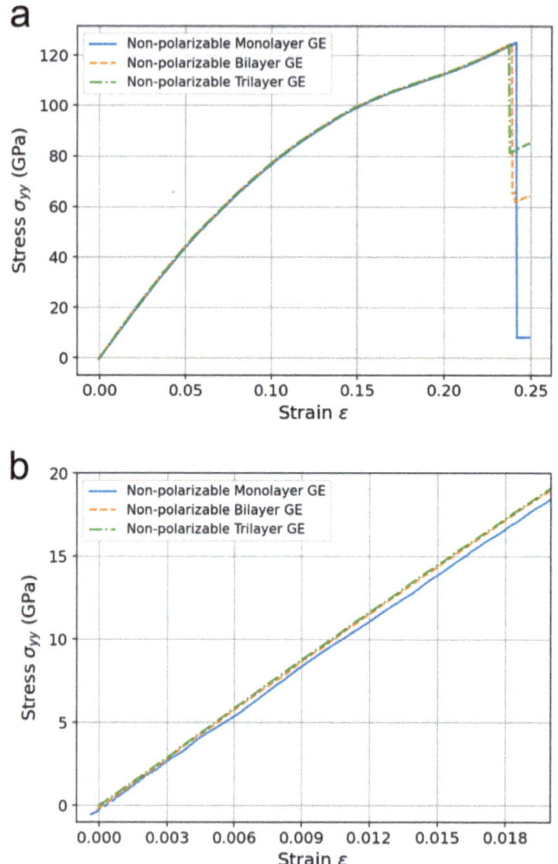

Fig. 4. (a) Stress-strain curve for non-polarizable graphene with 1, 2, and 3 layers under uniaxial deformation in the ZZ direction. (b) Zoomed-in view of the elastic region. Source: Own elaboration.

Table 1. Elastic modulus for non-polarizable graphene systems.

Number of layers	Direction	Elastic Modulus (GPa)
1	ZZ	942.8
1	AC	923.1
2	ZZ	956.5
2	AC	948.1
3	ZZ	957.5
3	AC	948.5

Table 2. Elastic modulus for polarizable graphene systems.

Number of layers	Direction	Elastic Modulus (GPa)
1	ZZ	909.0
1	AC	899.6
2	ZZ	919.6
2	AC	915.3
3	ZZ	932.1
3	AC	922.7

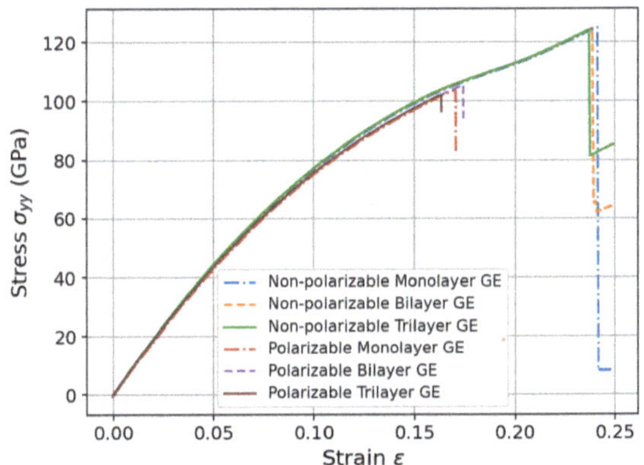

Fig. 5. Stress-strain curves for polarizable and non-polarizable graphene with 1, 2, and 3 layers under uniaxial deformation in the ZZ direction. Source: Own elaboration.

2.2 Conclusions

In this work, molecular dynamics simulations were carried out to evaluate the effect of polarizability on the mechanical properties of mono- and multilayer graphene. Through simulations mimicking tensile tests, σ–ϵ curves were obtained for graphene with 1, 2, and 3 layers along the two principal in-plane directions of the material (armchair, AC, and zigzag, ZZ). Two models were compared: nonpolarizable graphene and polarizable graphene. From the linear region of the σ–ϵ curves, the elastic modulus was calculated, yielding values between 923.1 and 957.5 GPa for non-polarizable graphene, and between 899.6 and 932.1 GPa for polarizable graphene. Based on the calculated values, it is established that while polarizability introduces a softening effect in the material, its impact is minimal.

In addition to evaluating the elastic modulus, the simulations revealed significant differences in tensile strength and fracture behavior between polarizable and non-polarizable graphene. Polarizability reduces the tensile strength of graphene, resulting in earlier fractures in the polarizable systems compared to their non-polarizable counterparts. This behavior is associated with the destabilization of C–C bonds due to the presence of non-permanent dipoles on each carbon atom.

Future research is encouraged to further explore the atomistic mechanisms behind C–C bond destabilization caused by polarizability and to assess how different graphene configurations, including variations in layer number and orientation, may influence the observed properties.

Acknowledgments. This work was funded by ANID under project FONDECYT de iniciación no. 11240551. Powered@NLHPC: This research was partially supported by the supercomputing infrastructure of the NLHPC (CCSS210001).

References

1. Geim, A.K.: Graphene: Status and prospects (2009). https://doi.org/10.1126/science.1158877
2. Balandin, A.A., et al.: Superior thermal conductivity of single-layer graphene. Nano Lett. **8**(3), 902–907 (2008). https://doi.org/10.1021/nl0731872
3. Alemour, B., Yaacob, M.H., Lim, H.N., Hassan, M.R.: Review of electrical properties of graphene conductive composites (2018)
4. Papageorgiou, D.G., Kinloch, I.A., Young, R.J.: Mechanical properties of graphene and graphene-based nanocomposites. Prog. Mater. Sci. **90**, 75–127 (2017). https://doi.org/10.1016/j.pmatsci.2017.07.004
5. Singh, E., Meyyappan, M., Nalwa, H.S.: Flexible graphene-based wearable gas and chemical sensors. ACS Appl. Mater. Interfaces **9**(40), 34544–34586 (2017). https://doi.org/10.1021/acsami.7b07063
6. Han, T.H., Kim, H., Kwon, S.J., Lee, T.W.: Graphene-based flexible electronic devices. Mater. Sci. Eng. R. Rep. **118**, 1–43 (2017). https://doi.org/10.1016/j.mser.2017.05.001
7. Castellanos-Gomez, A., et al.: Van der Waals heterostructures. Nat. Rev. Methods Primers **2**(1), 58 (2022). https://doi.org/10.1038/s43586-022-00139-1
8. Olabi, A.G., Abdelkareem, M.A., Wilberforce, T., Sayed, E.T.: Application of graphene in energy storage device–a review. Renew. Sustain. Energy Rev. **135**, 110026 (2021). https://doi.org/10.1016/j.rser.2020.110026
9. Thompson, A.P., et al.: LAMMPS - a flexible simulation tool for particle-based materials modeling at the atomic, meso, and continuum scales. Comput. Phys. Commun. **271**, 108171 (2022). https://doi.org/10.1016/j.cpc.2021.108171
10. Stukowski, A.: Visualization and analysis of atomistic simulation data with OVITO–the open visualization tool. Model. Simul. Mater. Sci. Eng. **18**(1), 015012 (2010)
11. Lindsay, L., Broido, D.A.: Optimized Tersoff and Brenner empirical potential parameters for lattice dynamics and phonon thermal transport in carbon nanotubes and graphene. Phys. Rev. B Condens. Matter. Mater. Phys. **81**(20), 205441 (2010). https://doi.org/10.1103/PhysRevB.81.205441
12. Girifalco, L.A., Hodak, M., Lee, R.S.: Carbon nanotubes, buckyballs, ropes, and a universal graphitic potential. Phys. Rev. B Condens. Matter. Mater. Phys. **62**(19), 13104 (2000). https://doi.org/10.1103/PhysRevB.62.13104
13. MacKerell, A.D.: Empirical force fields for biological macromolecules: overview and issues. J. Comput. Chem. **25**(13), 1584–1604 (2004). https://doi.org/10.1002/jcc.20082

14. Misra, R.P., Blankschtein, D.: Insights on the role of many-body polarization effects in the wetting of graphitic surfaces by water. J. Phys. Chem. C **121**(50), 28166–28179 (2017). https://doi.org/10.1021/acs.jpcc.7b08891

Evaluation of Few-Layer Graphene Addition on the Hardness and Fracture Toughness of WC-10Co

Jhonatan Dantas dos Santos Rosa[1], Daniele Rodrigues das Neves[1], Marcello Filgueira[2], Marcelo Bertolete[1](✉), and Patrícia Alves Barbosa[1]

[1] Mechanical Technology Laboratory, Federal University of Espírito Santo, Vitória, ES 29075-910, Brazil
marcelo.b.carneiro@ufes.br

[2] Advanced Materials Laboratory, Northern Fluminense State University, Campos dos Goytacazes, RJ 28013-602, Brazil

Abstract. Cemented carbide (WC-Co) is a powder metallurgy product that exhibits an excellent combination of hardness and fracture toughness. This work aims to evaluate the influence of graphene addition on the densification, hardness and fracture toughness of WC-10Co. For this purpose, 0.1% by weight of few-layer graphene (FLG) was added to a mixture of ultrafine and fine powders of WC and Co, respectively. The samples were sintered by spark plasma sintering (SPS) at 1400 °C, with uniaxial pressure of 40 MPa and a low vacuum atmosphere. Density (Archimedes method, mass/volume ratio, and image analysis), Vickers hardness and fracture toughness (K_{IC} – indentation method) were evaluated. The results showed little influence of graphene on densification and hardness. However, there was a statistically significant increase of 14.9% in K_{IC} compared to the reference sample. It is concluded that graphene can potentially improve the materials' resistance; however, the sintering conditions must be adjusted to highlight the results.

Keywords: Cemented Carbide · Graphene · Vickers hardness · Fracture Toughness

1 Introduction

Cemented carbide is one of the most widespread powder metallurgy products [1]. Basically, it comprises tungsten carbide (WC) particles, which are hard and brittle, and a metallic binder, mainly cobalt (Co), as a soft and ductile phase. This material exhibits an excellent combination of mechanical properties, such as hardness and fracture toughness [2, 3]. For this reason, cemented carbide has been successfully used as a tool material in forming, machining, and mining processes for over 100 years. Garcia et al. [1] mention that about 65% of cemented carbide production is destined for machining and 15% for mining. Bobzin [4] cites that cemented carbide occupies around 53% of the global cutting tool market, high-speed steel 20%, ceramics, PcBN and PCD 21%, and cermets 6%.

O. F. Farías Fuentes et al. (Eds.): CIBIM 2024, *Proceedings of the XVI Ibero-American Congress of Mechanical Engineering*, pp. 230–243, 2026.
https://doi.org/10.1007/978-3-032-22823-9_17

This material is in constant development. First with the addition of refractory carbides (TiC or Ti(C,N)), then with the application of coatings utilising chemical or vapour deposition (CVD or PVD), functionally gradient material [5, 6], and now studies involving Co replacement due to scarcity [7], and graphene addition [8–11]. Graphene is commonly referred to as the mother of all graphitic forms. It is a monolayer of carbon atoms bonded by sp^2 hybridisation, arranged in a 2D hexagonal pattern, resembling a honeycomb lattice. A monolayer of carbon atoms may form a 0D Buckminster fullerenes, a 1D carbon nanotube (CNT), and 3D graphite, which is the stacking of graphene sheets, in general more than 10 layers. Goswami et al. [12] and Wang et al. [13] mention that graphene is not only a material with extraordinary properties, but also a nanomaterial additive with the possibility to increase the performance of the conventional materials, such as metals, polymers, and ceramics. In general, the addition of graphene can improve mechanical and lubrication properties.

Su et al. [8] studied the addition of few-layer graphene (up to 10 layers) at mass contents of 0.05% to 0.20% in WC-6Co. They observed a decrease in hardness values with higher graphene content. However, the rupture strength, fracture toughness, and thermal conductivity were improved for the 0.05% content. Sun et al. [9] add 0.1% in mass of graphene monolayer and multilayer, as a reinforcement phase in WC-15TiC-6Co. They noted that multilayer graphene allowed higher hardness and fracture toughness values than monolayer graphene and reference samples. Regarding fracture toughness, the graphene samples presented similar values; however, higher than the reference. He et al. [14] manufactured WC-11Co with gradient of graphene, WC grain size gradient, and consequently Co gradient from the surface. The results showed high hardness and fracture toughness on the sample surface.

Zhang et al. [10] studied the addition effect of the 0.1 wt.% graphene oxide (GO) on the mechanical and tribological properties of WC-Co. The results showed that GO addition had little effect on densification and hardness; however, it provided greater fracture toughness, as well as a lower friction coefficient and wear rate. Shang et al. [11] introduced multilayer graphene to improve the mechanical and tribological performance of coarse WC-8Co powders. The graphene addition provided better hardness and fracture toughness, as well as a decrease in friction coefficient and wear rate compared to the sample without graphene addition.

The work aims to investigate the influence of few-layer graphene on the density, hardness, and fracture toughness of cemented carbide (WC-10Co) with a mostly ultrafine particle size, sintered by spark plasma sintering (SPS).

2 Materials and Methods

The materials used were ultrafine powders of WC (average particle size of 0.35 μm, density of 15.63 g/cm^3, Shangai Xinglu Chemical Co.), fine powders of Co (average particle size of 1.2 μm, density of 8.8 g/cm^3, Umicore-Cobalt & Specialty Materials), and few-layer graphene – FLG (thickness up to 5 nm, lateral dimension of 1–5 μm, number of layers 1 to 10, density of 2.25 g/cm^3, CTNano).

The powders were weighted using a precision scale with a 0.001 g resolution (model 200 CE, Marte) to obtain samples with 0.1 wt.% addition of FLG in WC-10Co, named

WC-10Co+0.1FLG; and without graphene addition, WC-10Co, also called reference. The graphene was deagglomerated in cyclohexane for one hour in the ultrasonic device (Cristófoli). Then, the powders were mixed in a high-energy ball mill (8000 Mixer/Mill, SPEX), at 1500 rpm, with a ball milling ratio of 10:1 (ball:powder) for 2 h. Next, the powders were dried in a laboratory drying oven (NL80/42, New Lab) at 70 °C for 24 h. Finally, the powders were placed in a graphite die (grade MBIS60X, Morganite) to be sintered in a spark plasma sintering machine (SPS 211 LX, Dr. Sinter). The sintering conditions were heating rate of 77 °C/min, uniaxial pressure of 40 MPa, dwell temperature of 1400 °C, dwell time of 10 min, vacuum around 10 Pa and DC pulse pattern of 12 On 2 Off lasting 3.3 ms/pulse.

After the sintering operation, the samples were sectioned with a precision saw (CPT25, Teclago) and a diamond wheel (15HC, Buehler). One section was used for density evaluation, and the other to microscopy and mechanical tests.

Experimental density (ρ_{exp}) was conducted based on the ASTM B962-23 [15] standard, following the Archimedes' principle and using a precision scale (model 200 CE, Marte) and its hydrostatic kit, Eq. (1). The samples were dried in a laboratory drying oven (NL80/42, New Lab) at 110 °C for 24 h, and then, the dry mass (m_1) was measured. Next, the samples were boiled in distilled water for 2 h, and the masses of the samples impregnated and suspended in water, respectively (m_2) and (m_3), were measured.

$$\rho_{exp} = \frac{m_1}{m_2 - m_3}\rho_{H2O} \tag{1}$$

Equation (2) calculated the distilled water density (ρ_{H2O}) as function of temperature in °C [16]. During the suspended mass step, the water temperature was measured with a spit digital thermometer (TP300, Liba).

$$\rho_{H2O} = 1.0017 - 0.0002235T \tag{2}$$

The experimental density (ρ_{exp}) was also evaluated considering mass (m) and volume (v), Eq. (3). The last was determined by an outside micrometre 0–25 mm with a resolution of 0.001 mm (110.284, DIGIMESS).

$$\rho_{exp} = \frac{m}{v} \tag{3}$$

The theoretical density (ρ_{th}) was determined by the inverse rule of mixtures [17], Eq. (4).

$$\frac{1}{\rho_{th}} = \frac{wt\%WC}{\rho_{WC}} + \frac{wt\%Co}{\rho_{Co}} + \frac{wt\%FLG}{\rho_{FLG}} \tag{4}$$

The relative density (ρ_{rel}), which is a percentage of the solid theoretical density (porosity-free), was obtained based on Eq. (5).

$$\rho_{rel} = \frac{\rho_{exp}}{\rho_{th}} \tag{5}$$

The samples were prepared metallographically before to microscopic analysis and mechanical testing. They were mounted on phenolic thermosetting resin using a manual compression-mounting machine (EFD 30, Fortel). The samples were ground and polished in a semi-automatic grinder-polisher (PFLDV, Fortel) using nylon clothes (PSA, Bueler) on flat and rounded glass plates with diamond paste with particle sizes of 100, 70, 50, 30, 15, 6, 1 e ¼ μm.

Images recording and chemical mapping of samples were made using scanning electron microscopy – SEM (Quanta FEG650, FEI) equipped with an energy-dispersive spectrometer – EDS (Quantax, Bruker). Image analyses of two random regions of the samples were performed at magnifications of 2000, 5000, and 10000x to evaluate solid phase fraction and, consequently, the porosity. For this purpose, the ImageJ software (IJ1.54g, NIH) was used. The crop function was selected to limit the analysis area; additionally, the images were processed in binary function to facilitate the identification of the solid phase and porosity. Finally, the fraction area analysis was performed.

To evaluate Vickers hardness and fracture toughness (K_{IC}) by indentation, a Vickers hardness tester (400.316, DIGIMESS) was used with a load of 10 kgf. The hardness test was based on the ASTM C1327-15 [18] standard and the Vickers hardness number (HV) was calculated in agreement with Eq. (6), in which P is the load in kgf and d is the average diagonal length in mm.

$$HV = 1.8544\frac{P}{d^2} \tag{6}$$

K_{IC} in MPa.m$^{1/2}$ was estimated based on literature [19, 20], Eq. (7), in which E is the elasticity modulus in GPa, H is the Vickers hardness value in GPa, P is the load in N, and c is the crack length in m. The load was adjusted so that the crack length was higher than the average diagonal length of the impressions.

$$K_{IC} = 0.016\left(\frac{E}{H}\right)^{\frac{1}{2}}\frac{P}{c^{\frac{3}{2}}} \tag{7}$$

The elasticity modulus of the sample E_a was calculated based on the rule of mixtures [17], Eq. (8). Where, $E_{WC} = 682.5$ GPa [21], $E_{Co} = 211$ GPa [22], $E_{FLG} = 750$ GPa [23, 24] and f is the volume fraction.

$$E_\alpha = E_{WC}.f_{WC} + E_{Co}.f_{Co} + E_{FLG}.f_{FLG} \tag{8}$$

All quantitative data generated underwent variance analysis. The assumptions made were that the sample data reflected the behaviour of a population with a normal distribution and that systematic errors were eliminated.

3 Results and Discussion

Figures 1(a) and (b) show SEM images of WC-10Co and WC-10Co+0.1FLG samples. WC grains in grey contrast with polygonal features and porosity in black contrast can be seen. Chemical analysis by EDS confirms the presence of the element W, which is related to WC in blue, and the metallic binder Co in orange.

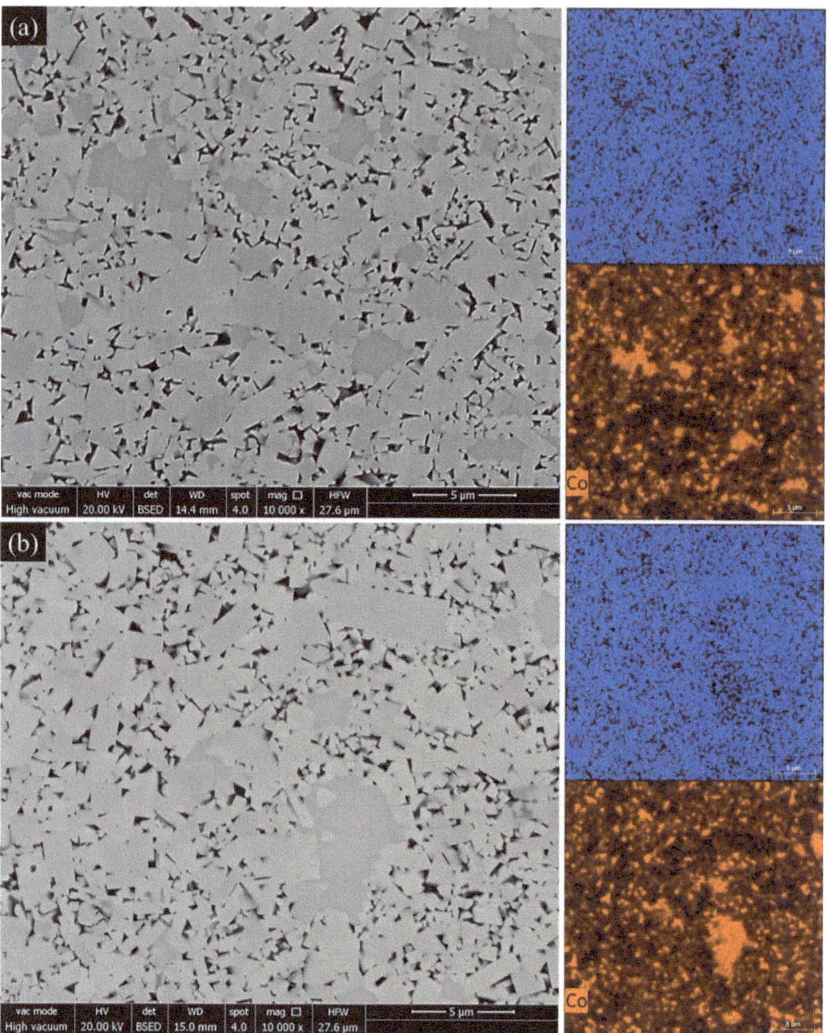

Fig. 1. (a) WC-10Co (reference), WC grains in grey, porosity in black, EDS identifying W in blue and Co in orange; (b) WC-10Co+0.1FLG, WC grains in grey, porosity in black, EDS identifying W in blue and Co in orange. Source: own production.

Figure 2 shows a representative image of few-layer graphene after it has been exfoliated by ultrasonication. A very thin layer can be seen compared to a large lateral dimension. For an addition of 0.1% in mass of FLG in cemented carbide, finding it in the microstructure is very difficult.

Relative density results, considering Archimedes' principle, suggested that the samples were completely densified. WC-10Co obtained $100 \pm 1{,}08\%$ and WC-10Co + 0.1FLG $100 \pm 0.5\%$.

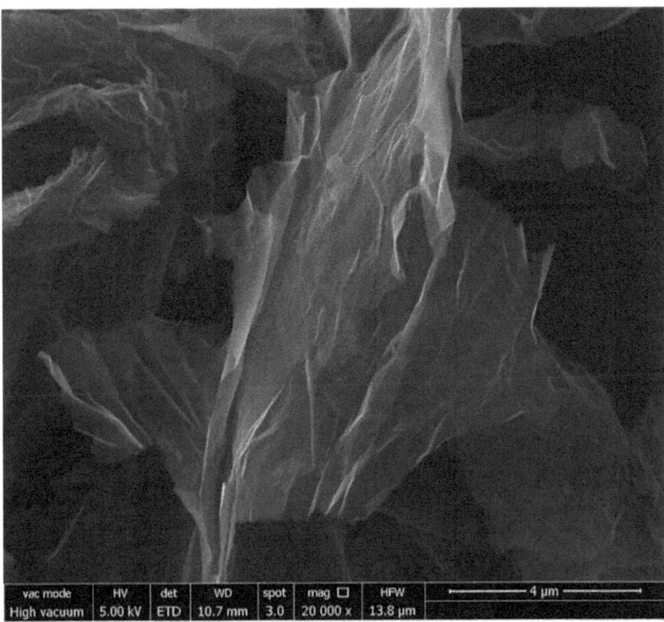

Fig. 2. Representative image of few-layer graphene (FLG) after exfoliation by ultrasonic device. Source: own production.

Table 1 presents the analysis of variance (ANOVA) of raised data from relative density tests based on mass/volume ratio for a statistical significance level (α) of 5%. The Treatment variable indicates if there is a difference between WC-10Co and WC-10Co+FLG samples on the output variable, SS represents the residual sum of squares, DF degree of freedom, MS mean square (or variance), F calculated value for Fisher's distribution, and p probability or significance level calculated.

Table 1. ANOVA for relative density data.

Effect	SS	DF	MS	F	p
Intercept	196408.6	1	196408.6	7950311	0.00
Treatment	6.0	1	6.0	244	0.00
Error	0.4	18	0.02		

Source: own production

The statistical analysis indicates a significant difference between the samples for the output variable evaluated, as the significance level calculated (p) was lower than the criteria (α) used in the hypothesis test.

Figure 3 illustrates the effect of treatment (graphene addition or not) on relative density. The percentage difference in relative density between the treatment samples is 1.1%. According to the analysis, WC-10Co is statistically denser than that with graphene addition.

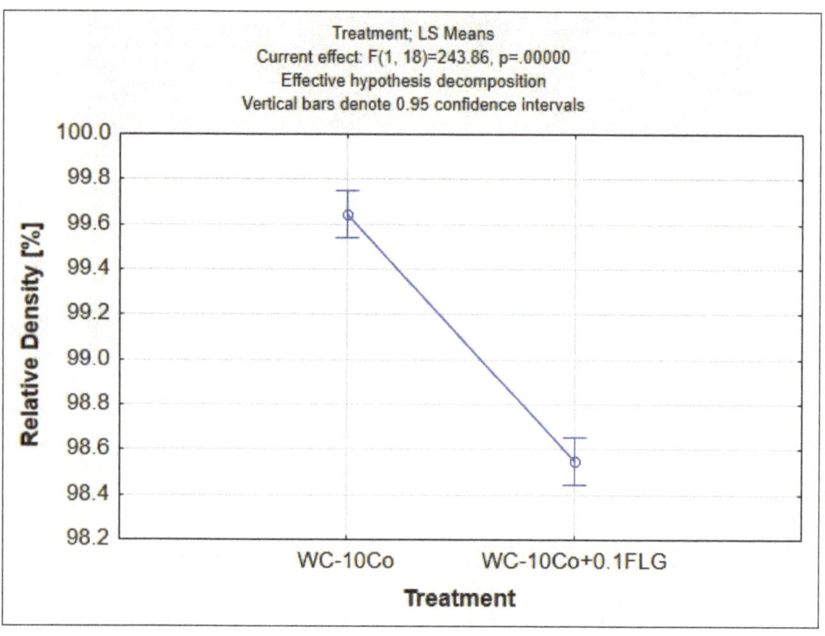

Fig. 3. Effect of treatment on the relative density. Source: own production.

Figures 4(a) and (c) present illustrative SEM images at 5000x magnification of samples to evaluate solid phase fraction and porosity. Figures 4(b) and (d) show binary processed images, in which black and white colour represent the solid phase and the porosity, respectively.

Table 2 presents ANOVA for the image processed data for (α) of 5% regarding solid phase fraction.

In this case, the statistical analysis (Table 2) indicates that no significant difference exists between the samples ($p > \alpha$) for solid phase fraction and, consequently, for porosity.

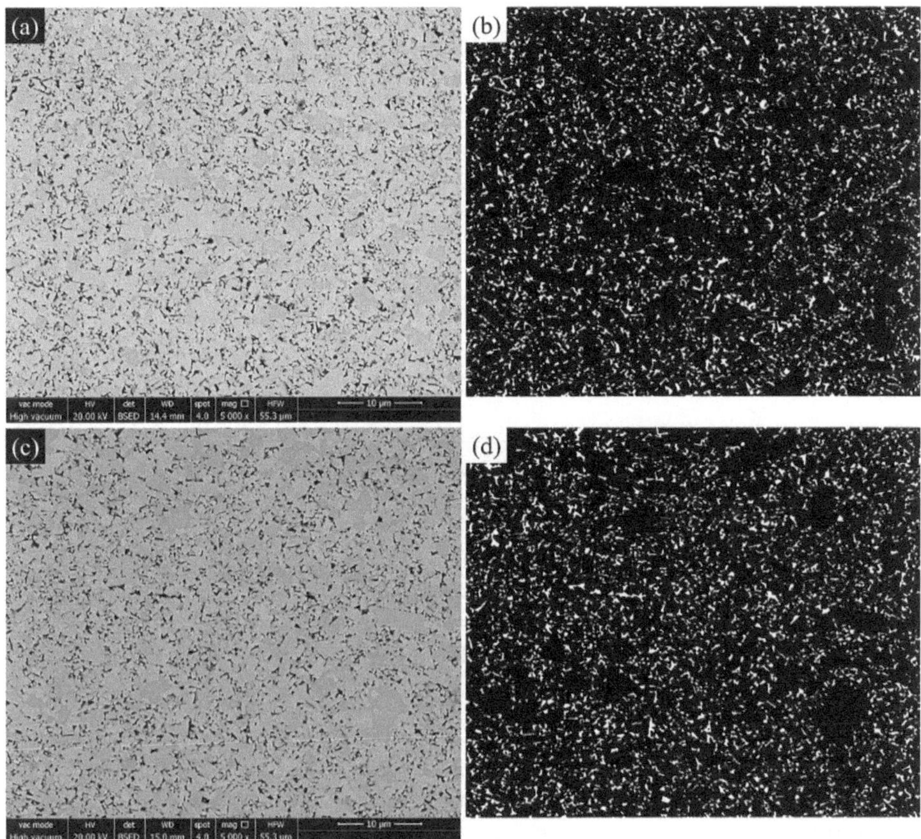

Fig. 4. (a) WC-10Co (reference) SEM image; (b) WC-10Co binary processed image; (c) WC-10Co+0.1FLG SEM image; (d) WC-10Co+FLG binary processed image. Source: own production.

Table 2. ANOVA for image processed data.

Effect	SS	DF	MS	F	p
Intercept	96341.15	1	96341.15	89443.76	0.00
Treatment	0.48	1	0.48	0.44	0.52
Error	10.77	10	1.08		

Source: own production

Figure 5 shows the effect of treatment (materials) on the solid phase fraction, which has a similar meaning to relative density.

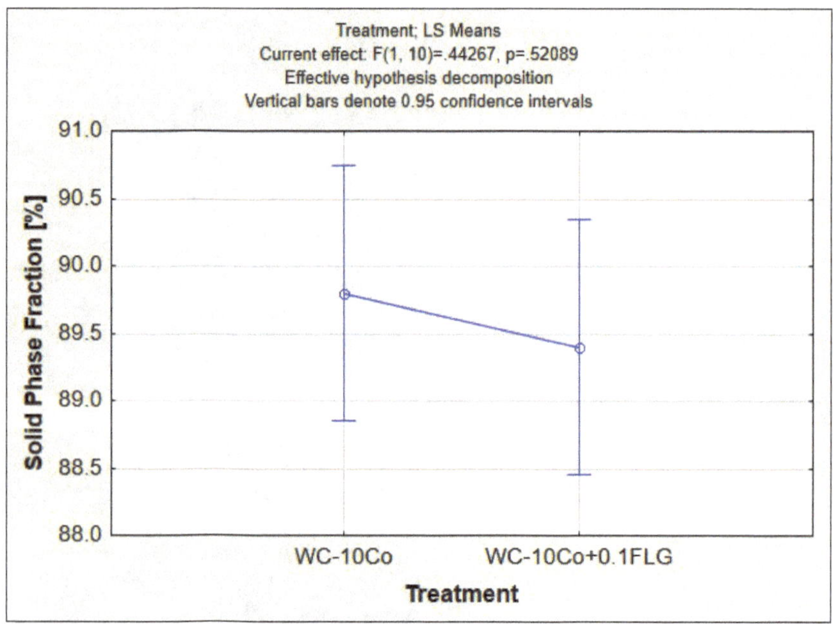

Fig. 5. Effect of treatment on the solid phase fraction. Source: own production.

For Fig. 5 results, the percentage difference between the samples is only 0.4%, with no statistically significant difference. The image processing analysis showed more sensitivity in detecting solid phase fraction or porosity. The results indicated around 10% porosity in the samples, whereas the previous technique indicated between 0.3 and 1.5%.

It is understood that the analysis technique influences the evaluated outcome. In Archimedes' principle the sources of error may be associated with the procedure to measure humid mass, which is subjective and depends on the fluid impregnation (viscosity) in the porous channels. In the same way, the suspended mass measurement depends on the kit's hydrostatic quality, air bubbles and contact among system parts elimination, as well as the operator's skills to put and remove the samples without impact to the basket. The main error source in the experimental density determination by mass/volume ratio is related to the geometric deviations of the samples. In the processing image technique, specifically for this work, the Co fraction may alias with the porosity fraction, influencing the mean results and dispersion. On the other hand, independent of the analysis technique, it is noted that the amount of graphene added had little influence on densification, which is in agrees with the literature. Sun et al. [9] noted an increase of just 1% in relative density when 0.1% of multilayer graphene was added to WC-15TiC-6Co. Zhang et al. [10] cited that 0.1% of GO had little effect on WC-Co densification. Shang et al. [11] reported 0.8% and 1.7% improvement in the relative density of WC-8Co samples by adding 0.1% and 0.2% of multilayer graphene, respectively.

Figure 6 presents representative images of indentations performed with a Vickers indenter, load of 10 kgf, with cracks at the tips of the impressions to evaluate the materials' fracture toughness (K_{IC}). According to ASTM C1327-15 [18], the presented indentations are accepted.

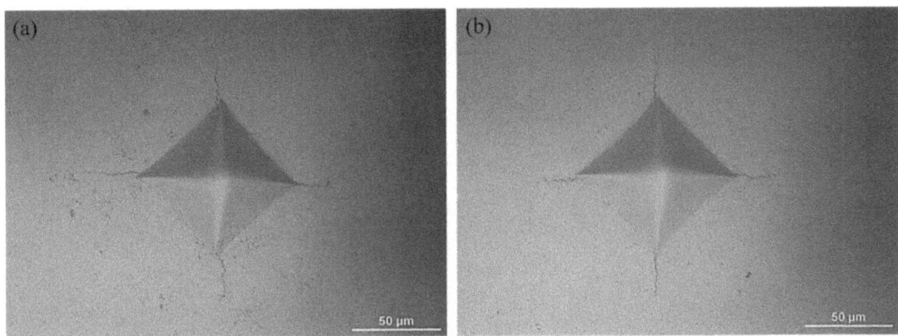

Fig. 6. Representative SEM images of Vickers indentation. (a) WC-10Co sample; (b) WC-10Co+0.1FLG sample. Source: own production.

Table 3 presents ANOVA for the Vickers hardness number (HV10) data for the evaluated samples, considering (α) of 5%.

Table 3. ANOVA for Vickers hardness number (HV10) data.

Effect	SS	DF	MS	F	p
Intercept	54296902	1	54296902	20463.28	0.00
Treatment	2960	1	2960	1.12	0.30
Error	53068	20	2653		

Source: own production

The results in Table 3 show no statistical difference between the evaluated materials (Treatment), since $p > \alpha$.

Figure 7 shows the effect of materials on the Vickers hardness number (HV). Adding 0.1 wt.% FLG caused a 1.5% decrease in the Vickers hardness number. Su et al. [8] noted that adding multilayer graphene from 0.05% to 0.20% in WC-6Co caused a continuous decrease in hardness, although not significant. Sun et al. [9] reported a negligible influence on the Vickers hardness when adding 0.1% of monolayer graphene in WC-15TiC-6Co. When they added 0.1% of multilayer graphene, they noted a 6.2% increase in hardness values. Zhang et al. [10] mentioned that adding 0.1% GO had little influence on Vickers hardness, decreasing the outcome by 4%. Shang et al. [11] noted an increase in Vickers hardness of around 7% when adding 0.1% multilayer graphene in WC-8Co, with WC particle size of 25 µm.

Table 4 has ANOVA with (α) of 5% for the fracture toughness (K_{IC}) data.

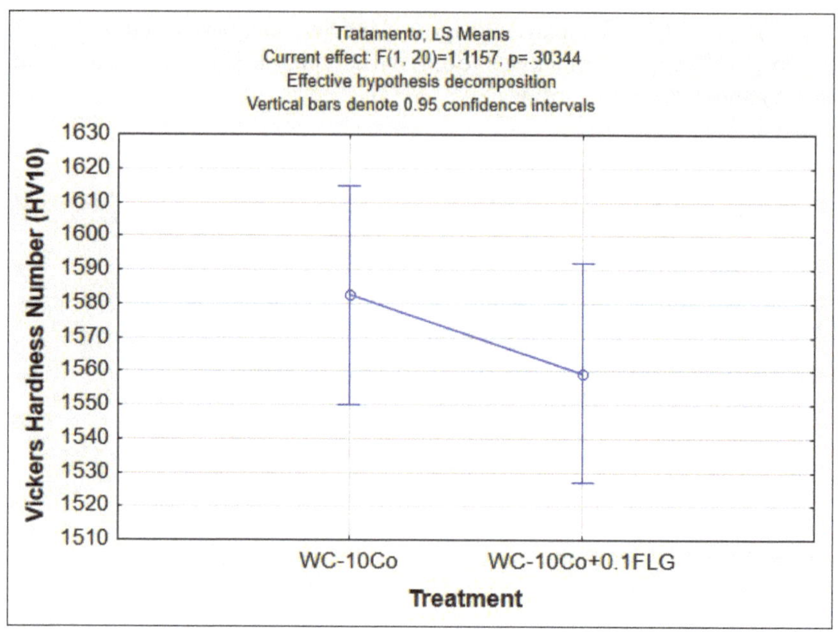

Fig. 7. Effect of treatment on Vickers hardness number (HV). Source: own production.

Table 4. ANOVA for fracture toughness (K_{IC}) data.

Effect	SS	DF	MS	F	p
Intercept	5097.71	1	5097.71	5097.41	0.00
Treatment	24.39	1	24.39	28.57	0.00
Error	17.07	20	0.85		

Source: own production

From the statistical analysis (Table 4) of K_{IC} data, it is observed that there is a significant difference between WC-10Co and WC-10Co+0.1FLG samples, since $p < \alpha$.

Figure 8 presents the effect of treatment (materials evaluated) on the response. It is noted that the addition of 0.1% FLG caused an average increase of 14.9% compared to the reference sample. Su et al. [8] mentioned a rise of 29% in K_{IC} with conservation of hardness values when adding 0.05% multilayer graphene in WC-6Co. Sun et al. [9] observed an increase of 27.4% and 24.5% on K_{IC} when adding 0.1% monolayer and multilayer graphene, respectively, in WC-15TiC-6Co. Zhang et al. [10] noted a 29% increase in K_{IC} when adding 0.1% GO in WC-Co. Shang et al. [11] reported a 7% increase in K_{IC} with a similar increase in hardness by adding 0.1% multilayer graphene in WC-8Co with coarse WC particle size.

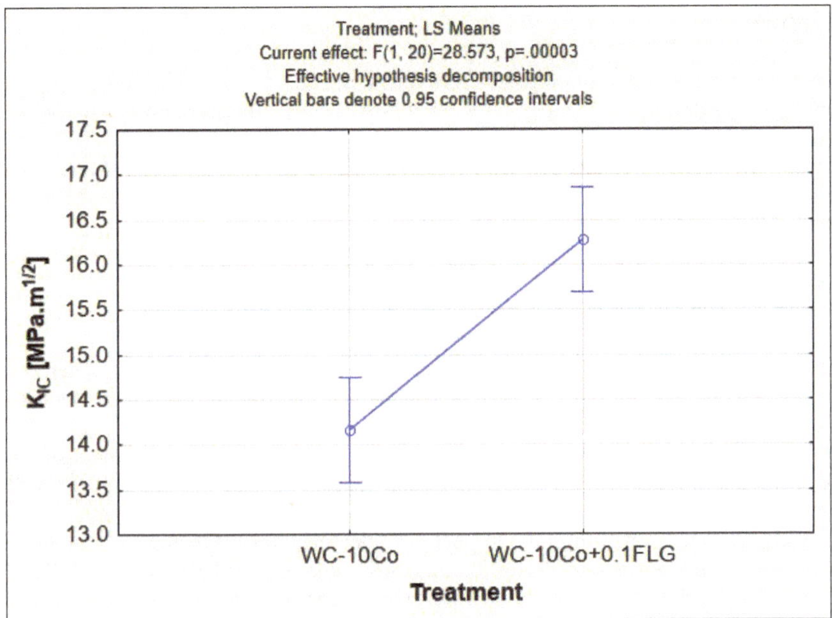

Fig. 8. Effect of treatment on fracture toughness (K_{IC}). Source: own production.

The positive effect of graphene on fracture toughness may be associated with its capacity to act as matrix reinforcement, preventing grain growth, anchoring grain boundaries, increasing the elasticity modulus, and making crack propagation difficult by consuming fracture energy. The toughening mechanisms commonly reported in composites with graphene addition are crack bridging, branching, deflection and pull-out [8, 12, 25].

4 Conclusions

Regardless of the relative density analysis technique, the results indicate that the addition of few layers graphene (<10) in the proportion of 0.1 wt.% has little influence on the sample densification. This is probably because graphene acts to hide or inhibit the WC grain growth.

In the same way, the graphene addition did not have a statistical influence on the hardness values for the evaluated conditions, and it is common to find a negative effect of the graphene addition in WC-Co in the literature for this property.

However, analysing fracture toughness results, it was observed that the graphene had a statistical influence, increasing the outcome by an average of 14.9%. This increase in the property is commonly reported in the literature for graphene addition of 0.1 wt.% in WC-Co while maintaining high hardness values.

Acknowledgements. The authors thank to FAPES (Grants 1066/2022; 970/2024), CNPq (Grant 405483/2021-0), and CAPES for supporting research.

References

1. García, J., Ciprés, V.C., Blomqvist, A., Kaplan, B.: Cemented carbide microstructures: a review. Int. J. Refractory Metals Mater. **80**, 40–68 (2019). https://doi.org/10.1016/j.ijrmhm.2018.12.004
2. Upadhyaya, G.S.: Materials science of cemented carbide – an overview. Mater. Des. **22**, 483–489 (2001). https://doi.org/10.1016/S0261-3069(01)00007-3
3. Sheikh, S., et al.: Fracture toughness of cemented carbide: testing method and microstructural effects. Int. J. Refractory Metals Mater. **49**, 153–160 (2015). https://doi.org/10.1016/j.ijrmhm.2014.08.018
4. Bobzin, K.: High-performance coatings for cutting tools. CIRP J. Manuf. Sci. Technol. **18**, 1–9 (2017). https://doi.org/10.1016/j.cirpj.2016.11.004
5. Nomura, T., Moriguchi, H., Tsuda, K., Isobe, K., Ikegaya, A., Moriyama, K.: Material design method for the functionally graded cemented carbide tool. Int. J. Refractory Metals Mater. **17**, 397–404 (1999). https://doi.org/10.1016/S0263-4368(99)00029-3
6. Frykholm, R., Andrén, H.O.: Development of the microstructure during gradient sintering of a cemented carbide. Mater. Chem. Phys. **67**, 203–208 (2001). https://doi.org/10.1016/S0263-4368(99)00029-3
7. Tarraste, M., et al.: Ferritic chromium steel as binder metal for WC cemented carbides. Int. J. Refractory Metals Mater. **73**, 183–191 (2018). https://doi.org/10.1016/j.ijrmhm.2018.02.010
8. Su, W., Li, S., Sun, L.: Effect of multilayer graphene as a reinforcement on mechanical properties of WC-6Co cemented carbide. Ceram. Int. **46**, 15392–15399 (2020). https://doi.org/10.1016/j.ceramint.2020.03.084
9. Sun, J., Huang, Z., Zhao, J.: High-hard and high-tough WC-TiC-Co cemented carbide reinforced with graphene. Mater. Today Commun. **29**, 102841 (2021). https://doi.org/10.1016/j.mtcomm.2021.102841
10. Zhang, X., Zhang, J., Ding, J.: Effect of the additive graphene oxide on tribological properties of WC-Co cemented carbide. Int. J. Refractory Metals Mater. **109**, 105962 (2022). https://doi.org/10.1016/j.ijrmhm.2022.105962
11. Shang, Z., Sun, Z., Wang, Z., Chen, Y., Zhao, W.: Effect of the additive multilayer graphene oxide on the enhanced tribological properties of ultra-coarse WC-Co cemented carbide. Mater. Lett. **360**, 135996 (2024). https://doi.org/10.1016/j.matlet.2024.135996
12. Goswami, S., Glosh, R., Hirani, H., Mandal, N.: Mechano-tribological performance of Graphene/CNT reinforced alumina nanocomposites – review and quantitative insights. Ceram. Int. **48**, 11879–11908 (2022). https://doi.org/10.1016/j.ceramint.2022.02.214
13. Wang, X., Zhao, J., Cui, E., Song, S., Liu, H., Song, W.: Microstructure, mechanical properties and toughening mechanisms of graphene reinforced Al_2O_3-WC-TiC composite ceramic tool material. Ceram. Int. **45**, 10321–10329 (2019). https://doi.org/10.1016/j.ceramint.2019.02.087
14. He, G., Liang, Y., Chen, P.: Fabrication of cemented carbides with bidirectional gradient structure via graphene deposition to simultaneously enhance surface hardness and toughness. Ceram. Int. **50**, 23643–23655 (2024). https://doi.org/10.1016/j.ceramint.2024.04.088
15. American Society for Testing and Materials. ASTM International B962-23: Standard test methods for density of compacted or sintered powder metallurgy (PM) products using Archimedes' principle. West Conshohocken: ASTM (2023)
16. German, R.M.: Sintering Theory and Practice. Wiley, New York (1996)
17. German, R.M., Park, S.J.: Mathematical Relations in Particulate Materials Processing. Wiley, New York (2008)
18. American Society for Testing and Materials. ASTM International C1327-15: Standard test method for Vickers indentation hardness of advanced ceramics. West Conshohocken: ASTM (2019)

19. Meyers, A.M., Chawla, K.K.: Mechanical Behavior of Materials. Cambridge University Press, Cambridge (2009)
20. Chychko, A., García, J., Ciprés, V.C., Holmström, E., Blomqvist, A.: HV-K_{IC} property charts pf cemented carbide: A comprehensive data collection. Int. J. Refractory Metals Mater. **103**, 105763 (2022). https://doi.org/10.1016/j.ijrmhm.2021.105763
21. MatWeb. Tungsten Carbide, WC. https://www.matweb.com/search/DataSheet.aspx?MatGUID=e68b647b86104478a32012cbbd5ad3ea&ckck=1. Accessed 20 May 2025
22. MatWeb. Cobalt, Co. https://www.matweb.com/search/DataSheet.aspx?MatGUID=4602449fc566494ab3efa286e8827c99. Accessed 20 May 2025
23. Frank, I.W., Tanenbaum, D.M., van der Zandle, A.M., McEuen, P.L.: Mechanical properties of suspended graphene sheets. J. Vaccum Sci. Technol. B **25**, 2558–2561 (2007). https://doi.org/10.1116/1.2789446
24. Mohan, V.B., Lau, K., Hui, D., Bhattacharyya, D.: Graphene-based materials and their composites: a review on production, applications and product limitations. Compos. B Eng. **142**, 200–220 (2018). https://doi.org/10.1016/j.compositesb.2018.01.013
25. Boniecki, M., et al.: Alumina/zirconia composites toughened by the addition of graphene flakes. Ceram. Int. **43**, 10066–10070 (2017). https://doi.org/10.1016/j.ceramint.2017.05.025

Feasibility of Using Acoustic Emission Signals to Monitor Abrasive Condition in Robotic Sanding Operations

Pablo Sanhueza-Carvajal, Ricardo Alzugaray-Franz, and Eduardo Diez-Cifuentes(✉)

Departamento de Ingeniería Mecánica, Universidad de La Frontera, Temuco, Chile
eduardo.diez@ufrontera.cl

Abstract. In the woodworking industry, sanding is essential for achieving smooth surfaces and high-quality finishes. However, abrasive tool performance degrades progressively due to wear mechanisms that depend on material properties and operating conditions. Despite advancements in robotic sanding, which have improved the efficiency and uniformity of the process, the development of an effective solution to automate the evaluation of the abrasive condition remains an open challenge.

This study investigates a robotic sanding operation using a rotary sanding tool to machine medium-density-fiberboard (MDF) and solid oak specimens. The analysis focuses on the measurement of the acoustic emission (AE) and motor current signals as process variables for indirect estimation of the abrasive condition. The findings reveal that amplitude-based characteristics extracted from both signals are effective for monitoring the evolution of the sandpaper performance over time.

Keywords: robotic sanding · abrasive wear monitoring · tool condition monitoring · acoustic emission · wood sanding

1 Introduction

Sanding is a subtractive machining process that smooths rough surfaces using a coated abrasive, removing imperfections and preparing the surface for subsequent finishing operations [1]. Coated abrasives are composed of mineral grains bonded to a backing material, acting as multiple cutting edges. The characteristics of the backing material determine the strength, stiffness, and material removal capacity of the abrasive. The effectiveness and tool life of the abrasive are influenced not only by the properties of the wood [2], but also by the morphology and spatial distribution of the abrasive grains, which directly affect the cutting forces and local velocities acting on individual grains during the process [3].

In the wood processing industry, the condition of the abrasive is typically evaluated manually by the machine operator. Since abrasive wear has a direct impact on product

O. F. Farías Fuentes et al. (Eds.): CIBIM 2024, *Proceedings of the XVI Ibero-American Congress of Mechanical Engineering*, pp. 244–253, 2026.
https://doi.org/10.1007/978-3-032-22823-9_18

quality, monitoring the evolution of abrasive performance can significantly enhance process efficiency [3].

In the 1990s, Matsumoto and Murase [4] investigated the influence of sanding time, grit size, and applied pressure on acoustic emission (AE) signals. To evaluate process efficiency, they used the material removal rate (MRR), defined as the amount of material removed within a given period, as a key performance metric. Their findings indicate that the MRR increases with larger grit sizes but decreases over time as sanding progresses. This behavior can be related with the acoustic emission count rate, which declines as abrasive grains become worn and increasingly saturated with dust, reducing their ability to penetrate the workpiece surface.

Carrano et al. [5], on the other hand, implemented a monitoring system for abrasive belt saturation during wood sanding by combining machine vision techniques with neural networks. Their approach enabled the estimation of abrasive tool life with a confidence level exceeding 90%.

Saloni and Lemaster [6, 7] further developed an online monitoring system designed to detect abrasive belt wear and clogging. The system integrated optical sensing, AE monitoring, and temperature measurement to enable real-time assessment of abrasive condition. Sensor data were used to schedule periodic cleaning operations, thereby extending the effective service life of the abrasive. Moreover, the system supported the timely replacement of the sanding belt when excessive grain wear was detected, preventing surface quality degradation.

Recent research has focused on understanding abrasive wear and failure mechanisms, as well as on optimizing the sanding process through offline optical measurement techniques. Zhang [8,9] investigated the wear behavior of abrasive belts and its influence on material removal during the sanding of medium-density fiberboard (MDF). The study identified grain fracture and abrasion as the predominant wear modes. In a related contribution, Du [10] reported two principal failure mechanisms affecting abrasive belts in MDF sanding: progressive grain wear, which reduces the maximum protrusion height of abrasive particles, and dust accumulation between grains, which leads to loading and eventual belt blockage.

The adoption of industrial robots in sanding operations has increased in recent years due to their ability to improve both efficiency and process consistency compared to manual sanding. Several recent developments include end-of-arm tooling specifically designed for robotic sanding applications [11, 12]. Despite this progress, the automated assessment of abrasive condition remains largely underexplore. Furthermore, existing studies on abrasive condition monitoring has focused predominantly on belt sanding processes, with little attention given to rotary sanding tools.

This study evaluates the feasibility of using acoustic emission (AE) signals and the current measurements from the end-effector motor as process variables to monitor abrasive condition in robotic sanding with a rotary tool. The aim is to employ these signals to enable timely abrasive maintenance, through cleaning in cases of clogging, or replacement in cases of wear, thereby improving both the autonomy and efficiency of the process.

2 Methodology

2.1 Experimental Design

The experimental design involved two types of workpiece materials: medium-density fiberboard (MDF) and solid oak wood. MDF was selected due to its structural and compositional characteristics, which produce a high volume of fine dust particles that adhere to the abrasive surface, thereby accelerating abrasive loading. In contrast, oak wood, being significatively harder than MDF, induces abrasive grain wear with minimal dust accumulation, allowing the analysis of wear mechanisms in the absence of substantial loading effects.

The specimens used in the experiments had nominal dimensions of 180 mm in length, 110 mm in width, and 18 mm in thickness. They were held in place using a vacuum fixture. The experimental factors summarized in Table 1 were determined based on the results of preliminary trials.

Table 1. Experimental factors for loading and abrasive wear tests.

Factor	Abrasive loading tests	Abrasive wear tests
Spindle speed	3000 RPM	3000 RPM
Feed rate	10 mm/s	10 mm/s
Normal force	10 N	10 N
Abrasive grit size	P150	P220

2.2 Experimental Setup

A custom end-effector was designed and fabricated for the experiments. It was driven by an AKM-23D servomotor with a maximum torque of 1.1 Nm and mounted on the wrist of an industrial robot as illustrated in Fig. 1. The AKD-P00306 servo drive was configured to control the spindle speed and provide an analog output signal corresponding to the motor current, which was used as an indirect measure of the sanding torque.

The tool holder and dust extraction system were 3D-printed using PLA, resulting in a lightweight structure. The sanding trajectory was executed using a force control strategy to achieve uniform normal force throughout the operation.

2.3 Data Acquisition, Communication, and Control System

The process variables – acoustic emission (AE), normal sanding force, and motor current – were measured using a National Instruments PXI-1062Q platform. The AE signal was acquired at a sampling rate of 2 MHz using a NI-6132 module, while the normal force and motor current signals were sampled at 100 kHz using the NI-4472B module.

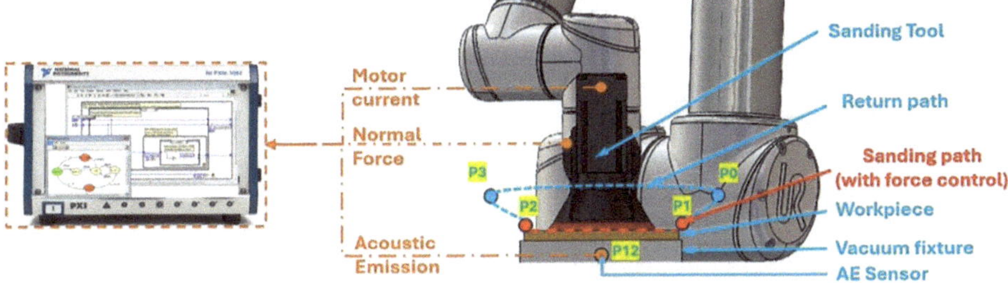

Fig. 1. Schematic of the experimental setup, including the sanding station, tool trajectories, measured signals and data acquisition system.

The data acquisition routine is initiated by a digital signal sent from the industrial robot when the sanding tool trajectory passes point P12, as indicated in Fig. 1. This signal acts as a trigger for both the NI-4472B and NI-6132 modules, enabling simultaneous data acquisition over a one-second interval. This duration corresponds to approximately fifty complete revolutions of the abrasive disc over both the workpiece and the AE sensor.

Communication between the control computer and the industrial robot was implemented via the Real-Time Data Exchange (RTDE) interface provided by Universal Robots. A queued message handler (QMH) architecture was developed in LabVIEW to coordinate communication tasks, manage image acquisition, and record the mass of the processed specimens.

2.4 Data Processing and Analysis

The acoustic emission (AE) signal, motor current, and normal force were stored on the NI PXI-1062Q system in TDMS format and subsequently processed in LabVIEW. A fifth-order Butterworth bandpass filter was applied to the signals, with cutoff frequencies set at 50 kHz and 400 kHz, to eliminate components outside the operational bandwidth of the AE sensor. The peak-to-peak amplitude of the AE signal was then extracted, and this value, along with the average motor current, was logged in an Excel spreadsheet for further analysis.

Abrasive performance was evaluated using the normalized material removal ratio (MRR), defined as the ratio of the removal rate at a given time (R) to the initial removal rate (R_0), following the approach described by Matsumoto and Murase [4]. The mass of each specimen was measured every five sanding passes, corresponding to 90 s of cutting time.

3 Results and Discussion

3.1 Abrasive Loading Tests

To investigate the effect of abrasive loading, experiments were conducted using two workpiece materials with contrasting characteristics: (1) medium-density fiberboard (MDF), a material that promotes rapid accumulation of fine particles between abrasive grains due to its composition and structure; and (2) oak wood, a harder material with a significantly lower tendency to cause abrasive loading.

Both materials were sanded using abrasive discs of the same grit type and under identical process parameters, as detailed in Table 1.

Abrasive Loading in MDF

To investigate abrasive loading, a total of 125 sanding passes were performed using a P150 grit disc on two MDF specimens, resulting in a total cutting time of 37.5 min. After 35 passes, the first workpiece was replaced due to excessive thickness reduction, which approached the minimum limit for stable vacuum fixation.

Throughout the test, visual inspection of the sanding disc revealed a clear evolution in its surface condition. As shown in Fig. 2(a) and 2(b), the abrasive surface was initially clean and unobstructed. However, by the end of the sanding on the first workpiece, substantial accumulation of material can be observed between the abrasive grains, indicating an advanced loading state.

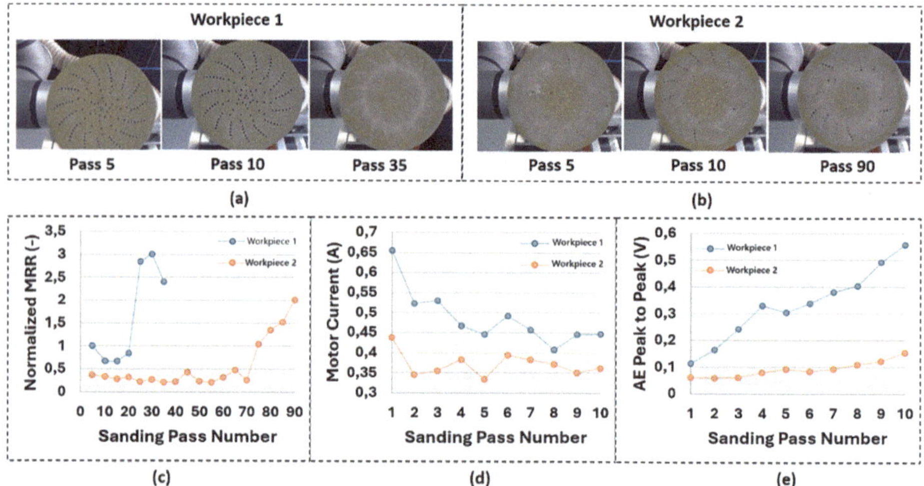

Fig. 2. Evolution of process variables during abrasive loading tests on MDF. (a) and (b) show visual inspections of the sanding disc during the test, (c) material removal ratio (MRR), (d) peak-to-peak amplitude of the acoustic emission (AE) signal, and (e) average motor current.

Figure 2(c) illustrates the evolution of the material removal ratio (MRR) over the entire test. Figure 2(d) and 2(e) capture the early-stage behavior of the process during the first ten sanding passes on each workpiece, showing the peak-to-peak amplitude of the acoustic emission signal and the average motor current, respectively. These initial values offer the most consistent representation of the process, as data from subsequent passes become increasingly scattered due to the inherent density variability within the internal layers of the MDF.

The acoustic emission (AE) signal shows an increasing trend in peak-to-peak amplitude as the number of sanding passes progresses, similar to the behavior observed in the material removal ratio. This increase is more pronounced during the sanding of workpiece 1, indicating greater AE activity for low levels of loading. These results confirm that AE is sensitive to changes in abrasive condition and can serve as an indicator of the loading state of the sanding disc. In contrast, the motor current exhibits a decreasing trend, consistent with a gradual loss of cutting efficiency caused by the accumulation of debris on the abrasive surface. As the disc becomes increasingly loaded, the mechanical resistance during sanding is reduced, resulting in lower current consumption.

Abrasive Loading in Oak

Saturation tests were also performed on solid wood specimens to evaluate the evolution of process variables in a material with a more homogeneous structure and density than MDF. The experiment consisted of 250 consecutive sanding passes distributed across two oak workpieces. The first 150 passes were carried out on workpiece 1, followed by 100 passes on workpiece 2, resulting in a total cutting time of approximately 75 min. Workpiece 1 was sanded using a fresh P150 grit abrasive, whereas workpiece 2 was processed with the same disc after it had been artificially loaded through prior MDF sanding, thereby simulating a saturated abrasive condition.

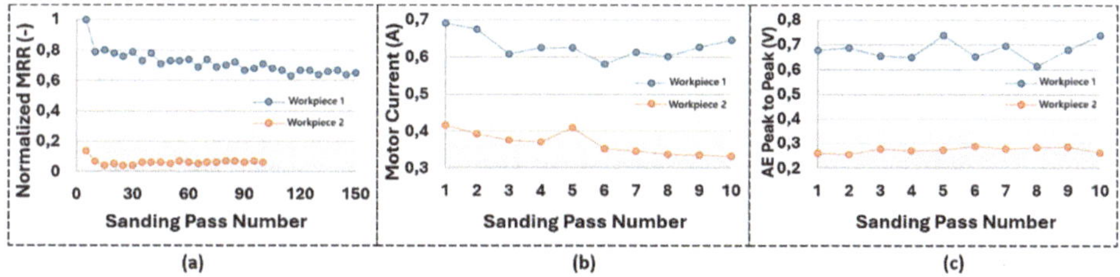

Fig. 3. Evolution of process variables during abrasive loading tests on oak wood. (a) Material removal ratio (MRR), (b) average motor current, and (c) peak-to-peak amplitude of the acoustic emission (AE) signal.

Figure 3 summarizes the main findings from the test. In all cases, the process variables exhibit consistent amplitude trend for both the fresh and preloaded abrasive conditions. Figure 3(a) shows that the material removal ratio with a loaded disc drops to approximately 10% of the value observed with a fresh abrasive, indicating a significant loss in cutting efficiency. The motor current shown in Fig. 3(b), also decreases under the loaded condition, reflecting reduced mechanical resistance as the abrasive surface becomes less effective. Similarly, the acoustic emission (AE) signal in Fig. 3(c), shows diminished activity, further supporting the correlation between abrasive loading and attenuation of amplitude-based features.

These results confirm that the acoustic emission and motor current are sensitive indicators of abrasive condition and can be used to enable timely maintenance actions, such as cleaning, that help restore cutting performance and extend tool service life.

3.2 Abrasive Wear Tests

Wear tests were carried out on solid oak specimens using a P220 grit abrasive. A total of 1400 consecutive sanding passes were performed across three workpieces. As in the abrasive loading tests, each specimen was replaced before reaching the minimum allowable thickness for stable vacuum clamping. The first 300 passes were carried out on workpiece 1, followed by 450 passes on workpiece 2, and the remaining 650 passes on workpiece 3, resulting in a total cutting time of approximately seven hours. The test was concluded once the abrasive disc could no longer reduce the thickness of the third workpiece, indicating a critical loss in cutting capability.

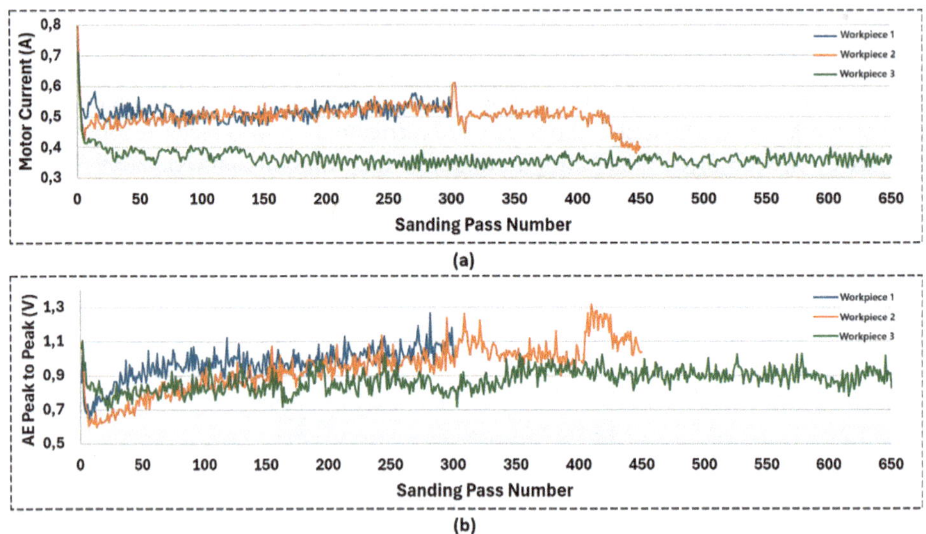

Fig. 4. Evolution of process variables during abrasive wear tests on oak specimens: (a) average motor current, (b) peak-to-peak amplitude of the acoustic emission signal.

Given the extended duration of the test and the previously established correlation between motor current and material removal rate, neither mass measurements nor photographic documentation of the sanding disc were performed during this phase.

During the test, visual inspections of the abrasive disc confirmed the absence of loading. Therefore, the reduction in cutting efficiency is attributed exclusively to abrasive wear. The latter is supported by the evolution of the process variables over time, as shown in Fig. 4.

Figure 4(a) shows that the motor current remains relatively stable during the initial phase of the test, extending over approximately 700 sanding passes. After this period, the signal undergoes a noticeable decline, settling into a second plateau at a lower amplitude, which persists until the end of the test. This shift indicates a substantial loss of cutting capability, suggesting that the abrasive reached the end of its effective service life after approximately 3.5 h of operation.

Figure 5(a) shows that even within the transient region associated with the settling of new workpieces, the motor current does not reach the threshold identified as the end of abrasive life. This observation further supports the reliability of motor current as a valid indicator of abrasive wear.

Regarding the acoustic emission (AE) signal, the peak-to-peak amplitude shown in Fig. 4(b) exhibits a transient phase during the initial passes on each workpiece, likely resulting from the initial surface roughness of the fresh workpiece. Once uniform cutting conditions are established, the signal amplitude shows a slight upward trend, which may be attributed to the progressive reduction in specimen thickness, effectively shortening the distance between the AE source and the sensor. After approximately 1000 sanding passes, the signal stabilizes at a plateau, suggesting that the abrasive disc has reached the end of its effective service life.

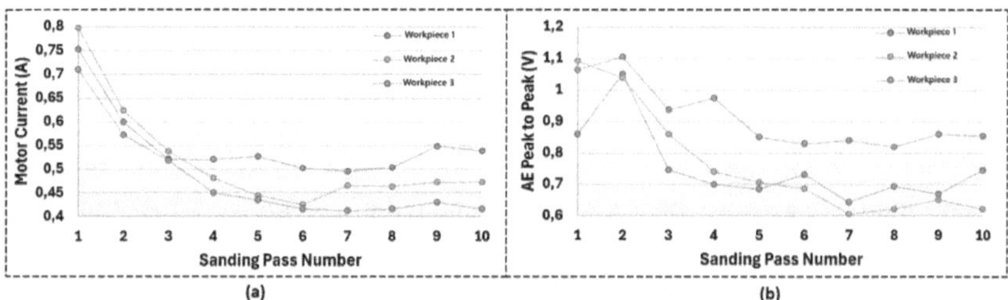

Fig. 5. Evolution of process variables during the initial ten sanding passes of the abrasive wear test on oak: (a) average motor current and (b) peak-to-peak amplitude of the acoustic emission signal.

On the other hand, Fig. 5 presents the evolution of process variables during the first ten sanding passes on each workpiece. These results are representative of finishing operations, where the primary objective is to improve surface quality rather than modify the workpiece dimensions. Excluding the first three passes, which are strongly influenced by the initial surface condition of the material, a worn abrasive is associated with lower motor current consumption. In contrast, the acoustic emission (AE) signal exhibits an increasing trend under abrasive wear conditions, likely due to the generation of new emission sources caused by the friction between the dull abrasive grains and the workpiece surface.

4 Conclusions

This study examined the two main failure mechanisms of abrasive tools: wear and loading. An experimental methodology was designed to isolate the influence of each mechanism and to evaluate the response of two indirect process variables: the acoustic emission (AE) signal and the servomotor current consumption. The findings contribute to the development of an online monitoring system for robotic sanding operations.

Loading experiments were conducted on two different materials, confirming that MDF specimens promote faster abrasive loading compared to harder solid wood such as oak. In both materials, loading was found to negatively impact abrasive performance, reducing the material removal rate, decreasing motor current consumption, and diminishing acoustic emission (AE) activity. This behavior highlights the potential of using these process variables as criteria for abrasive tool reconditioning, ultimately extending tool service life.

On the other hand, the results from the abrasive wear test indicate that motor current decreases progressively as the abrasive tool wears out. In contrast, the acoustic emission

(AE) signal tends to increase under wear conditions, likely due to a shift in the dominant generation mechanism, from cutting sources to frictional interactions caused by dull abrasive grains.

The results presented in this study support the feasibility of implementing an online monitoring system based on indirect process variables, specifically acoustic emission and motor current, to assess the condition of the abrasive tool during robotic sanding operations. Moreover, the proposed methodology can be adapted to other wood types, however, calibration experiments are required to validate the consistency of these findings across different materials.

Although the analysis was performed as a post-processing step, the proposed method is compatible with real-time monitoring systems. Implementing this approach requires the development of an algorithm capable of integrating with the robotic sanding program to continuously acquire, process, and evaluate the selected variables during operation. This would enable real-time tracking of the abrasive condition and support timely decisions for tool reconditioning or replacement.

Acknowledgments. The authors would like to thank the Universidad de La Frontera for promoting and supporting research in intelligent manufacturing technologies. This work was funded by the National Agency for Research and Development ANID (Chile) through the FONDEF projects ID18i10042 and IT21i0069, focused on the development of intelligent robotic sanding processes. The authors also acknowledge the contributions of Maderas Nativas Woodnic and 3M Chile for providing tools and consumables essential to this research.

References

1. Ratnasingam, J.: Sanding process. In: Furniture Manufacturing: A Production Engineering Approach. Springer, Singapore (2022)
2. Sydor, M., Mirski, R., Stuper-Szablewska, K., Rogoziński, T.: Efficiency of machine sanding of wood. Appl. Sci. (Switzerland) **11**(6), Article no. 2860 (2021). https://doi.org/10.3390/app11062860
3. Ratnasingam, J., Reid, H.F., Perkins, M.C.: The productivity imperatives in coated abrasives – Application in furniture manufacturing. Holz Als Roh- Und Werkstoff **57**, 117–120 (1999)
4. Matsumoto, H., Murase, Y.: Acoustic emission characteristics in wood sanding: amplitude distribution of acoustic emission in disc sanding process. J. Faculty Agric. Kyushu Univ. **43**, 257–268 (1998). https://doi.org/10.5109/24270
5. Carrano, A., Vora, B., Sahin, F., Lemaster, R.: Monitoring of abrasive loading for optimal belt cleaning or replacement. For. Prod. J. **57**, 78–83 (2007)
6. Saloni, D.E., Lemaster, R.L., Jackson, S.D.: Process monitoring evaluation and implementation for the wood abrasive machining process. Sensors **10**(11), 10401–10412 (2010). https://doi.org/10.3390/s101110401
7. Saloni, D.E., Lemaster, R., Jackson, S.: Control system evaluation and implementation for the abrasive machining process on wood. BioResources **6**(3), 3116–3131 (2011)
8. Zhang, J., Ying, J., Cheng, F., Liu, H., Luo, B., Li, L.: Investigating the sanding process of medium-density fiberboard and Korean Pine for material removal and surface creation. Coatings **8**(12), Article no. 416 (2018). https://doi.org/10.3390/coatings8120416
9. Zhang, J., Yang, Y., Luo, B., Liu, H., Li, L.: Research on wear characteristics of abrasive belt and the effect on material removal during sanding of medium density fiberboard (MDF). Eur. J. Wood Wood Products **79**, 1563–1576 (2021). https://doi.org/10.1007/s00107-021-01718-x

10. Du, Y., Sun, X., Luo, B., Li, L., Liu, H.: Research on failure mechanism of abrasive belt and effect on sanding of medium-density fiberboard (MDF). Coatings **12**(5), Article no. 621 (2022). https://doi.org/10.3390/coatings12050621
11. Robotiq: Sanding kit. https://robotiq.com/products/sanding-kit. Accessed 30 May 2025
12. Onrobot: Onrobot sander. https://onrobot.com/us/products/onrobot-sander. Accessed 30 May 2025

Machinability Evaluation of Free-Cutting Steel with Lead and/or Bismuth Addition

João Paulo Luiz Grisotto Alves[1], João Batista Ribeiro Martins[2,3], Mariane Gonçalves de Miranda Salustre[3], Marcelo Bertolete[1], and Patrícia Alves Barbosa[1](✉)

[1] Mechanical Technology Laboratory, Department of Mechanical Engineering, Federal University of Espírito Santo, Vitoria, Brazil
patricia.a.barbosa@ufes.br

[2] Federal Institute of Education, Science and Technology of Espírito Santo, Metallurgy Coordination, Rio de Janeiro, Brazil

[3] ArcelorMittal, Global R&D Brazil, Belo Horizonte, Brazil

Abstract. Free-cutting steels are defined by the element alloying additions interrupting the matrix and facilitating machining, with sulphur and lead as the main elements. In the automotive industry, its application is known to increase productivity and reduce costs. However, adding lead presents several adversities, and its use has been banned in several countries. Alternatively, "friendly" free-cutting steel developments are replacing lead with bismuth. This work aims to evaluate the machinability of two manganese sulphide free-cutting steels: (i) with lead and bismuth addition (MnS+PbBi) and (ii) with bismuth addition (MnS+Bi). These materials were characterized by microstructure and hardness. Turning tests were carried out with varying feed rates in 0.1, 0.2, and 0.3 mm/rev. Cutting force and roughness (Ra) were the machinability indexes assessed. No significant difference between materials on FU response from 0.2 mm/rev was evidenced. Adding Bi alone increased the FU response by 106% compared to MnS+PbBi steel to 0.1 mm/rev. Ra was influenced by feed rate, which increased the response proportionally. MnS+Bi cutting free steel may be promising in the rough turning process since no significant difference between materials on Ra response and FU response from 0.2 mm/rev was evidenced.

Keywords: Free-Cutting Steel · Lead · Bismuth · Machinability · Machining Force · Ra

1 Introduction

Free-cutting steels are a special steel grade defined by element alloying additions that, alone or compounds with sulphur (S) and manganese (Mn) forming inclusions. These inclusions interrupt the matrix and also have a lubricating effect. These inclusion functions improve the machinability, influencing the tool life, surface finish, machining force, and chip breaking [1].

O. F. Farías Fuentes et al. (Eds.): CIBIM 2024, *Proceedings of the XVI Ibero-American Congress of Mechanical Engineering*, pp. 254–267, 2026.
https://doi.org/10.1007/978-3-032-22823-9_19

Free-cutting steels, when compared with steels of similar structure, are mainly applied in the industry when seeking to increase productivity while simultaneously reducing costs [2].

Lead (Pb) was considered one of the most important free-cutting steel additive in the 20th century. It acts as a solid lubricant at the chip-tool interface and as a chip embrittlement mechanism [2–4]. However, lead addition has been restricted in several countries due to its toxicity to the environment and humans [3, 4].

Alternatively, "friendly" free-cutting steel developments are replacing lead with non-toxic alloying elements, considering acceptable levels in terms of environmental impact and production costs [1, 5].

Iwamoto and Murakami [6] investigated two proposals for lead-free free-cutting steels: (i) graphite-precipitation steel and (ii) high-Cr steel. Changing hard cementite to graphite makes the steel structure softer, and graphite acts as a solid lubricant at the chip-tool interface, substantially improving machinability. In addition, graphite-precipitation steel exhibited excellent cold forgeability, satisfying steel's property requirements for machine structural use. Cr addition with increasing sulphur content crystallizes large sulphide inclusions. These results in equal or superior machinability compared to SAE 12L14 steel regarding tool life and roughness criteria.

Wu and Li [7] investigated a new Pb-free high-machinability austenitic stainless steel with bismuth addition. They showed the importance of Bi addition to austenitic stainless steel machinability, promoting longer tool life and lower cutting forces.

Xu et al. [8] investigated the feasibility of producing microalloyed free-cutting steel with tin (Sn) instead of lead addition. They concluded that increasing Sn content improves steel machinability by decreasing the surface roughness of the machined part and making the chips more brittle.

In this context, this work aims to evaluate the machinability in turning process of two free-cutting steels with a ferritic matrix containing manganese sulphide (MnS): (i) with the addition of lead and bismuth (Pb+Bi) and (ii) with the addition of only bismuth (Bi).

2 Materials and Methods

The materials investigated were two free-cutting steel alloys with manganese sulphide (MnS) and ferritic matrix, coded as (i) MnS+PbBi (MnS-free-cutting steel with lead and bismuth additions) and (ii) MnS+Bi (MnS-free-cutting steel with bismuth addition). These materials were supplied as rolled bars with circular cross-sections with a nominal diameter of 25.4 mm. The chemical composition of MnS-free cutting steels is shown in Table 1, as the supplier indicates.

The materials were characterized by microstructure, highlighting the free-cutting elements and Vickers hardness number. Microstructure characterization and free-cut elements analysis were performed using an Eclipse MA200 inverted optical microscope (NIKON) and a Quanta 650 FEG scanning electron microscope (FEI) with a Quantax energy dispersive spectrum (EDS) detector (Bruker). Vickers Hardness was evaluated in a 200HBRV-187.5S hardness tester (Lleida) according to ASTM E92-23 [9]. Ten indentations were made on each sample, both in transverse and longitudinal sections, applying a load of 30 kgf for 10 s.

Table 1. Chemical composition (wt%).

Material	Element							
	Mn	S	Si	P	Al	Mg+Pb+Bi	Cr+Ni+Cu+Mo	C
MnS+PbBi	0.6 1.5	0.5 max	0.1 0.4	0.06 max	0.06 max	0.3 max	0.3 max	0.06 0.1
MnS+Bi	0.6 1.5	0.5 max	0.1 0.4	0.06 max	0.06 max	0.7 max	0.03 max	0.06 0.1

Source: own production

Dry turning tests were carried out on a Centur 35D CNC lathe (ROMI). Uncoated carbide tools, TPUN 160304, ISO P30 grade (BRASSINTER), were used. The insert was mounted on a CTGPR 2525 M16 support (SANDVIK), which provided a semi-orthogonal cutting geometry with an inclination angle (λs) of 0°, a positive rake angle (γo) of 6° and a side-cutting edge angle (χr) of 91°. Cutting speed (Vc) and depth of cut (ap) were kept constant at 150 m/min and 1 mm, respectively. Feed rate (f) was varied by 0.2, 0.3, and 0.3 mm/rev. Cutting speed was kept constant due to the small bar diameter (25.4 mm) and the maximum machine-tool rotation limitation of 2000 rpm when using a 250 mm diameter plate. Machining force and surface roughness were the machinability indexes evaluated.

The machining force components were monitored through a machining force measurement system (KISTLER), consisting of a piezoelectric dynamometer 9129 AA model, signal amplifier 5080A1030001, acquisition system 5697A1 with DynoWare analysis software, as shown in Fig. 1. The system captures the signal of the three orthogonal components of the force acting during machining: Fx, Fy and Fz, for a measurement range of ±10 kN; sensitivity of ≈ −8 pC/N for the x and z components, and ≈ −4.1 pC/N for the y component; and system measurement uncertainty of 2%. The acquisition rate was 1 kHz. The machining force is determined by Eq. (1).

$$FU = \sqrt{{F_x}^2 + {F_y}^2 + {F_z}^2} \tag{1}$$

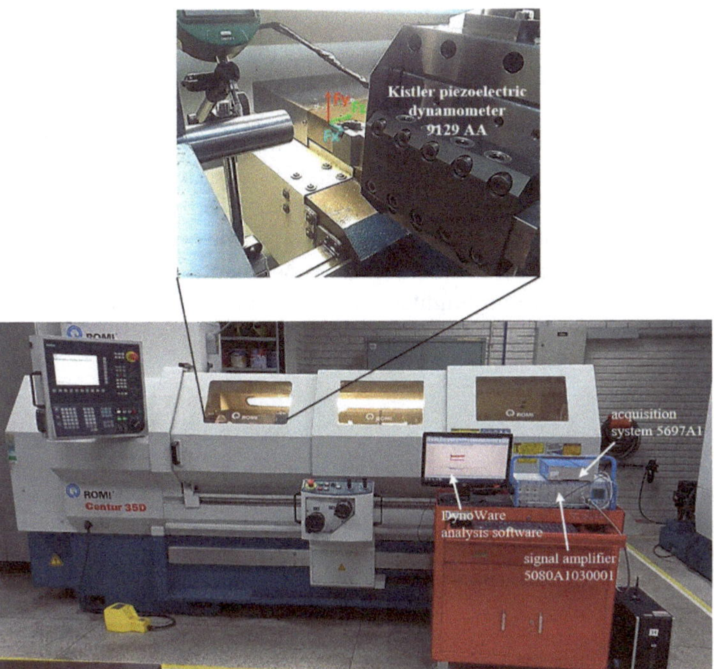

Fig. 1. Kistler machining force measurement system for monitoring force orthogonal components. Source: own production.

Arithmetic average surface roughness (Ra) was measured via a portable surface roughness tester, Surftest SJ-210 (MITUTOYO). The cut-off was set according to ABNT NBR ISO 4287:2002 [10] and ABNT NBR ISO 4288:2008 [11]. Roughness measurements were performed for each machined feed length (Lf) of 15 mm by testing (T) and replications (R1 and R2) in four measurement planes spaced at 90°, as shown in Fig. 2. Ra results were obtained from the average values of the three experimental regions (T, R1 and R2).

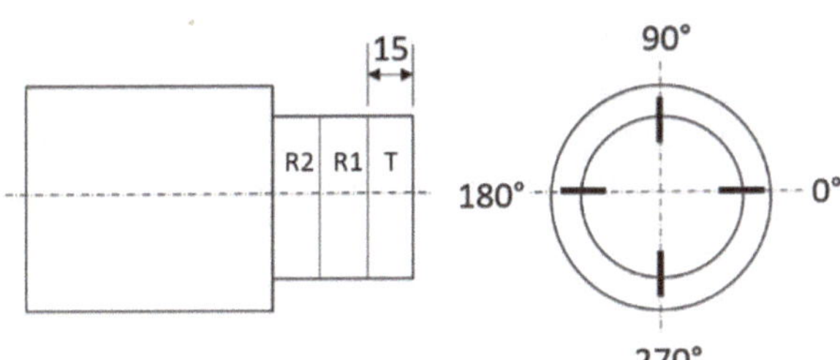

Fig. 2. Positioning schematic configuration for measuring roughness on the machined bar. The machined feed length (Lf) is highlighted on the left, and the measurement plane identification is on the right. Source: own production.

3 Results

3.1 Characterization

Figure 3 and Fig. 4 show representative images of the microstructure of the MnS+PbBi and MnS+Bi free-cutting steels, etched with Nital 2%, respectively. These images reveal that both materials have a ferritic matrix with pearlitic islands and small MnS inclusions dispersed in the matrix. The longitudinal section shows the pearlitic phase and inclusions stretched, indicating the bar rolling direction. SEM images show the presence of PbBi and Bi additions, which are highlighted as a white envelope surrounding some MnS inclusions.

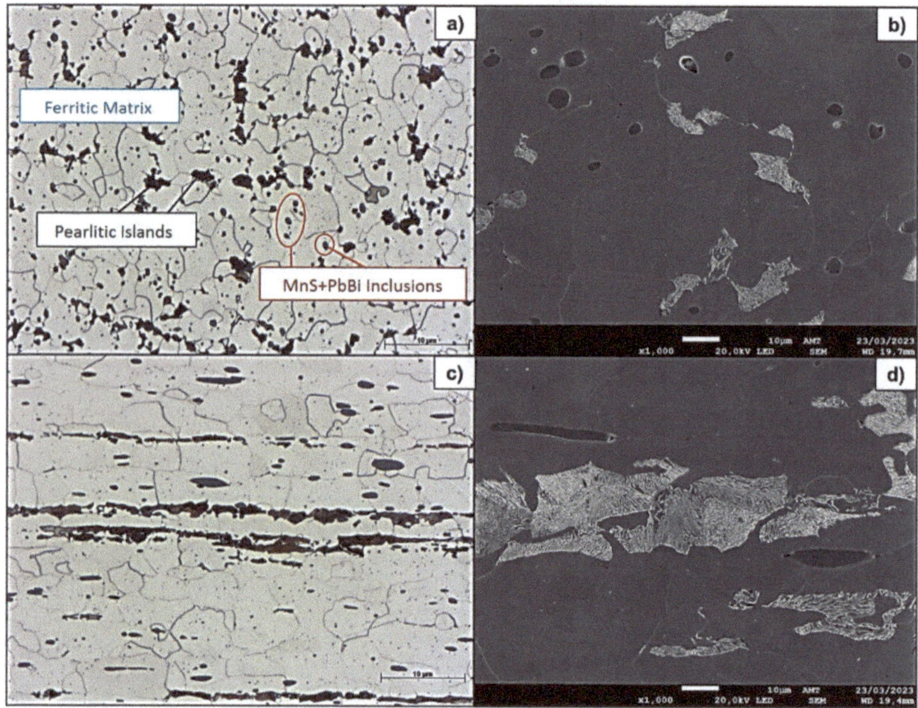

Fig. 3. Microstructure images of MnS+PbBi free-cutting steel etched with 2% Nital. a) MO of transverse section (x100). b) SEM of transverse section (x1000). c) MO of longitudinal section (x100). d) SEM of longitudinal section (x1000). Source: own production.

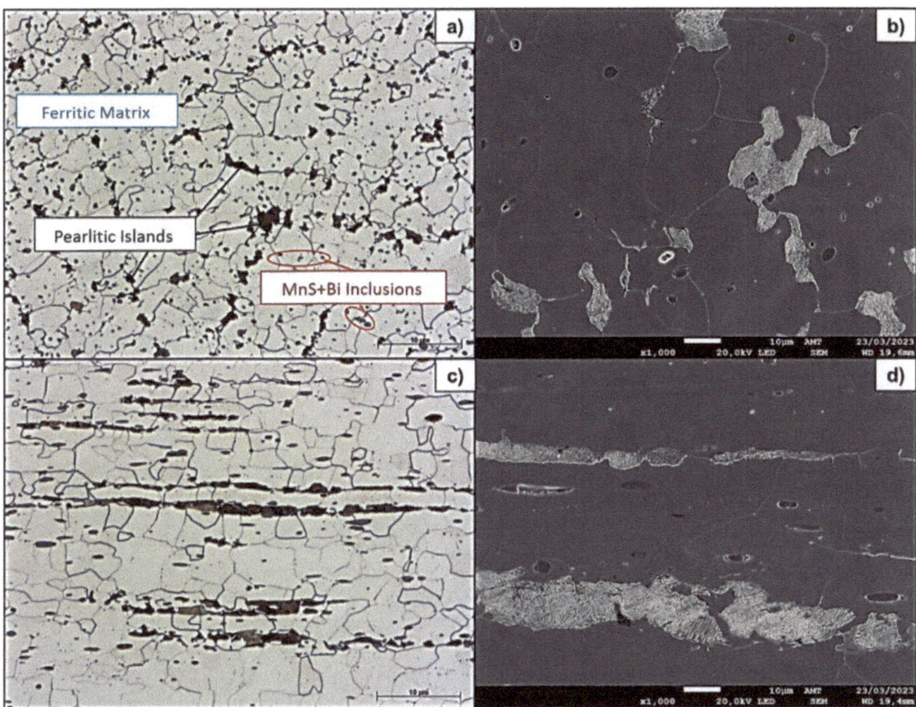

Fig. 4. Microstructure images of MnS+Bi free-cutting steel etched with 2% Nital. a) MO of transverse section (x100). b) SEM of transverse section (x1000). c) MO of longitudinal section (x100). d) SEM of longitudinal section (x1000). Source: own production.

Figure 5 and Fig. 6 show the EDS analysis of the free-cutting steel inclusions. The chemical mapping and the chemical analysis spectra indicate the presence of manganese sulphide in both materials, surrounded by varying elements specific according to each material under study. Lead and bismuth were detected in the MnS+PbBi free-cutting steel (Fig. 5b and Fig. 5d), and bismuth in the MnS+Bi free-cutting steel (Fig. 6b and Fig. 6d). According to Xi et al. [12], Bi can be present in an isolated form dispersed in the matrix, forming a cavity bridge between manganese sulphide inclusions, semi-encapsulating MnS, and/or completely encapsulating MnS. Similar behaviour was also observed for PbBi (Fig. 5a and Fig. 5b).

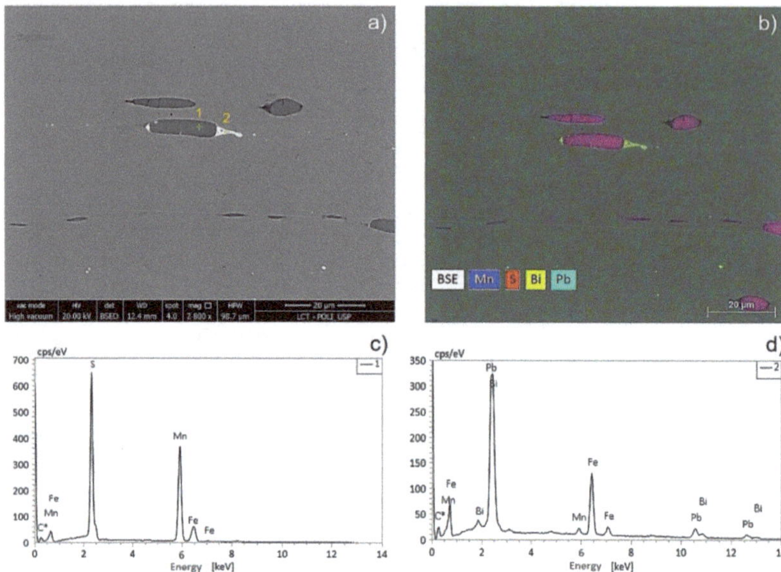

Fig. 5. SEM/EDS images of MnS+PbBi free-cutting steel inclusion. a) SEM of longitudinal section highlighting EDS analysis points. b) Chemical mapping by EDS. c) Chemical analysis spectrum within the MnS inclusion (Point 1). d) Chemical analysis spectrum within the white phase (PbBi) involving MnS inclusion (Point 2). Source: own production.

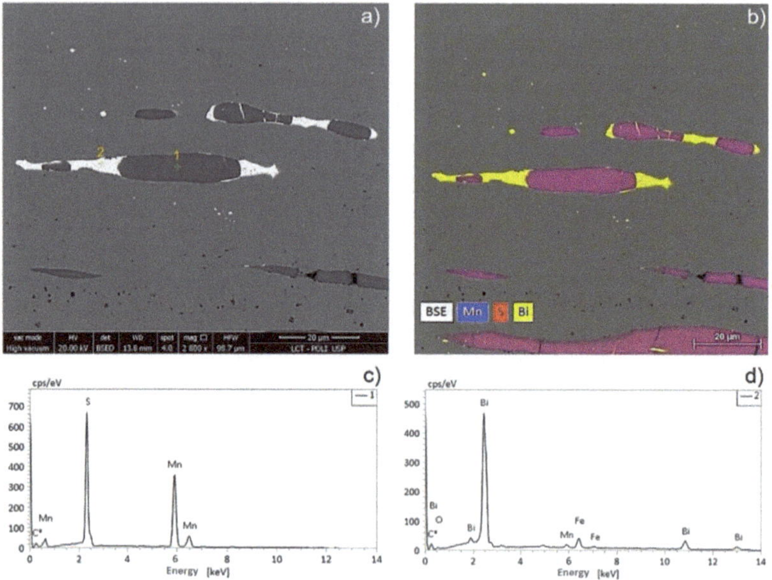

Fig. 6. SEM/EDS images of MnS+Pb free-cutting steel inclusion. a) SEM of longitudinal section highlighting EDS analysis points. b) Chemical mapping by EDS. c) Chemical analysis spectrum within the MnS inclusion (Point 1). d) Chemical analysis spectrum within the white phase (Bi) involving MnS inclusion (Point 2). Source: own production.

Table 2 shows Vickers hardness number results (average HV30) for transverse and longitudinal sections of the evaluated materials.

Table 2. HV30 Vickers hardness number of free-cutting steels with lead and/or bismuth additions.

MnS+PbBi		MnS+Bi	
Transverse section	Longitudinal section	Transverse section	Longitudinal section
188.3 ± 5.3	185.0 ± 4.9	180.5 ± 6.4	180.6 ± 4.1

Source: own production

Table 3 presents the Vickers hardness number analysis of variance (ANOVA), with replicates, for a significance level (α) of 5%, considering the different materials and sample sections. SS means the sum of squares, GL degrees of freedom, MS denotes the mean square (or variance), F is the value calculated for Fisher distribution, and p is the calculated significance level. The results showed that only the primary variable – Material – significantly influenced the hardness results since the calculated significance level (p) is lower than the criterion (α) used in the hypothesis test ($p < 0.05$).

Table 3. ANOVA for Vickers hardness number (HV30).

Effect	SS	GL	MS	F	p
Intercept	1348248	1	1348248	44167.09	0.000000
Material	369	1	369	12.09	0.001342
Section	25	1	25	0.83	0.367366
Material x Section	31	1	31	1.02	0.319139
Error	1099	36	31		

Source: own production

Figure 7 shows the effect graph of the significant difference between different free-cutting steels on hardness response. It can be observed that MnS+PbBi free-cutting steel hardness is, on average, 3.3% higher than MnS+Bi ones. The slight difference in hardness response between the MnS+PbBi and MnS+Bi free-cutting steels shown in Fig. 7 may have been influenced by different chemical compositions and inclusion sizes.

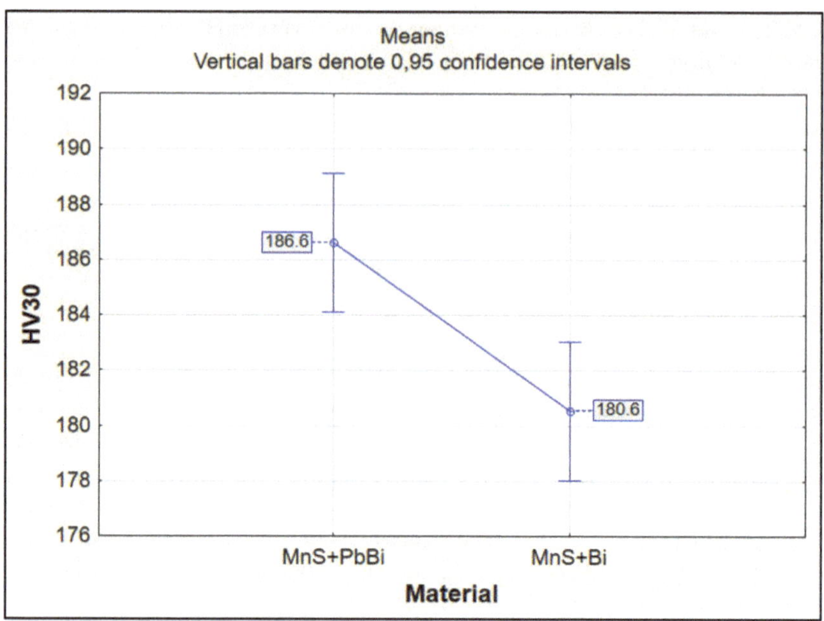

Fig. 7. Material's effect graph on Vickers hardness number response (HV30). Source: own production.

3.2 Machinability

Figure 8 shows the representative machining force (FU) signals over cutting time for different free-cutting steels, MnS+PbBi and MnS+Bi, in three feed rate (f) conditions evaluated. The higher the feed rate, the shorter the cutting time.

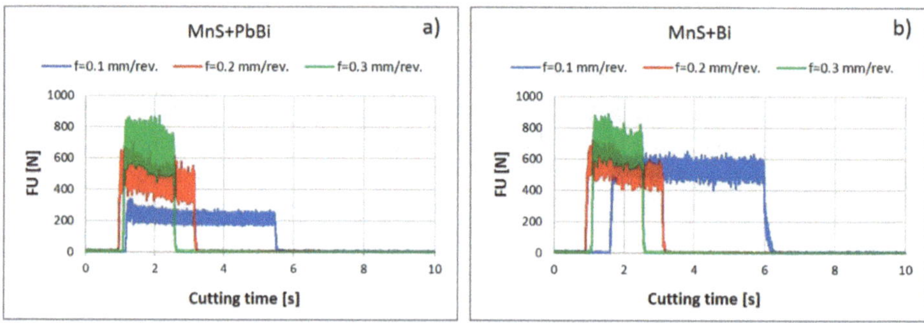

Fig. 8. Representative machining force (FU) signals over cutting time for three different feed rate conditions evaluated. a) MnS+PbBi free-cutting steel. b) MnS+Bi free-cutting steel. Source: own production.

Table 4 presents the analysis of variance (ANOVA) of machining force data for a significance level (α) of 5%, considering the different materials and feed rates (f). In this case, the statistical analysis indicates the main variables - material and feed rate – significantly influence FU response ($p < \alpha$). However, these variables have a strong interaction ($p < 0.05$).

Table 4. ANOVA for machining force (FU).

Effect	SS	GL	MS	F	p
Intercept	4834078	1	4834078	5548.932	0.000000
Material	53005	1	53005	60.843	0.000005
f	268428	2	134214	154.061	0.000000
Material x f	54373	2	27186	31.207	0.000018
Error	10454	12	871		

Source: own production

Figure 9 shows the FU response effect graph for material (MnS+PbBi and MnS+Bi) and feed rate (0.1, 0.2 and 0.3 mm/rev) interaction. It can be seen there is a behaviour difference in the average machining force response for materials and different feed rate conditions. FU response of MnS+PbBi free-cutting steel showed an approximately linear average increase with increasing feed rate. Statistical analysis results of MnS+PbBi free-cutting steel indicated the increase in feed rate from 0.1 mm/rev to 0.2 mm/rev did not significantly influence FU response. In comparison, a significant influence was evidenced for the feed rate of 0.3 mm/rev. For the latter, FU increased, on average, by 26.75% compared to f of 0.2 mm/rev. Furthermore, it can be observed the free-cutting steels showed statistically different responses for FU at 0.1 mm/rev. MnS+Bi free-cutting steel exhibited significantly higher machining force than MnS+PbBi ones in this described feed rate condition, averaging 106% greater. With the feed rate increase within the investigated range, this difference was reduced so that the two free-cutting steels did not present a statistical difference in FU response from f of 0.2 mm/rev.

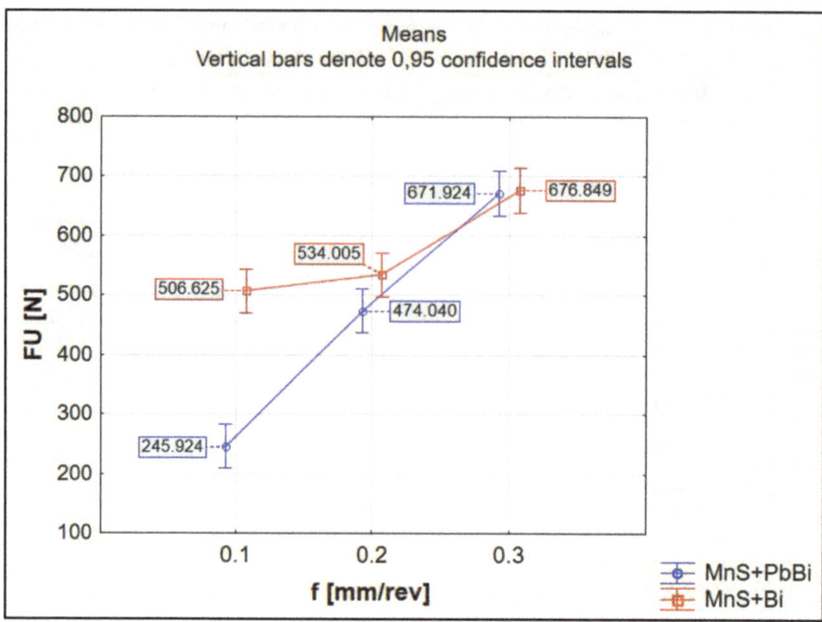

Fig. 9. Material and feed rate interaction effect graph on machining force (FU) response. Source: own production.

Figure 10 shows the arithmetic average roughness (Ra) results for the different free-cutting steels under the evaluated feed rate (f) conditions. The materials exhibited similar behaviour, with an increasing Ra response as f increased.

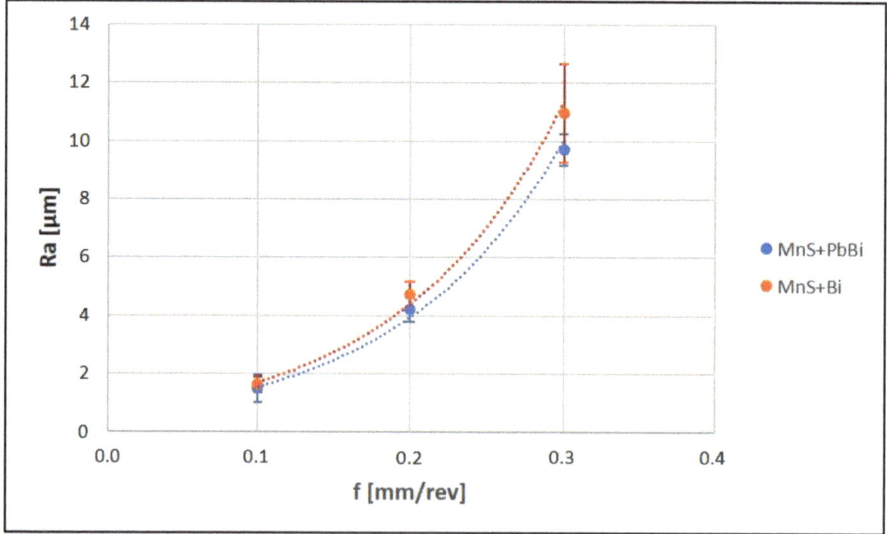

Fig. 10. Ra results as a function of feed rate for MnS+PbBi and MnS+Bi free-cutting steels. Source: own production.

Table 5 presents the analysis of variance (ANOVA) of arithmetic average roughness data for a significance level (α) of 5%, considering the different materials and feed rates (f). In this case, only the primary input variable – feed rate – significantly influenced Ra response ($p < 0.05$), corroborating with visually representative results in Fig. 10.

Table 5. ANOVA for arithmetic average roughness (Ra).

Effect	SS	GL	MS	F	p
Intercept	537.1768	1	537.1768	860.4552	0.000000
Material	1.7534	1	1.7534	2.8087	0.119593
f	239.8150	2	119.9075	192.0690	0.000000
Material x f	0.9481	2	0.4741	0.7494	0.489190
Error	7.4915	12	0.6243		

Source: own production

Figure 11 displays the effect graph highlighting the increase of Ra increase with feed rate. Ra response increased, on average, 184.7% and 131.3% when f increased from 0.1 mm/rev to 0.2 mm/rev and from 0.2 mm/rev to 0.3 mm/rev, respectively. The results of feed rate influence on arithmetic average roughness follow a similar behaviour to the theoretical average roughness proposed by Boothroyd and Knight [13], which presents Ra as a function directly proportional to f2.

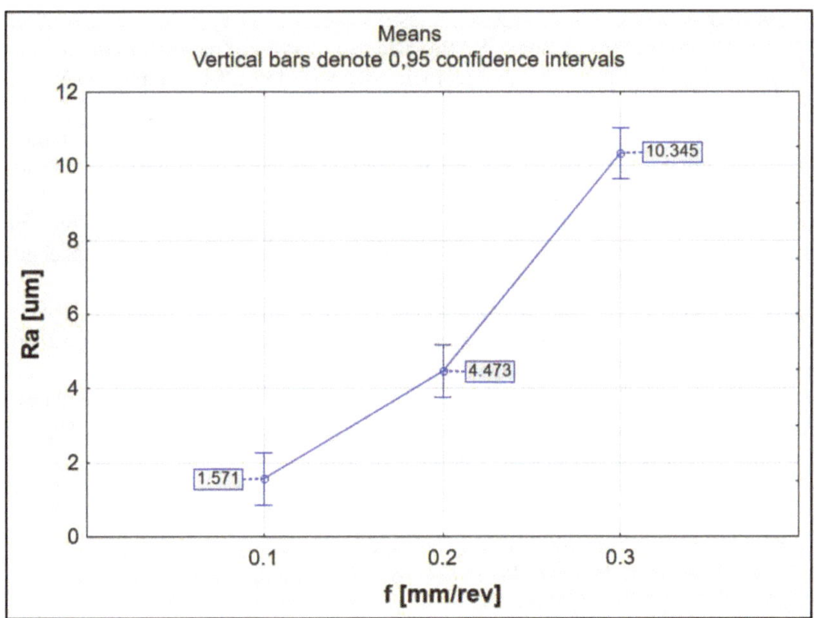

Fig. 11. Feed rate's effect graph on arithmetic average surface roughness (Ra). Source: own production.

4 Conclusions

Free-cutting steel with lead and bismuth additions (MnS+PbBi) exhibits Vickers hardness, on average, 3.3% higher than free-cutting steel with only bismuth additions (MnS+Bi).

In the turning process, MnS+Bi free-cutting steel exhibits significantly higher machining force than MnS+PbBi free-cutting steel, on average 106% greater at a feed rate of 0.1 mm/rev.

PbBi addition or Bi alone to MnS free-cutting steel does not change surface finish response.

This study presents MnS+Bi free-cutting steel as a promising substitute for lead-based free-cutting steels, particularly in rough turning processes. The main finding is that the bismuth presence in MnS free-cutting steel during the turning process at high feed rates does not significantly affect the machinability criteria based on machining force and surface roughness, making the MnS+Bi additive a promising lead-free alternative.

Acknowledgements. The authors thank to FAPES (144/2020), FAPES (995/2024), CAPES, Embrapii-IFES, and ArcelorMittal Brazil.

References

1. Desaigues, J.-E., Lescalier, C., Bomont-Arzur, A., Dudzinski, D., Bomont, O.: Experimental study of built-up layer formation during machining of high strength free-cutting steel. J. Mater. Process. Technol. **236**, 204–215 (2016). https://doi.org/10.1016/j.jmatprotec.2016.05.016
2. Živković, D., Štrbac, N., Ekinović, S., Begović, E.: Lead-free free-cutting steels as modern environmentally friendly materials. Ecologica **18**, 451–456 (2011). https://www.researchgate.net/publication/281812810
3. Garcia, C.I., Hua, M.J., Miller, M.K., Deardo, A.J.: Application of grain boundary engineering in lead-free green steel. ISIJ Int. **43**, 2023–2027 (2003). https://doi.org/10.2355/isijinternational.43.2023
4. Kurka, V., Kuboň, Z., Kander, L., Jonšta, P., Kotásek, O.: The effect of bismuth on techn logical and material characteristics of low-alloyed automotive steels with a good machinability. Metals **12**, 1–14 (2022). https://doi.org/10.3390/met12020301
5. Watari, K., Iwasaki, T., Aiso, T.: Development of lead-free free-cutting steel and cutting technology. Nippon Steel Sumitomo Metal Tech. Rep. **116**, 32–37 (2017). https://www.nipponsteel.com/en/tech/report/nssmc/pdf/116-07.pdf
6. Iwamoto, T., Murakami, T.: Bar and wire steels for gears and valves of automobiles - eco-friendly free cutting steel without lead addition. JFE Tech. Rep. **4**, 74–80 (2004). https://www.jfe-steel.co.jp/en/research/report/004/pdf/004-13.pdf
7. Wu, D., Li, Z.: A new Pb-free machinable austenitic stainless steel. J. Iron Steel Res. Int. **17**, 59–63 (2010). https://doi.org/10.1016/S1006-706X(10)60046-5
8. Xu, X., et al.: Effect of tin microalloying on the microstructure of low-carbon free-machining steels. J. Mater. Res. Technol. **20**, 1172–1185 (2022). https://doi.org/10.1016/j.jmrt.2022.07.153
9. American Society for Testing and Materials. ASTM International E92-23: Standard Test Methods for Vickers Hardness and Knoop Hardness of Metallic Materials1. West Co shohocken: ASTM (2023). https://www.astm.org/e0092-23.html

10. Associação Brasileira De Normas Técnicas. ABNT NBR ISO 4287: Especificações geométricas do produto (GPS) - Rugosidade: Método do perfil - Termos, definições e parâmetros da rugosidade. Rio de Janeiro: ABNT (2002). https://www.normas.com.br/visualizar/abnt-nbrnm/22323/abnt-nbriso4287
11. Associação Brasileira De Normas Técnicas. ABNT NBR ISO 4288: Especificações geométricas do produto (GPS)—Rugosidade: Método do perfil - Regras e procedimentos para avaliação de rugosidade. ABNT, Rio de Janeiro (2008). https://www.normas.com.br/visualizar/abnt-nbrnm/27793/nbriso4288
12. Xie, J., Fan, T., Zeng, Z., Sun, H., Fu, J.: Bi-sulfide existence in 0Cr18Ni9 steel: correlation with machinability and mechanical properties. J. Mater. Res. Technol. **9**(4), 9142–9152 (2020). https://doi.org/10.1016/j.jmrt.2020.06.043
13. Boothroyd, G., Knight, W.A.: Fundamentals of Machining and Machine Tools. Marcel Dekker Inc., New York (1989)

Methodological Approach to the Development of a Robotic Additive Manufacturing Cell Based on the LMD-Wire Process

Brayan S. Figueroa(✉) and Alberto Alvares

Department of Mechanical and Mechatronics Engineering, University of Brasilia, Brasilia, Brazil
brayan.s.figueroa47@gmail.com

Abstract. This study presents a comprehensive methodology for the development a robotic additive manufacturing cell based on Laser Metal Deposition with wire feed (LMD-wire). The proposed system integrates a KUKA KR70 R2100 industrial robot with a Meltio Engine deposition head and is supported by a CAD/CAPP/CAM architecture structured around a Digital Twin (DT) framework, compliant with ISO 23247 standards. The CAD component generates the 3D geometry of the parts, which is then processed by the CAPP system to define optimized deposition strategies and generate KUKA KRL machine instructions via Rhino/Grasshopper and Meltio Space. These instructions are transferred to the CAM environment, which incorporates a virtual representation of the physical manufacturing cell for realistic simulation and real-time monitoring. The DT enables continuous synchronization between physical operations and the digital model. The proposed methodology is validated through the fabrication of a planar metallic part under various process parameters, demonstrating the effectiveness of the integrated system for process control and operational reliability.

Keywords: Additive Manufacturing · LMD · Robotics · Digital Twin

1 Introduction

Additive manufacturing has significantly transformed traditional production methods by enabling the creation of three-dimensional objects through the controlled, layer-by-layer addition of material based on digital models [1,2]. This technological advancement has gained increasing relevance in sectors such as aerospace, automotive, medical, and consumer goods industries due to its capacity for extreme customization, high geometric precision, and substantial reductions in both operational costs and production times [3,4].

Among the various advanced additive manufacturing techniques, Laser Metal Deposition with wire feed (LMD-wire) has emerged as a highly efficient method

O. F. Farías Fuentes et al. (Eds.): CIBIM 2024, *Proceedings of the XVI Ibero-American Congress of Mechanical Engineering*, pp. 268–280, 2026.
https://doi.org/10.1007/978-3-032-22823-9_20

for producing metallic components with excellent mechanical properties and dimensional accuracy [5–7]. This process utilizes a high-power laser to melt a metal wire, which is then precisely deposited onto pre-formed substrates. This controlled deposition facilitates the fabrication of complex and adaptive geometries, optimizes material usage, and minimizes waste throughout the production cycle [8,9]. To maximize the advantages of LMD-wire, Robotic Additive Manufacturing Cells (RAMC) have been developed. These systems integrate robotics, motion planning, and process monitoring to ensure repeatability, quality, and operational flexibility [10,11].

In parallel, Digital Twins (DTs) have emerged as essential tools for enhancing control and optimization in additive manufacturing environments [12,13]. By integrating real-time sensor data, DTs replicate the behavior of physical systems under dynamic conditions [14,15]. In LMD-wire processes, DTs enable accurate simulation of critical parameters such as deposition speed, laser power, and material flow, supporting defect prediction and mitigation. This leads to improved part quality and consistency, while reducing production time and operational costs.

The integration of Digital Twins with additive manufacturing opens new opportunities for innovation and efficiency, particularly in the production of functional and structural components [16]. Incorporating LMD-wire technology into robotic manufacturing cells represents a major step forward in producing complex, high-performance metallic parts. The synergy between additive manufacturing and DTs allows for real-time simulation and adaptive control of critical process variables, enhancing both process quality and overall system performance [17].

Despite these advancements, several challenges remain in the deployment of robotic additive manufacturing systems based on LMD-wire. Conventional robotic cells often suffer from limited interoperability and inadequate integration for real-time monitoring, adaptive control, and predictive fault detection. These shortcomings reduce process flexibility, compromise part quality, and hinder the dynamic optimization of deposition parameters [18]. Moreover, the absence of intelligent systems to synchronize design, planning, and execution impedes broader industrial adoption. Addressing these challenges requires integrated solutions that combine automation, simulation, and real-time feedback [19].

This work proposes a methodological approach for the development of a robotic additive manufacturing cell based on LMD-wire, supported by a CAD/CAPP/CAM system specifically adapted for additive manufacturing and integrated with a Digital Twin architecture. The methodology details the modules and their interactions using IDEF0 diagrams, emphasizing both the physical and logical integration of components. The CAD/CAPP/CAM framework is structured to align design (CAD), process planning (CAPP), and fabrication (CAM) within a DT environment compliant with ISO 23247 standards. The proposed approach is validated through a case study involving the production of two metallic parts using different deposition parameters, demonstrating the system's adaptability, precision, and real-time monitoring capabilities.

2 Methodology

The Robotic Additive Manufacturing Cell (RAMC), based on a wire-based metal deposition process, was developed using a structured methodology grounded in the IDEF0 functional modeling approach. This systematic framework provides a comprehensive representation of the RAMC development process, encompassing the physical and logical integration of hardware components, their interactions with the control system, and the overall assembly of the manufacturing cell. In addition to these core elements, the methodology also addresses part design, process planning, and the execution of the additive manufacturing process for metallic components. Moreover, it includes detailed diagrams illustrating the development and deployment of the CAD/CAPP/CAM system, which is built upon a Digital Twin architecture specifically adapted to the requirements of metal additive manufacturing. This ensures precise control, real-time monitoring, and workflow optimization across the entire process.

Figure 1 illustrates the proposed methodology for the Robotic Additive Manufacturing Cell (RAMC) based on the Laser Metal Deposition (LMD) process. The methodology is structured into three interdependent stages: inputs, process, and outputs, establishing a clear and integrated framework for the additive manufacturing of metallic parts. The input stage comprises various essential parameters and system signals required to initiate and manage the manufacturing operation. These include logical signals from the control interface, the calibration weight of the robotic arm, argon gas pressure, cooling fluid conditions, the STL file defining the part geometry, and the metal wire feedstock for deposition. Collectively, these inputs establish the operational foundation for the RAMC, enabling seamless integration between physical components and digital systems for accurate and efficient manufacturing.

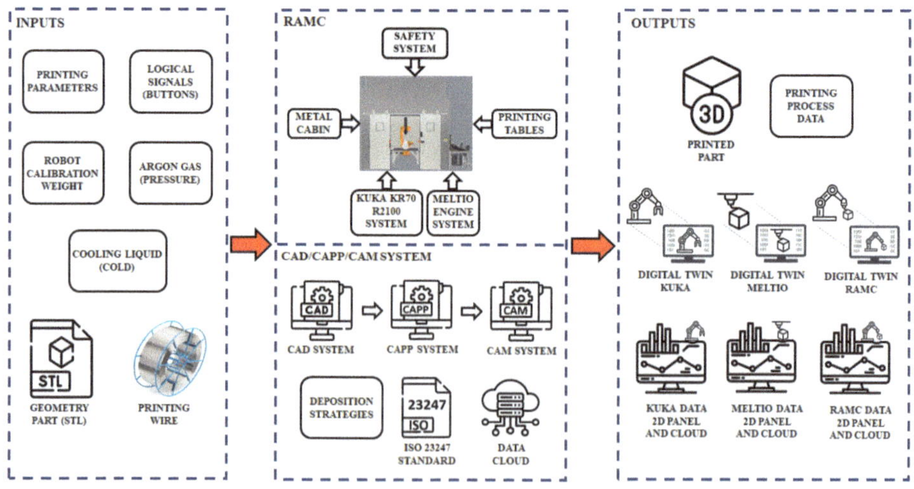

Fig. 1. Robotic Additive Manufacturing System.

The process stage involves the coordinated operation of the RAMC and the CAD/CAPP/CAM system, which jointly manage the full sequence from part design and process planning to physical manufacturing. The RAMC comprises a metal enclosure, integrated operator safety systems, precision printing tables, the KUKA KR70 R2100 robotic arm, and the MELTIO ENGINE deposition head. These components function together to ensure both the physical assembly and the logical control needed to maintain precision, stability, and efficiency throughout the LMD process.

The CAD/CAPP/CAM system forms a comprehensive framework integrating three main modules: CAD for part modeling, CAPP for process planning, and CAM for toolpath generation and execution. This system operates within a Digital Twin architecture aligned with ISO 23247 standards, enabling real-time monitoring and bidirectional data exchange with a cloud platform. Such architecture supports adaptive control, continuous optimization, and dynamic supervision, ensuring that manufacturing processes can be adjusted in real time to maintain high quality and performance standards.

The output stage encompasses both digital and physical outcomes generated throughout the manufacturing workflow. It comprises three Digital Twin (DT) implementations associated with the KUKA robot and the MELTIO ENGINE: (i) the Meltio Dashboard DT, responsible for continuous monitoring of deposition process variables; (ii) the KUKA iiQoT DT, which enables condition-based monitoring and predictive maintenance of the robotic system; and (iii) the RAMC DT, which provides general-level supervision, data acquisition, and simulation of the robotic additive manufacturing cell. Additional outputs include the 2D visualization and control interface, cloud-integrated production data, and the final printed metallic component. Collectively, these elements establish a robust framework for traceability, operational control, and high-performance outcomes in LMD-Wire processes.

Figure 2 shows the IDEF0 diagram representing the layered structure of the system, which is divided into two levels associated with the RAMC integration. The first level addresses physical and logical integration, covering the detailed interconnection of all components within the RAMC-including hardware, control systems, and data management units-designed to enable efficient and synchronized operation. The second level pertains to the CAD/CAPP/CAM system, which manages and optimizes the design, planning, and fabrication stages. This system operates in close coordination with the RAMC's physical components, ensuring a seamless flow of information and control from digital design to physical production.

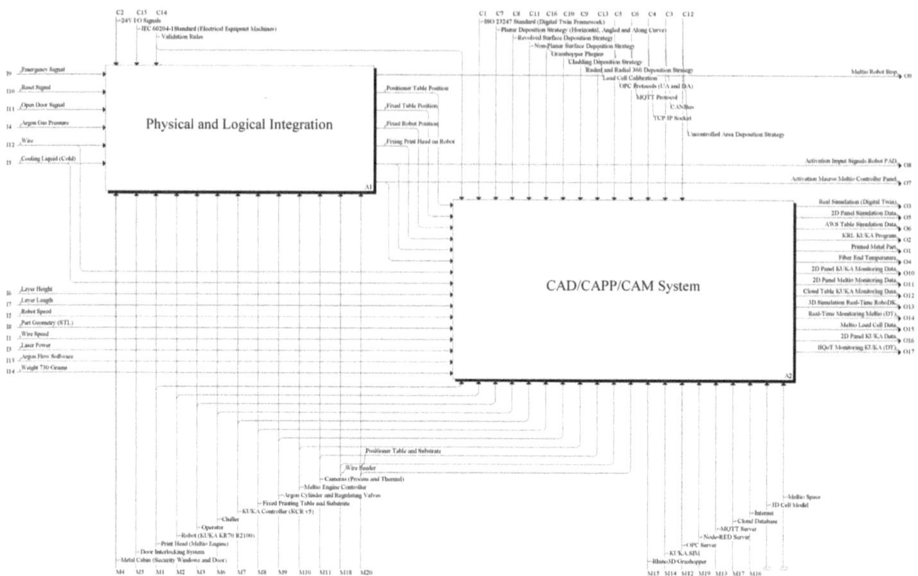

Fig. 2. IDEF0 Diagram of the Robotic Additive Manufacturing Cell.

The physical and logical integration activity handles the processing of system signals and operational inputs to coordinate interactions among the RAMC's components. It produces essential outputs such as safety system activation, robot spatial positioning, and alignment of the printing table-each serving as a prerequisite for the subsequent CAD/CAPP/CAM processes. This coordinated integration enables precise task execution, enhances synchronization among components, and improves responsiveness to operational disruptions.

The CAD/CAPP/CAM system activity processes input data such as part geometry, process parameters, and other key information necessary for manufacturing. Its outputs include a detailed simulation of the additive process, generation of robot-specific toolpaths and control code, physical part fabrication, and real-time monitoring of process variables and performance indicators. This integrated approach ensures consistent part quality, full traceability, and efficient operation throughout the production lifecycle.

2.1 Physical Integration

Figure 3 presents the IDEF0 model for Physical Integration, which outlines the interconnected processes that establish the physical, electrical, and mechanical connections of the components within the additive manufacturing cell. The process begins with the assembly of the metallic enclosure to ensure operator safety during printing, followed by the installation of the robot and its controller for optimal performance. The printing tables are then strategically positioned to maximize process efficiency. Afterward, the Meltio Engine system is installed, integrating the cooling and gas systems, print head, wire feeder, PCB electrical connections, and physical assembly. A safety system with door interlocking is implemented to enhance operator protection. Finally, the robot and Meltio Engine are integrated, ensuring that both systems operate in a coordinated and efficient manner.

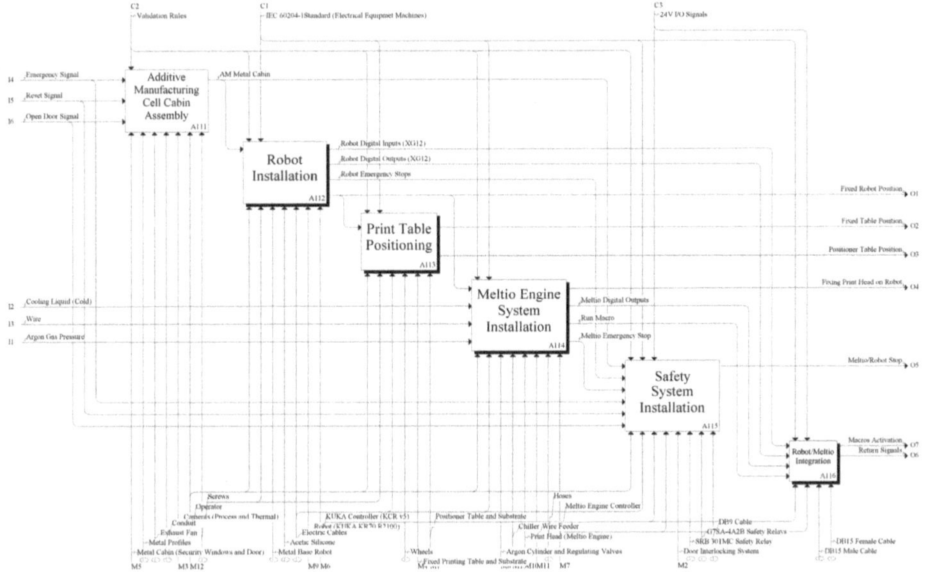

Fig. 3. IDEF0 Diagram Physical Integration Activities.

These activities are intricately interconnected within the physical structure of the additive manufacturing cell, governed by stringent validation rules, electrical standards, and the management of 24V input/output signals. The outputs generated not only fulfill the specific operational requirements of the cell but also provide essential inputs for other system functions, ensuring a seamless integration and optimal performance of the entire manufacturing process.

2.2 Logical Integration

Figure 4 presents the IDEF0 model of the Logical Integration activity, which is responsible for coordinating the interaction between the control logic and the physical components of the robotic additive manufacturing cell. This stage ensures proper synchronization between the robotic system and the Meltio deposition unit through reliable signal exchange and control validation. The process begins with the verification of input signals generated during the physical integration phase, ensuring that they are correctly processed to activate the appropriate functions within the Meltio controller. Simultaneously, the output signals from the robot are validated to confirm accurate reception and interpretation by the Meltio system. This bidirectional validation enables consistent coordination between both systems, reducing communication errors and supporting precise execution of the deposition process. Logical Integration is thus essential to ensure the operational coherence and responsiveness required for efficient LMD-Wire manufacturing.

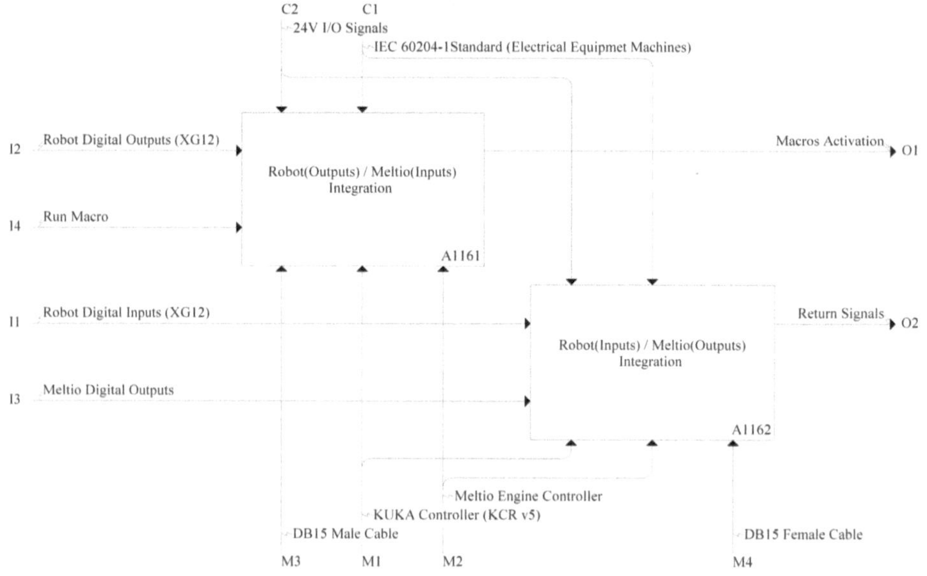

Fig. 4. IDEF0 Diagram Logical Integration Activities.

This validation process triggers the activation of input signals on the robot's control panel and initiates macros on the Meltio controller panel, ensuring full interoperability between both systems. Logical integration plays a critical role in establishing a seamless connection between control processes and physical components, enabling precise communication between subsystems and ensuring the efficient and reliable execution of tasks within the additive manufacturing environment.

3 CAD/CAPP/CAM System

Figure 5 presents the IDEF0 model representing the development of the CAD/ CAPP/CAM system, which is composed of three main activities: Computer-Aided Design (CAD), Computer-Aided Process Planning (CAPP), and Computer-Aided Manufacturing (CAM). These modules collectively support the design, planning, and execution stages of the additive manufacturing workflow, ensuring the integration of digital models with physical production.

The CAD activity is organized into three phases: importing the STL file, generating the base model, and refining it. The imported STL file is processed to create an initial geometry, which is subsequently optimized through dimensional and structural adjustments within the CAD environment to ensure compatibility with the deposition strategy and manufacturing constraints.

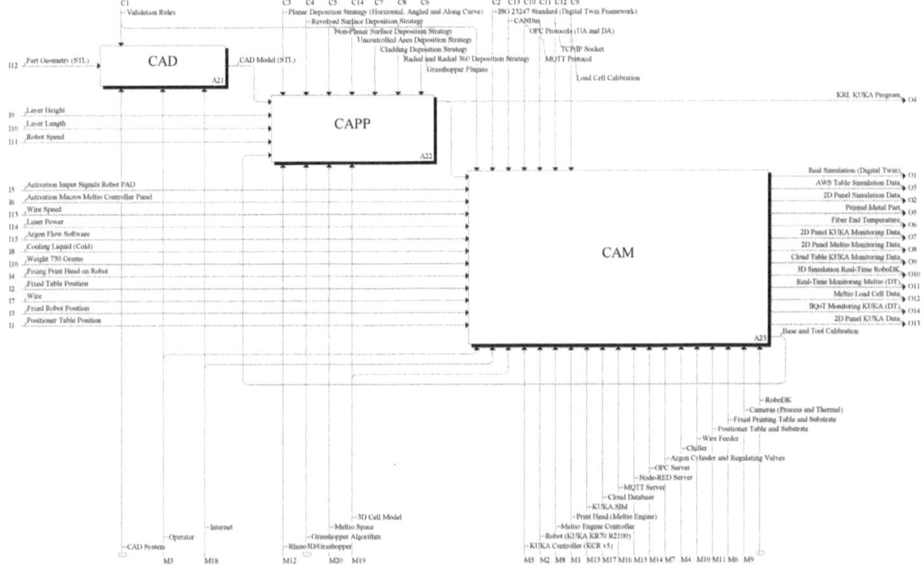

Fig. 5. IDEF0 Diagram CAD/CAPP/CAM System Activities.

The CAPP activity includes three sub-steps: importing the CAD model, slicing the geometry into layers, and generating the KUKA Robot Language (KRL) program. It processes the STL geometry from CAD, incorporates tool origin adjustments from CAM, and integrates external parameters such as robot speed, layer height, and width. The result is a KRL code that governs the additive manufacturing process. The CAM activity consists of two sub-activities: Digital Twin Simulation (CAM Planning) and 3D Printing Execution (CAM Execution). The KRL program generated by CAPP is used in CAM Planning to simulate the deposition process using a digital twin, with real-time monitoring through a 2D interface and an AWS-based dashboard. CAM Execution applies the printing parameters, outputs from both physical and logical integration stages, and external inputs, while CAM Monitoring ensures continuous tracking of the printing process through the digital twin.

4 Results

Figure 6 shows the printing head, the MELTIO ENGINE system, the KUKA KR70 R2100 robot, the metallic enclosure, the integrated safety system, and the computer units used for process monitoring. The enclosure is equipped with an advanced safety infrastructure, including interlocks on doors and windows, as well as a 980 nm laser filter to ensure operator protection. Inside the enclosure, the robot, the printing table, and the MELTIO ENGINE head are arranged to promote a safe and efficient deposition environment. The Robotic Additive Manufacturing Cell (RAMC), based on the Laser Metal Deposition (LMD) process, was developed following the proposed methodology. This RAMC is currently installed at the GRACO Laboratory of the University of Brasília. Its configuration reflects the careful integration of components and systems as defined in the methodology and provides a dedicated experimental platform for advanced research and development in the field of additive manufacturing.

Fig. 6. Robotic Additive Manufacturing Cell Based on the LMD Process.

A video demonstration of the metal part printing process using the Robotic Additive Manufacturing Cell (RAMC) and the Laser Metal Deposition (LMD) system is available at: https://youtu.be/tVQH4ARsGFU?si=grPeRUZSnkye7nu. This video highlights the coordinated operation of the robotic system and the deposition head during fabrication. The CAD/CAPP/CAM system, integrated with a Digital Twin (DT) architecture, is embedded in the workflow to support part design, process planning, and automated execution. A second video, available at https://youtu.be/EZgtQ_zMuaI, showcases the Digital

Twin implementation. It features simulation via *KUKA.Sim* for print code validation, real-time data visualization using a 2D interface built with *Node-RED*, and cloud-based data storage through the *AWS* platform.

5 Case Studies

Figure 7 presents the printed components corresponding to two case studies, both consisting of 20 × 20 × 20 mm cubes. Case Study Part 1 (Fig. 7 (A)) was modeled using Grasshopper/Rhino3D, enabling the import of STL geometry, the definition of a deposition strategy, and the generation of KUKA Robot Language (KRL) code. Case Study Part 2 (Fig. 7 (B)) was developed using Meltio Space, which likewise imported STL geometry but employed different process parameters to generate the corresponding KRL code. A 2 mm raft was included to facilitate part removal; however, the reported measurements refer exclusively to the part excluding the raft. Key printing parameters, part dimensions, processing temperatures, materials, and slicing software used in both cases are summarized in Table 1.

A) Printed Part of Case Study 1

B) Printed Part of Case Study 2

Fig. 7. Printed parts of Case Studies.

Table 1. Parameters and Dimensions of Printed Parts.

Parameter	Case Study Part 1	Case Study Part 2
Laser Power (W)	800	845
Wire Feed Speed (mm/s)	15.30	15.28
Print Speed (mm/s)	10	10
Wire Diameter (mm)	1	1
Gas Flow (ml/min)	12000	10000
Material	316 LSi	316 LSi
Process Temperature (°C)	216.5	216.5
Layer Width (mm)	1	1
Layer Height (mm)	1.2	1.2
Part Width (mm)	19.82	20.50
Part Height (mm)	21.90	20.70
Part Length (mm)	19.85	20.10
3D Slicing Software	Grasshopper/Rhino3D	Meltio Space

6 Conclusions

The methodology proposed in this study for the development of a robotic additive manufacturing cell based on the LMD-wire process has demonstrated its effectiveness in the fabrication of complex, large-scale metallic components. The integration of a Digital Twin with the CAD/CAPP/CAM system enables seamless synchronization between physical operations and digital models, thereby enhancing precision, efficiency, and control throughout the manufacturing process. This approach not only contributes to the advancement of automation in additive manufacturing but also establishes a standardized framework that supports future system development, fostering greater interoperability and scalability. Additionally, the implementation of specialized software tools and the optimization of deposition parameters resulted in notable improvements in the dimensional accuracy of the second case study, highlighting the importance of selecting appropriate tools and process settings to ensure consistency and high-quality outcomes in wire-based additive manufacturing.

Acknowledgments. The authors would like to acknowledge the financial support provided by the National Council for Scientific and Technological Development (CNPq) and the Research Support Foundation of the Federal District (FAPDF). They also express their appreciation to the University of Brasília and the Graduate Program in Mechatronic Systems within the Department of Mechanical Engineering at the Faculty of Technology.

References

1. Liu, G., et al.: Additive manufacturing of structural materials. Mater. Sci. Eng. R. Rep. **145**, 100596 (2021). https://doi.org/10.1016/j.mser.2020.100596
2. Figueroa, B.S., Riaño, C.I., Peña C., C.A., Alvares, A.: Artificial vision system for digital inspection of dimensional and geometric measurements in closed-loop manufacturing architecture. In: 2024 IEEE Latin American Conference on Computational Intelligence (LA-CCI), pp. 1–6 (2024). IEEE. https://doi.org/10.1109/LA-CCI62337.2024.10814821
3. Maleki, E., Bagherifard, S., Bandini, M., Guagliano, M.: Surface posttreatments for metal additive manufacturing: progress, challenges, and opportunities. Addit. Manuf. **37**, 101619 (2021). https://doi.org/10.1016/j.addma.2020.101619
4. Blakey-Milner, B., et al.: Metal additive manufacturing in aerospace: a review. Mater. Des. **209**, 110008 (2021). https://doi.org/10.1016/j.matdes.2021.110008
5. Bhatt, P.M., Kabir, A.M., Peralta, M., Bruck, H.A., Gupta, S.K.: A robotic cell for performing sheet lamination-based additive manufacturing. Addit. Manuf. **27**, 278–289 (2019). https://doi.org/10.1016/j.addma.2019.02.002
6. Álvarez, P., et al.: Model-based input energy control for reproducible AISI 316L laser deposited tracks. Finite Elem. Anal. Des. **236** (2024). https://doi.org/10.1016/j.finel.2024.104184
7. Gu, D., Shi, X., Poprawe, R., Bourell, D.L., Setchi, R., Zhu, J.: Material-structure-performance integrated laser-metal additive manufacturing. Science **372**(6545), eabg1487 (2021). https://doi.org/10.1126/science.abg1487
8. Alvares, A.J., Lacroix, I., Maron, M.A. de L., Figueroa, B.S.: Desenvolvimento de uma célula de manufatura aditiva robotizada baseada no processo de deposição de metal à laser usando arame de soldagem. PRW **5**(21), 17–39 (2023)
9. Yan, L., Chen, Y., Liou, F.: Additive manufacturing of functionally graded metallic materials using laser metal deposition. Addit. Manuf. **31**, 100901 (2020). https://doi.org/10.1016/j.addma.2019.100901
10. Alvares, A., Lacroix, I., Maron, M., Figueroa, B.: Robotic additive manufacturing by laser metal deposition in the context of industry 4.0. CLIUM **23**(23), 79–103 (2023). https://10.53660/CLM-2571-23U13
11. Figueroa, B.S., Alvares, A.: Methodology for the development of a robotic additive manufacturing cell based on the laser metal deposition (LMD) system. In: Proceedings of Congreso Iberoamericano de Ingeniería Mecánica (2024). [in Spanish]
12. Phua, A., Davies, C., Delaney, G.: A digital twin hierarchy for metal additive manufacturing. Comput. Ind. **140**, 103667 (2022). https://doi.org/10.1016/j.compind.2022.103667
13. Stavropoulos, P., Papacharalampopoulos, A., Michail, C.K., Chryssolouris, G.: Robust additive manufacturing performance through a control oriented digital twin. Metals **11**(5), 708 (2021). https://doi.org/10.3390/met11050708
14. ISO 23247-1: Automation systems and integration - Digital twin framework for manufacturing - Part 1: Overview and general principles. International Organization for Standardization (2021)
15. Glaessgen, E., Stargel, D.: The digital twin paradigm for future of NASA and U.S. Air Force vehicles. In: 53rd AIAA/ASME/ASCE/AHS/ASC Structures, Structural Dynamics and Materials Conference (2012). https://doi.org/10.2514/6.2012-1818
16. Cabral, J., Gasca, E., Alvares, A.: Digital twin implementation for machining center based on ISO 23247 standard. IEEE Lat. Am. Trans. (2023). https://doi.org/10.1109/TLA.2023.10130834

17. Figueroa, B.S., Araújo, L., Alvares, A.: Development of a digital twin for a laser metal deposition (LMD) additive manufacturing cell. Adv. Autom. Robot. Res., 68–76 (2024). https://doi.org/10.1007/978-3-031-54763-8_7
18. Alvares, A.J., Figueroa, B.S., Cabral, J.V.A., Lacroix, I.: Automated defect classification in additive manufacturing LMD-wire using deep learning. J. Braz. Soc. Mech. Sci. Eng. **47**, 432 (2025). https://doi.org/10.1007/s40430-025-05740-5
19. Alvares, A.J., Rodriguez, E., Figueroa, B.: Digital-twin-enabled process monitoring for a robotic additive manufacturing cell using wire-based laser metal deposition. Processes **13**(8), 2335 (2025). https://doi.org/10.3390/pr13082335

Design Improvements and Evaluation of an Inverted Delta 3D Printer

Vinicius Camilo da Rocha, Luiz Fernando Segalin de Andrade, and Aurélio da Costa Sabino Netto(✉)

Instituto Federal de Santa Catarina, Florianópolis, SC 88020300, Brazil
asabino@ifsc.edu.br

Abstract. Additive Manufacturing continues to advance, with Fused Filament Fabrication (FFF) being one of the most widespread technologies. The availability of filaments made from new polymeric materials, including high-performance polymers, has increased significantly. These materials offer a wide range of applications but are difficult to process using conventional 3D printers. A prototype printer employing inverted delta kinematics was developed to allow research with these materials. However, initial validation revealed surface waviness and excessive dimensional variation in printed parts. This study aimed to redesign a print bed and improve the linear motion modules to improve print quality and process stability. Following these modifications, the print bed achieved higher temperatures, and when printing with ABS filament, the dimensional repeatability of printed parts improved from ±0.27 mm to ±0.09 mm. These enhancements are crucial for further machine development and enabling studies with high-performance polymers.

Keywords: Additive Manufacturing · Inverted Delta 3D Printer · ABS

1 Introduction

Additive Manufacturing (AM) has increasingly taken a prominent position as a manufacturing process. Among AM processes, Fused Filament Fabrication (FFF) remains the most widely adopted globally [1].

Recent years have witnessed a substantial expansion in the range of polymeric materials available for FFF, notably with the emergence of high-performance polymers such as polysulfone (PSU), polyetherimide (PEI), and polyetheretherketone (PEEK). Generally, these materials exhibit high thermal decomposition temperatures, long-term durability, and elevated glass transition temperatures, which are relevant characteristics for applications in the healthcare, automotive, and aerospace sectors [2, 3].

The processing of these materials requires elevated extrusion and bed temperatures, along with stringent control of the chamber temperature—factors that significantly hinder their use in low-cost 3D printing systems.

O. F. Farías Fuentes et al. (Eds.): CIBIM 2024, *Proceedings of the XVI Ibero-American Congress of Mechanical Engineering*, pp. 281–296, 2026.
https://doi.org/10.1007/978-3-032-22823-9_21

With the aim of conducting research with high-performance polymers, a 3D printer with inverted delta kinematics was developed. Thomazetti [4] carried out the electro-electronic integration and conducted tests to validate the prototype. Overall, successful printing of parts was achieved; however, the occurrence of waviness patterns on the surface and limitations in heating the print bed were identified. Based on the initial tests, Feistauer [5] conducted a study in which a series of adjustments were suggested to mitigate the quality issues of the printed parts.

This work presents targeted enhancements to the print bed and linear motion modules, with the primary objective of mitigating the quality deficiencies initially identified in the printed parts. These modifications represent a novelty and constitute a fundamental step in the continued development of the inverted delta 3D printer, aiming to enable its future application in the processing of high-performance polymers.

2 Background

2.1 Delta Kinematics for 3D Printers

Several kinematic configurations are used in 3D printers, each affecting speed, precision, and mechanical complexity. Among these, architectures cartesian and delta are the most widely adopted in Fused Filament Fabrication (FFF) due to their proven reliability, ease of control, and availability of open-source designs (Fig. 1).

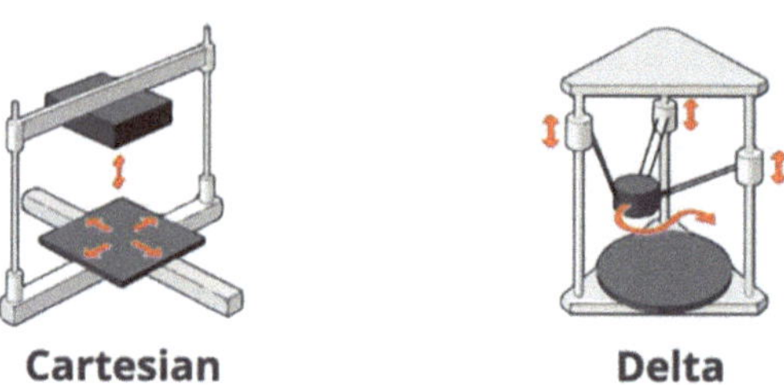

Fig. 1. Schematic representations of cartesian and delta printers [6].

The delta-configured robot was developed by Reymond Clavel in the 1980s at the École Polytechnique Fédérale de Lausanne (EPFL) [7]. The idea was to use parallelograms to construct a parallel robot with four degrees of freedom: three translational and one rotational [8]. The architecture of a robot is defined by how its kinematic structures and respective actuators move. Delta robots possess a parallel kinematic chain where the actuators must work together to achieve the desired movement, as one actuator influences the others [9].

Delta printers have three actuators positioned vertically and equidistantly around a circle, unlike Cartesian printers, where the actuators are parallel to each of the three orthogonal planes. By controlling the actuators, it is possible to direct the extruder nozzle to any point in the cylindrical printing volume. The movement, illustrated in Fig. 2, involves the positioning of each axis, forming a right triangle between the fixed rod (c) and the imaginary vertical and horizontal lines serving as the legs, with the horizontal lines (b) of the three triangles moving together linked to the end effector. Only the vertical lengths (a) are variable, providing three degrees of freedom to the machine [9].

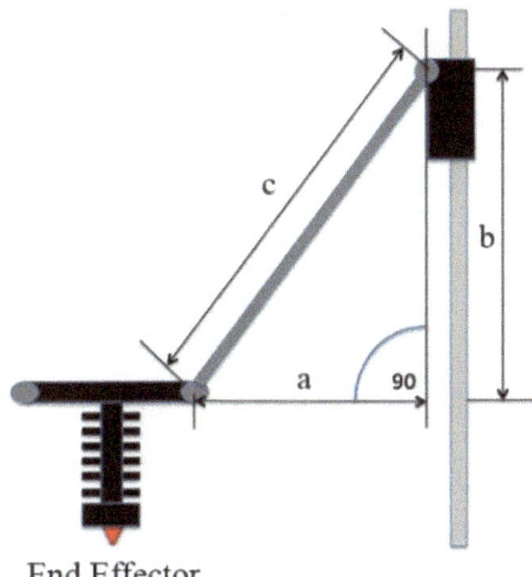

Fig. 2. Movements of the axes of a delta machine [10].

The prototype developed for 3D printing high-performance polymers was inspired by an open-source project from the Diyouware group called Twin Teeth PCB Mini-Factory [11]. The project consists of a machine with an inverted delta configuration capable of performing various processes, such as machining, 3D printing, and the deposition of paste-like materials, among others (Fig. 3). The inverted configuration allows the table to move while the tool remains stationary. This solution proved to be advantageous for applications with high-performance polymers, as it isolates the motors and the extrusion module, preventing motor heating and the filament buckling effect.

Fig. 3. TwinTeeth PCB Mini-Factory [11].

3 Reconfiguration of the Inverted Delta Printer

3.1 Print Bed

The initial print bed, incorporating a MK2Y heater (12V/90W), was replaced due to thermal issues. Its direct attachment to the end effector resulted in overheating of the delta support rods (Fig. 4).

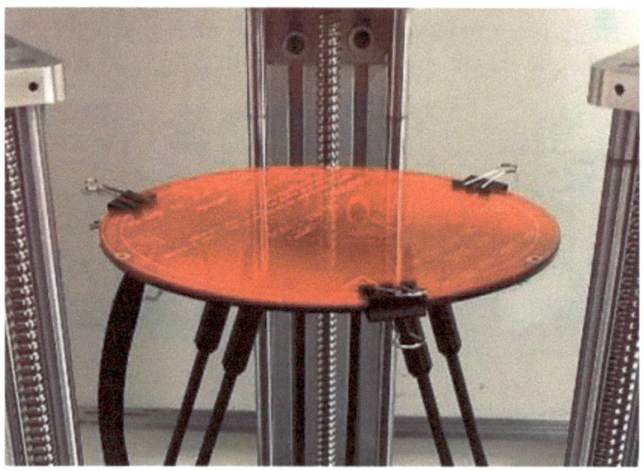

Fig. 4. Initial print bed directly assembled at the end effector [4].

The redesigned print bed incorporated a 150 mm diameter Keenovo self-adhesive silicone heater (12V/300W), which was affixed to an aluminum plate positioned away from the end effector. A leveling mechanism—comprising screws, springs, and adjustment nuts—was implemented to ensure proper alignment of the build surface. The main components and the final assembly are presented in Fig. 5.

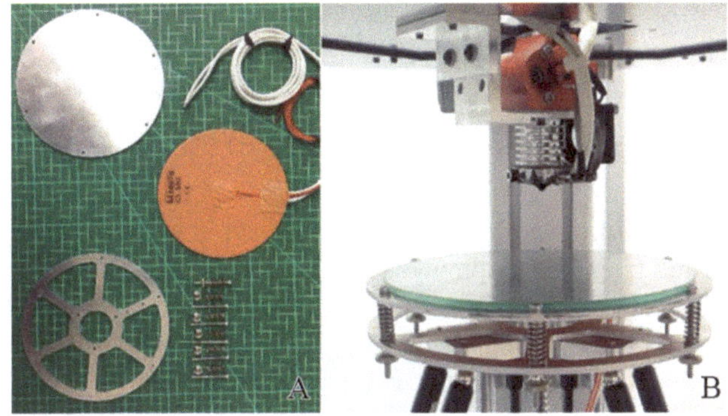

Fig. 5. Improved print bed. (A) main components and (B) final assembly at the end effector.

3.2 Linear Motion Modules

Printing tests revealed that surface waviness in the fabricated parts was likely attributable to mechanical play in the carriage guidance system. To mitigate this, the LME12UU linear bearings (Fig. 6A) were replaced with extended LMEK12LUU bearings (Fig. 6B), offering improved stability and alignment.

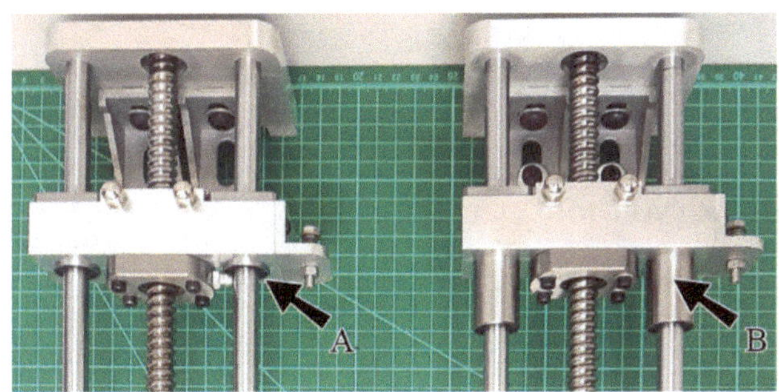

Fig. 6. Replacement of the linear bearing LME12UU (A) with LME12UUU (B).

The previous configuration employed multi-beam aluminum flexible couplings (Fig. 7A), which exhibited limited performance under frequent direction reversals during printing trajectories. These were replaced with JAW-type couplings (Fig. 7B), offering quicker response and improved resistance to deformation during motion reversals.

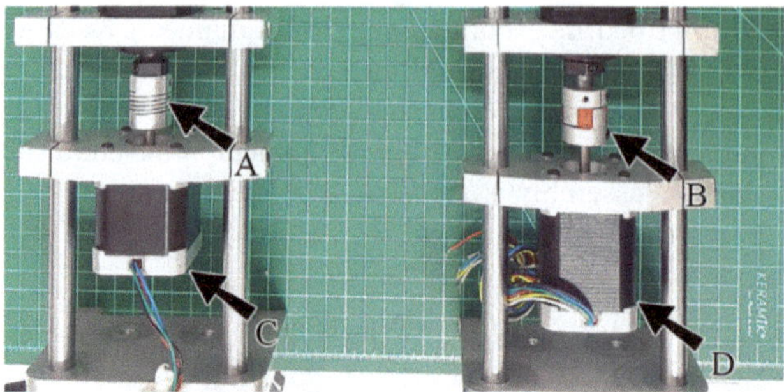

Fig. 7. Replacement of the flexible couplings (from A to B) and stepper motors (from C to D).

In the movement tests, step losses were identified at speeds above 50 mm/s due to insufficient torque from the employed stepper motors (Fig. 7C). Therefore, stepper motors with 6.5 kgf.cm torque from Vurtz Motors, model VZS1760-065-1206, were installed as shown in Fig. 7D. Since the same Pololu A4988 stepper motor drivers were maintained, adjustments to the reference voltage (Vref) of the driver were necessary for the use of the new motors (Fig. 8).

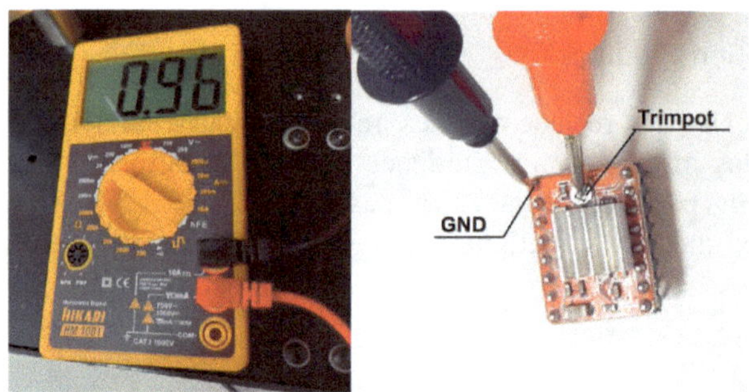

Fig. 8. Adjustment of the reference voltage of the A4988 driver

4 Materials and Methods

4.1 Materials

For printing the test specimens, ABS filaments with 1.75 mm diameter in blue and yellow colors were used, both from manufacturer 3D Fila.

4.2 Printing Parameters

The following parameters listed in Table 1 were used for the fabrication of the parts. Before starting the process, the print bed was prepared with liquid adhesive from manufacturer SLIM 3D, composed of non-ionic polymer and ethyl alcohol, specifically designed to aid adhesion of PLA and ABS filaments on glass surfaces.

Table 1. Printing parameters used in the fabrication of the parts.

Parameter	Value
Perimeter printing speed	30 mm/s
Infill printing speed	50 mm/s
Layer height	0,20 mm
First layer height	0,35 mm
Layer width	100%
Base/top solid layers	2
Number of perimeters	2
Infill density	10%
Printing temperature	230 °C
Bed temperature	100 °C
Print flow	100%

4.3 Machine Improvements Evaluation

Print Bed Performance

To evaluate the heating behavior of the print bed, a K-type thermocouple was fixed to the central region of the glass plate. Through the manual control interface in Repetier-Host, a study was conducted with six setpoints, with 20 °C increments between them. The heating curve was monitored using a Picolog TC-08 model datalogger. Throughout the study, thermal imaging was performed using a FLUKE VT04A visual infrared thermometer.

Print Quality

Print quality assessment was performed using the xyzCalibration_cube model, sourced from Thingiverse.com (Fig. 9). This standardized 20 mm cube is extensively used in dimensional accuracy evaluations in 3D printing. The inclusion of axis markings (X, Y, and Z) on its faces enables clear identification of each axis and aids in diagnosing direction-specific deviations.

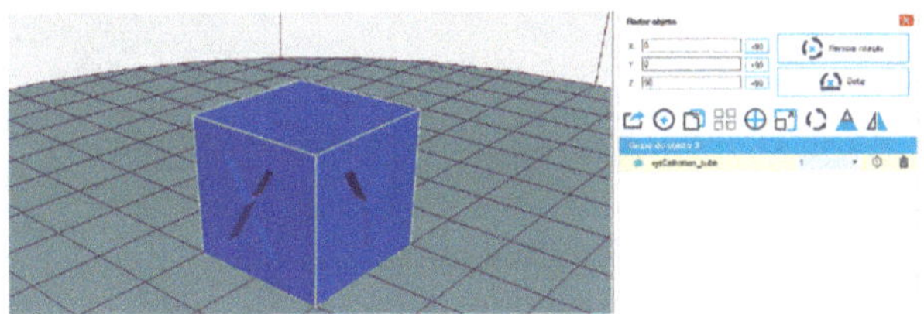

Fig. 9. "xyzCalibration_cube" STL model.

Print Scale

In order to investigate the progression of dimensional errors along each axis, a series of calibration cubes with side lengths increasing in 5 mm increments was printed. In total, 10 cubes were fabricated, ranging from 5 mm to 50 mm in size.

Multiple Parts Printing on the Platform

To evaluate the printing of more slender parts on the build platform, a set of 6 test specimens was printed according to ASTM D638 Type V standard, with 4 dimensions selected for evaluation (Fig. 10).

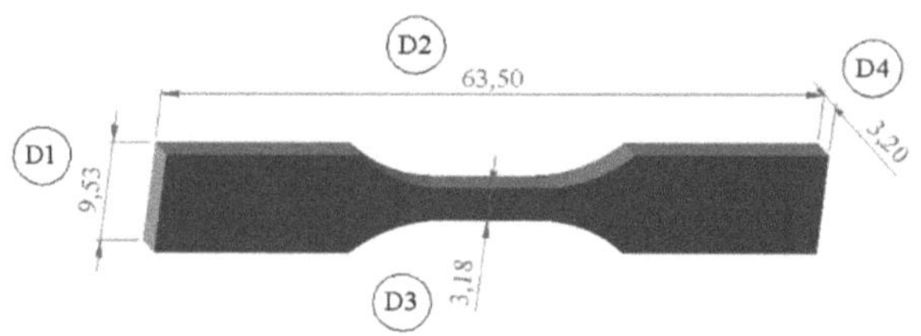

Fig. 10. Dimensions evaluated in the ASTM D638 type V specimen.

Repeatability Study

To assess whether the modifications enhanced equipment repeatability, the same L-shaped model used by Thomazetti [4] was employed. This model includes five predefined dimensions for measurement, as illustrated in Fig. 11.

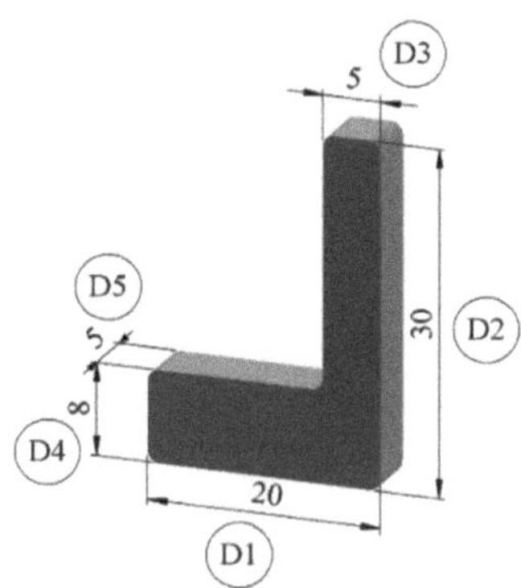

Fig. 11. Geometry used for evaluating repeatability.

The models were positioned according to the X+, X−, Y+, and Y− printing directions, with 3 parts printed for each direction. To evaluate systematic error, or bias (b), first the arithmetic mean ($\overline{x}$) of the samples for each dimension was calculated and then subtracted from the nominal value (x), as shown in Eq. 1:

$$b = \overline{x} - x \tag{1}$$

For calculating repeatability, a Student's t-Distribution was used, considering a 95% confidence interval and the number of degrees of freedom (v) according to Eq. 2:

$$v = n - 1 \quad (2)$$

Thus, the repeatability (r) given by Eq. 3 is the symmetric value of t multiplied by the standard deviation (s):

$$r = t \cdot s \quad (3)$$

5 Results and Discussion

5.1 Evaluation of the Print Bed

Figure 12 presents the heating curve relative to the evaluated setpoints. The previous print bed was unable to reach temperatures above 100 °C, which hindered print tests using ABS. The redesign allowed the base to reach temperatures of up to 140 °C, with a heating time of approximately 5 min for stabilization.

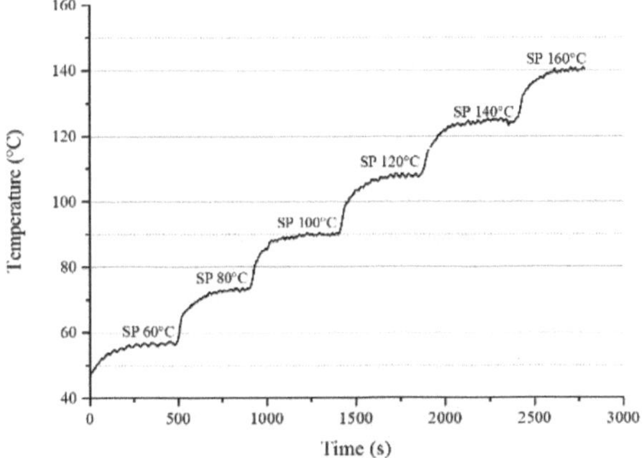

Fig. 12. Heating curves for the different setpoints.

Temperature measurements taken at the center of the platform averaged around 90% of the programmed setpoint values. This discrepancy is attributed to the positioning of the thermistor (embedded in the heating element) relative to the thermocouple on the glass surface. Future tests will be conducted using more accurate thermocouples, followed by firmware calibration, to ensure that the displayed temperature accurately reflects the temperature at the glass surface.

Figure 13 presents a thermal image of the print bed assembly during the heating process. It can be qualitatively observed that the new design prevents excessive heat from reaching the effector area and support rods—an issue identified in the previous platform version, which contributed to the failure of one of the rods.

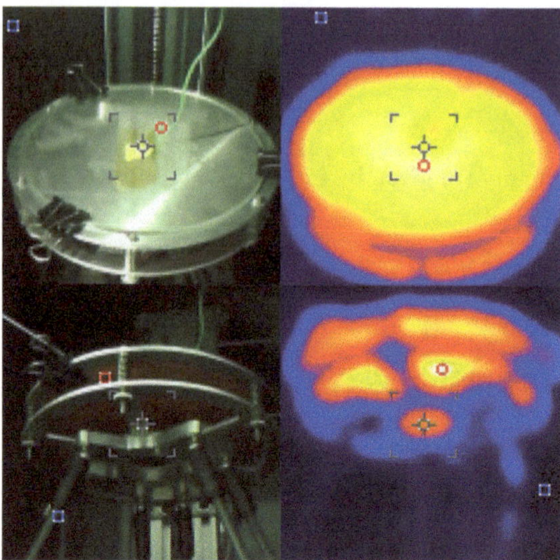

Fig. 13. Heat distribution on the print bed.

5.2 Print Quality Assessment

Figure 14 presents the printed calibration cube results. Model A reflects the part produced following the design improvements, whereas Model B corresponds to the earlier configuration. The substantial reduction in surface waviness on the vertical X-axis face of Model A highlights the positive impact of the implemented modifications.

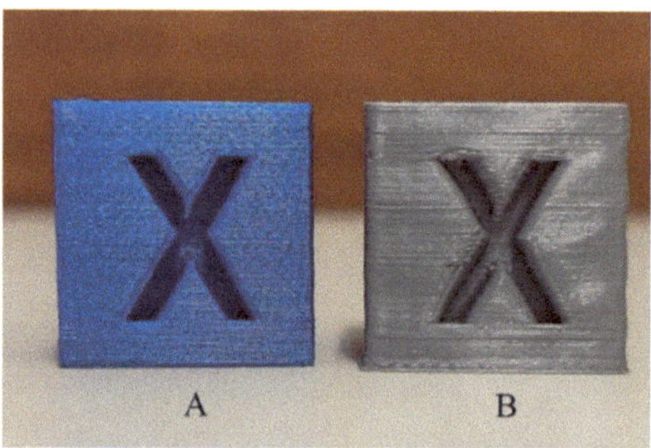

Fig. 14. Printed calibration cubes: (A) after the improvements and (B) earlier configuration.

5.3 Dimensional Evaluation

Figure 15 presents the measurement results along the X, Y, and Z axes of the printed cubes as a function of scale increase. While the Z-axis values show good agreement with the nominal dimensions, the X and Y axis values exhibit proportional deviation from the nominal values, suggesting the presence of a systematic error. Figure 16 displays a graph of the calculated error relative to the nominal dimension for each printed cube.

The graph indicates that the most significant errors are concentrated along the X and Y axes, with an average deviation of approximately -3%.

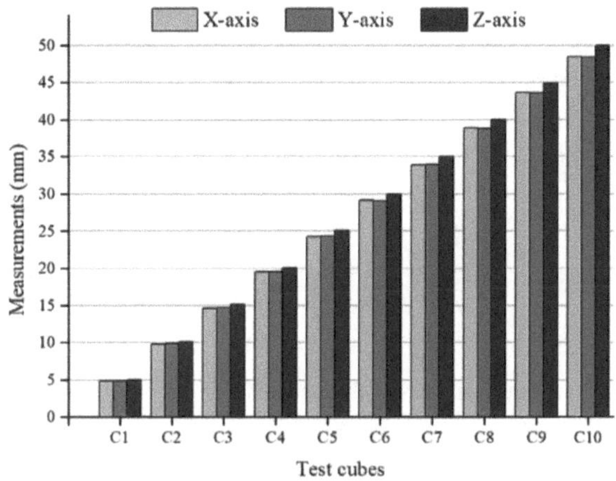

Fig. 15. Measurements of the test cubes for the three axes.

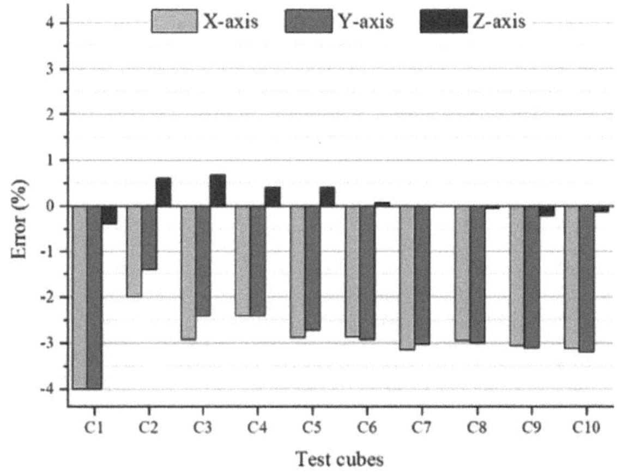

Fig. 16. Calculation of the error of the test cubes for the three axes.

Based on the calculated average deviation, firmware settings were adjusted, and the cubes from C1 to C10 were reprinted for evaluation. Figure 17 shows the measurement results along the X, Y, and Z axes for the newly printed cubes.

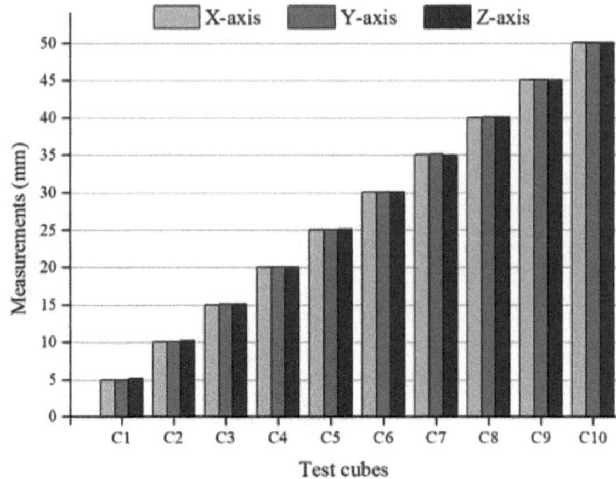

Fig. 17. Measurements of the new test cubes for the three axes.

Figure 18 illustrates the error calculations for these new cubes. For dimensions starting at 15 mm, the errors along the X and Y axes were below 0.5%, which is considered acceptable for the equipment.

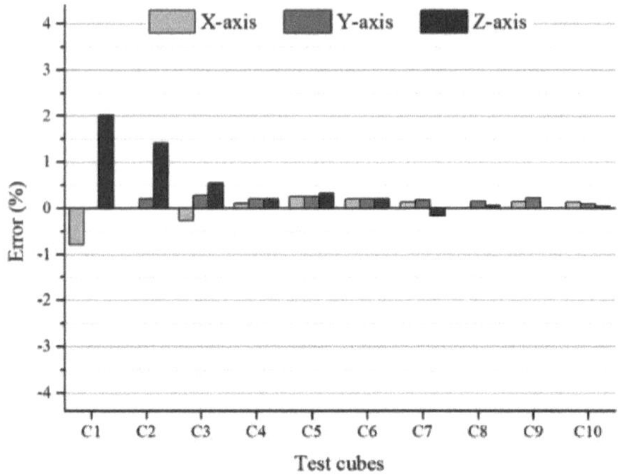

Fig. 18. Calculation of the error of the new test cubes for the three axes.

5.4 Evaluation of Multi-Part Printing

Good adhesion was observed across a larger surface area of the print bed. After printing, manual force was applied to remove the parts from the glass; however, they only detached after the platform cooled down. Percentage errors were calculated in relation to the nominal dimensions. Analysis of the data in Table 2 reveals that the parts exhibited minimal dimensional variation and closely matched the nominal dimensions, corroborating the findings of the previous study (Fig. 19).

Table 2. Result of the measurements of the test specimens.

Measurements	Nominal (mm)	Average (mm)	Error (%)
D1	9,53	9,53	0,00
D2	63,5	63,51	0,01
D3	3,18	3,23	1,57
D4	3,2	3,21	0,42

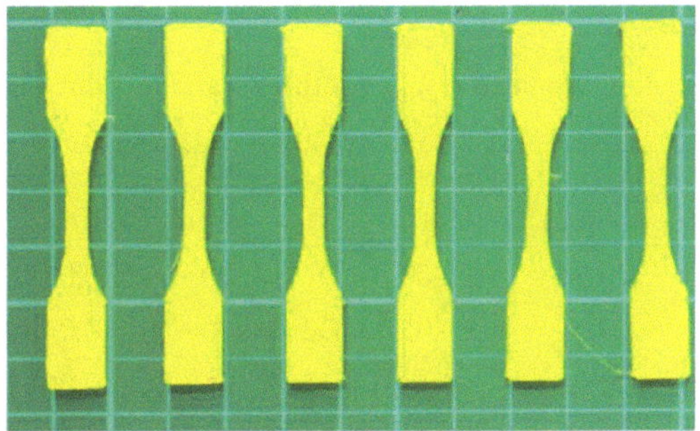

Fig. 19. ASTM D638 Type V printed test specimens.

5.5 Repeatability Evaluation

One of the sets of "L" shaped parts produced for repeatability assessment is shown in Fig. 20. Repeatability calculations for the X+, X−, Y+, and Y− directions are presented in Tables 3 and 4, respectively.

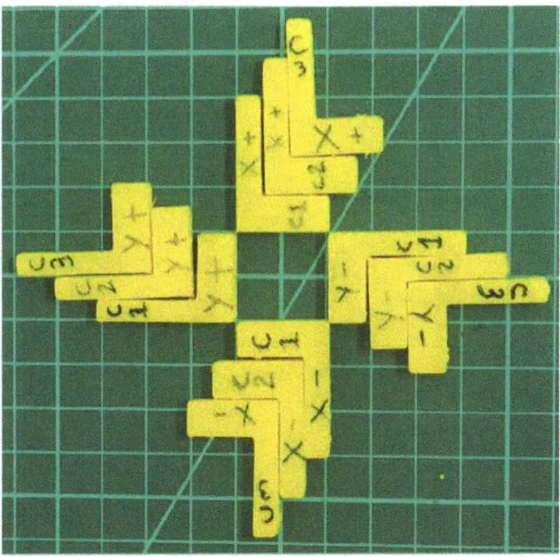

Fig. 20. Set of printed L parts for evaluating repeatability.

Table 3. Calculated results for the X+ and X− directions.

X+	x (mm)	b (mm)	s (mm)	r (mm)	X−	x (mm)	b (mm)	s (mm)	r (mm)
D1	20	0,05	0,01	±0,05	D1	20	0,07	0,01	±0,05
D2	30	0,02	0,02	±0,09	D2	30	0,07	0,03	±0,13
D3	5	0,07	0,03	±0,13	D3	5	0,06	0,02	±0,09
D4	8	0,01	0,01	±0,05	D4	8	0,02	0,02	±0,09
D5	5	0,03	0,01	±0,05	D5	5	0,08	0,02	±0,09

Table 4. Calculated results for the Y+ and Y− directions.

Y+	x (mm)	b (mm)	s (mm)	r (mm)	Y−	x (mm)	b (mm)	s (mm)	r (mm)
D1	20	0,00	0,02	±0,09	D1	20	0,03	0,03	±0,13
D2	30	0,03	0,01	±0,05	D2	30	0,03	0,04	±0,18
D3	5	−0,01	0,03	±0,13	D3	5	0,06	0,00	±0,00
D4	8	0,04	0,02	±0,09	D4	8	0,05	0,05	±0,22
D5	5	0,07	0,04	±0,18	D5	5	0,03	0,01	±0,05

Measurement results show that, following the firmware adjustments, the equipment consistently produced parts with dimensions approximately 0.05 mm above the nominal values (Fig. 21). Analysis across build directions indicated a random distribution, suggesting that no alignment issues are affecting the machine's performance.

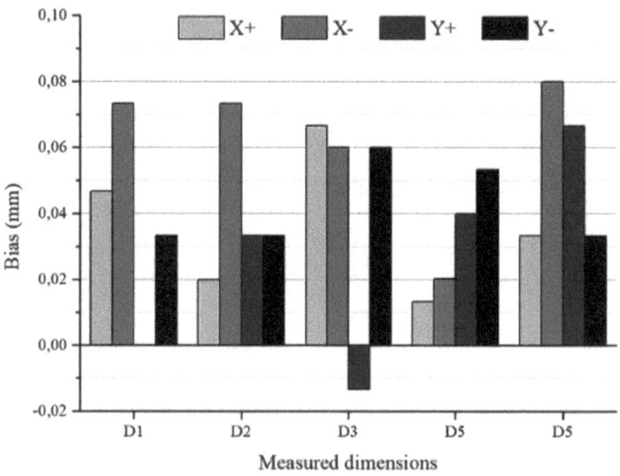

Fig. 21. Bias of the measurements for the directions.

To obtain a comprehensive assessment of the machine's repeatability, calculations were performed considering all measurements per section. With a sample size of n = 12, degrees of freedom of 11, and a t-value of 2.201, the results are summarized in Table 5. The maximum repeatability obtained was ±0.09 mm—three times lower than the value reported in initial tests conducted by Thomazetti (±0.27 mm).

Table 5. Repeatability values considering the global set of parts.

Global Measurements	x (mm)	b (mm)	s (mm)	r (mm)
D1	20	0,04	0,03	±0,07
D2	30	0,04	0,03	±0,07
D3	5	0,04	0,04	±0,09
D4	8	0,03	0,03	±0,07
D5	5	0,05	0,03	±0,07

6 Conclusions

This study proposed and implemented a set of structural and functional improvements to the print bed and linear motion modules of a delta 3D printer, with the objective of enhancing its performance and reliability, particularly for applications involving high-performance polymers in future.

Thermal performance tests demonstrated that the redesigned print bed was capable of reaching 140 °C in approximately 5 min, a temperature and heating rate suitable for processing engineering-grade thermoplastics. Furthermore, the new design effectively isolated the heat source, preventing undesired thermal transfer to the effector and motion rods that was one of the critical shortcomings identified in the initial bed prototype.

Modifications to the linear motion system, including the replacement of key components, resulted in a marked improvement in the surface quality of printed parts, eliminating previously observed surface waviness. Dimensional accuracy also benefited significantly: based on measurement data and statistical analysis, the random error was reduced from ±0.27 mm to ±0.09 mm in 95% of the cases. This represents a threefold improvement in repeatability, underscoring the robustness of the proposed enhancements and their relevance for research and applications requiring high dimensional precision and material performance.

These results highlight the substantial contribution of this work toward enabling reliable, high-temperature-capable additive manufacturing platforms, and reinforce the potential of the inverted delta printer architecture for advanced material processing.

Future work will focus on printing high-performance polymers such as PSU, PEI, and PEEK, aiming to evaluate the dimensional accuracy, surface quality, and overall integrity of the fabricated parts.

Acknowledgments. The authors gratefully acknowledge the support provided for this research through funding from IFSC, CNPq, and FAPESC.

References

1. Gibson, I., Rosen, D.W., Stucker, B.: Additive Manufacturing Technologies: 3D Printing, Rapid Prototyping, and Direct Digital Manufacturing, 2nd edn. Springer, Cham (2015)
2. Weyhrich, C.W., Long, E.T.: Additive manufacturing of high-performance engineering polymers: present and future. Polym. Int. **71**(5), 532–536 (2022)
3. Haleem, A., Javaid, M.: Polyether ether ketone (PEEK) and its manufacturing of customised 3D printed dentistry parts using additive manufacturing. Clin. Epidemiol. Glob. Health **7**(4), 654–660 (2019)
4. Thomazetti, J.L.S.: Integração eletroeletrônica em impressora 3D com cinemática delta invertida [Bachelor's thesis]. Florianópolis: IFSC, p. 74 (2019)
5. Feistauer, G.B.: Aprimoramento de impressora 3D com cinemática delta invertida [Bachelor's thesis]. Florianópolis: IFSC, p. 81 (2020)
6. Types of 3D Printers, the definitive guide 2020 (2020). https://bitfab.io/blog/types-of-3d-printers/
7. Bonev, I.: Delta Parallel Robot – The Story of Success (2001). http://parallelmic.org
8. Pessina, L.: Reymond Clavel, creator of the Delta Robot reflects on his career (2013). Available from: https://sti.epfl.ch/reymond-clavel-creator-of-the-delta-robot-reflects-on-his-career/
9. Tomei GPS. Desenvolvimento de um protótipo de um robô de cinemática paralela do tipo delta para impressão tridimensional de peças [Bachelor's thesis]. Lajeado: UNIVATES, p. 88 (2015)
10. Bell, C.: 3D Printing with Delta Printers. Apress Berkeley (2015)
11. DIYOUWARE. Diyouware TwinTeeth: The PCB mini-factory (2019). http://diyouware.com/

Exploratory Analysis of Plastic Waste Utilization as Fine Aggregates in Cement and Geopolymer Mortar

Rakesh Kuchana[1], Samatha Bairi[2], Cristian Canales[3](✉), Siva Avudaiappan[4], and Krishna Prakash Arunachalam[4]

[1] Guntur Engineering College, Opp. Katuri Medical College, Yanamadala 522019, Andhra Pradesh, India

[2] Universal College of Engineering and Technology, Guntur 522001, India

[3] Mechanical Engineering Department, Universidad de Concepción, Concepción, Chile
cristcanales@udec.cl

[4] Departamento de Ciencias de la Construcción, Facultad de Ciencias de la Construcción Ordenamiento Territorial, Universidad Tecnológica Metropolitana, Santiago, Chile
{s.avudaiappan,k.prakash}@utem.cl

Abstract. Chile is currently dealing with a growing problem related to plastic waste, amounting to 1.2 million tonnes per year. Experts are currently examining the possibility of using solid waste resources as construction materials. This study explored the utilization of plastic waste (PW) polyethylene terephthalate and polyolefin waste as a replacement for fine aggregate in geopolymer mortar. PW was utilised as a substitute for natural fine aggregate in geopolymer mortar, replacing 0%, 10%, 20%, 30%, and 40% by volume. The resulting mixtures were then compared to a standard cement mortar. The mortars were subjected to laboratory experiments to determine their density, water absorption, porosity, flow, thermal conductivity, UPV and mechanical strengths. The findings of these tests are then compared to conventional Portland cement mortar. The findings demonstrated that as the amount of plastic trash increased, the mechanical characteristics of the geopolymer mortar declined. Increases in Plastic Waste concentration from 0% to 40% resulted in a drop in compressive strength 70% for geopolymer mortars and for 53% cement mortars. Cement mortars and geopolymer that included PW showed an improvement in the strength of flexural comparing to compressive strength ratio. Furthermore, as the PW concentration increased, mortars thermal conductivity and density are noticeably decreased; for geopolymer mortar, density ranged from 2222 to 1886 kg/m and for cement mortar, from 2389 to 1957 kg/m^3, and the thermal conductivity for geopolymer mortar ranges from 1.4354 to 0.7219 W/m·K, and for cement mortar it ranges from 2.1953 to 1.2190 W/m·K. Plastic waste lower mortar's weight and improve its thermal performance. This novel approach reduces plastic waste and makes concrete lighter and better for thermal insulation.

Keywords: Geopolymer Mortar · Plastic Waste Utilization · Mechanical Properties · Fine aggregate replacement · Thermal conductivity

O. F. Farías Fuentes et al. (Eds.): CIBIM 2024, *Proceedings of the XVI Ibero-American Congress of Mechanical Engineering*, pp. 297–308, 2026.
https://doi.org/10.1007/978-3-032-22823-9_22

1 Introduction

The waste plastic generation has increased worldwide day by day [1]. The USA's Environment Programme reports that the amount of trash plastic created each year is increasing at a rate of 4.6%, reaching 400 million tons. Nine percent of the world's plastic trash gets recycled, while twelve percent is incinerated [2], most of the remaining plastic 79% ends up in landfills or water bodies. The expel of harmful air pollutants like furan, dioxins, and particulate matter, as well as an increase in the concentrations of microplastics and heavy metals in aquatic systems due to degradation processes, and a decrease in water permeability and soil fertility of agricultural yields, are some of the major environmental problems caused by careless disposal and open burning of plastic waste.

There is an immediate need for researchers to develop safe, effective, affordable, and environmentally friendly ways to deal with the growing amount of single-use plastic waste (PW) [3]. The desirable attributes, such as low water absorption capacity, low electrical and thermal conductivity, corrosion resistance, and high impact resistance capacity, of waste plastic indicate that it could be a promising efficient substitute and environmentally friendly for natural coarse and fine aggregate in the production of cement concrete. This utilisation of waste plastic aligns with the principles of a circular economy.

Aggregate makes up about 60–80% of concrete the total annual consumption of aggregates needed to meet this demand is around 13.12 billion tons [4, 5]. The mining and processing of natural aggregates via a quarry is an ecologically detrimental activity that results in increased energy consumption and the release of greenhouse gases (GHGs). It also leads to landscape degradation, destabilisation of land, contamination of water resources, atmospheric pollution, and adverse social effects such as vandalism. Utilising plastic waste as a replacement for natural aggregates in recycling efforts has the potential to effectively decrease the occurrence of environmental dangers and social problems due to mining of natural aggregates [6].

Research on the topic of determining the strength of concrete based on waste plastic aggregate has been done in the past [7]. Concrete made with recycled plastic aggregate has a lot of potential, but there is a dearth of research on its endurance [8]. More water was absorbed by specimens of mortar that included only PET aggregates as opposed to a mixture of PET and sand [9]. When 75% of polyethylene (LDPE) aggregate of low density is utilized in replacement of fine natural aggregate there is a drop in sorptivity coefficient of 31% [10]. Another study looked at the possibility of using HDPE as a suitable substitute for up to 12% CA with regard of water adsorption capacity [11]. In addition, it is mentioned that adding an extra 6% HDPE increased the water adsorption capacity by 75%.

Building materials that include WP have shown favourable material qualities in terms of decreased density, thermal conductivity and unit weight, in comparison to concrete that do not incorporate WP substitution. Nevertheless, increasing in the WP quantity leads to a decline in the mechanical qualities of building materials. Some applications of WP in building materials are asphalt mixtures, bitumen concrete, paver blocks, and bricks [12]. Sodium silicate is the key ingredient in geopolymer concrete, which has the added advantages of lower energy consumption and carbon dioxide emissions compared to conventional concrete [13]. Geopolymers are inorganic alumino-silicate polymers of three-dimensions [14] made by reacting aluminosilicate-rich agricultural wastes, municipal, industrial, and byproducts like rice husk ash, GGBS, fly ash, etc. with alkaline aqueous solution [15, 16].

In the research conducted by Hama and Hilal [17] the results of using plastic waste as a partial sand replacement in concrete is observed, the strength of compression values ranged from around 65 MPa to 37 MPa, and it was shown that the strength decreased with increasing plastic percentage compared to the control mix without WP. It was noted that the plastic debris was softer than the natural aggregate, which led to reduction in strength. In addition, Pacheco-Torgal [18] used polyethylene terephthalate (PET) from plastic bottles to partially replace sand in OPC concrete, with a maximum plastic replacement of 50%. The effects of this substitution were studied. Substituting recycled PET into environmentally friendly concrete at specified rates was determined to be feasible by the research. This decision was made in part because using PET instead of sand reduces concrete's self-weight, making it better suited for use in non-structural parts that don't need a lot of compressive strength.

In the studies conducted by, Shaikh [19] he utilised recycled PET fibre of length 12–19 mm to strengthen OPC-fly ash-based concrete, and geopolymer concrete. The strength of compression of geopolymer composites was increased to 57 MPa and 47 MPa, respectively, with the addition of 1.0 & 1.5% of PET fibres. In comparison to the plastic-free control mix, geopolymer concretes made of PET waste, fly ash and blast-furnace slag reduced the strength of tensile to 1.2 MPa and the compressive strength to 13.8 MPa. Consequently, the mechanical characteristics were marginally enhanced due to the 10% substitution of plastic granules for sand in the fine aggregate [20]. In their evaluation of innovative geopolymer composites for sustainable construction, Lazorenko et al. [21] used WP as fillers and included waste plastic of varying sizes and forms. Findings demonstrated that geopolymer concrete containing small WP particles exhibited excellent strength, however geopolymer workability declined with increasing WP particle size. The authors recommended that geopolymer concrete employ recycled plastics at a rate of 50% as a substitute for coarse aggregates. The utilization of WP as a fine aggregate in geopolymer concrete has not been extensively studied.

The purpose of this research is to find the mechanical and microstructural properties of the geopolymer that results from using WP as fine aggregate replacement at geopolymer mortar production in a sustainable manner and comparing with normal mortar. The present study's goals are to strengthen what is already known about geopolymer concrete research, promote a more environmentally friendly and sustainable future in the building industry, and inspire more environmentally conscious practices in the creation of geopolymer concrete.

2 Experimental Study

2.1 Mix Production and Test Method

As aluminosilicate materials, dry fly ash with a specific gravity of 2.4 and about 40% of particles smaller than 7 μm was utilised. Figure 1 shows the plastic waste fine aggregates utilised in this research. For this work, we used plastic cullet that was between 0.75 and

1 mm in size. The geopolymer mortar was made using the following ingredients: 10 M NaOH, a 2.75:1 aggregate to binder ratio, a 0.70:1 ratio of liquid alkali solution to fly ash, and a 1.00:1 ratio of sodium hydroxide to sodium silicate. The properties of fly ash and Cement are given in Table 1.

Fig. 1. Plastic waste fine aggregate and casted specimens from plastic aggregate.

Table 1. Chemical Composition of OPC and Fly ash.

Chemical Composition	Cement	Fly ash
Si02	20.4	42.6
Al2O3	6.55	24.4
Fe2O3	3.56	12.8
CaO	66	14.3
MgO	1.75	2.1
K2O	0.54	2.6
Na2O	0.25	0.21
SO3	0.42	0.53

A total of five different mixtures were created, with corresponding volume ratios of 0%, 10%, 20%, 30%, and 40% PW to sand: GP0, GP10, GP20, GP30, GP40. The three cement mortar mixes were OPC0, OPC10, and OPC20, where OPC0 represents 0% PW to the volume of sand replacement, OPC10 represents 10%, and OPC20 represents 20%. The sand substitute level was limited to 20–40% in the present investigation owing to workability concerns. The water by cement ratio employed was 0.50, while the binder to aggregate ratio was 2.64. The specific ratios for the cement and geopolymer mortar mixtures can be found in Table 2.

The process of mixing the geopolymer began by combining fly ash and a solution of sodium hydroxide for a duration of 5 min, then the aggregates were added and thoroughly

Table 2. Mix Proportions of cement mortar and geopolymer mortar.

Mix	OPC	FA	NH	NS	Sand	PW	Water
OPCO	546	-	-	-	1624	-	294
OPC10	546	-	-	-	1462	83	294
OPC20	546	-	-	-	1299	132	294
GP0	-	493	159	159	1624	-	-
GP10	-	493	159	159	1462	83	-
GP20	-	493	159	159	1299	132	-
GP30	-	493	159	159	1137	204	-
GP40	-	493	159	159	974	298	-

blend for a duration of 5 min, solution of sodium silicate was introduced and well blended for an additional duration of 5 min. The geopolymer mortar is then poured into moulds and subjected to vibration to a duration of 10 s to ensure effective compaction for minimising the segregation. Then samples were enveloped on clingfilm to avoid moisture losses after that the specimens will be subjected to curing in an oven at a temperature of 60ºC for a duration of 48 h.

After curing, the specimens are removed from the moulds and placed in a controlled environment with a relative humidity of 50% and a temperature of 22ºC for a period of 28 days. The cement mortar was casted as per the guidelines of ASTM C305.

Following the process of mixing, the specimens were subjected to vibration and cured as per the procedure followed for geopolymer; after drying the specimens were removed from the mould and it is subjected to immersion in water for the curing prior to testing. The physical properties and mechanical strength of specimens with 28 days of hardened mortar are assessed using various tests. These tests included measuring the water absorption, porosity, density, UPV, splitting tensile strength, flexural strength and compressive strength. The thermal conductivity was determined as specified in the ASTM D5930-17.

The specimens, which had dimensions of 50 mm × 50 mm were subjected to compressive strength. A prism measuring 160 × 40 × 40 mm was subjected to a strength of flexural test while the cylinder of 100 mm diameter and 200 mm height was subjected to strength of splitting tensile test. The specimens having dimensions of 100 mm × 100 mm were subjected to tests to measure thermal conductivity and ultrasonic pulse velocity. The findings are the mean values obtained from three samples in each combination.

3 Result and Discussion

3.1 Flow Value

The cement mortar and geopolymer mortar's flow values with varying percentages of PW are illustrated in Fig. 2. As the volume of PW increased, the flow value dropped sharply. With a 40% increase in PW, the geopolymer mortar's flow dropped from 141%

to 72%. Adding 20% PW trash to OPC reduced the flow rate from 121% to 111%. The particles have high angularity and cause the mortar's flow to decrease. Other research has reached similar conclusions [21].

In addition, according to Prakash et al. [22], local masons would prescribe mortar flow rates of 65%–75% for brick joint mortar and 100%–110% for clay brick plastering. Flow values ranging from 99% to 112% were recorded for both cement and geopolymer mortar when sand was replaced with 10% and 20% PW by volume, respectively. This range is suitable for plastering operations. Brick joint work was successfully accomplished by substituting 30% to 40% of the volume of PW trash for natural fine aggregate in geopolymer mortar.

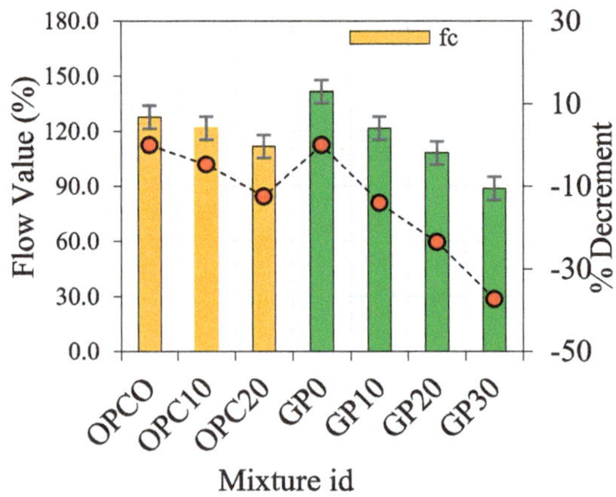

Fig. 2. Flow Value of Cement mortar and Geopolymer mortar.

3.2 Water Absorption, Porosity, Density, UPV, Density of Cement and Geopolymer Motors

The findings for the Water Absorption, Porosity, Density, UPV, Density of mortars is illustrated in Fig. 3. The density of cement mortar and geopolymer mortar, which included PW, dropped as the amount of PW increased. Other researchers have also shown PW causes a decrease in the dry density of the mortar [23]. Furthermore, substituting sand with PW enhanced permeability and decreased density. Table 3 displays the decrease in density resulting from substituting of sand with plastic waste.

The findings demonstrated that the utilisation of various lightweight plastic waste aggregates resulted in a decrease in the density of the mortar. While both the geopolymer and OPC mortars had the same amount of PW, the density comparison revealed that the geopolymer mortar was a little less dense than the cement mortar.

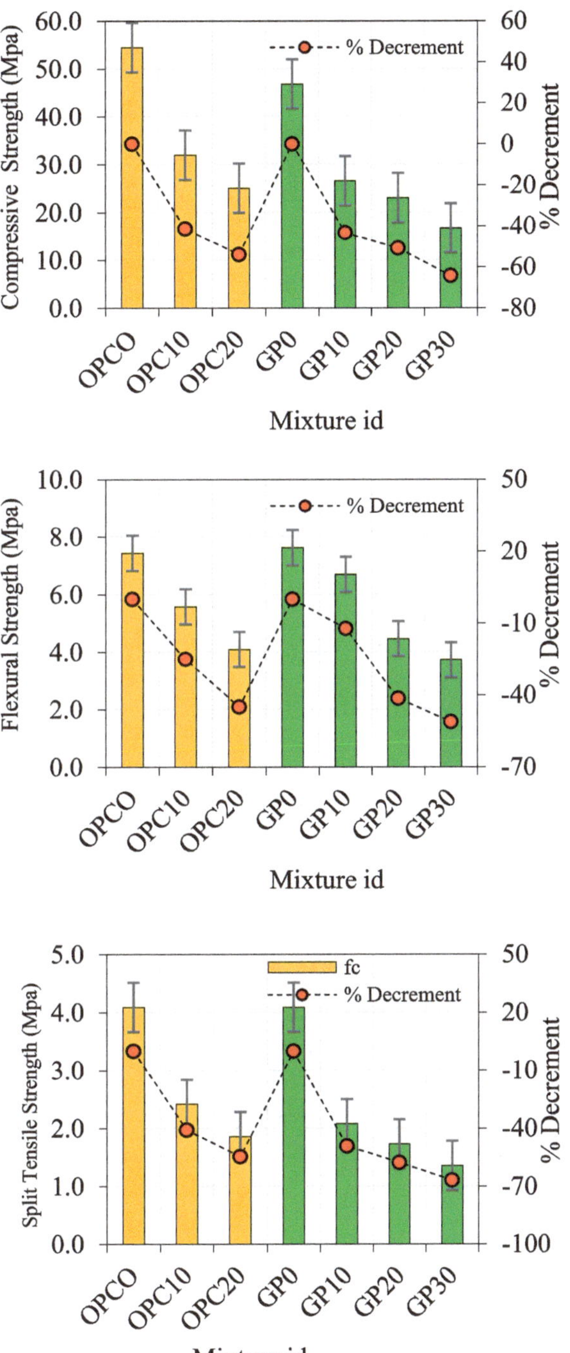

Fig. 3. Compressive, flexure and split tensile strength of cement mortar and geopolymer mortar.

The mortars containing 40% PW exhibited significantly low-density values of 1886 kg/m. Utilising mortar of geopolymer with PW can substantially decrease the structural dead load. Unlike porosity and density, water absorption exhibited a rise with more PW content. The geopolymer mortar exhibited an increase in porosity from 24.7% to 30.1% and an increase in water absorption from 8.7% to 9.9%. The elevated water absorption was caused by the heightened porosity of the mortar, facilitating the infiltration of water into the specimen with greater ease. Because of the very weak bond among the cement matrix and PW, pores formed at the interfacial transition zone, leading to increase of porosity in the PW-containing mortar [24].

Table 3. Water Absorption, Porosity, Density, UPV, Density and Thermal Conductivity of Cement and Geopolymer Motors.

Mix	Water Absorption	Porosity (%)	UPV (m/s)	Density (Kg/m^3)	Thermal Conductivity (W/m·K)
OPCO	6.3	13.01	4143	2389	2.1953
OPC10	7.6	18.6	3352	2168	1.6254
OPC20	8.2	25.9	2692	1957	1.2190
GP0	8.1	20.1	2749	2222	1.4354
GP10	8.7	24.7	2554	2206	1.1029
GP20	9.1	28.0	2395	2062	0.9130
GP30	9.2	29.6	1990	1904	0.8180
GP40	9.9	30.1	1785	1886	0.7219

The data shown in Table 3 shows that the inclusion of a more amount of PW in the mortars resulted in a decrease of value in UPV test results. The UPV test results of the control geopolymer mortar was initially measured as 2749 m/s. When 40% PW was added to the mortar mix, the UPV decreased to 1785 m/s. The incorporation of PW in cement composites led to a decrease in the ultrasonic pulse velocity due to the lower density of PW comparing to the natural aggregate and increases the porosity. The UPV test is a method of non-destructiveness employed to identify voids and cracks in concrete. Additionally, it could be utilised for assessing the concrete quality [25].

3.3 Thermal Conductivity

The GP0 geopolymer specimen has thermal conductivity of 1.4354 W/m·K, but there's a drop in thermal conductivity to a range of 1.1029–0.7219 W/m·K after adding 10–40% PW materials. The reduction in conductivity aligns with the findings by Hacini et al. [26], who investigated the potential of incorporating polyethylene plastic waste in mortar. His study also reveals when the amount of plastic content increased the conductivity of thermal-ness of mortar will gets decreased. The decline in thermal conductivity of mortar containing PW exhibited a comparable pattern to that observed in lightweight concrete [27, 28].

The addition of PW resulted in an increase in the number of pores in the material. The thermal conductivity of the entrapped air, which is 0.024 W/m·K, is significantly lower than that of other materials [29], resulting in a further reduction of thermal conductivity. In addition to the decreased heat conductivity when compared to natural aggregate, PW acts as an insulator. During PW replacement process, the geopolymer mortar exhibited a reduced thermal conductivity compared to the cement mortar. Geopolymer mortar has a significant effect on thermal conductivity due to its decreased density and higher porosity in comparison to cement mortar. Table 3 presents the values of thermal conductivity.

3.4 Mechanical Strength of Cement and Geopolymer Mortar

The strength parameters of geopolymer as well as the cement mortar that contains PW is illustrated in Fig. 3. Based to the study outcomes, the utilisation of PW resulted in a reduction in the characteristics of both the cement mortar and geopolymer mortar. The geopolymer mortars exhibited compressive strengths ranging from 46.85 to 13.76 MPa when mixed with 0–40% PW, while the cement mortar had compressive strengths ranging from 54.47 to 25.10 MPa when mixed with 0–20% PW. This finding aligns with earlier studies that have indicated that substituting sand with PW results in a reduction in the strength of mortar or concrete [23]. Reduced compressive strength is the result of increasing the PW-to-sand ratio. One major issue that reduces the compressive strength of mortar with PW is the weak bond strength between the cement matrix and PW [30]. The compressive strength of the geopolymer and cement mortar containing PW is reduced because of the increased porosity caused by the existence of several pores in the interfacial transition zone. Types N, S, and M of masonry mortar are required to have minimum compressive strengths of 6.2, 14.5 and 20 MPa, respectively, according to ASTM C91. Type M mortar specifications were satisfied by combinations containing 10 to 20% volume of PW in geopolymer and cement mortars, since their compressive strengths were greater than 20 MPa. Also, the Type S mortar specifications were met by mixture PG30 with a compressive strength of 16.73 and the Type N mortar specifications by mixture PG40 with strength of compressiveness 13.76 MPa.

Density should range from 1140 to 1850 kg/m3, and strength of compressive is 17.0 to 41.0 MPa in order to meet the structural lightweight requirement, in accordance with ACI 213R-14, This means that structural lightweight concrete is the outcome that is produced when we add 10–20% PW into mortar.

In terms of the geopolymer mortar's tension response, increasing the PW content resulted in lower flexural and tensile strengths. The splitting tensile and flexural and strengths were 7.62–3.16 and 4.09–1.21 MPa, respectively, when 0–40% PW was incorporated, as illustrated in Fig. 3. The flexural and cracking strengths of cement mortar were also decreased by PW. Because PW has a smooth surface, it forms a weak bond with the paste, which reduces its tensile strength [31]. Nevertheless, compared to compressive strength, the flexural strength drop was smaller. The behaviour can be attributed to the matrix's load interacted to the PW before the specimen breaking and the PW's flexibility [32]. Geopolymer and cement mortars containing PW had a less impact on flexural strength as compared to compressive strength.

4 Conclusions

The substitution of a portion of sand with discarded polyethylene terephthalate (PW) resulted in a deterioration of the strength characteristics of geopolymer mortar containing fly ash. There was a notable drop in both the compressive strength and splitting tensile strength as the quantity of PW increased. The PW exhibited a morphology characterised by flaky and angular shapes. Additionally, it possessed flexible qualities, enabling it to bear the stress transferred from the matrix until it fractured. Consequently, the inclusion of PW in the mortar led to an increased ratio of flexural strength to compressive strength.

The mortar incorporating PW exhibited increased water absorption and porosity because of broader ITZ between the cement paste and PW. Utilization of PW resulted in decrease in density, that has the potential to be advantageous in the creation of lightweight materials, thereby lessening the burden of weight on structures.

Furthermore, substituting a portion of PW has the potential to improve the thermal efficiency of both cement mortar and geopolymer mortar. The incorporation of 30% and 40% PW by volume into mortar of geopolymer led to significant reduction in thermal conductivity, transforming into an effective material of insulation. This benefit could be utilised in mortar of masonry to decrease heat conductivity in the structure, leading to reduced energy usage.

References

1. Gondal, A.H., Bhat, R.A., Gómez, R.L., Areche, F.O., Huaman, J.T.: Advances in plastic pollution prevention and their fragile effects on soil, water, and air continuums. Int. J. Environ. Sci. Technol. **20**(6), 6897–6912 (2023). https://doi.org/10.1007/s13762-022-04607-9
2. da Luz Garcia, M., Oliveira, M.R., Silva, T.N., Castro, A.C.M.: Performance of mortars with PET. J. Mater. Cycles Waste Manage. **23**(2), 699–706 (2021). https://doi.org/10.1007/s10163-020-01160-w
3. Zhang, C., et al.: Life cycle assessment of material footprint in recycling: a case of concrete recycling. Waste Manage. **155**, 311–319 (2023). https://doi.org/10.1016/j.wasman.2022.10.035
4. Islam, M.J., Shahjalal, M.: Effect of polypropylene plastic on concrete properties as a partial replacement of stone and brick aggregate. Case Stud. Constr. Mater. **15**, e00627 (2021). https://doi.org/10.1016/j.cscm.2021.e00627
5. Naderi, M., Kaboudan, A.: Experimental study of the effect of aggregate type on concrete strength and permeability. J. Build. Eng. **37**, 101928 (2021). https://doi.org/10.1016/j.jobe.2020.101928
6. Jain, D., Bhadauria, S.S., Kushwah, S.S.: An experimental study of utilization of plastic waste for manufacturing of composite construction material. Int. J. Environ. Sci. Technol. **20**(8), 8829–8838 (2023). https://doi.org/10.1007/s13762-022-04447-7
7. Sim, J., Park, C.: Compressive strength and resistance to chloride ion penetration and carbonation of recycled aggregate concrete with varying amount of fly ash and fine recycled aggregate. Waste Manage. **31**(11), 2352–2360 (2011). https://doi.org/10.1016/j.wasman.2011.06.014
8. Zega, C.J., Di Maio, Á.A.: Use of recycled fine aggregate in concretes with durable requirements. Waste Manage. **31**(11), 2336–2340 (2011). https://doi.org/10.1016/j.wasman.2011.06.011

9. Akçaözoğlu, S., Atiş, C.D., Akçaözoğlu, K.: An investigation on the use of shredded waste PET bottles as aggregate in lightweight concrete. Waste Manage. **30**(2), 285–290 (2010). https://doi.org/10.1016/j.wasman.2009.09.033
10. Ohemeng, E.A., Ekolu, S.O.: Strength prediction model for cement mortar made with waste LDPE plastic as fine aggregate. J. Sustain. Cement-Based Mater. **8**(4), 228–243 (2019). https://doi.org/10.1080/21650373.2019.1625826
11. Majeed, A.Z., Kurian, T., Davis, B., Alex, S.T., Fernandez, K.S., Mathew, A.V.: Partial replacement of aggregates with granulated waste plastic in solid concrete blocks--an intensive study. In: Drück, H., Pillai, R.G., Tharian, M.G., Majeed, A.Z. (eds.) Green Buildings and Sustainable Engineering, pp. 405–415. Springer, Singapore (2019)
12. Lazorenko, G., Kasprzhitskii, A., Fini, E.H.: Sustainable construction via novel geopolymer composites incorporating waste plastic of different sizes and shapes. Constr. Build. Mater. **324**, 126697 (2022). https://doi.org/10.1016/j.conbuildmat.2022.126697
13. Sun, K., Peng, X., Wang, S., Zeng, L., Ran, P., Ji, G.: Effect of nano-SiO2 on the efflorescence of an alkali-activated metakaolin mortar. Constr. Build. Mater. **253**, 118952 (2020). https://doi.org/10.1016/j.conbuildmat.2020.118952
14. Simão, L., et al.: Controlling efflorescence in geopolymers: a new approach. Case Stud. Constr. Mater. **15**, e00740 (2021). https://doi.org/10.1016/j.cscm.2021.e00740
15. Skariah Thomas, B., et al.: Geopolymer concrete incorporating recycled aggregates: a comprehensive review. Clean. Mater. **3**, 100056 (2022). https://doi.org/10.1016/j.clema.2022.100056
16. Nath, P., Sarker, P.K.: Effect of GGBFS on setting, workability and early strength properties of fly ash geopolymer concrete cured in ambient condition. Constr. Build. Mater. **66**, 163–171 (2014). https://doi.org/10.1016/j.conbuildmat.2014.05.080
17. Hama, S.M., Hilal, N.N.: Fresh properties of self-compacting concrete with plastic waste as partial replacement of sand. Int. J. Sustain. Built Environ. **6**(2), 299–308 (2017). https://doi.org/10.1016/j.ijsbe.2017.01.001
18. Pacheco-Torgal, F.: Introduction to the use of recycled plastics in eco-efficient concrete. In: Use of Recycled Plastics in Eco-Efficient Concrete, pp. 1–8. Elsevier (2019)
19. Shaikh, F.U.A.: Tensile and flexural behaviour of recycled polyethylene terephthalate (PET) fibre reinforced geopolymer composites. Constr. Build. Mater. **245**, 118438 (2020). https://doi.org/10.1016/j.conbuildmat.2020.118438
20. Chithambar Ganesh, A., Deepak, N., Deepak, V., Ajay, S., Pandian, A., Karthik: Utilization of PET bottles and plastic granules in geopolymer concrete. Mater. Today: Proc. **42**, 444–449 (2021). https://doi.org/10.1016/j.matpr.2020.10.170
21. Batayneh, M., Marie, I., Asi, I.: Use of selected waste materials in concrete mixes. Waste Manage. **27**(12), 1870–1876 (2007). https://doi.org/10.1016/j.wasman.2006.07.026
22. Prakash, K., Jane, A., Henderson, H.: Experimental study on mechanical strength of vibro - compacted interlocking concrete blocks using image processing and microstructural analysis. Iranian J. Sci. Technol. Trans. Civil Eng. 0123456789 (2023). https://doi.org/10.1007/s40996-023-01194-8
23. Almeshal, I., Tayeh, B.A., Alyousef, R., Alabduljabbar, H., Mohamed, A.M.: Eco-friendly concrete containing recycled plastic as partial replacement for sand. J. Mater. Res. Technol. **9**(3), 4631–4643 (2020). https://doi.org/10.1016/j.jmrt.2020.02.090
24. Avudaiappan, S., et al.: Innovative use of single-use face mask fibers for the production of a sustainable cement mortar. J. Compos. Sci. **7**(6) (2023). https://doi.org/10.3390/jcs7060214
25. Avudaiappan, S., et al.: Experimental investigation on the physical, microstructural, and mechanical properties of hemp limecrete. Sci. Rep. **13**(1), 22650 (2023). https://doi.org/10.1038/s41598-023-48144-y

26. Hacini, M., et al.: Utilization and assessment of recycled polyethylene terephthalate strapping bands as lightweight aggregates in Eco-efficient composite mortars. Constr. Build. Mater. **270**, 121427 (2021). https://doi.org/10.1016/j.conbuildmat.2020.121427
27. Kavitha, S.A., Priya, R.K., Arunachalam, K.P., Avudaiappan, S., Saavedra Flores, E.I., Blanco, D.: Experimental investigation on strengthening of Zea mays root fibres for biodegradable composite materials using potassium permanganate treatment. Sci. Rep. **14**(1), 12754 (2024). https://doi.org/10.1038/s41598-024-58913-y
28. Patil, A., et al.: Performance analysis of self-compacting concrete with use of artificial aggregate and partial replacement of cement by fly ash. Buildings **14**(1) (2024). https://doi.org/10.3390/buildings14010143
29. Arunachalam, K.P., et al.: Exploring the potential of pumice stone as coarse aggregate: an experimental approach to reduce concrete self-weight. In: Saavedra Flores, E.I., Astroza, R., Das, R. (eds.) Recent Advances on the Mechanical Behaviour of Materials, pp. 305–314. Springer, Cham (2024)
30. Safi, B., Saidi, M., Aboutaleb, D., Maallem, M.: The use of plastic waste as fine aggregate in the self-compacting mortars: effect on physical and mechanical properties. Constr. Build. Mater. **43**, 436–442 (2013). https://doi.org/10.1016/j.conbuildmat.2013.02.049
31. Marzouk, O.Y., Dheilly, R.M., Queneudec, M.: Valorization of post-consumer waste plastic in cementitious concrete composites. Waste Manage. **27**(2), 310–318 (2007). https://doi.org/10.1016/j.wasman.2006.03.012
32. Hannawi, K., Kamali-Bernard, S., Prince, W.: Physical and mechanical properties of mortars containing PET and PC waste aggregates. Waste Manage. **30**(11), 2312–2320 (2010). https://doi.org/10.1016/j.wasman.2010.03.028

Experimental Investigation on Marble Powder and Metakaolin as Partial Replacement of Cement on Self-compacting Mortar

Samatha Bairi[1], Rakesh Kuchana[2], Cristian Canales[3](✉), Siva Avudaiappan[4], and Krishna Prakash Arunachalam[4]

[1] Universal College of Engineering and Technology, Guntur 522001, India
[2] Guntur Engineering College, Opp. Katuri Medical College, Yanamadala 522019, Andhra Pradesh, India
[3] Department of Mechanical Engineering, Universidad de Concepción, Concepción, Chile
cristcanales@udec.cl
[4] Departamento de Ciencias de la Construcción, Facultad de Ciencias de la Construcción Ordenamiento Territorial, Universidad Tecnológica Metropolitana, Santiago, Chile
{s.avudaiappan,k.prakash}@utem.cl

Abstract. Chile is currently dealing with a growing problem related to plastic waste, amounting to 1.2 million tonnes per year. Experts are currently examining the possibility of using solid waste resources as construction materials. This study explored the utilization of plastic waste (PW) polyethylene terephthalate and pol-yolefin waste as a replacement for fine aggregate in geopolymer mortar. PW was utilised as a substitute for natural fine aggregate in geopolymer mortar, replacing 0%, 10%, 20%, 30%, and 40% by volume. The resulting mixtures were then compared to a standard cement mortar. The mortars were subjected to laboratory experiments to determine their density, water absorption, porosity, flow, thermal conductivity, UPV and mechanical strengths. The findings of these tests are then compared to conventional Portland cement mortar. The findings demonstrated that as the amount of plastic trash increased, the mechanical characteristics of the geopolymer mortar declined. Increases in Plastic Waste concentration from 0% to 40% resulted in a drop in compressive strength 70% for geopolymer mortars and for 53% cement mortars. Cement mortars and geopolymer that included PW showed an improvement in the strength of flexural comparing to compressive strength ratio. Furthermore, as the PW concentration increased, mortars thermal conductivity and density are noticeably decreased; for geopolymer mortar, density ranged from 2222 to 1886 kg/m and for cement mortar, from 2389 to 1957 kg/m^3, and the thermal conductivity for geopolymer mortar ranges from 1.4354 to 0.7219 W/m·K, and for cement mortar it ranges from 2.1953 to 1.2190 W/m·K. Plastic waste lower mortar's weight and improve its thermal performance. This novel approach reduces plastic waste and makes concrete lighter and better for thermal insulation.

Keywords: Geopolymer Mortar · Plastic Waste Utilization · Mechanical Properties · Fine aggregate replacement · Thermal conductivity

O. F. Farías Fuentes et al. (Eds.): CIBIM 2024, *Proceedings of the XVI Ibero-American Congress of Mechanical Engineering*, pp. 309–321, 2026.
https://doi.org/10.1007/978-3-032-22823-9_23

1 Introduction

In the construction company, self-compacting concrete, also known as SCC, is frequently used because of its propensities of easy flow because of its self-weight, beyond the requirement of vibrations externally. The self-consolidating concrete (SCC) effectively occupies the narrow gaps between the bars of reinforcement and concrete cover. A sufficient amount of mortar is needed to distribute moisture and transfer aggregate particles achieve this. Higher quantities of Superplasticizer (SP), Viscosity Modifying Admixtures (VMA) and powder materials are used in SCC blends to makes sure the mixture is sufficiently cohesive and stable. This increases the likelihood of reducing bleeds, Settlement and segregation. SCC enhances uniformity in concrete, consequently improving quality of surface and preventing the formation of void holes [1–3]. The initial mix design approach for the creation of self-compacting concrete (SCC) was suggested by Okamura, Ozawa, and Maekawa in 1995 [4–6]. Subsequently, Ouchi et al. made more enhancements to this method in 1988 [7]. The University of Tokyo is credited with developing this approach, which is sometimes known as the rational or Japanese technique. By including admixtures of minerals into SCC, segregation and workability of the concrete are improved. Several studies indicate the utilization of admixtures of mineral to SCC as a means to expedite durability, mechanical properties and self-compatibility. Fly ash (FA), silica fume (SF), Metakaolin (MK), waste marble powder, lime stone powder, etc. are some of the mineral admixtures commonly utilized in the production of SCC. In addition to improving concrete performances, the environmental benefits of MK make it a desirable partial cement replacement, according to the researchers. It diminishes the reduction of CO2 emissions and decreases the consumption of regular Portland cement [7–9]. One mineral admixture that falls into the category of N pozzolana constitutes MK, which has a molecular size ranging from 1.5 to 2.5 um [10]. Due to its high reactivity and composition of (50–55% SiO2 and 40–45% A12O3), it reacts with Ca(OH)2 to form CaH2O4Si (C-S-H) gel, which improves the pore structure and, in turn, increases the durability and strength of concrete [11, 12]. C-S-H gel has a greater Ca: SI ratio in MK compared to blast furnace slag and fly ash, etc., due to larger amounts of alumina and silica in MK [13–15]. Metakaolin existed prior to 1960, but it became known for its application as SCMs or pozzolanic material in concrete after 1980 [16, 17]. For almost 20 years, metakaolin (MK) has been considered significant to the production of normally vibrated concrete (NVC) [18–22]. By utilizing MK, the pore structure of concrete has been greatly reduced, and the amount of Ca(OH)2 in the cement paste has been reduced. This occurs as a result of the profound effect of MK's high pozzolanic reactivity and purity level on the cement-hydration process. When combined with Ca(OH)2, MK generates more C-S-H gel that improves the micro-structure of concrete, which in turn increases the material's mechanical and durability properties [23, 24]. Previous researches have also investigated the effects of MK on the fresh, mechanical and durability properties of SCC [25–28]. Many varieties of stone, including granite, lime stone, slate, marble and many more are used in the construction industry. The utilization of modern technologies in stone processing, such as quarrying, sawing and cutting, continuously increases the rate utilization of natural stones. When the stone is cut into the required shapes using a saw, which results in the production of marble sludge's and waste marble powder. Since it is not biodegradable, the stone waste so generated poses a threat to human civilization and generates significant environmental difficulties. The disposal of stone waste in streams, and waterways leads to pollution of water. A wide variety of solutions exist for the disposal of the stone debris. It has multiple potential uses, such as filler for specific areas as an ingredient in a wide range of cement composites. With its practical and ecological advantages utilization in cement composites could be a great way to dispose of stone debris [29–32]. The production of SCC can make use of some stone waste, such as tiles, marble, or bricks, as a filler material to fill the spaces between the grains of sand [33–36]. The increased packing density of stone waste increases its strength, durability, resistance to abrasion, penetration coefficient, and thickness microstructure, as well as its compressive, tensile, and fracture strength [36–39]. The impact of filler materials on SCC properties has been the subject of several investigations. We found that the workability is improved when filler materials are used [40–42]. The crystalline structure is destroyed and chemically bonded water is expelled by the heat of hydration. Research has shown that the fineness of WMP can enhance the cohesiveness of concrete mortar. Adding WMP to concrete mixes that also contain SF or MK makes the mixture more cohesive. Because of its filler ability, WMP has a beneficial effect even at young ages [43]. It is possible to utilize mineral admixtures as viscosity enhancers in powder-type SCC. In addition, these mineral admixtures improve the workability of SCC [44–47]. The inclusion of mineral admixtures has been found to reduce the proportion of SP in SCC [48–50]. Several studies have explored the use of WMP in place of cement, in concrete and mortars as a mineral additive, and in studying the relationship between the fresh and hardened properties of SCC. To create self-compacting concrete, scientists have mixed WMP with cement mortar. WMP was ground in a variety of ratios with PC clinker, which stands for Portland cement. The strength of concrete was found to be significantly increased compared to a control mix when 5% marble waste powder and 10% diatomite by weight of cement were added. According to the latest study, WMP can improve the mechanical and physical qualities of concrete when used in place of cement or sand [51]. Compared to cement replacement, they found that replacing WMP with fine aggregate significantly enhanced the mechanical qualities of concrete. Blast furnace ash, fly ash, stone dust, and silica fume are some of the most typical industrial by-products utilized as filler materials in SCC. It is possible to lessen environmental damage and save money by using these filler materials.

Prior research has done comprehensive investigations into the strength and durability characteristics of marble powder as a potential replacement for sand or cement. However, substituting sand with marble powder yielded superior outcomes as a result of the filler characteristic of marble powder. However, when cement was replaced, no significant outcomes were achieved. The pozzolanic properties of concrete including marble powder have not yet been studied. This is because marble powder contains a relatively small amount of pozzolan, which acts as an inert ingredient and does not actively participate in the hydration process of concrete. In this study, the addition of pozzolan to marble powder is found to enhance the cementing capabilities of concrete. To achieve the purpose, the inclusion of metakaolin as a supplemental cementitious material can be implemented. Metakaolin will serve as micro fillers, leading to enhanced strength and durability attributes. This study suggests maximizing the usage of marble powder by partially replacing sand with an additional cementitious ingredient.

2 Materials and Parameters

In this study, metakaolin was used as a cement replacement at a consistent level in five different concrete mixes, maintaining a proportion of 10%. Marble powder was utilized as a replacement for fine aggregate in three different proportions: 10%, 15%, and 20%. The concrete mixtures were labelled as CS, MP05, MP10, MP15, MP20, and MP25. Here, CS stands for the control specimen; MP05 represents the mixture of cement having 10% metakaolin and 5% marble powder replaced with fine aggregates. Table 1 provides chemical characteristics in grade 43 cement, marble powder, metakaolin, and OPC according to IS 8112: 2013. Material physical parameters are presented in Table 2. The natural fine aggregates fall under zone II as per IS 383 2016 standards and the fine aggregates particle size distribution were determined per IS 383 2016, ensuring that the aggregates adhered to these standards through meticulous sieving. Normal tap water was used in both the manufacture and curing of the concrete. Various percentages of marble powder were employed to partially replace fine aggregates in different concrete mixes. The marble powder was sourced from a local supplier in Chennai. In all concrete mixtures cement is replaced partially with metakaolin by 10 percentages. To achieve the desired workability, a superplasticizer based on polycarboxylate ether, referred to as Sika, was used. Table 3 shows the mix proportions and percentage of replacements in the study.

Table 1. Chemical Properties of Cement, Metakaolin and Marble Powder

Chemical Composition	Cement	Metakaloine	Marble Powder
$Si0_2$	20.4	56	4.3
Al_2O_3	6.55	31	0.42
Fe_2O_3	3.56	6.3	0.21
CaO	64	2.4	56.1
MgO	1.75	0.12	0.1
K_2O	0.54	0.6	0.01
Na_2O	0.25	0.23	0
LOI	2.23	2.43	38.8

Table 2. Properties of Fine and Coarse aggregate

Properties	FA	CA
Bulk Density(compacted)(gm/cm^3)	1.63	1.84
Bulk Density(Loose)/gm/cm^3	1.52	1.79
Elongation Index(%)	-	10.24
Fineness Modulus(%)	2.53	6.76

(continued)

Table 2. *(continued)*

Properties	FA	CA
Specific Gravity	2.58	2.76
Flakiness Index(%)	-	11.83
Water absorption(%)	0.99	0.51

Table 3. Mix Proportions of the Concrete

Mix	CS	MP5	MP10	MP15	MP20	MP25
Water	165	165	165	165	165	165
Cement	364	328	328	328	328	328
Metakaolin	0	36	36	36	36	36
Marble Powder	0	41	82	123	164	205
Fine Aggregate	820	779	738	697	656	615
Coarse Aggregate	1126	1126	1126	1126	1126	1126

3 Experimental Results

3.1 Workablity

As a method of evaluating the workability of concrete mixes the slump cone was used to measure the consistency of the mixtures when fine aggregates and metakaolin are replaced to marble powder partially. The values of slump with the various mixtures were compared to the control mix to determine the impact of the replacements on the concrete consistency. Figure 1 shows a 37% decrease in workability for mix MP25 compared to the control mix. Reduction in workability of 6%, 23%, 27% and 31% for mixes MP5 to MP20 compared to the control mixture. The decrease in workability is observed in the medium range due to smaller particle sizes of marble powder and metakaolin, which possess specific surface area better than cement and fine aggregates. This increased surface area demands more water to achieve the desired workability in the concrete. It was observed that mix MP incorporated specimens exhibited an increase in fresh density comparing to CS specimen. This rise of wet density is likely due to the particle's fineness in metakaolin and marble powder, which contribute to a denser mix. The incorporation of metakaolin and marble powder affects concretes workability and density, with the finer particles necessitating more water for consistency while simultaneously enhancing the density of the fresh concrete.

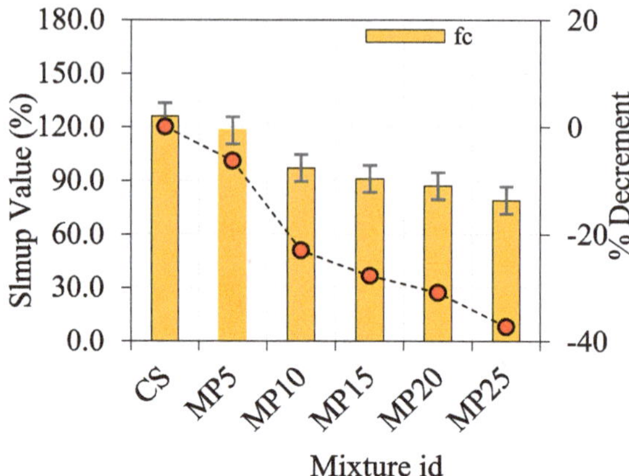

Fig. 1. Slump Value of Marble Powder Concrete

3.2 Compressive Strength

The concrete specimens were tested using a Universal testing machine, adhering to procedures outlined in Indian standard 516:1959. The test result is presented in Fig. 2. It was observed that MM20 mix exhibited the highest strength that is 38.32% greater than the CS specimen. The MM20 mix shows strength of increase of 32% compared to the CS specimen. This increase of strength is attributed to metakaolin's pozzolana effect. The increased proportion of marble powder enhances concrete strength due to its filler effect. Metakaolin at a concentration of 10% and marble powder at 20% significantly improved the concrete's strength, further strengthening its microstructure [23]. The results resemblance the findings of Ganesh who reported a 29% enhancement in compressive strength when replacing 20% of the aggregate of fine by marble powder. This consistency reinforces the beneficial impact of marble powder and metakaolin on concrete strength and durability.

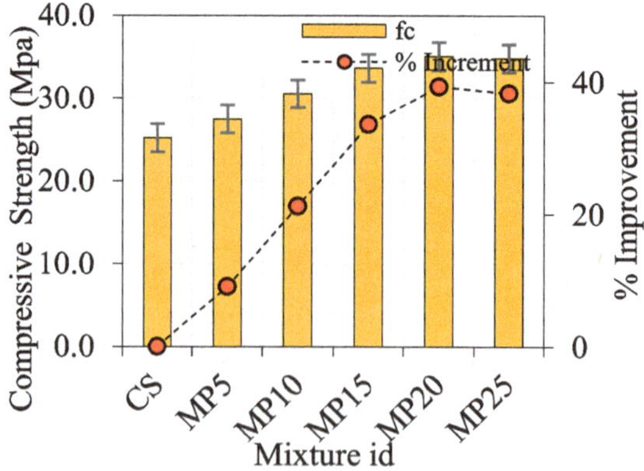

Fig. 2. Compressive Strength of Marble Powder Concrete

3.3 Split Tensile Strength Test

The testing process of split tensile strength (STS) is determined as per ASTM496-11. The test equipment and samples are shown in Fig. 3a, with values in Table 6. The STS was determined after 28 days, revealing increases of 3%, 9%,17%,26% and 27% for mixes MP5, MP10, MP15, MP20 and MP25 comparing to the control mix (CM). After 7 days, the strength of the concrete was enhanced due to the accelerated hydration rate caused by the addition of metakaolin [4]. The split tensile strength was further improved through the use of marble powder [9], as well as furnace slag and river sand. Additionally, the presence of metakaolin contributed to a decrease in dry density when using recycled aggregates [29]. The results are visually depicted in Fig. 3.

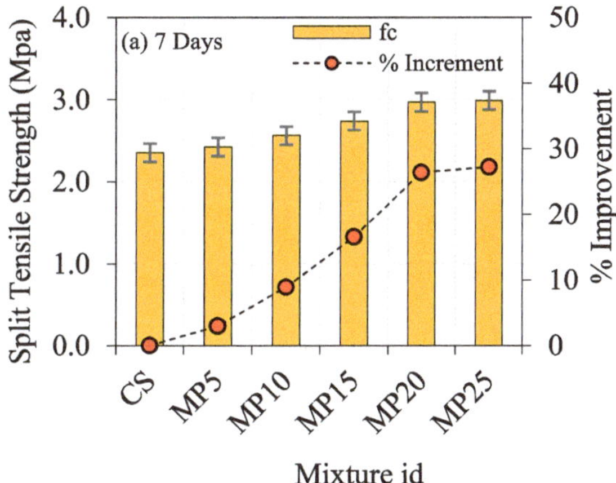

Fig. 3. Split Tensile Strength of Marble Powder Concrete

3.4 Flexural Strength

Following IS 516-1959 requirements, the prismatic beam were subjected to two-point load test. This testing setup is illustrated in Fig. 4. The findings indicated increases in flexural strength of 5%, 10%,14%,19% and 16% for mixes MP5, MP10, MP15, MP20 and MP25, respectively, comparing to control mix. This increase in flexural strength can be attributed to the fine nature of marble powder. Strength on flexural exhibited a progressive rise of up to 10% as the ratio of waste marble increased.

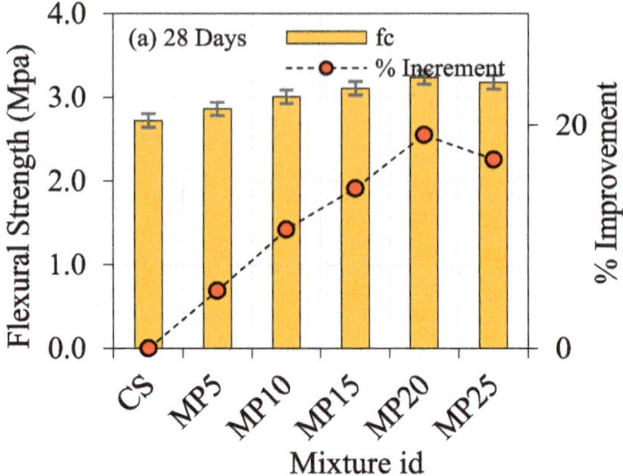

Fig. 4. Split Tensile Strength of Marble Powder Concrete

3.5 Water Absorption

The water absorption tests are carried out in accordance with ASTM C642 [42] following the concrete samples' 28-day curing period. A 100 mm × 100 mm test sample was prepared as per the specifications. Table 4. Results show that water absorption decreased smaller as marble powder was added to concrete mixes. Similar to the concrete mixture (CM), MM25 demonstrated a significant decrease in water absorption, this decrease in water absorption can be because of metakaolin's pozzolanic action. The reduction could be because of the pore reinforcement that occurs in mixes when using powders of marble as they ae very fine. It was found that SCC caused increased permeability, and discovered the same outcome with different waste marble dust powder-concrete mixtures.

Table 4. Water Absorption and Weight of Marble Powder Concrete

Mix	Wet Weight (kg)	Dry Weight(kg)	Water Absorption
CS	4.53	4.31	4.97
MP5	4.61	4.41	4.43
MP10	4.76	4.59	3.63
MP15	4.84	4.68	3.36
MP20	4.97	4.88	2.24
MP25	5.21	5.19	2.13

3.6 Sorptivity

The tests on Sorptivity test is conducted as per ASTM C 1585-13, which estimates the amount of water that is pulled up into a concrete mixture via capillary action. This test is conducted after the concrete sample has been cured for 28 days. Figure 5 illustrates the sorptivity rate reduction of marble powder concrete comparing to the CS specimens.

The decrease in water absorption rate [5] increases the pozzolanic activity of metakaolin. The incorporation of marble powder, a very fine substance, contributes to the reduction in water absorption by minimising pore structure of mixture. The incorporation of waste marble powder at 10% to 25% metakaolin significantly alters pore architectures of the concrete, enhancing its resistance to water absorption [29].

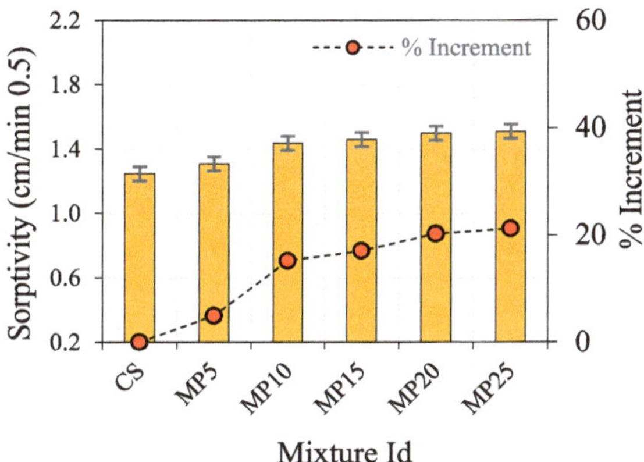

Fig. 5. Sorptivity of Marble Powder Concrete

4 Conclusions

The following observations and conclusions are stated based on the present studies:

1. The wet density increased when fine aggregates were replaced with marble powder and cement is replaced for metakaolin.
2. Workability decreased due to the larger surface area of metakaolin, necessitating a higher amount of superplasticizer.
3. Compressive strength increased when higher amount of aggregates of fine were replaced to marble powder. Notably, a combination of marble powder 20% and 10% metakaolin resulted in a significant rise in compressive strength. Split tensile strength also showed significant improvements under these conditions, with the optimal results achieved at 20% marble powder replacement.
4. Metakaolin and marble powder were added to cement it increases the flexural strength, the ideal amount of replacement for fine aggregate is found to be 20%.
5. Increased replacements with metakaolin and marble powder resulted in decreased water absorption and sorptivity, indicating enhanced durability of the concrete mix.

References

1. Khaleel, O.R., Al-Mishhadani, S.A., Razak, H.A.: The effect of coarse aggregate on fresh and hardened properties of self-compacting concrete (SCC). Procedia Eng. **14**, 805–813 (2011)
2. Ouchi, M., Nakamura, S., Osterberg, T., Hallberg, S., Lwin, M.: Applications of self-compacting concrete in Japan, Europe and the United States, Kochi University Technology Kochi, Japan (2003)

3. Zhu, W., Gibbs, J.C.: Use of different limestone and chalk powders in self-compacting concrete. Cem. Concr. Res. **35**, 1457–1462 (2005)
4. Domone, P.: Mix design, self-compacting concrete: state-of-the-art report of RILEM technical committee 174-SCC (2000)
5. Okamura, H., Ozawa, K., Ouchi, M.: Self-compacting concrete. Struct. Concr. **1**, 3–17 (2000). https://doi.org/10.1680/stco.2000.1.1.3
6. Nawa, T.: State-of-the-art report on materials and design of self-compacting concrete. In: Proceedings of the International Workshop on Self-Compacting Concrete, pp. 160–190 (1999)
7. Ouchi, M., Hibino, M., Ozawa, K., Okamura, H.: A rational mix-design method for mortar in self-compacting concrete. Struct. Eng. Constr. Tradit. Present Futur. **2**, 1307–1312 (1998)
8. Güneyisi, E., Gesoğlu, M., Mermerdaş, K.: Improving strength, drying shrinkage, and pore structure of concrete using metakaolin. Mater. Struct. **41**, 937–949 (2008)
9. Cassagnabère, F., Mouret, M., Escadeillas, G., Broilliard, P., Bertrand, A.: Metakaolin, a solution for the precast industry to limit the clinker content in concrete: mechanical aspects, Constr. Build. Mater. **24**, 1109–1118 (2010). https://doi.org/10.1016/j.conbuildmat.2009.12.032
10. Sabir, B.B., Wild, S., Bai, J.: Metakaolin and calcined clays as pozzolans for concrete: a review. Cem. Concr. Compos. **23**, 441–454 (2001). https://doi.org/10.1016/S0958-9465(00)00092-5
11. Jawahar, J.G., Sashidhar, C., Reddy, I.V.R., Peter, J.A.: Micro and macrolevel properties of fly ash blended self compacting concrete. Mater. Des. **46**, 696–705 (2013)
12. Poon, C.S., Kou, S.C., Lam, L.: Compressive strength, chloride diffusivity and pore structure of high performance metakaolin and silica fume concrete. Constr. Build. Mater. **20**, 858–865 (2006). https://doi.org/10.1016/j.conbuildmat.2005.07.001
13. Lakhani, R., Kumar, R., Tomar, P.: Utilization of stone waste in the development of value added products: a state of the art review. J. Eng. Sci. Technol. Rev. **7** (2014)
14. Poon, C.-S., Lam, L., Kou, S.C., Wong, Y.-L., Wong, R.: Rate of pozzolanic reaction of metakaolin in high-performance cement pastes. Cem. Concr. Res. **31**, 1301–1306 (2001). https://doi.org/10.1016/S0008-8846(01)00581-6
15. Zibara, H., Hooton, R.D., Thomas, M.D.A., Stanish, K.: Influence of the C/S and C/A ratios of hydration products on the chloride ion binding capacity of lime-SF and lime-MK mixtures. Cem. Concr. Res. **38**, 422–426 (2008)
16. Siddique, R., Klaus, J.: Influence of metakaolin on the properties of mortar and concrete: a review. Appl. Clay Sci. **43**, 392–400 (2009)
17. Gesoğlu, M., Güneyisi, E., Kocabağ, M.E., Bayram, V., Mermerdaş, K.: Fresh and hardened characteristics of self compacting concretes made with combined use of marble powder, limestone filler, and fly ash, Constr. Build. Mater. **37**, 160–170 (2012). https://doi.org/10.1016/j.conbuildmat.2012.07.092
18. Ding, J.-T., Li, Z.: Effects of metakaolin and silica fume on properties of concrete. ACI Mater. J. **99** (n.d.). https://doi.org/10.14359/12222
19. Li, Z., Ding, Z.: Property improvement of Portland cement by incorporating with metakaolin and slag. Cem. Concr. Res. **33**, 579–584 (2003). https://doi.org/10.1016/S0008-8846(02)01025-6
20. Qian, X., Li, Z.: The relationships between stress and strain for high-performance concrete with metakaolin. Cem. Concr. Res. **31**, 1607–1611 (2001). https://doi.org/10.1016/S0008-8846(01)00612-3
21. Badogiannis, E., Tsivilis, S.: Exploitation of poor Greek kaolins: durability of metakaolin concrete. Cem. Concr. Compos. **31**, 128–133 (2009). https://doi.org/10.1016/j.cemconcomp.2008.11.001
22. Ramezanianpour, A.A., Bahrami Jovein, H.: Influence of metakaolin as supplementary cementing material on strength and durability of concretes. Constr. Build. Mater. **30**, 470–479 (2012). https://doi.org/10.1016/j.conbuildmat.2011.12.050

23. Badogiannis, E., Kakali, G., Tsivilis, S.: Metakaolin as supplementary cementitious material. J. Therm. Anal. Calorim. **81**, 457–462 (2005). https://doi.org/10.1007/s10973-005-0806-3
24. Badogiannis, E., Papadakis, V.G., Chaniotakis, E., Tsivilis, S.: Exploitation of poor Greek kaolins: strength development of metakaolin concrete and evaluation by means of k-value. Cem. Concr. Res. **34**, 1035–1041 (2004). https://doi.org/10.1016/j.cemconres.2003.11.014
25. Hassan, A.A.A., Lachemi, M., Hossain, K.M.A.: Effect of metakaolin on the rheology of self-consolidating concrete. In: Design, Production and Placement of Self-Consolidating Concrete: Proceedings of SCC2010, Montreal, Canada, 26–29 September 2010, pp. 103–112. , Springer, Cham (2010)
26. Madandoust, R., Mousavi, S.Y.: Fresh and hardened properties of self-compacting concrete containing metakaolin. Constr. Build. Mater. **35**, 752–760 (2012). https://doi.org/10.1016/j.conbuildmat.2012.04.109
27. Sfikas, I.P., Kanellopoulos, A., Trezos, K.G., Petrou, M.F.: Durability of similar self-compacting concrete batches produced in two different EU laboratories. Constr. Build. Mater. **40**, 207–216 (2013). https://doi.org/10.1016/j.conbuildmat.2012.09.100
28. Behfarnia, K., Farshadfar, O.: The effects of pozzolanic binders and polypropylene fibers on durability of SCC to magnesium sulfate attack. Constr. Build. Mater. **38**, 64–71 (2013). https://doi.org/10.1016/j.conbuildmat.2012.08.035
29. Rekha, M.S., Sumathy, S.R., Arunachalam, K.P., Avudaiappan, S., Abbas, M., Fernande, D.B.: Effects of alkaline concentration on workability and strength properties of ambient cured green geopolymer concrete. Asian J. Civ. Eng. (2024). https://doi.org/10.1007/s42107-024-01087-9
30. Shirule, P.A., Rahman, A., Gupta, R.D.: Partial replacement of cement with marble dust powder. Int. J. Adv. Eng. Res. Stud. **1**, 2249 (2012)
31. Hacini, M., et al.: Utilization and assessment of recycled polyethylene terephthalate strapping bands as lightweight aggregates in Eco-efficient composite mortars. Constr. Build. Mater. **270**, 121427 (2021). https://doi.org/10.1016/j.conbuildmat.2020.121427
32. Arunachalam, K.P., et al.: Exploring the potential of pumice stone as coarse aggregate: an experimental approach to reduce concrete self-weight. In: Saavedra Flores, E.I., Astroza, R., Das, R. (eds.) Recent Advances Mechanical Behaviour of Materials, pp. 305–314. Springer, Cham (2024)
33. Patil, A., et al.: Performance analysis of self-compacting concrete with use of artificial aggregate and partial replacement of cement by fly ash. Buildings **14** (2024). https://doi.org/10.3390/buildings14010143
34. Samatha, B., et al.: Experimental study of nanosilica based concrete with nano silica gel BT-recent advances on the mechanical behaviour of materials. In: Saavedra Flores, E.I., Astroza, R., Das, R. (eds.) ICM 2023, pp. 315–330. Springer, Cham (2024)
35. Kavitha, S.A., Priya, R.K., Arunachalam, K.P., Avudaiappan, S., Saavedra Flores, E.I., Blanco, D.: Experimental investigation on strengthening of Zea mays root fibres for biodegradable composite materials using potassium permanganate treatment. Sci. Rep. **14**, 12754 (2024). https://doi.org/10.1038/s41598-024-58913-y
36. Arunachalam, K.P., et al.: Exploring the potential of pumice stone as coarse aggregate: an experimental approach to reduce concrete self-weight BT - recent advances on the mechanical behaviour of materials. In: Saavedra Flores, E.I., Astroza, R., Das, R. (eds.) ICM 2023, pp. 305–314. Springer, Cham (2024)
37. Avudaiappan, S., et al.: Experimental investigation on the physical, microstructural, and mechanical properties of hemp limecrete. Sci. Rep. **13**, 22650 (2023). https://doi.org/10.1038/s41598-023-48144-y
38. Sheeba, K.R.J., Priya, R.K., Arunachalam, K.P., Shobana, S., Avudaiappan, S., Flores, E.S.: Examining the physico-chemical, structural and thermo-mechanical properties of naturally

occurring Acacia pennata fibres treated with KMnO4. Sci. Rep. **13**, 20643 (2023). https://doi.org/10.1038/s41598-023-46989-x
39. Mohan, A., Priya, R.K., Arunachalam, K.P., Avudaiappan, S., Maureira-Carsalade, N., Roco-Videla, A.: Investigating the mechanical, thermal, and crystalline properties of raw and potassium hydroxide treated butea parviflora fibers for green polymer composites. Polymers (Basel) **15** (2023). https://doi.org/10.3390/polym15173522
40. Sheeba, K.R.J., Priya, R.K., Arunachalam, K.P., Avudaiappan, S., Flores, E.S., Kozlov, P.: Enhancing structural, thermal, and mechanical properties of Acacia pennata natural fibers through benzoyl chloride treatment for construction applications. Case Stud. Constr. Mater. **19**, e02443 (2023). https://doi.org/10.1016/j.cscm.2023.e02443
41. Monteiro, P.: Concrete: Microstructure, Properties, and Materials. McGraw-Hill Publishing (2006)
42. Patel, A.N., Pitroda, J.: Stone waste: effective replacement of cement for establishing green concrete. Inter. J. Innov. Tech. Explor. Eng. **2**, 24–27 (2013)
43. Farhana, S., Bhumika, P., Jayesh, P., Pitroda, J.: A study on utilization aspects of stone chips as an aggregate replacement in concrete in Indian context. Int. J. Eng. Trends Technol. **4**, 3500–3505 (2013)
44. Almeida, N., Branco, F., Santos, J.R.: Recycling of stone slurry in industrial activities: application to concrete mixtures. Build. Environ. **42**, 810–819 (2007). https://doi.org/10.1016/j.buildenv.2005.09.018
45. Hebhoub, H., Aoun, H., Belachia, M., Houari, H., Ghorbel, E.: Use of waste marble aggregates in concrete. Constr. Build. Mater. **25**, 1167–1171 (2011). https://doi.org/10.1016/j.conbuildmat.2010.09.037
46. Binici, H., Kaplan, H., Yilmaz, S.: Influence of marble and limestone dusts as additives on some mechanical properties of concrete. Sci. Res. Essay. **2**, 372–379 (2007)
47. Demirel, B.: The effect of the using waste marble dust as fine sand on the mechanical properties of the concrete. Int. J. Phys. Sci. **5**, 1372–1380 (2010)
48. Baboo, R., Khan, H.N., Kr, A., Rushad, S.T., Duggal, S.K.: Influence of Marble powder/granules in Concrete mix. Int. J. Civ. Struct. Eng. **1**, 827–834 (2011)
49. Gürü, M., Tekeli, S., Akin, E.: Manufacturing of polymer matrix composite material using marble dust and fly ash. Key Eng. Mater. **336**, 1353–1356 (2007)
50. Terzi, S., Karashin, M.: Evaluation of marble waste dust in the mixture of asphaltic concrete. J. Constr. Build. Mater. **21**, 616–620 (2007)
51. Poppe, A.-M., De Schutter, G.: Cement hydration in the presence of high filler contents. Cem. Concr. Res. **35**, 2290–2299 (2005). https://doi.org/10.1016/j.cemconres.2005.03.008
52. Grim, R.E.: Applied clay mineralogy. Geol. Föreningen i Stock. Förhandlingar. **84**, 533 (1962)
53. Ye, G., Liu, X., De Schutter, G., Poppe, A.-M., Taerwe, L.: Influence of limestone powder used as filler in SCC on hydration and microstructure of cement pastes. Cem. Concr. Compos. **29**, 94–102 (2007)
54. Corinaldesi, V., Moriconi, G., Naik, T.R.: Characterization of a marble powder for its use in mortar and concrete. In: Sustainable Development of Cement, Concrete and Concrete Structures, Toronto, Canada, pp. 313–336. American Concrete Institute (2005)
55. Aliabdo, A.A., Abd Elmoaty, M., Auda, E.M.: Re-use of waste marble dust in the production of cement and concrete. Constr. Build. Mater. **50**, 28–41 (2014)

Industrial Maintenance

Identification of the Most Prominent Frequency Components in the Vibration Signature of a High-Speed Train

Alejandro Bustos Caballero[1](✉), Rodrigo Gómez Iglesias[1], Marta Zamorano Garzón[2], and Juan Carlos García Prada[1]

[1] MAQLAB, Mechanics Department, Universidad Nacional de Educación a Distancia, Madrid, Spain
{albustos,jcgprada}@ind.uned.es, rodgomez@madrid.uned.es
[2] Higher Polytechnic School, Universidad Francisco de Vitoria, Crta. Pozuelo-Majadahonda Km 1, 800, 28223 Pozuelo de Alarcón, Madrid, Spain
marta.zamorano@ufv.es

Abstract. At present, European industry is continuing its particular Industrial Revolution towards improved productivity and decarbonisation, with high-speed rail transport being a real alternative to air transport and its dependence on oil. Traditionally, the elements of the railway sector have been subject to preventive maintenance through periodic inspections, based on the recommendations of the different manufacturers of rolling stock, lubricants, etc. Vibration analysis of a railway system is a very useful tool for determining the condition of mechanical elements if it is carried out in real-time. This work studies the data of a High-Speed Train on the Madrid-Seville route, with the aim of characterising its vibration signature by counting the most important frequency components. To do this, a user-variable reference level is established, which allows the frequency components that exceed the previously established level to be quickly located. By relating these more repetitive frequency components with their probable physical origin, it would be possible to establish an indicator of the wear of the different elements of the system, and thus be able to carry out the consequent maintenance operations before failure occurs.

Keywords: Significant frequency components · vibrations · maintenance · high-speed train

1 Introduction

It will soon be 200 years since the Stockton and Darlington Railway company inaugurated its first line for transporting coal from the mines to the towns of Darlingnton and Stockton-on-Tees son September 27, 1825, so that it could then be shipped by boat to various corners of the world in need of fossil energy. Since then, railway transport has continuously evolved. Today, this progress remains ongoing, both in freight and passenger transport. Not surprisingly, rail transport in Spain consumes half as much energy as

O. F. Farías Fuentes et al. (Eds.): CIBIM 2024, *Proceedings of the XVI Ibero-American Congress of Mechanical Engineering*, pp. 325–339, 2026.
https://doi.org/10.1007/978-3-032-22823-9_24

road transport in passenger transport per transport unit (passenger-km) and less than a third in freight transport (tonne-km) [1]. Today, the rail sector is still in full swing. By 2022, passenger rail transport will have surpassed the levels of passenger numbers that existed before the COVID-19 pandemic [2]. High Speed Rail, which was born 60 years ago in Japan, is playing a key role in this trend and is spreading all over the world. The success of High Speed as a means of transport is evidenced by the fact that China used its High Speed train, Fuxing, as the flagship of its Beijing 2022 Winter Olympic Games. This train linked the cities of Zhangijakoy and Yanqing to the capital Beijing and transported the athletes quickly, comfortably and safely.

Maintaining the safety standards required by the authorities requires special attention to maintenance of both rolling stock and infrastructure. Industrial maintenance was initially corrective maintenance, which can lead to unplanned downtime. Later, it evolved to preventive maintenance, which produces unnecessary expenses in most cases, but does not prevent railway accidents such as Viareggio [3] and Eschede [4, 5] from occurring.

Industrial technology, taken to the railway sector, means that maintenance is evolving towards predictive maintenance, with the aim of avoiding events such as those mentioned above and, in addition, reducing unnecessary maintenance costs.

The industry in general needs non-destructive flaw detection techniques [6], so that it does not involve the loss of units. The railway industry, due to the magnitude of its elements, uses these techniques for the detection of possible defects. One of the most studied and used is vibration and noise analysis, which are highly legislated [7].

Normally, the scientific literature dealing with vibration analysis in railway rolling stock focuses on bogies and their elements, mainly: transmissions [8], axle boxes [9], axles [10], wheel defects [11], such as cracks [12] and primary suspension elements [13]. These investigations are usually carried out in laboratory conditions, without reaching operating speeds, creating numerical models, for subsequent tests.

In recent years, studies have begun to be carried out at normal operating speeds. Thus, Wang et al. [14] aims to study the vibrations recorded in the axle box of a High Speed Train in order to lengthen the periods between wheel re-profiling. Bustos et al. [15] studies the dynamic behaviour of a High Speed Train after a reprofiling operation, obtaining frequency indicators of the condition of the train. Lebel et al. [16] carry out multibody dynamics simulations with in situ measurements taken on a French High Speed Train, performing a mathematical model based on Bayesian methods to detect how the train suspension elements evolve. Bayesian methods have also been used by Zhang et al. [17] and Wang et al. [18], in these cases to determine the effect of wheel re-profiling. Regarding the treatment of the vibration signal data, Bustos et al. [19] propose the application of EGRSC and ECBF techniques.

This work proposes the normalization of the power spectral density of the signals with respect to the maximum amplitude of the spectrum, determining the most prominent components in the frequency spectrum based on a user-variable reference level. Once the most prominent components have been identified, their repeatability is studied in order to characterize the operational status of the train under standard conditions, thus being able to determine the need for maintenance operations on the train elements.

2 Methods

The data under study were obtained through a system configured for this purpose, which is located in the extreme tourist class trailer car, and this section describes this system, and the subsequent techniques used to process the signals.

2.1 Data Acquisition System

The subject of this study is a High Speed train, which runs on the Madrid-Seville route, on a section with a uniform speed of around 270 km/h. The train is made up of two powerheads and eight articulated cars that are located between the powerheads.

The system used consists of four main phases: acquisition, conditioning, recording and analysis. Acquisition is carried out by means of sensors placed in the axle boxes of the shaft to be studied; the parameters of the measurement signals are shown in Table 1. The signals are conditioned by means of two IMx-R units from SKF, and the measurements are automatically recorded by means of the @ptitude Observer software. These measurements are then processed with MATLAB®. Figure 1 shows the schematic of the measurement system.

Table 1. Parameters of the measurement signals

Parameter	Value
Sampling frequency	5,120 Hz
Measurement time	3.2 s
Speed range	75–2,000 rpm (13–347 km/h)
Data per measurement	16,384

Fig. 1. Schematic of the measurement system

The measuring accelerometers are located in the axle box covers, as well as the speed sensor, in order to correctly monitor the vibrations of the axle bearings and the interaction with the track. The accelerometer is a piezoelectric, ICP, SKF model CMSS-RAIL-9100, the main characteristics of which are given in Table 2.

Table 2. CMSS-RAIL 9100 accelerometer characteristics

Parameter	Value
Sensitivity (±20%)	10.2 mV/(m/s^2)
Acceleration range	±490 m/s^2
Frequency range (±3dB)	0.52 Hz to 8 kHz
Resonance frequency	25 kHz
Amplitude linearity	±1%
Transversal sensitivity	≤7%

2.2 Signal Processing

Signal processing is carried out using the Power Spectral Density (PSD). The PSD is based on the Fourier transform, which decomposes complex signals into a sum of harmonic signals. Its analysis in the frequency domain makes it possible to express the power at each frequency.

Given a time-dependent function, $x(t)$, its Fourier transform, Eq. (1) is defined as:

$$X(f) = \int_{-\infty}^{+\infty} x(t) \cdot e^{-j2\pi ft} dt \quad (1)$$

where t is the time, f is the frequency, and j is the imaginary unit. From a real function $x(t)$ it is obtained $X(f)$, a transform which is a complex function of the frequency with its own spectrum.

Provided that the integral (1) exists, it is possible to return to the initial function $x(t)$, starting from the transform $X(f)$, using the anti-transform Eq. (2):

$$x(t) = \int_{-\infty}^{+\infty} X(f) \cdot e^{-j2\pi ft} dt \quad (2)$$

Both functions $X(f)$ and $x(t)$ are two ways of representing the same data, and the properties of the former are verified by the latter and vice versa.

Digitization of the signal for further processing on computer equipment requires discretisation of a continuous signal, so the discrete Fourier transform given by Eq. (3) and Eq. (4) is used.

$$X(k) = \sum_{n=0}^{N-1} x(n) \cdot e^{-j\frac{2\pi}{N}kn} \quad (3)$$

$$x(n) = \frac{1}{N}\sum_{k=0}^{N-1} X(k)e^{-\frac{j2\pi}{N}kn} \quad (4)$$

In this transform, N frequency components ($N/2$ positive and $N/2$ negative) are obtained from N components in time with N^2 operations in the complex plane. The discrete Fourier transform relates the time function of the monitored signals to a function in the frequency spectrum.

The characterization of random signals is performed by means of the power density distribution function. The definition of the PSD is done through the general distribution function S, Eq. (5), as its integral, for the continuous case, and as its summation, for the discrete case. Over the entire frequency range, the PSD must be equal to the signal power P, Eq. (6) [20].

$$S(k)_{1side} = \begin{cases} 2S(k)k = 1 \ldots \frac{N}{2} - 1 \\ S(k)k = 0, k = \frac{N}{2} \end{cases} \tag{5}$$

$$P = \sum\nolimits_{k=0}^{N-1} S(fk)\Delta f = \sum\nolimits_{k=0}^{N/2} S(k)_{1side}\Delta f \tag{6}$$

The calculation of the PSD is performed through the Fast Fourier Transform, FFT, as indicated in Eq. (7).

$$S(k)_{1side} = \frac{1}{N\Delta t} \sum x^2 \Delta t \tag{7}$$

And must be equal to the total power, as indicated in Eq. (8):

$$\sum S\Delta f = \frac{1}{N\Delta t} \sum S \tag{8}$$

Considering Paserval's Law, Eq. (9) is obtained:

$$S(k) = \frac{\Delta t}{N} |X(k)|^2 \tag{9}$$

where Δt is the sampling time, $S(k)$ is the PSD, N is the amount of data in the signal and $X(k)$ is the Fourier Transform of the signal.

In a moving bearing, each time a defective rolling element passes over one of its defects, a series of periodic frequencies called fundamental fault frequencies are produced. These frequencies must be considered in the power spectral density. The faults to be considered in this study, from Eq. (10) to Eq. (13) are the following:

BPFI: Ball Pass Frequency Inner Race.

$$BPFI = \frac{N}{2} F_r \left(1 + \frac{d}{D} \cos\beta\right) \tag{10}$$

BPFO: Ball Pass Frequency Outer Race.

$$BPFO = \frac{N}{2} F_r \left(1 - \frac{d}{D} \cos\beta\right) \tag{11}$$

BSF: Ball Spin Frequency.

$$BSF = \frac{D}{2d} F_r \left(1 - \left(\frac{d}{D}\right)^2 \cos^2\beta\right) \tag{12}$$

FTF: Fundamental Train Frequency.

$$FTF = \frac{F_r}{2}\left(1 - \frac{d}{D}\cos\beta\right) \quad (13)$$

where N is the number of rolling elements in the bearing, F_r is the rotation frequency of the inner race, d is the diameter of the rolling elements, D is the mean diameter of the bearing and β is the contact angle.

It is also relevant to know the sleeper passing frequency (SPF), which is given by Eq. (14).

$$SPF = v/\lambda \quad (14)$$

where λ is the distance between the sleepers and v is the linear speed at which the train is running.

2.3 Data Analysis

Once the signals measured by the IMx-R system have been processed with MATLAB®, the power spectral densities are studied in three ways: unnormalized, normalized with respect to the total power and normalized with respect to the maximum power. For the characterization of the most important frequency components, the power spectral density normalized with respect to the maximum power will be used, setting a reference level or threshold as a percentage of the maximum power, which allows the frequency components whose amplitude is above this threshold to be selected.

In order to address different levels of detail, four reference levels are initially established: 50% (0.5%), 25% (0.25), 10% (0.1) and 5% (0.05). In addition, the frequency components above the reference value are grouped into bands of 1 Hz, 3 Hz and 5 Hz in order to subsequently compare the results obtained and establish the most significant grouping for the analysis of the vibration signals.

3 Results

Previous works [15, 19] have shown that vertical vibrations are the most representative of the operational status of the bogie, so this study will focus on the analysis of the vertical accelerometer signals. The signals were collected during normal train operation, on a journey between Madrid and Seville, at an average speed of approximately 270 km/h.

Before starting the analysis of the signals, it is necessary to fix a series of parameters to obtain some of the frequencies that characterise the system, calculated for a speed of 270 km/h. The parameters of interest are listed in Table 3 and are the rotation frequency, the bearing failure frequencies and the frequency of passage through the sleeper.

Table 3. Characteristic frequencies for v = 270 km/h

Phenomenon	Frequency (Hz)
Rotation frequency (F_r)	25.95
BPFI	329.14
BPFO	267.69
BSF	122.84
FTF	11.65
SPF	125

Figure 2 shows one of the time signals recorded on-board the train at 270 km/h. The power spectral density of this signal is also shown. The result of normalizing the power spectral density with the methods discussed above (with respect to the total power and the maximum amplitude of the spectrum) are shown in Fig. 3.

Several active zones can be distinguished in the spectra: the first one is from 0 Hz to about 350 Hz; the second zone is between 400 Hz and 650 Hz; the third zone comprises frequencies between 800 Hz and 900 Hz. There is a final, much smaller zone of activity between 2000 Hz and 2560 Hz, which can be associated with corrugation phenomena in the wheel [21].

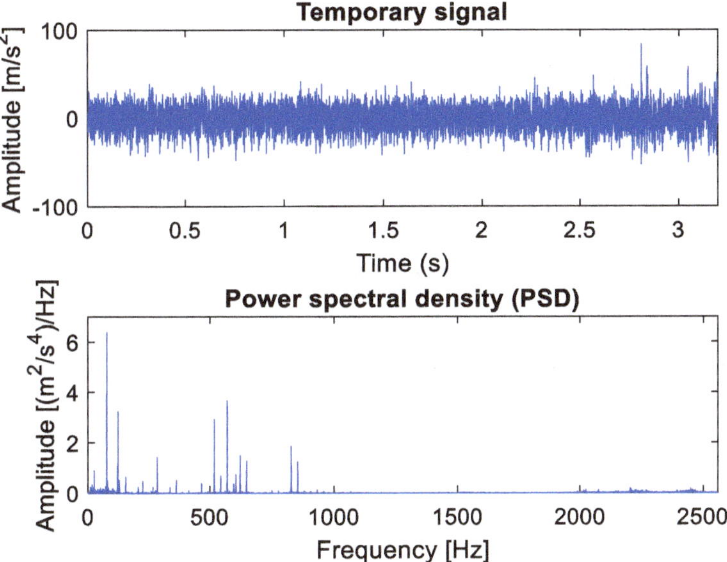

Fig. 2. Representation of the signal and power spectral density (PSD)

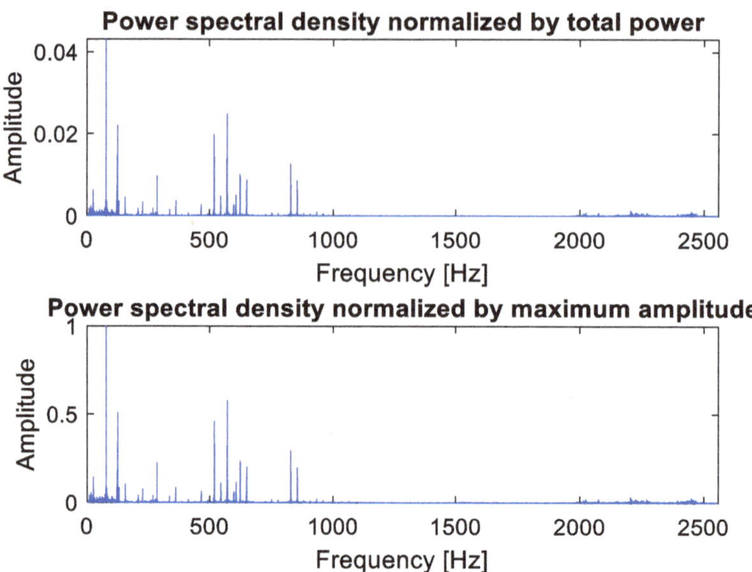

Fig. 3. Representation of power spectral density normalized by the two methods

The normalization of the power spectral density offers the possibility to establish a common reference value for all analysed signals. Both spectra offer similar information, however, the amplitude values obtained by normalizing the spectrum with respect to the total power are very small (in the order of a hundredth), while the other normalization offers amplitude values in the range from 0 to 1. This larger working range makes it possible to establish a greater number of reference values and, therefore, the spectrum normalized with respect to the maximum amplitude will be used.

Four reference levels have been established, namely 50%, 25%, 10% and 5% of the maximum power of the spectrum. Based on these thresholds, a count was made of the number of repetitions of the most prominent frequency components of the spectrum in a given frequency band in all the signals analysed. The frequency bands used are 1 Hz, 3 Hz and 5 Hz. The 1 Hz band results from rounding the frequency of the prominent components to the nearest integer. The 3 Hz band represents approximately 10 times the frequency resolution of the spectrum ($\Delta f = 0.3125$ Hz). Both the 3 Hz band and the 5 Hz band allow to group in the same band frequency components that are strongly dependent on the wheel rotation speed and that therefore vary as the speed of the train varies.

The results obtained are shown from Figs. 4, 5, 6 and 7. In all cases, the most prominent components recorded are between 0 Hz and 1000 Hz, so these limits are specified for the abscissa axes in all figures. The number of repetitions of the most prominent frequency components are represented by blue bar diagrams. Above and in red, the average normalized power spectral density (with respect to the total power) of all analysed signals has been superimposed for comparison.

By setting the reference value to 50% (Fig. 4), the number of frequency components that meet this criterion is limited, which is reflected in the number of repetitions. Regardless of the grouping by frequency bands, it can be seen that the most repeated frequency component is around 77 Hz, which corresponds to the third harmonic of the wheel rotation frequency Fr. In addition to this particularly prominent frequency component, significant components can also be seen around 26 Hz, 123 Hz, 278 Hz, 513 Hz, between 550 Hz and 650 Hz, and some more components between 800 Hz and 900 Hz.

Setting the reference value to 25% (Fig. 5) causes frequency components that were not visible with the previous reference level are identified as prominent. In particular, prominent frequency components appear around 12 Hz and 13 Hz, which is half of the wheel rotation frequency. The rest of the prominent components of the spectrum look similar. Changing the threshold value has a direct effect on the number of repetitions of the most prominent frequency components The lowering of the reference value leads to a higher number of components being identified and thus to an increase in the number of repetitions. For this reason, the number of repetitions of the significant frequency components is slightly higher than when the reference level was set at 50% of the maximum power of the spectrum.

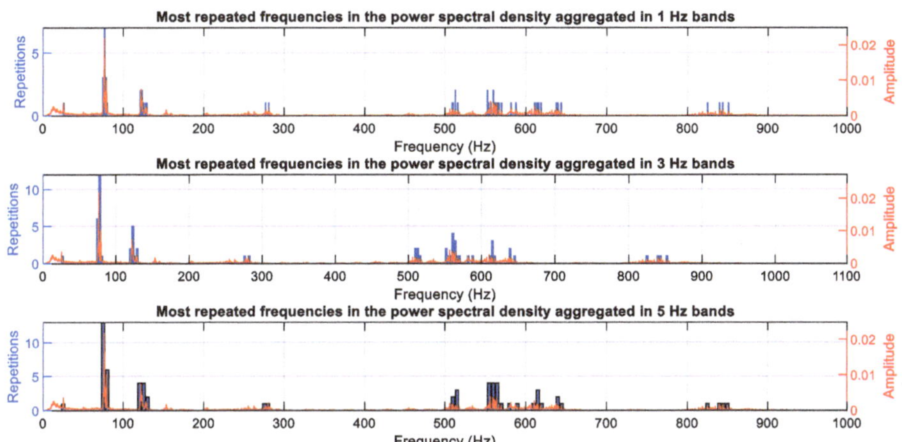

Fig. 4. Representation of the most repeated frequency components in power spectral density normalized by the maximum amplitude, aggregated in bands of 1 Hz, 3 Hz and 5 Hz. Threshold set at 50% of the maximum amplitude of the power spectrum.

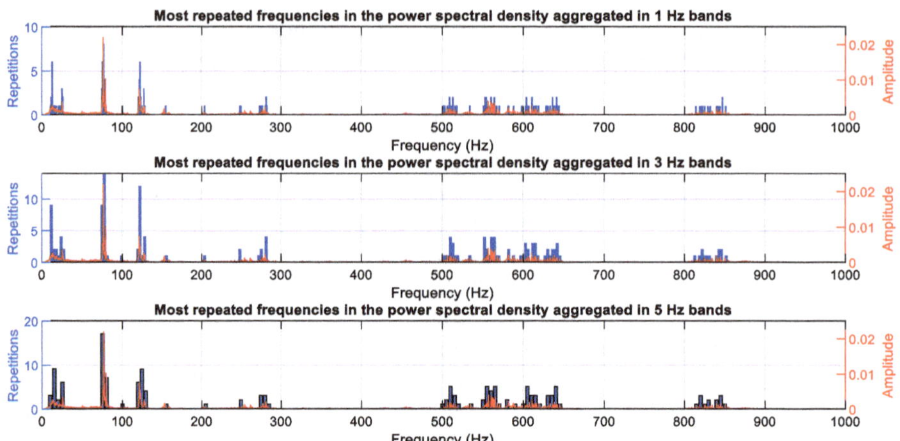

Fig. 5. Representation of the most repeated frequency components in the power spectral density normalized by the maximum amplitude, aggregated in bands of 1 Hz, 3 Hz and 5 Hz. Threshold set at 25% of the maximum amplitude of the power spectrum.

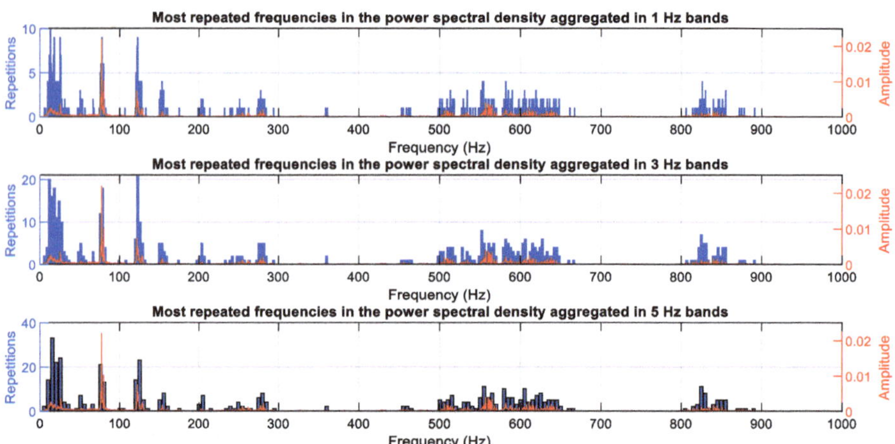

Fig. 6. Representation of the most repeated frequency components in the power spectral density normalized by the maximum amplitude, aggregated in bands of 1 Hz, 3 Hz and 5 Hz. Threshold set at 10% of the maximum amplitude of the power spectrum.

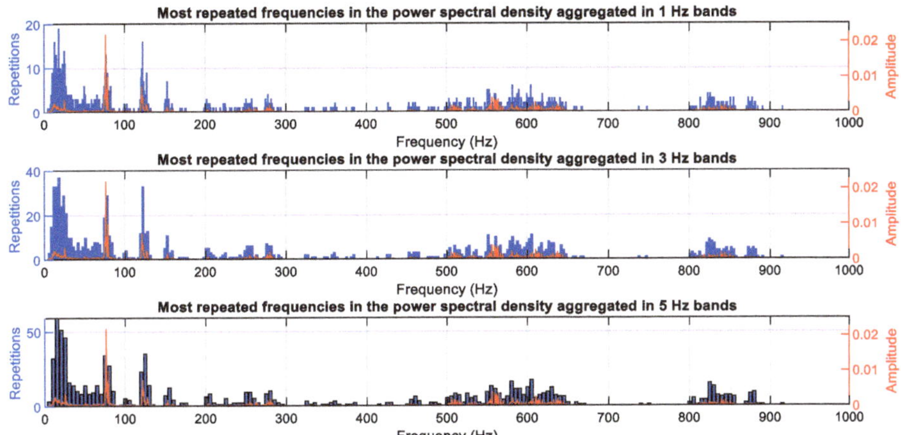

Fig. 7. Representation of the most repeated frequency components in the power spectral density normalized by the maximum amplitude, aggregated in bands of 1 Hz, 3 Hz and 5 Hz. Threshold set at 5% of the maximum amplitude of the power spectrum.

In Fig. 6 the result of setting the reference level to 10% of the maximum power is illustrated. The number of prominent frequency components exceeding the threshold is considerably higher than in the previous cases. At low frequencies, the frequency components around 13 Hz become more prominent, masking those around 26 Hz. The components around 76 Hz and 123 Hz are still very well identified. Additionally, new frequency components between 400 Hz and 500 Hz are observed. Logically, grouping the components into wider bands increases the number of repetitions in each of the bands, which is especially reflected in the 5 Hz bands. Compared with the averaged power spectral density of all the signals, it can be seen that the proposed method gives a prominent role to the low frequency components.

Finally, the reference value is set to 5% of the maximum power of the spectrum (Fig. 7). In this case, the number of identified frequency components is such that the information shown in the plots is hardly simplified with respect to that of the power spectral density. Even so, it can be seen that the selection of frequency components according to the proposed method particularly emphasises the low-frequency area. In fact, the highest number of repetitions is located between 10 Hz and 30 Hz.

In light of these results, it can be concluded that the reference level set at 50% of the maximum power of the spectrum reduces the information extracted from the power spectra too much, while the 10% and 5% thresholds show too much information. Setting the reference level at 25% seems to offer the best compromise between synthesising the information from the power spectra and not losing relevant information from the spectra.

As for the bandwidth, although the idea of widening the bandwidth is to group in the same band frequency components with the same likely physical origin that vary slightly with the wheel rotation speed, this is not always possible, even if the 5 Hz width is used. Therefore, 3 Hz also seems to be the best option.

Based on the data in Table 3, several of the identified prominent frequency components can be related to their possible physical origin. The values presented below relate to 3 Hz bandwidths and a reference level of 25% of the maximum spectrum power.

In the first zone of activity, between 0 Hz and 350 Hz, the following frequency bands stand out as the most repeated:

- 12 Hz: would roughly correspond to the rolling element failure frequency, FTF, whose value is 11.65 Hz at 270 km/h.
- 24 Hz and 27 Hz: would correspond to the wheel rotation frequency Fr.
- 75 Hz and 78 Hz: would correspond approximately to the third harmonic of the wheel rotation frequency Fr.
- 123 Hz: the physical origin of this frequency band may be due to two phenomena very close to each other: the BSF of the bearing and the frequency of passage through the sleepers.
- 282 ± 1.5 Hz: this would correspond to the eleventh harmonic of the wheel rotation frequency Fr.

Within the second activity zone (between 400 Hz and 650 Hz), the most significant components are located between 500 Hz and 650 Hz. The following frequency bands stand out as the most repeated:

- 510 Hz: approximately coincides with the 20th harmonic of the wheel's rotational frequency.
- 552 Hz, 561 Hz and 564 Hz: the probable physical origin is related to the 22nd harmonic of the wheel rotation frequency. Train speed variations during the recording of vibration signals cause the same phenomenon to be classified in different bands.
- The frequency components grouped in the bands observed between 600 Hz and 650 Hz are probably due to the 23rd and 24th harmonics of the wheel rotation frequency, with the band in which they are grouped varying according to the variation of the train speed.

In the third activity zone, located between 800 Hz and 900 Hz, the most prominent bands are located at 822 Hz, 843 Hz and 846 Hz. All of them seem to originate from the seventh harmonic of the BSF of the bearing, being classified in one or another band depending on the speed of the train when the vibrational measurements were taken.

4 Conclusions

This work presents a method based on the selection of significant frequency components to extract relevant information from the spectra of vibration signals, with the aim of characterizing the mechanical system of interest under specific conditions.

The method is applied to vibration signals recorded in the vertical direction by accelerometers installed on a high-speed train operating between Madrid and Seville. To ensure comparable measurement conditions, the signals are recorded over a specific section of the route where the train travels at approximately 270 km/h.

The application of the method involves defining a reference level or threshold relative to the maximum power of the spectrum, in order to select the most prominent frequency components. The reference levels proposed in this work are 50% (0.5), 25% (0.25), 10% (0.1), and 5% (0.05). In addition, the selected frequency components are grouped into bands with widths of 1 Hz, 3 Hz, and 5 Hz.

The results show that the best compromise between conciseness and the retention of relevant information is achieved with a reference level of 25% and a bandwidth of 3 Hz.

The proposed method is useful for identifying the most frequently occurring frequency components in the spectra of the collected measurements. In other words, it enables the identification of characteristic frequency components of the mechanical system under specific operating conditions, thereby obtaining a system signature. Variations in this signature under identical conditions may indicate the presence of a potential fault.

In most cases, it was possible to associate frequency components with their probable physical origins (typically harmonics of the wheel rotation frequency).

In future work, it is intended to establish a reference average power spectrum under normal operating conditions, from which other measurements can be normalized. The reference levels will then be determined based on this averaged spectrum, making it easier to detect changes in the operating condition of the mechanical system.

Acknowledgments. This publication is part of the R&D Project MC4.0 PID2020-116984RB-C21, funded by MCIN/AEI/10.13039/501100011033. This publication is part of the R&D Project Intelligent diagnosis of critical railway components, funded by the call “Ayudas a Investigadores Tempranos UNED-Santander 2024”.

Competing Interests. The authors have no conflicts of interest to declare that are relevant to the content of this chapter.

References

1. Informe anual 2022. Observatorio del Transporte y la logística en España, OTLE. Ministerio de Transportes y Movilidad Sostenible (2023)
2. Martín del Moral, J.M.: Estadísticas del INE de transporte por ferrocarril. Indice: Revista de Estadística y Sociedad 22–25 (2024)
3. Landucci, G., Tugnoli, A., Busini, V., Derudi, M., Rota, R., Cozzani, V.: The Viareggio LPG accident: lessons learnt. J. Loss Prev. Process Ind. **24**, 466–476 (2011). https://doi.org/10.1016/j.jlp.2011.04.001
4. Esslinger, V., Kieselbach, R., Koller, R., Weisse, B.: The railway accident of Eschede – technical background. Eng. Fail. Anal. **11**, 515–535 (2004). https://doi.org/10.1016/j.engfailanal.2003.11.001
5. Oestern, H.J., Huels, B., Quirini, W., Pohlemann, T.: Facts about the disaster at Eschede. J. Orthop. Trauma **14**, 287–290; discussion 277 (2000). https://doi.org/10.1097/00005131-200005000-00011
6. Mafla, C., Castejon, C., Rubio, H.: Mantenimiento predictivo en tractores agrícolas. Propuesta de metodología orientada al mantenimiento conectado. Revista Iberoamerica de Ingeniería Mecánica **26**, 63–75 (2022)
7. Pertuz Argumedo, J.M.: Investigación científica sobre normas aplicables al mantenimiento predictivo en máquinas rotativas usando análisis de vibraciones mecánicas (2021). http://repositorioslatinoamericanos.uchile.cl/handle/2250/6225819
8. Ye, L., Xia, X., Chang, Z.: Dynamic prediction of the performance reliability of high-speed railway bearings. J. Braz. Soc. Mech. Sci. Eng. **41**, 532 (2019). https://doi.org/10.1007/s40430-019-2041-z

9. Li, Y., Liang, X., Chen, Y., Chen, Z., Lin, J.: Wheelset bearing fault detection using morphological signal and image analysis. Struct. Control. Health Monit. **27**, e2619 (2020). https://doi.org/10.1002/stc.2619
10. Hassan, M., Bruni, S.: Experimental and numerical investigation of the possibilities for the structural health monitoring of railway axles based on acceleration measurements. Struct. Health Monit. **18**, 902–919 (2018). https://doi.org/10.1177/1475921718786427
11. Liu, P., Yang, S., Liu, Y.: Full-scale test and numerical simulation of wheelset-gear box vibration excited by wheel polygon wear and track irregularity. Mech. Syst. Signal Process. **167**, 108515 (2022). https://doi.org/10.1016/j.ymssp.2021.108515
12. Kalengayi Tshilumbu, Z., Rubio Alonso, H., Bustos Caballero, A., Castejón Sisamón, C., Meneses Alonso, J., Garcia Prada, J.C.: Nueva metodología para el análisis de la evolución de las frecuencias naturales con la presencia de grietas en ruedas ferroviarias. In: Cruz Machado, V. and Navas, H. (eds.) 13o Congresso Ibero-americano de Engenharia Mecânica/Ingeniería Mecánica. Livro de actas, Lisboa, Portugal (2017)
13. Huang, D., Li, S., Qin, N., Zhang, Y.: Fault diagnosis of high-speed train bogie based on the improved-CEEMDAN and 1-D CNN algorithms. IEEE Trans. Instrum. Meas. **70**, 1–11 (2021). https://doi.org/10.1109/TIM.2020.3047922
14. Wang, J., Song, C., Wu, P., Dai, H.: Wheel reprofiling interval optimization based on dynamic behavior evolution for high speed trains. Wear **366–367**, 316–324 (2016). https://doi.org/10.1016/j.wear.2016.06.016
15. Bustos, A., Rubio, H., Castejon, C., Garcia-Prada, J.C.: Enhancement of chromatographic spectral technique applied to a high-speed train. Struct. Control. Health Monit. **28**, e2842 (2021). https://doi.org/10.1002/stc.2842
16. Lebel, D., Soize, C., Funfschilling, C., Perrin, G.: High-speed train suspension health monitoring using computational dynamics and acceleration measurements. Veh. Syst. Dyn. 1–22 (2019). https://doi.org/10.1080/00423114.2019.1601744
17. Zhang, L.-H., Wang, Y.-W., Ni, Y.-Q., Lai, S.-K.: Online condition assessment of high-speed trains based on Bayesian forecasting approach and time series analysis. Smart Struct. Syst. **21**, 705–713 (2018). https://doi.org/10.12989/sss.2018.21.5.705
18. Wang, Y.W., Ni, Y.Q., Wang, X.: Real-time defect detection of high-speed train wheels by using Bayesian forecasting and dynamic model. Mech. Syst. Signal Process. **139**, 106654 (2020). https://doi.org/10.1016/j.ymssp.2020.106654
19. Bustos, A., del Castillo ZasBustos Caballero, L., Artés, M.L. del, García Prada, J.C.: Estudio del comportamiento vibratorio de un tren de alta velocidad en operación real mediante EGRSC y ECBF. In: Actas del XV Congreso Iberoamericano de Ingeniería Mecánica. Universidad Nacional de Educación a Distancia (España), Universidad Politécnica de Madrid. Departamento de Ingeniería Mecánica, Madrid, Spain (2022)
20. Braun, S.: Discover signal processing: an interactive guide for engineers. John Wiley & Sons, Chicester, West Sussex (England) (2008)
21. Kouroussis, G., Connolly, D.P., Verlinden, O.: Railway-induced ground vibrations – a review of vehicle effects. Int. J. Rail Transp. **2**, 69–110 (2014). https://doi.org/10.1080/23248378.2014.897791

Correlation Between Microstructure and Hardness of AISI/SAE 1045 Steel Subjected to Different Thermal Cycles Using a Modified Jominy End-Quench Test

Carlos Fosca Pastor and Victor Cupe Missa(✉)

Pontificia Universidad Católica del Perú, Lima, Peru
{cfosca,vcupe}@pucp.edu.pe

Abstract. The present study focuses on the correlation between microstructure and hardness in AISI/SAE 1045 steel subjected to different thermal cycles, including within the range of partial austenitization. For that purpose, a modified Jominy End-Quench Test was applied, cooling rates were estimated, hardness measurement and microstructural analyses for each thermal cycle condition were performed. The outcomes establish a microstructure-to-hardness that may assist in understanding the thermal cycling conditions to which steel may have been exposed, particularly under scenarios simulating severe fire exposure. Moreover, for specific thermal cycles and at equivalent hardness values, significant mixture phases were observed, leading to the conclusion that, under such conditions, hardness testing by itself is insufficient to accurately assess the extent of material degradation resulting from high-temperature exposure.

Keywords: fire damage assessment · steel transformations · Jominy test · microstructural changes of steel

1 Introduction

Steels are the most relevant metallic materials in the field of engineering due to their wide range of microstructural modifications achievable through the application of thermal or thermomechanical treatments. This leads to the development of superior mechanical properties, such as high strength and toughness [1]. The controlled application of temperature (including temperature within the intercritical region) and deformation leads to the development of advanced steels, particularly Dual Phase (DP) and Advanced High Strength Steels (AHSS) [2]. Under these conditions, the nucleation of new grains is enhanced, and in conjunction with swift cooling, austinite transforms into martensite [3].

O. F. Farías Fuentes et al. (Eds.): CIBIM 2024, *Proceedings of the XVI Ibero-American Congress of Mechanical Engineering*, pp. 340–357, 2026.
https://doi.org/10.1007/978-3-032-22823-9_25

Nonetheless, when steel undergo elevated temperatures and thermal cycling under unforeseen conditions, such as exposure to fire during a conflagration, develop microstructures which are not commensurate with those typically associated with conventional manufacturing processes or heat treatments, potentially affecting their mechanical properties significantly.

Therefore, within the scope of Fitness for Service (FFS) evaluation of components exposed to fire-induced damage, on-site hardness measurements and metallographic replica analyses, serve to evaluate the material degradation level [4]. Nonetheless, discrepancies often arise between hardness measurement and the corresponding microstructural changes.

Exposure of steel to fire induces thermal cycles that produce a variety of phases, such as allotriomorphic ferrite, Widmanstätten ferrite, fine perlite, globular perlite, bainite and martensite in varying combinations. Accurate observation of these microstructures using replica metallography is inherent difficult, which diminishes the capacity to reach definitive conclusions and increases the risk of inaccuracies in inspection and analysis of in-situ results.

Consequently, in order to determine microstructural and hardness correlations within the intercritical and complete austenitization temperature ranges, the standard ASTM A255 for Jominy end-quench test were implemented, with variations in heating temperature conditions. Specimens were heated to three different temperatures (890 °C, 760 °C and 740 °C) and exposed to diverse cooling rates via the quasi-unidirectional cooling characteristic of the Jominy specimen, thereby generating a range of microstructure and hardness values along its length, with the purpose of establishing a correlation between them for AISI/SAE 1045 steel.

Establishing the relationship between mechanical proprieties and microstructure will greatly enhance the interpretation of on-site findings, including the identification of phases formed and their corresponding hardness values and the consolidation of definitive conclusions to support crucial determinations concerning the decommissioning of the component, its repair or its ongoing operational viability.

2 Research Methodology

Seven Jominy specimens manufactured in accordance with the geometry established by AST A255-20a [5] were employed and their chemical composition by weight is summarized in Table 1.

2.1 Determination of Critical Temperature

The Eqs. (1) and (2) established by Kasatkin (1984) [6] were employed to compute the upper and lower critical temperatures, with the results summarized in Table 2.

$$\begin{aligned} A_1 = {} & 723 - 7.08 \times Mn + 37.7 \times Si + 18.1 \times Cr + 44.2 \times Mo + 8.95 \times Ni + 50.1 \times V + 21.7 \times Al \\ & + 3.18 \times W + 297 \times S - 830 \times N - 11.5 \times C \times Si - 14 \times Mn \times Si - 3.10 \times Si \times Cr - 57.9 \times C \times Mo \\ & - 15.5 \times Mn \times Mo - 5.28 \times C \times Ni - 6 \times Mn \times Ni + 6.77 \times Si \times Ni - 0.80 \times Cr \times Ni - 27.4 \times V \\ & + 30.8 \times Mo \times V - 0.84 \times Cr^2 - 3.46 \times Mo^2 - 0.46 \times Ni^2 - 28 \times V^2 \end{aligned} \tag{1}$$

Table 1. Chemical composition of the Jominy specimens.

%C	%Mn	%P	%Si	%Cr	%Ni	%Fe	%S
0.46	0.71	0.01	0.19	0.02	0.02	98.4	<0.05
0.46	0.71	0.01	0.19	0.02	0.02	98.4	<0.05
0.46	0.71	0.01	0.19	0.02	0.02	98.4	<0.05
0.45	0.70	0.01	0.21	0.03	0.03	98.4	0.04
0.46	0.70	0.01	0.20	0.03	0.04	98.4	0.04
0.46	0.70	0.01	0.20	0.03	0.04	98.4	0.04
0.46	0.70	0.01	0.20	0.03	0.03	98.4	0.04

$$\begin{aligned} A_3 =\ & 912 - 370 \times C - 27.4 \times Mn + 27.3 \times Si - 6.35 \times Cr - 32.7 \times Ni + 95.2 \times V + 190 \times T \\ & + 72 \times Al + 64.5 \times Nb + 5.57 \times W + 332 \times S + 276 \times P + 486 \times N - 900 \times B + 16.2 \times C \times Mn \\ & + 32.3 \times C \times Si + 15.4 \times C \times Cr + 48 \times C \times Ni + 4.32 \times Si \times Cr - 17.3 \times Si \times Mo - 18.6 \times Si \times Ni \\ & + 4.8 \times Mn \times Ni + 40.5 \times Mo \times V + 174 \times C^2 + 2.46 \times Mn^2 - 6.86 \times Si^2 + 0.322 \times Cr^2 \\ & + 9.90 \times Mo^2 + 1.24 \times Ni^2 - 60.2 \times V^2 \end{aligned} \tag{2}$$

Table 2. Critical temperatures (A_1 y A_3) for the Jominy specimens under investigation.

Specimen	A_1 [°C]	A_3 [°C]
1	731.55	790.02
2	731.49	789.6
3	731.5	789.25
4	729.52	789.6
5	728.52	786.58
6	729.78	787.15
7	728.85	786.26

Thus, the temperatures considered in the intercritical region are 760 °C and 740 °C.

2.2 Jominy Test

Specimens were subjected to a one-hour thermal exposure within the furnace chamber as follows: three specimens were heated to 890 °C, corresponding to complete austenitization, and two specimens were treated under each intercritical condition (760 °C y 740 °C). Subsequently, each specimen war was carefully extracted and expedited within a maximin interval of five seconds to the water-jet cooling and the water-quenched device was adapted to maintain the water jet at a free height of 63.5 mm employing water maintained between 20 °C and 30 °C. Finally, the Jominy specimen was maintained under controlled cooling to room temperature for a period of ten minutes.

2.3 Computational Model Development

Finite element analysis (FEA) performed with Ansys Student 2022 R1 was employed to estimate the cooling curves. The model parameters implemented in the model are summarized below.

Initial Conditions

A three-dimensional model of Jominy specimen was developed in accordance with geometry specified by ASTM A255-20a [5]. Correspondingly, the constant thermal properties applied during cooling are detailed in Table 3 and thermal conductivity was considered variable in accordance with the provisions set forth by M. Drake (1972) [7].

Table 3. Thermal properties of AISI/SAE 1045 steel [1, 8].

Parameter	Notation	Value
Coefficient of thermal expansion	λ	$1.13\ 10^{-5}$/K
Specific heat capacity	Cp	465 J/kg·K
Density	ρ	7800 kg/m^3

Boundary Conditions

The heat transfer coefficients for convection between water jet and the base of the Jominy specimen's end, and between the air and its lateral surface, were incorporated and are detailed in Table 4.

Table 4. Boundary conditions of the computational model [1, 8].

Parameter	Notation	Value
Coefficient of forced convection	hc	10000 W/m^2·K
Fluid temperature	T	20 °C
Coefficient of natural convection and radiation	hr	35 W/m^2·K
Environmental temperature	T∞	20 °C

Mesh Configuration and Time Steps

The meshing properties for the model were as follows: a parabolic tetrahedral element type with size of 3.2mm, and for the refined element type was 1mm. Consequenly, 48,856 elements and 83,245 nodes were generated. Time increments for data acquisition were in three stages: 0–10 s with 0.02-s steps, 10–100 s with 0.5-s steps and 100–600 s with 5-s steps.

2.4 Cooling Rates

The cooling rates derived from temperature-time plots obtained though the computational model were calculated by applying Eq. (3) proposed by Cesar Nunura [8].

$$\dot{T} = \frac{\Delta T}{\Delta t} = \frac{T_a - T_{MI}}{\Delta t} \tag{3}$$

where:

- T_a: austenitization temperature
- T_{MI}: Martensite start temperature
- Δt: duration between T_a and T_{MI}

Under intercritical heating conditions, T_a correspond to the highest heating temperature (760 °C or 740 °C) and according to the ASM Handbook (2020) [9] T_{MI} in computed using Eq. (4):

$$\begin{aligned} T_{MI} = & 512 - 453 \times C - 16.9 \times Ni + 15 \times Cr - 9.5 \times Mo \\ & + 217 \times C^2 - 71.5 \times C \times Mn - 67 \times C \times Cr \end{aligned} \tag{4}$$

2.5 Hardness Measurement and Metallography Evaluation

Hardness testing was performed pursuant to the procedures stipulated in ASTM E18-22 (2022) [10], while metallographic specimen preparation was carried out in strict accordance with the protocols set forth in ASTM E3-11 [11].

Samples underwent precise griding and polishing to obtain a roughness of 1 um. Chemical etching was subsequently executed in conformity with ASTM E407-7 [12].

Samples were procured though cutting at 60 mm measured from the base of the cooled extremely of the specimen. Thereafter, two parallel trims were manufactured to produce a support base 15 mm in width. Figure 1 illustrates the geometric schematic of the sample used for hardness measurement and metallographic evaluation.

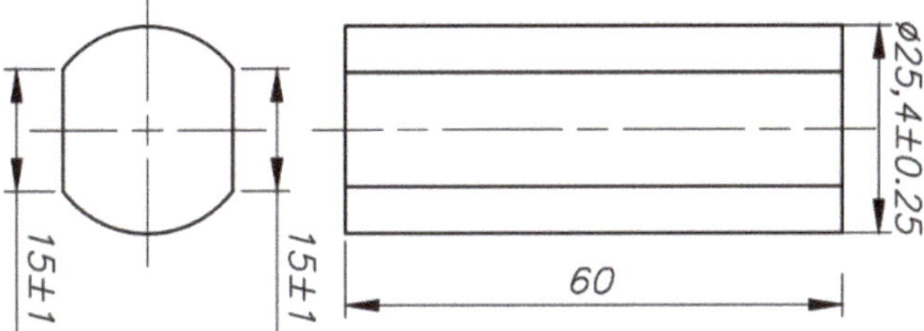

Fig. 1. Sample designated for hardness assessment and metallographic evaluation (units in mm)

3 Experimental Results

3.1 Hardness Evaluation

The hardness profile was performed from the cooled extremity employing Rockwell LCR-500 tester® on the seven Jominy specimens. For the complete austenitization condition (890 °C), the hardness values obtained along the longitudinal axis of the specimen within the steel hardenability range illustrated in Fig. 2, thereby indicating that the established procedure and resultant data conform closely to widely accepted standards.

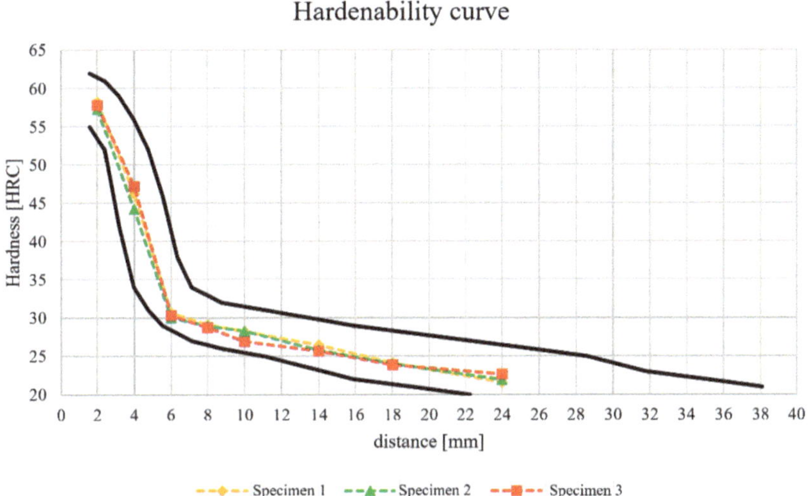

Fig. 2. Hardness measurements on Jominy specimens 1, 2 and 3 corresponding to full austenitization (890 °C)

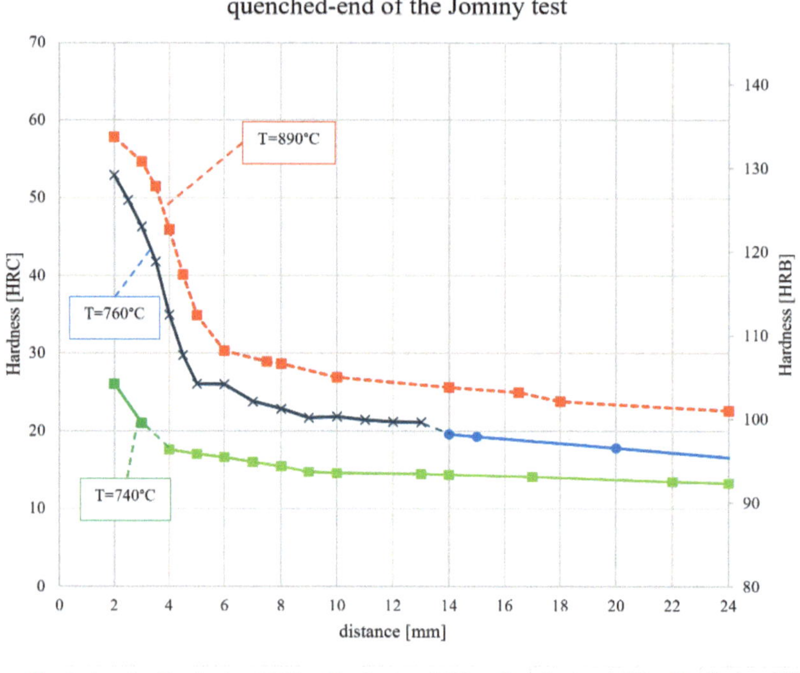

Fig. 3. Hardness profile relative to distance from the quenched end of the Jominy specimen

Therefore, the equivalent heating and cooling procedure was implemented for the two intercritical temperature conditions (760 °C y 740 °C). Subsequently, hardness measurements of the remaining specimens were conducted, demonstrating excellent reproducibility throughout the tested samples.

Figure 3 illustrates the hardness profile as a function of distance from the quenched end of the Jominy specimen for the three analyzed conditions. As demonstrated, a maximum exposure temperature exerts a direct influence on its hardness, owing to the microstructural changes that occur in the steel throughout the cooling process. Additionally, the Jominy Test cooling procedure enables a single specimen to display a continuum of microstructural variations linked to the distinct cooling rates attained throughout its length.

3.2 Cooling Rates

A finite element analysis was conducted to establish the cooling rates occurring throughout the length of a Jominy specimen.

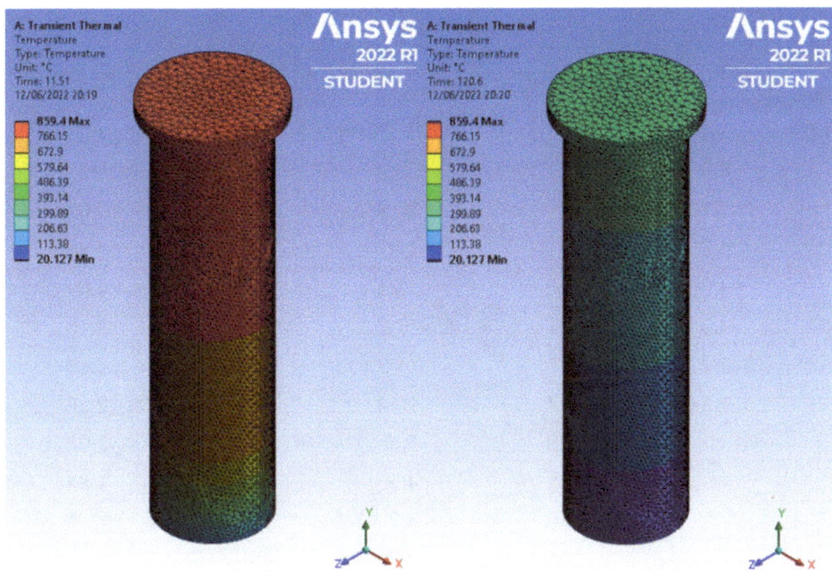

Fig. 4. Temperature distribution for condition 1 (890 °C). (a) 10 s and (b) 120 s

According to the methodology conditions, the computational model was simulated in Ansys Student 2022 R1 software®, as illustrated in Fig. 4, resulting in the acquisition of cooling curves.

Subsequently, cooling rates were computed based on Eq. (3) for each position and condition, detailed in Table 5 and showing a decline in cooling rates with increasing distance from the waster-quenched end.

Table 5. Cooling rates for each condition (890 °C, 760 °C and 740 °C) obtained by numerical simulation with Ansys Student 2022 R1 software®

Tc = 890 °C		Tc = 760 °C		Tc = 740 °C	
d [mm]	$\dot{T}$ [°C/s]	d [mm]	$\dot{T}$ [°C/s]	d [mm]	$\dot{T}$ [°C/s]
2	100	2	106.4	2	108.52
3	64.5	2.5	86.1	3	72.61
3.5	54.77	3	71.6	4	52.53
4	47.39	3.5	60.3	5	39.82
4.5	40.52	4	51.7	6	31.6
5	35.16	4.5	44.8	7	25.48
6	28.68	5	39.5	8	20.79
7.5	21.37	5.5	34.6	9	17.95
8	19.46	6	30.7	10	15.19
10	14.34	6.5	27.7	13	10.39

(*continued*)

Table 5. *(continued)*

Tc = 890 °C		Tc = 760 °C		Tc = 740 °C	
d [mm]	T˙[°C/s]	d [mm]	T˙[°C/s]	d [mm]	T˙[°C/s]
14	8.93	7	25.2	14	9.29
16.5	7.12	8	20.8	17	7.05
18	6.3	9	17.7	20	4.88
24	4.36	10	15.1	25	4.09
30	3.3	11	13.2	30	3.16
45	2.22	12	11.7	35	2.63
50	1.58	13	10.4	40	2.26
-	-	20	5.6	45	2.03
-	-	25	4.2	50	1.84
-	-	30	3.2	-	-
-	-	35	2.7	-	-
-	-	40	2.3	-	-
-	-	45	2	-	-
-	-	50	1.8	-	-

3.3 Microstructural Analysis at Tc = 890 °C

Microstructural analysis was undertaken for various distances measured from the quenched end, as illustrated in Fig. 5. The microstructures in Fig. 5(a) through (f) depict zones near the quenched extremely, where the martensitic matrix initiates, with increasing distance from the quenched end, accompanied by a progressive reduction in cooling rates, there is a noticeable decline in martensitic fraction, replaced by bainitic (acicular), pearlitic (laminar), and ferritic (Widmanstätten and allotriomorphic) morphologies, reflecting the diminished cooling rates and resulting in lower hardness values, as demonstrated in Fig. 3.

In areas further from the quenched end, the microstructure is typified by ferritic (allotriomorphic o idiomorphic) and pearlitic (laminar) morphologies associated with reduced cooling velocities, as evidenced in Fig. 5(g) to (j).

3.4 Microstructural Analysis at Tc = 760 °C

Figure 6(a) thorough (e) depict regions proximal to the quenched end, where martensitic matrix forms due to elevated cooling rates, coexisting with primary pro-eutectoid ferrite as a consequence of the steel exposure to intercritical temperatures (partial austenitization) with the microstructure exhibiting banded morphology indicative of the hot rolling process. Correspondingly, the identical pattern observed in the previous case was identified, whereby decreasing cooling rates favour the development of bainitic and ferritic-pearlitic structure, resulting in a reduction in hardness. The lower hardness measurements in Fig. 3 are linked to the presence of pro-eutectoid ferrite produced at elevated cooling rates owing to the steel undergoing incomplete austenitization at 760 °C. In contrast, at further distances, the matrix present is ferritic-pearlitic composition, as evidence in Fig. 6(f) through (j).

3.5 Microstructural Analysis at Tc = 740 °C

Figures 7(a) to (e) illustrate areas near the quenched end with significant pro-eutectoid ferrite presence, in conjunction with martensitic (resulting from elevated cooling rates) and lamellar pearlitic phases. This behaviour is attributable to exposure to a lower intercritical temperature, which reduced austenitic transformation and consequently decreased hardness along the longitudinal axis, as illustrated in Fig. 3. A decrease in cooling rates results in diminished martensitic formation, culminating in a microstructure dominated by ferritic-pearlitic phases.

Figure 7(f) to (h) illustrated pearlitic undergoing spheroidization, a process attributed in the literature [13] to Divorced Eutectoid Transformation (DET). This transformation is facilitated by heterogeneous austenite formation, typically characterized by pre-existing cementite particles or nuclei within the austenitic matrix, resulting from heating near the lower critical temperature A1 = 731.5 °C (ascertained via Differential Scanning Calorimetry, DSC) and slowing cooling rates of approximately 20 °C/s.

3.6 Microstructure-Hardness Correlation Analisys

Microstructural differences along the Jominy test specimen under the analyzed conditions are a direct consequence of exposure to non-uniform cooling rates varying with distance from the quenched end. For this reason, the formation of various phases occurs, including martensite, bainite, lamellar pearlite, allotriomorphic ferrite and Widmanstätten ferrite. Consequenly, a significant variation in mechanical properties arises, particularly hardness, which is directly correlated with the phases present and their proportions within the microstructure.

In this way, the microstructural and hardness pattern depicted in Fig. 8 was meticulously developed to emphasize that a single hardness value may correspond to a range of distinct microstructural states in steel subjected to varying thermal cycling regimes. For instance, under the analyzed conditions, a hardness of 30 ± 1 HRC (Fig. 8) may correspond to two distinct microstructures, whereas a hardness of 25 ± 1 HRC can be attained though five different microstructures states.

These microstructural conditions may exert differential effects on other critical properties, including material toughness and the Ductile Brittle Transition Temperature (DBTT) notwithstanding identical hardness values across all instances.

The criticality of this aspect lies in its application to assessing Fitness for Service of fire-exposed components, whereby microstructural analysis provides a robust basis for correlating diverse mechanical properties beyond the limitations on site hardness testing, thus advancing the precision of the Fitness for Service evaluations.

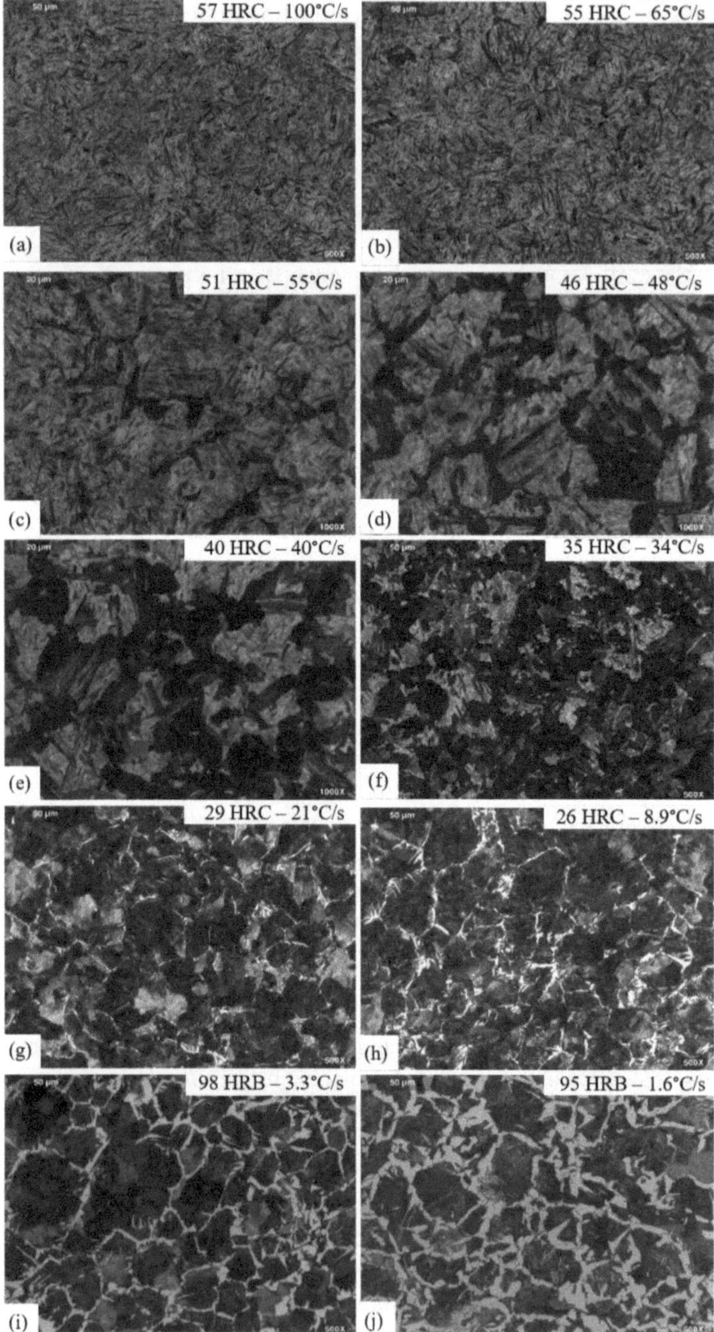

Fig. 5. Optical metallography images on AISI/SAE 1045 Jominy specimen heated to 890 °C illustrating areas corresponding to distance from the quenched end. (a) 2 mm; (b) 3 mm; (c) 3.5 mm; (d) 4 mm; (e) 4.5 mm; (f) 5 mm; (g) 7.5 mm; (h) 14 mm; (i) 30 mm; (j) 50 mm.

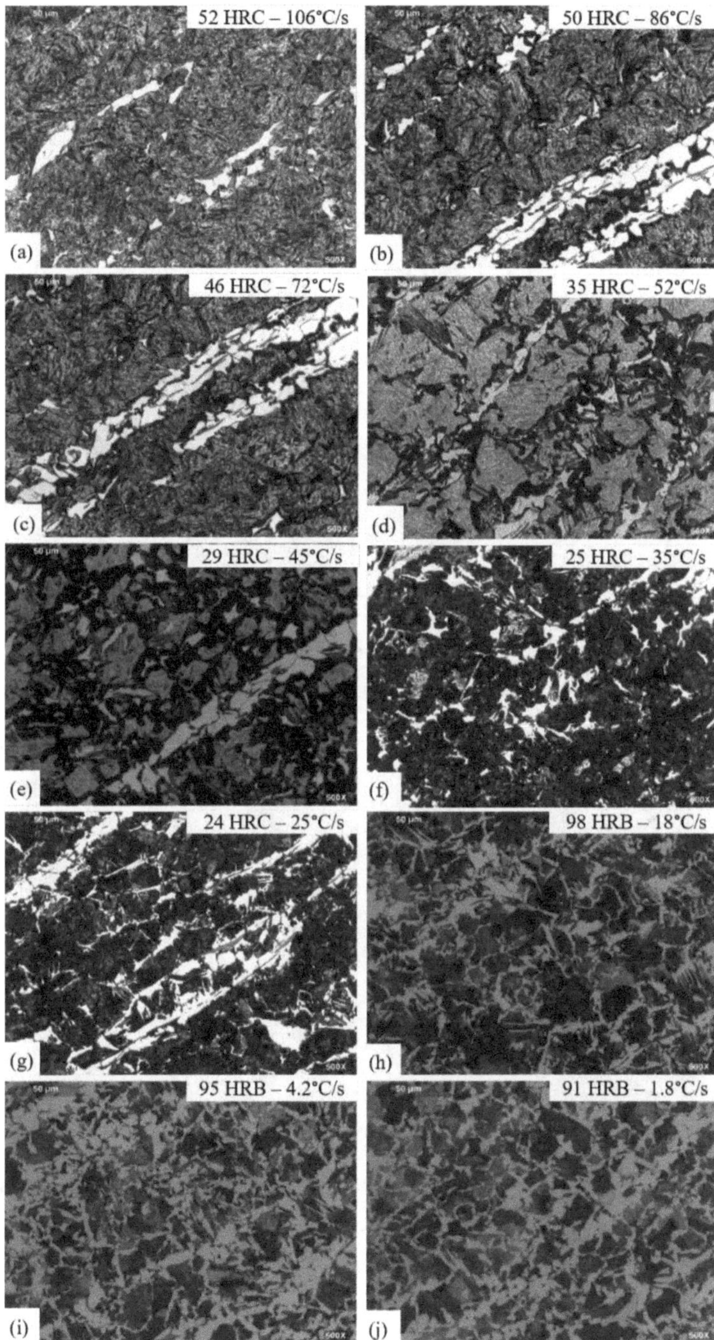

Fig. 6. Optical metallography images on AISI/SAE 1045 Jominy specimen heated to 760 °C illustrating areas corresponding to distance from the quenched end. (a) 2 mm; (b) 2.5 mm; (c) 3 mm; (d) 4 mm; (e) 4.5 mm; (f) 5.5 mm; (g) 7 mm; (h) 15 mm; (i) 25 mm; (j) 50 mm.

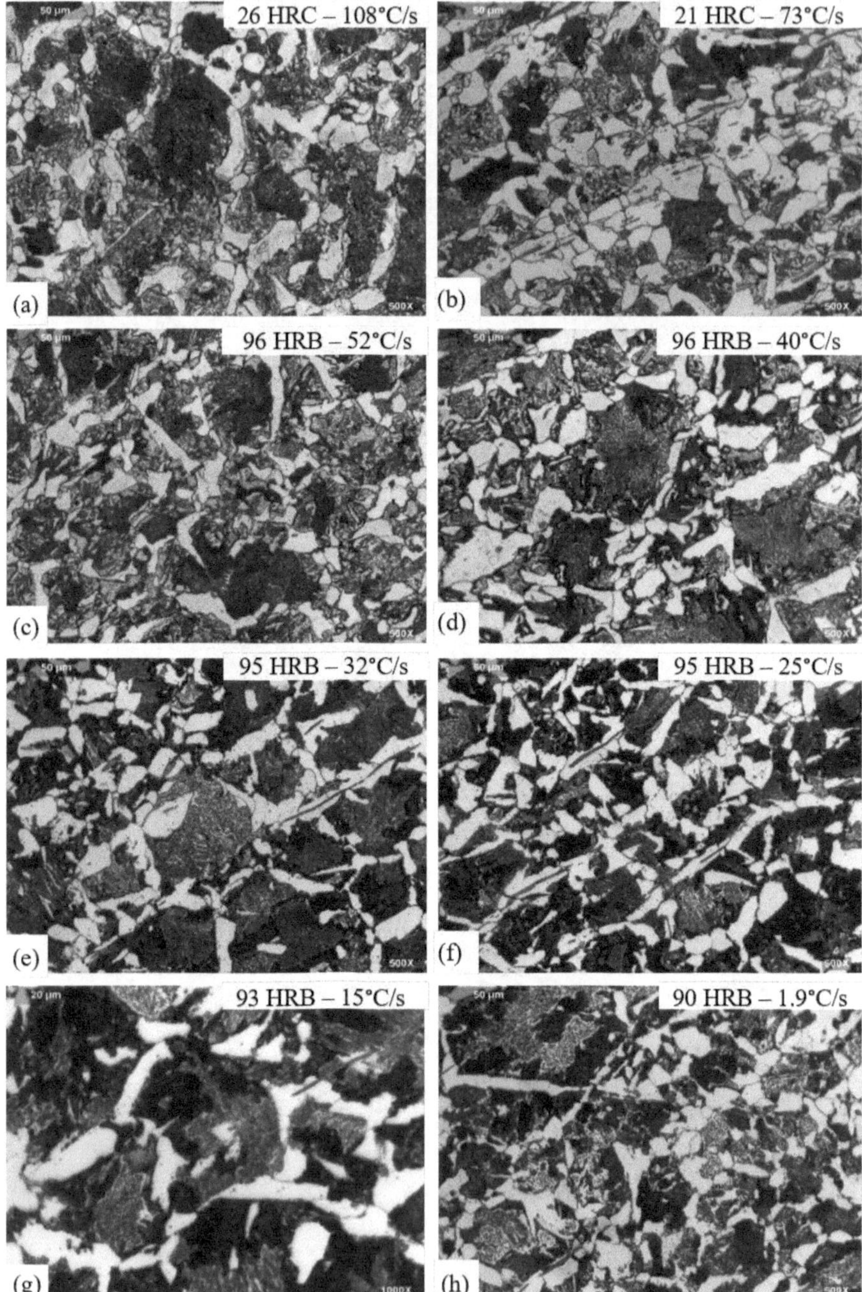

Fig. 7. Optical metallography images on AISI/SAE 1045 Jominy specimen heated to 760 °C illustrating areas corresponding to distance from the quenched end. (a) 2 mm; (b) 2.5 mm; (c) 3 mm; (d) 4 mm; (e) 4.5 mm; (f) 5.5 mm; (g) 7 mm; (h) 15 mm; (i) 25 mm; (j) 50 mm.

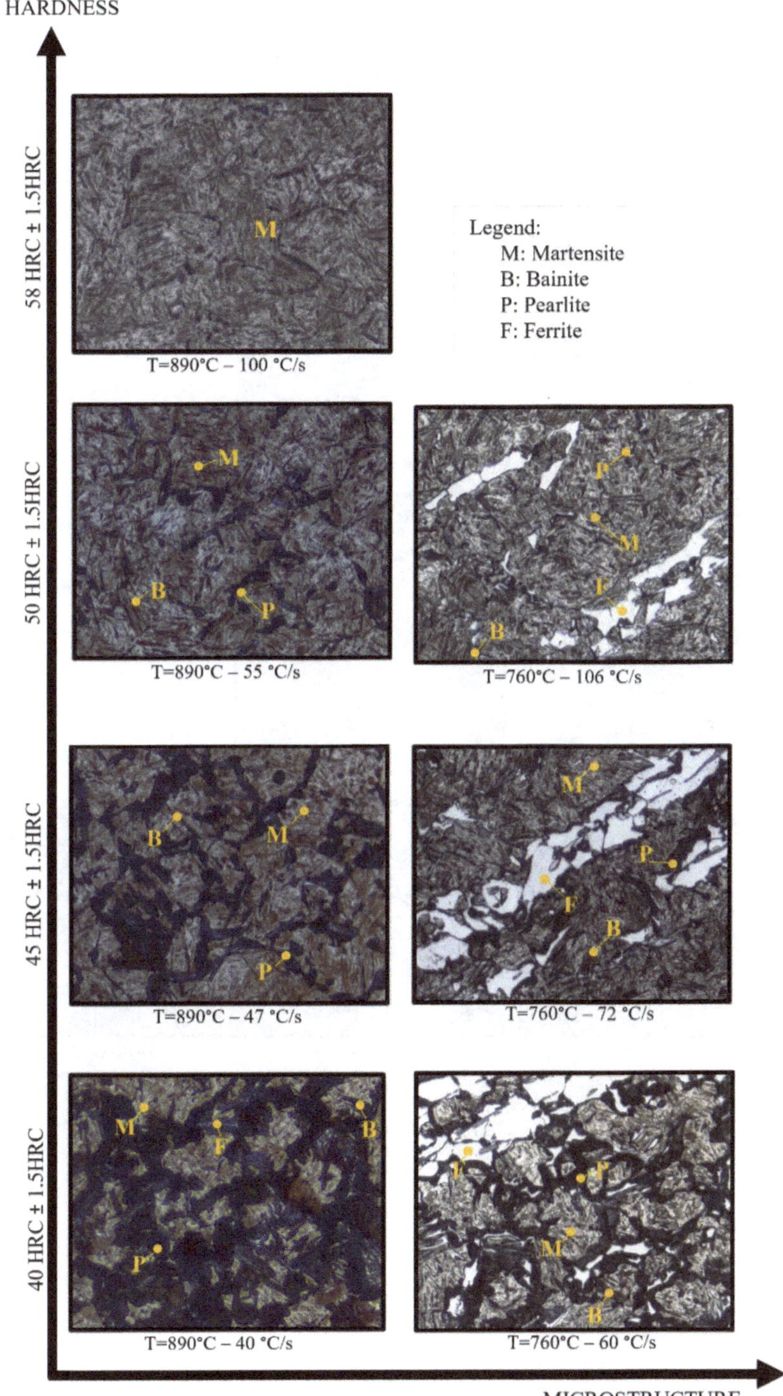

Fig. 8. Microstructural pattern corresponding to three heating conditions (890 °C, 760 °C, 740 °C)

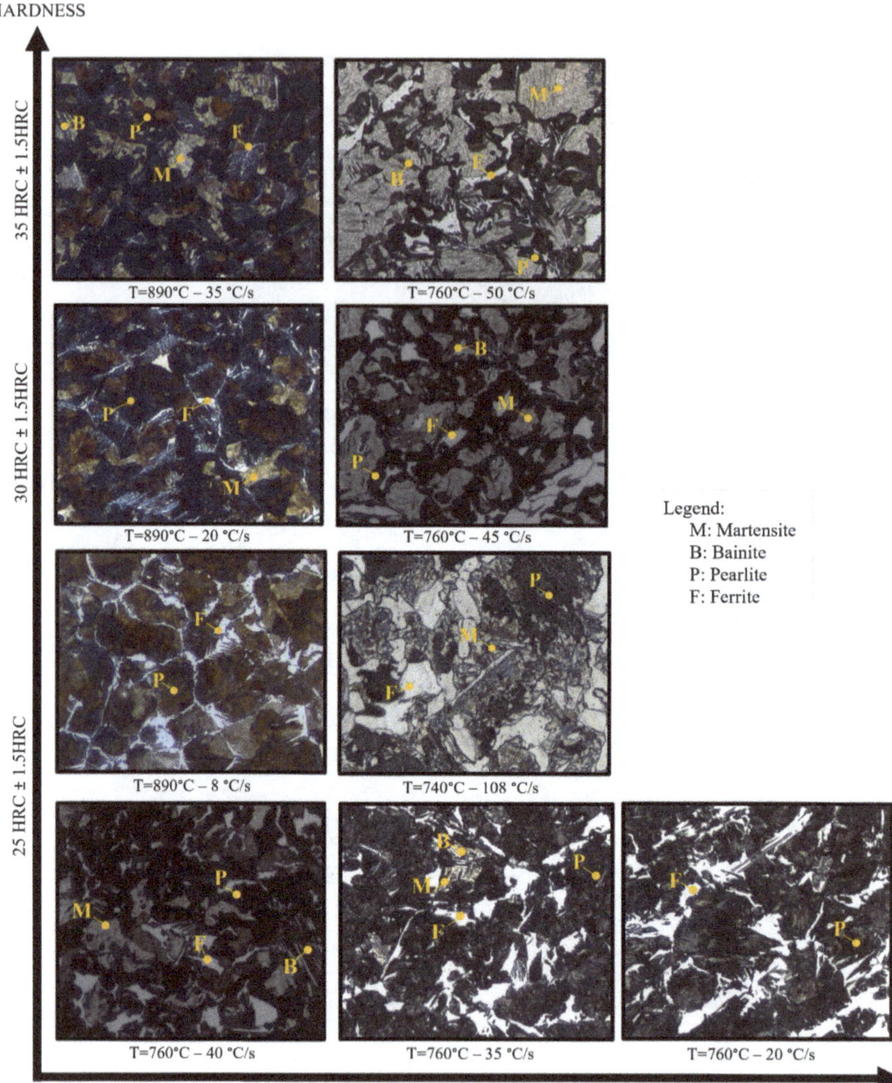

Fig. 8. (*continued*)

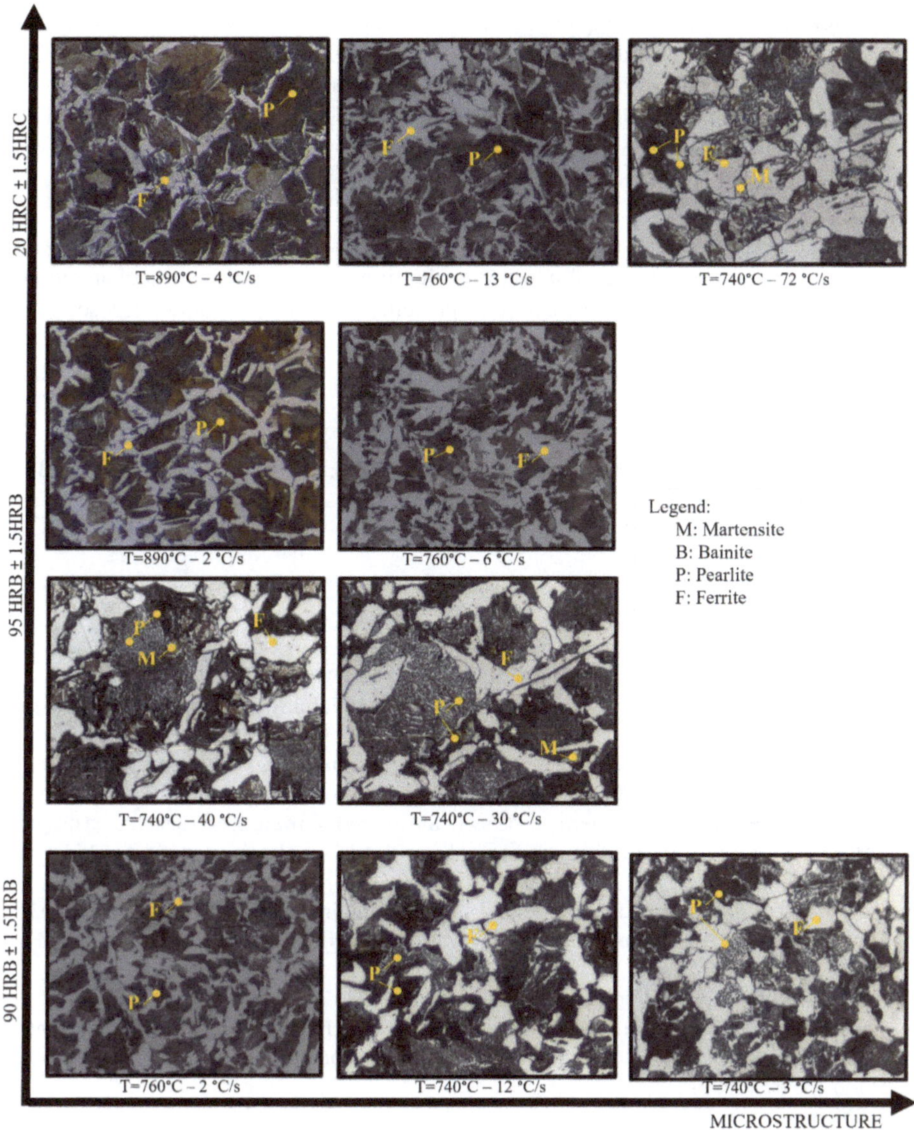

Fig. 8. (*continued*)

4 Conclusions

The implementation of the modified Jominy test facilitated a comprehensive investigation into the correlation between microstructure and hardness by subjecting AISI/SAE 1045 steel exposed to distinct thermal cycles compressing heating at 890 °C (complete austenitization) and 760 °C and 740 °C (biphasic phase range).

In regions proximal to the quenched extremely, exposure to elevated cooling rates (100 °C/s) led primarily to the development of a martensitic matrix, accompanied by pro-eutectoid ferrite and pearlitic phases during intercritical heating. Increasing distance from the quenched end results in reduced hardness and the formation of microstructures with lower hardness values, namely bainite, pearlite and ferrite. From the quenched end, hardness diminishes swiftly up to 5 mm and thereafter decreases progressively up to 50 mm.

A comprehensive microstructure-hardness pattern was established based on extensive hardness measurements, cooling rates evaluation and microstructural characterization, facilitating the precise identification of principal phases via optical microscopy for AISI/SAE 1045 subjected to three maximum exposure temperatures (890 °C; 760 °C, 740 °C) and the cooling conditions produced by Jominy test. The hardness values considered for the development of these patterns are: 58 HRC, 55 HRC, 50 HRC, 45 HRC, 40 HRC, 35 HRC, 30 HRC, 25 HRC, 20 HRC, 95 HRB and 90 HRB. Additionally, the proposed hardness values possess an accuracy of ± 1.5 HRC or ± 1.5 HRB, in accordance with the applicable scale.

References

1. Higuera, O.: Thermal simulation in Cosmoworks of Steel put under test of hardenability Jominy. Scientia et Technica Año XII **35**, 231–236 (2007)
2. Lopez, E.: Predicción del perfil de dureza en probetas Jominy de aceros de medio y bajo carbono. Revista Mexicana de Ingeniería Química **12**(3), 609–619 (2013)
3. Ramos, R.: Estudio de la soldabilidad de la union disimar de un cobre con 5% de Zn con acero estructural ASTM A36. Pontificia Universidad Católica del Perú, Lima, Perú (2013)
4. Quiros, G.: Nuevo tratamiento termico de recocido intercrítico de aceros resistentes al desgaste con Boro de bajo impacto medioamabiental. Universidad Complutense Madrid, Madrid, Spain (2021)
5. ASTM A255-20a: Standard Teset Methods for Determining Hardenability of Steels (2020)
6. Kasatkin, O.G.: Calculation models for determining the critical point of steel. Met. Sci. Heat Treat. **26**(1), 27–31 (1984)
7. Drake, R.M.: Analysis of heat and mass transfer (1972)
8. Nunura, C.: Coorelação numérica e experimental da microestrutura, taxa de resfriamento e características mecânicas do aço ABNT 1045. Universidade Federal Do Rio Grande Do Sul, Porto Alegre (2009)
9. ASM Handbook: Volume 4D: Heat Treating Irons & Steels (2020)
10. ASTM E18-22: Standard Test Methods for Rockwell Hardness of Metallic Materials (2022)
11. ASTM E3-11: Standard Test Methods for Preparation of Metallographic Specimens (2011)
12. ASTM E407-07: Standard Test Methods for Microetching Metals and Alloys (2007)
13. Verhoeven, J.D.: The divorced Eutectoid transformation in steel. Metall. and Mater. Trans. A. **29A**, 1181–1189 (1998)

Statistical Characterization of an Aluminium Foam Using Compression Tests

Melissa Alpízar-Arce[1], Bruno Chinè[2], Francisco Rodríguez-Méndez[2], and Marcela Meneses-Guzmán[3](✉)

[1] Medical Device Engineering, Technological Institute of Costa Rica, Cartago, Costa Rica
[2] Materials Science and Engineering School, Technological Institute of Costa Rica, Cartago, Costa Rica
[3] Industrial Production Engineering School, Technological Institute of Costa Rica, Cartago, Costa Rica
mameneses@itcr.ac.cr

Abstract. In this work, the mechanical behaviour of aluminium foam specimens is studied through compression tests in accordance with the ISO13314 standard. For this purpose, 36 specimens with two geometries, cylindrical and parallelepipedal, were extracted near the edge and the central region of a single rectangular block of foam. To verify whether the geometry of the specimens and the area of their extraction influence the mechanical properties, six hypotheses are proposed, which are then analysed by statistical methods. The stress-strain diagrams obtained exhibit the three expected regions for cellular materials: the elastic zone, the Plateau zone, and the densification zone. It is observed that, regardless of the shape and location of the samples, the densities, and the mechanical properties (Plateau stress, maximum compression stress, Young's modulus and energy absorption capacity) are statistically similar.

Keywords: Aluminium Foam · Mechanical Properties · Statistical Analysis

1 Introduction

Banhart [1] defines a metal foam as: "a special class of cellular metals that originate from liquid-metal foams and, therefore, have a restricted morphology. The cells are closed, round or polyhedral and are separated from each other by thin films". These materials have attracted significant research interest due to their exceptional properties, e.g., low density, high stiffness, low thermal conductivity, and excellent capability to absorb impacts and noise. As a result, they are used in several applications such as load-bearing and protective structures in buildings, automobiles, aircrafts, ships, and satellites. Additionally, their biocompatibility makes them suitable for medical prosthetics, while their thermal properties lend them to use in heat exchangers.

Metal foams can be manufactured using various methods [2-4]: injecting a gas (air, nitrogen, or argon) into a molten metal, introducing a gas-releasing blowing agent (e.g.,

O. F. Farías Fuentes et al. (Eds.): CIBIM 2024, *Proceedings of the XVI Ibero-American Congress of Mechanical Engineering*, pp. 358–368, 2026.
https://doi.org/10.1007/978-3-032-22823-9_26

1–2% TiH_2), or by precipitating a previously dissolved gas in a molten metal under pressure. However, these foaming processes produce non-homogeneous cellular materials characterized by density gradients, randomly shaped cells, and structural imperfections such as pore size variation, non-uniform cell wall thickness, and cracks in the cell walls [5]. These "defects" compromise the mechanical properties of the metallic foam and lead to significant variability in the experimental results.

For this reason, characterizing these materials after manufacturing is essential to ensure their mechanical stability and integrity at any point within their structure, regardless of the internal or spatial distribution of defects. Usually, to determine the mechanical behaviour and properties of metallic foams, an experimental stress-strain curve is obtained under compressive loading [5-7].

Following this approach, mechanical properties of aluminium foams have been extensively studied in recent years. E.g., powder metallurgy methods produce foams (628–713 kg/m^3 density) with high Young's Modulus (~5000 MPa), but moderate compressive strength (5–22 MPa), demonstrating excellent energy absorption (29.27 J/g at 77% strain) [8]. Direct foaming methods show comparable densities (627–717 kg/m^3), but lower stiffness (500–1295 MPa) and plateau stresses of 7–8 MPa [9].

Notably, sandwich structures with solid aluminum sheets achieve exceptional yield strength (88–320 MPa) and energy absorption (13–60 J/m^3) [10]. High-speed testing (284 mm/s) reveals aluminum foams can absorb 15 MJ/m^3 energy [11]. Ultra-light foams made with TiH_2 additives (243 kg/m^3 density) exhibit lower mechanical properties (1.19 MPa yield strength) [12, 13], while Alporas/Alcan foams show significant stiffness variations (105–700 MPa) [14]. These results show that manufacturing methods critically influence pore structure and mechanical performance, with trade-offs between density, strength, and energy absorption capabilities [2, 5, 15].

It is also important to note that in the literature, it is common to find that different compression tests follow standards such as ISO 13314: 2011, ASTM C393, or ASTM C365-05 [16, 17], where a variety of geometries and specimen sizes were reported, as well as several configurations used during the tests and the statistically reduced number of samples (5 repetitions, according to the different standards).

Those differences make an experimental characterization, with an approach that ensures the statistical significance of the obtained data, both interesting and necessary. For this reason, the main contribution of this work is to provide mechanical characterization data from compression tests based on statistically representative experiments, using specimens with two different geometries extracted from a single block of aluminium foam. With this numerical representativeness of samples, which have been taken from different positions within the matrix of the material, the aim is to address the hypothesis of whether the geometry and extraction site of the specimens affect their response, and the mechanical characteristics obtained through compression tests.

2 Methodology

2.1 Specimen Preparation

The original piece of aluminium foam, which had dimensions of 166 × 155 × 40 mm^3, was manufactured via a powder metallurgy process, employing precursors of an AlSi10 alloy mixed with titanium hydride (TiH_2) powder at 0.80 wt%. These last materials were produced by the Austrian company Aluligth® GmbH (Ranshofen, Austria), and they were subsequently foamed in a Nabertherm® GmbH convection furnace (Lilienthal, Germany).

According to ISO 13314 [18], specimens with either cylindrical or parallelepipedal geometry can be used for compression tests; therefore, 18 specimens of each shape were extracted by cutting them using an Electrical Discharge Machine (EDM). Figure 1 illustrates a spatial diagram of the cut samples, where a 25 mm wide section has been discarded from both the width and the length of the foam. This was done to avoid the effect of the higher density near the edges. Cylindrical samples are 20 mm in diameter and 40 mm in length, while the parallelepipedal specimens are 20 × 20 × 40 mm^3. The samples extracted from the central region (Fig. 1) are labelled as internal specimens (IS), while those outlined by the two boxes are referred to as external specimens (ES). Each sample is identified by a number that provides traceability based on its shape and position.

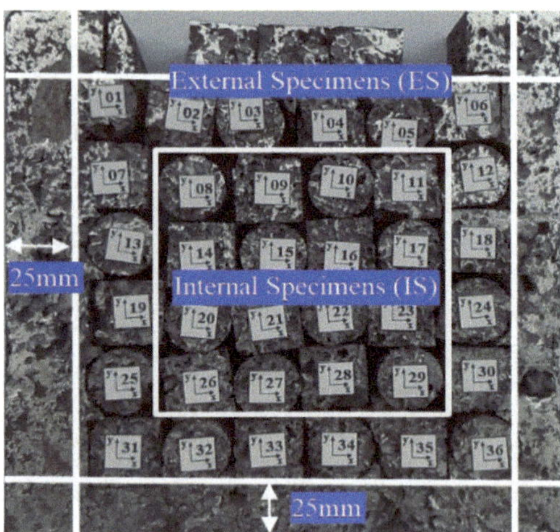

Fig. 1. Top view of the aluminium foam sample used, and the various cut test specimens.

2.2 Hypothesis Formulation

Hypotheses are formulated to address whether the geometry and extraction site of the specimens affect the mechanical properties obtained through compression tests. Table 1 shows all these possibilities. Subsequently, with the experimental data collected, the t-Student statistical test is used to compare means, then normality is verified using the Anderson-Darling test, while variance comparison is performed using the F-test.

2.3 Equipment and Specimen Characterization

The mass of each specimen was measured using an analytical balance, and its height and diameter with a Vernier calliper. The density of each sample (ρ_{foam}) and the relative density ($\rho_{Relative}$) were calculated with the following relation:

$$\rho_{Relative} = \frac{\rho_{foam}}{\rho_{Material}} \tag{1}$$

where $\rho_{Material}$ is the density of the bulk material that constitutes the foam, i.e., aluminium. Then, the porosity percentage of each sample was calculated using:

$$\%\ Porosity = \frac{\rho_{Material} - \rho_{foam}}{\rho_{Material}} * 100 \tag{2}$$

Table 1. Hypotheses for Mechanical Properties

Hypotheses (H)	Statement
1. ES: Cylindrical – Parallelepipedal	The mechanical properties of the specimens are independent of the geometry
2. IS: Cylindrical – Parallelepipedal	The mechanical properties of the specimens are independent of the geometry
3. Cylindrical: ES-IS	The mechanical properties of cylindrical specimens are independent of the extraction zone
4. Parallelepipedal: ES-IS	The mechanical properties of parallelepipedal specimens are independent of the extraction zone
5. Geometry	The mechanical properties of any specimen vary according to geometry but are independent of the extraction zone
6. Location	The mechanical properties vary according to the extraction zone, denoted as 1: PE, 2: PI, and are independent of the shape

Compression tests were conducted using an MTS Bionix equipment (Eden Prairie, MN, USA) with a 25kN load cell, 0.1 mm/s speed, and a preload of 150N, following the standard ISO 13314 [18].

The specimens were tested randomly to minimize errors. Force and displacement were measured by the Bionix machine, and from these measurements, the stress-strain curves for each sample were plotted. To visually identify the deformation bands, knowing that an exothermic reaction occurs in the material due to plastic deformation during the compression process, the tests were filmed using both a standard camera and a FLIR I50 thermographic system (Wilsonville, OR, USA) with an emissivity of 0.96 [11, 19, 20].

Finally, the energy absorption capacity (EA) of the foam is equal to the area under the stress-strain curve, and can be calculated as follows:

$$EA = \int_{\varepsilon_1}^{\varepsilon_2} \sigma(\varepsilon) d\varepsilon \tag{3}$$

where ε_1 and ε_2 are the strain values within an interval of interest, and $\sigma(\varepsilon)$ is the stress response of the material, which is dependent on the strain.

3 Results and Discussion

Nine of 36 specimens were discarded: cylindrical specimens 1, 15, 34, and 36, and parallelepipedal samples 2, 14, 16, 31, and 33. The main reason for this elimination was that the mechanical properties of these specimens could not be calculated according to the ISO 13314 standard, since the behaviour obtained for the stress-strain curve did not include the appropriate phases. The results of some of these samples were probably influenced by internal defects in the foam, while others had already undergone deformation prior to testing, as was the case for specimens 15 and 33.

In average, the densities of the 36 specimens are 0.4256 g/cm^3. However, the average density of the remaining 27 samples, after the discard process, is 0.4144 g/cm^3, with a standard deviation of 0.0104 g/cm^3. Using the two-sample t-Student test, the average densities for both geometries were compared, resulting in a p-value of 0.932, indicating that the densities of the parallelepipedal and cylindrical specimens are statistically equal. The same conclusion is drawn when comparing specimens from the internal and external zones, with a p-value of 0.238; in both cases, the samples exhibit normal behaviour and statistically equivalent variances.

Figure 2 illustrates the deformation of a parallelepipedal specimen during the compression test. It can be noted that the first deformation bands occur at the top of the sample (Fig. 2a, grey box), which are a direct consequence of the previously mentioned defects. From the IR images (Fig. 2b), it is confirmed that most of the deformation bands appear at the top of the sample, near the applied load. Additionally, different deformation bands are observed, oriented parallel to the load and distributed throughout the foam. Then, as the specimen is compressed further, the porous structure of the sample collapses layer by layer, which is attributed to the continuous evolution of the existing deformation bands and the formation of new ones [5].

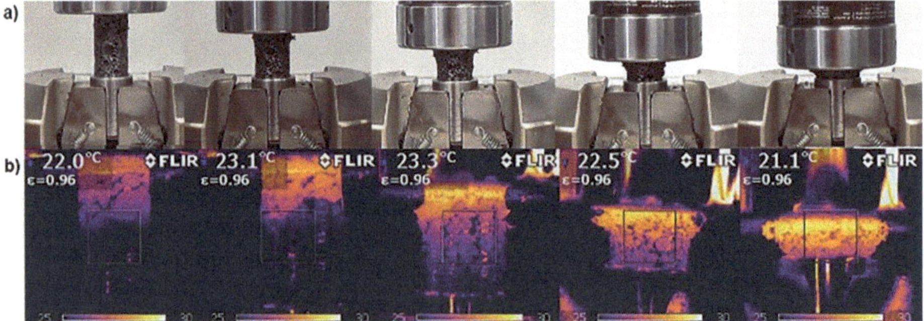

Fig. 2. Deformation of the aluminium foam specimen recorded with a) a standard camera, and b) a thermographic camera.

Figure 3 shows the stress-strain diagrams obtained for the cylindrical (Fig. 3a) and parallelepipedal (Fig. 3b) specimens, respectively. These curves reveal the three expected regions for this type of material: the elastic region, the Plateau region, and the densification region. The resulting yield strength is, on average, 240 MPa. This value is lower compared to those reported in the literature, and it is due to the relative low density of the foams. The elastic region of the foams is small owing to the anisotropic, irregular, and heterogeneous cellular structure. This yields an uneven distribution of stresses and strain, with local plastic deformation even at low loads [21, 22]. After surpassing the initial elastic region, the cell walls begin to bend and buckle, resulting in the Plateau stress. This region is responsible for the high energy absorption capacity exhibited by cellular materials [23, 24]. Finally, as deformation increases, the cell walls fracture and gradually collapse, leading to the densification of the cellular structure.

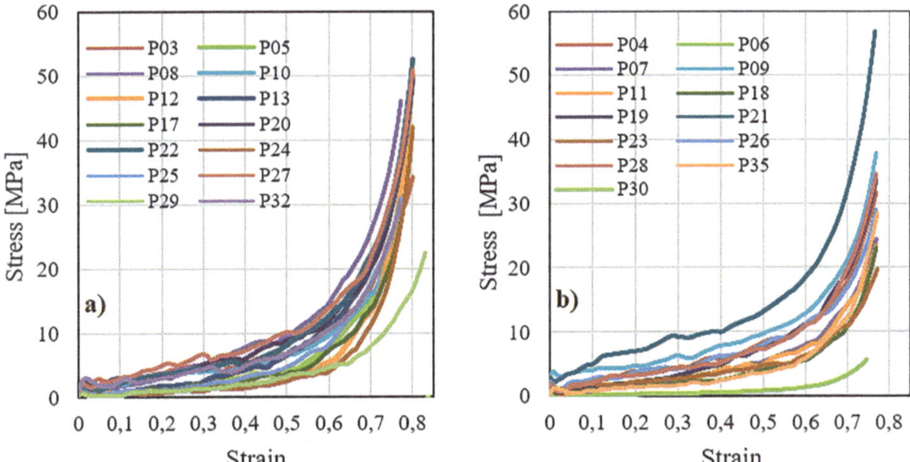

Fig. 3. Stress-strain diagrams for a) cylindrical, and b) parallelepipedal foam specimens.

Figure 4 plots the energy absorption capacity (EA) for the cylindrical specimens (Fig. 4a) and the parallelepipedal samples (Fig. 4b). It is observed that the parallelepipedal specimens exhibit a higher energy absorption capacity (EA) than the cylindrical ones.

Figure 5 exhibits the influence of material density on the energy absorption capacity of the aluminium foams. Both samples show a tendency to increase the EA value as the foam density increases. This behaviour is evident in both the regression equations: $-0.75 + 5.50\rho$ for the cylindrical samples, and $1.47 + 3.1\rho$ for the parallelepipedal pieces, but with very high variability. This instability in the data is reflected by the observation that specimen P9 reports the highest and best absorption capacity (7.63 MJ/m^3), but with a density of 0.410 g/cm^3. The literature reports energy absorption capacities ranging from 0.9 to 13 MJ/m^3 for aluminium foams with porosities between 40 and 80% [11, 17, 23], which agrees with the obtained results.

The Young's modulus measurements for both sample geometries are above 50 MPa and below 400 MPa. It was observed that the cylindrical specimens have more consistent values than the parallelepipedal specimens. Theoretically, the higher the density, the greater the stiffness of the sample, and thus, the higher the Young's modulus. However, specimen 6 exhibits high density and a very low Young's modulus, while specimens 21 and 25 show low density and high Young's modulus. Figure 6 displays these behaviours and the relationship between Young's modulus and density. Due to the high instability introduced by specimens 6, 7, 21, 25, 30, and 35, they were excluded from this regression analysis. For the cylindrical specimens, there is a slight trend for the Young's modulus to increase, which is confirmed by the regression equation, following the relationship $82 + 341\rho$, while for the parallelepipedal specimens, the equation is $19 + 342\rho$, with both equations showing a similar slope.

Table 2 summarizes the mean and the standard deviation of the mean for the aluminium specimens tested, according to the hypotheses analysed. Table 3 shows the p-values of the statistical tests for normality, homogeneity of variance, and equality of means.

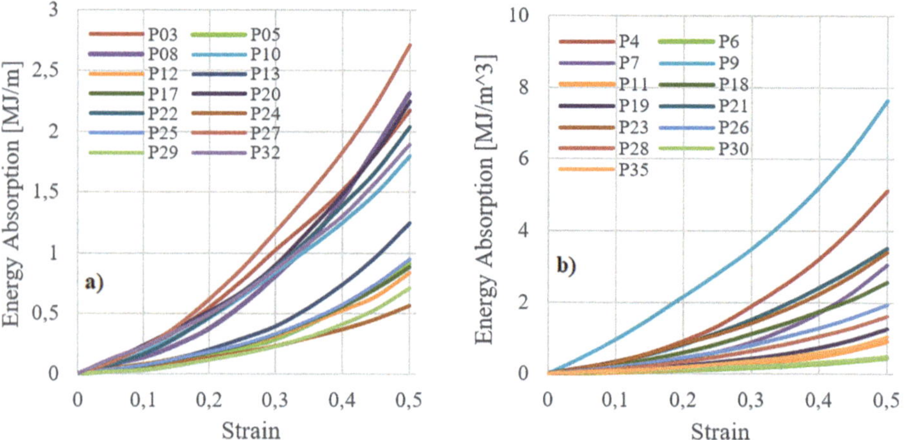

Fig. 4. Energy absorption capacity for a) cylindrical, and b) parallelepipedal foam samples.

Normality that does not pass the Anderson-Darling test is evaluated with the Ryan Joiner (RJ) test. The resulting p-values are contrasted against a type I error of 0.05. For the Young's modulus property, the samples located in the external zone with cylindrical shape (H1) show 190 units higher than those of parallelepipedal shape, leading to a rejection of the hypothesis of equal means. H6 also rejects the hypothesis that the Young's modulus is the same for specimens of both geometries, regardless of the extraction zone. The compressive yield strength for parallelepipedal specimens in H5 differs according to the extraction zone; the same result is obtained when comparing the yield strength of specimens located in both zones (H6), regardless of the shape. In all hypotheses, the mean density values fall within the range of 0.40 to 0.43 g/cm^3, with moderate standard deviation values. It is observed that the specimens with a parallelepipedal geometry located near the edge have a higher density than the cylindrical and all the centred samples. This is because, during the manufacturing process, the pores are compacted near the edges due to the metal solidification starting from the outside, which produces

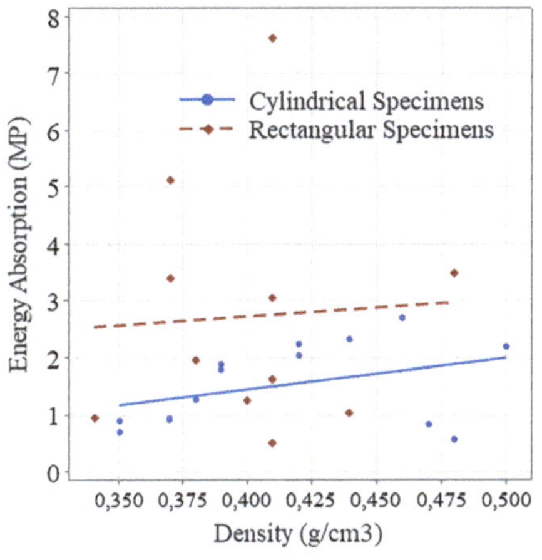

Fig. 5. Relationship between energy absorption capacity and specimen density.

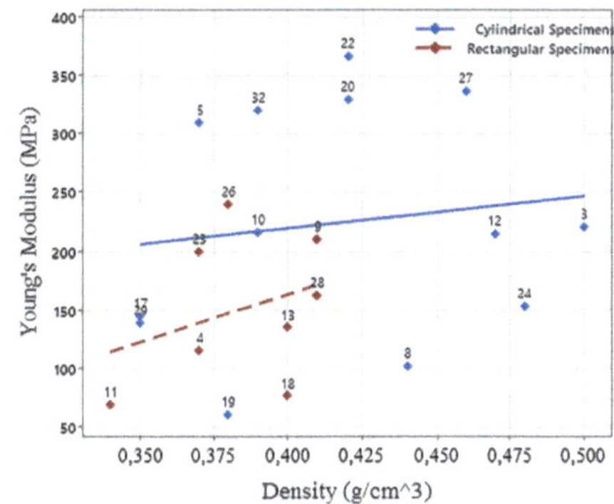

Fig. 6. Relationship between Young's Modulus and specimen density.

an increase in the foam density in these places [24]. Finally, for the Plateau region, both the internal and external specimens, regardless of the geometry, have different Plateau stresses according to H6; moreover, H1 suggests that specimens with different stress values from both geometries are those coming from the external section.

Table 2. Mechanical properties of aluminium foam obtained through compression tests.

		Young's Modulus [MPa]		Compressive yield strength [MPa]		Density [g/cm^3]		Plateau [MPa]	
H	Geometry/ Location	mean	standard error of the mean	mean	standard error of the mean	mean	standard error of the mean	mean	standard error of the mean
1	C	255,900	54,100	1,360	0,305	0,420	0,022	2,500	0,480
ES	P	64,800	17,600	0,910	0,151	0,430	0,026	4,580	2,670
2	C	233,600	41,500	1,770	0,250	0,400	0,016	3,990	0,650
IS	P	176,2*	29,500	2,150	0,420	0,400	0,02	3,980	0,880
3	ES	255,900	54,100	1,360	0,310	0,420	0,022	2,500	0,480
C	IS	233,600	41,500	1,770	0,250	0,400	0,016	3,990	0,650
4	ES	64,800	17,600	0,910	0,150	0,430	0,026	4,580	2,670
P	IS	340,000	165,000	2,150	0,420	0,400	0,02	3,980	0,880
5	C	244,800	32,900	1,560	0,200	0,410	0,013	3,250	0,440
G	P	191,700	83,100	1,480	0,270	0,420	0,017	4,300	1,440
6	ES	160,400	38,100	1,130	0,180	0,430	0,016	3,540	1,340
L	IS	282,600	77,100	1,940	0,230	0,400	0,012	3,990	0,510

H: Hypothesis, ES: External Specimen, IS: Internal Specimen, C: Cylindrical, P: Parallelepipedal.
*An outlier was removed from the dataset.

Table 3. p-values for normality tests using the Anderson-Darling test, F-test for equality of variances, and t-test for equality of means, for the 6 established hypotheses.

		Young's Modulus [MPa]			Compressive yield strength [MPa]			Density [g/cm3]			Plateau [MPa]		
H	Geometry/ Location	normal test	homog. of variance	equality of means	normal test	homog. of variance	equality of means	normal test	homog. of variance	equality of means	normal test	homog. of variance	equality of means
1	C	0,658	0,015	0,006	0,010*	0,110	0,212	0,054	0,067	0,837	0,221	0,001	0,047
ES	P	0,756			0,100*			0,010*			0,010		
2	C	0,195	0,000	0,561	0,142	0,313	0,435	0,440	0,778	0,817	0,367	0,429	0,987
IS	P	0,010*			0,906			0,513			0,563		
3	ES	0,685	0,535	0,750	0,010*	0,636	0,322	0,054	0,486	0,507	0,221	0,491	0,089
C	IS	0,195			0,142			0,440			0,367		
4	ES	0,756	0,000	0,096	0,376	0,040	0,013	0,010*	0,436	0,367	0,010	0,020	0,844
P	IS	0,010*			0,906			0,513			0,563		
5	C	0,510	0,136	0,003	0,017	0,355	0,807	0,391	0,502	0,932	0,100	0,000	0,476
G	P	0,531**			0,084			0,011			0,01*		
6	ES	0,114	0,187	0,318	0,010*	0,399	0,009	0,035	0,239	0,238	0,518	0,189	0,010
L	IS	0,520*			0,591			0,837			0,749		

* RJ test ** exclusion of an outlier

In general, the results regarding the mechanical characteristics of elasticity, compression strength, density, and Plateau stress are statistically independent of the shape and location of the sample. This indicates a certain degree of homogeneity in the mechanical properties within the piece of metal foam, despite the exceptions found in hypotheses H1 and H6, where the values of the Young's modulus are statistically different, while the compressive yield strengths are different for hypothesis H5 and H6.

4 Conclusions

In this work, a statistical analysis of the mechanical behaviour and properties of an aluminium foam was conducted, performing experimental compression tests using representative samples of two specimen geometries, extracted from a single piece of cellular material. It is observed that, regardless of the geometry and location of the tested specimens, the densities are statistically equivalent. According to the IR images, most of the deformation bands appear in the upper part of the sample, close to the applied load, while the remaining bands are distributed throughout the foam and oriented parallel to the compression load.

The stress–strain diagrams show the three expected regions for cellular materials: the elastic region, the Plateau region, and the densification region. The various mechanical properties studied, such as Young's modulus and energy absorption capacity, are proportional to the material's density for both cylindrical and parallelepipedal specimens. It is also demonstrated that the tested samples exhibit good energy absorption capacity, ranging from 1 to 9 MJ/m^3. The average values for Young's modulus, compressive yield strength, density, and Plateau stress are 244.8 MPa, 0.2 MPa, 0.41 g/cm^3, and 3.25 MPa for cylindrical specimens; and 191.7 MPa, 1.48 MPa, 0.42 g/cm^3, and 4.3 MPa for parallelepipedal specimens, respectively.

The results of the statistical tests show significant differences for Young's modulus with respect to the specimen shape. Correspondingly, the compressive yield strength is also statistically different but in relation to the extraction location. For the remaining indicators, the tests reveal statistical independence from both specimen shape and location. We conclude that there is a certain degree of homogeneity in the mechanical properties of the metal foam piece, which are generally independent of the specimen's geometry and extraction zone.

During the process of cutting the specimens and subsequent execution of the compression tests, it was observed that certain samples must be discarded for reasons related to structural defects, sudden failure during the experiment, or anomalous results that cannot be used as reliable values in the statistical analysis. This highlights that the number of samples, which must be considered and prepared for experimental design and execution requires greater attention than the guidelines specified in the ISO 13314:2011 standard. For instance, in this study, 25% of the specimens had to be discarded for the reasons mentioned above. This indicates that the number of specimens must always ensure the representativeness of the process and the reliability of the results as a valid reference.

References

1. Banhart, J.: Manufacturing routes for metallic foams. JOM **52**(12), 22–27 (2000)
2. Polmear, I., StJohn, D., Jian-Feng, N., Qian, M.: Light Alloys - Metallurgy of the Light Metals. Butterworth-Heinemann, UK (2017)
3. Ji, C., Huang, H., Wang, T., Huang, Q.: Recent advances and future trends in processing methods and characterization technologies of aluminum foam composite structures: a review. J. Manuf. Process. **93**, 116–152 (2023)
4. Madgule, M., Sreenivasa, C.G., Borgaonkar, A.: Aluminium metal foam production methods, properties and applications - a review. Mater. Today Proc. **77**, 673–679 (2023)
5. Nisa, S., Pandey, S., Pandey, P.: A review of the compressive properties of closed-cell aluminum metal foams. J. Process Mech. Eng. Proc. Inst. Mech. Eng. Part E **237**(2), 531–545 (2023)
6. Hanssen, A.G., Reyes, A., Langseth, M.: Design and finite element simulations of aluminium foam-filled thin-walled tubes. J. Veh. Des. **37**, 126–155 (2005)
7. Mangipudi, K.R.: Multiscale Modelling of Deformation and Fracture in Metal Foams. University of Groningen (2012)
8. Novak, N., et al.: Compressive behaviour of closed-cell aluminium foam at different strain rates. Rev. Mater. **12**(24) (2019)
9. Mankovits, T., Antal, T., Manó, S., Kocsis, I.: Compressive response determination of closed-cell aluminium foam and linear-elastic finite element simulation of μCT-based directly reconstructed geometrical models. J. Mech. Eng. **64**(2), 105–113 (2018)
10. Endut, N., Hazza, M.A., Sidek, A., Adesta, E., Ibrahim, N.: Compressive behaviour and energy absorption of aluminium foam sandwich. In: Conference Series: Materials Science and Engineering, vol. 290, no. 1 (2018)
11. Duarte, I., Vesenjak, M., Krstulović-Opara, L.: Compressive behaviour of unconstrained and constrained integral-skin closed-cell aluminium foam. Compos. Struct. **154**, 231–238 (2016)
12. Wang, X., Guangtao, Z.: The static compressive behavior of aluminum foam. J. Adv. Mater. Sci. **33**, 316–321 (2013)

13. Haag, F., Galio, A., SchaeVer, L.: Uniaxial compression tests of aluminium foams. J. Eng. Manufact. Part B. IMechE **216**(4) (2001)
14. Sugimura, Y., Meyer, J., Bart-Smith, H., Evans, J.A.: On the mechanical performance of closed cell Al alloy foams. Acta Mater. **45**(12), 5245–5259 (1997)
15. Faruk, O., Tjong, J., Sain, M.: Lightweight and Sustainable Materials for Automotive Applications. CRC Press, Boca Raton (2017)
16. Ramamurty, U., Paul, A.: Variability in mechanical properties of a metal foam. Acta Mater. **52**, 869–876 (2004)
17. Grilec, K., Maric, G., Suzana, J.: A study on energy absorption of aluminium foam. Berg-Und Huttenmannische Monatshefte **155**(5), 231–234 (2010)
18. International Organization for Standardization: ISO 13314:2011 - Mechanical testing of metals Ductility testing Compression test for porous and cellular metals (2011)
19. FLIR: FLIR i50 (2010 model) (2010). http://userequip.com/files/specs/5593/FLIR%20i50%20(2010%20model).pdf
20. FLIR: The Ultimate Infrared Handbook for R&D Professionals (2015). https://vault.flir.com/file/asset/18320/original?token=8df9f33c-ebf4-43ce-897e-4fdeb4672a18
21. Hassan, S., Sadek, M.: Aluminum Foam Sandwich with Adhesive Bonding: Computational Modelling. Universidad de Porto, Portugal (2016)
22. Gibson, L.J., Ashby, M.F.: Cellular Solids: Structure ad Properties. Cambridge University Press, Cambridge (1997)
23. Degischer, H.P., Kriszt, B.: Metallic Foams: Their production, Properties and Applications. Wiley-VCH, Weinheim, Germany (2002)
24. Koerner, C.: Integral Foam Molding of Light Metals: Technology, Foam Physics and Foam Simulation. Springer, Heidelberg (2008)

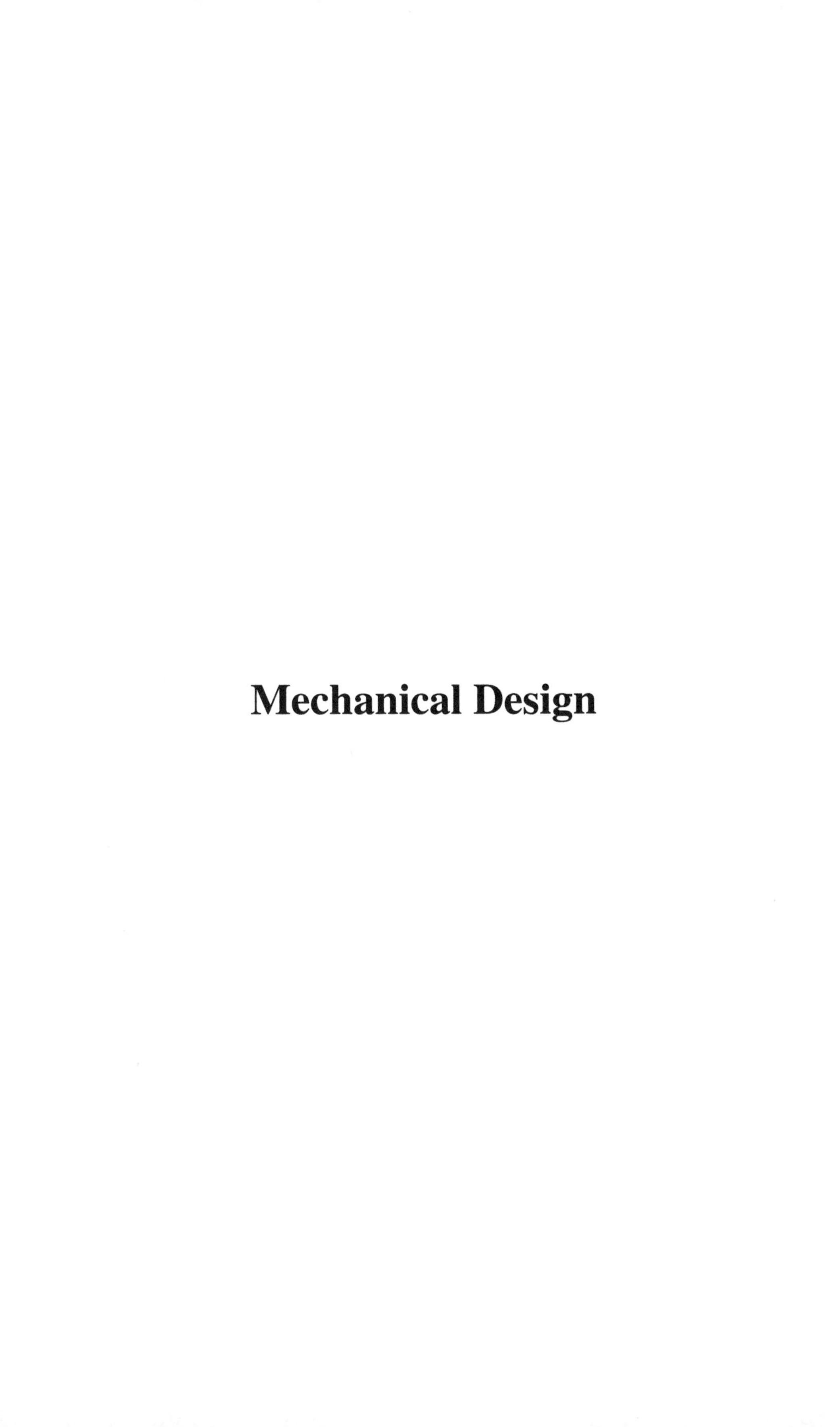

Mechanical Design

Mathematical Modeling of a Wave Energy Converter Equipped with a Negative Spring Mechanism

Braulio Neira-Verdugo[1](✉), Danilo Pastrana-Mendoza[1], Claudio Villegas-Ulloa[1], Fabián Pierart-Vásquez[1], and Thomas Knobloch[2]

[1] Department of Mechanical Engineering, University of Bío-Bío, Concepción, Chile
braulio.neira2001@alumnos.ubiobio.cl,
{dpastrana,cvillegas,fpierart}@ubiobio.cl
[2] Institute for Gear Technology, Machine Dynamics, and Robotics, RWTH Aachen University, Aachen, Germany
knobloch@igmr.rwth-aachen.de

Abstract. A renewable source of energy with great potential is wave energy, and Chile offers favorable conditions for its exploitation. To convert wave energy into a usable form of energy, wave energy converters are used. In this context, the University of Bío-Bío has designed and constructed a wave energy converter called *Lafkenewen*. Numerous wave energy converters have been developed over time; however, their efficiency remains low due to operation away from resonance. To address this issue, a spring mechanism with regions of negative stiffness has been proposed, which can enhance power output and broaden frequency bandwidth. This work presents a mathematical model of a scaled version of the Lafkenewen converter, developed using ANSYS software, under regular wave conditions. The results indicate that the integration of the negative stiffness mechanism leads to a 280% increase in the generated power.

Keywords: Wave energy · Nonlinear stiffness · Negative spring · Energy conversion

1 Introduction

The consumption of electrical energy is increasing and will continue to increase [10]. However, a significant portion of this electricity is still generated using fossil fuels. A well-known solution to these issues is the use of renewable energy sources that are accessible to everyone. In addition to addressing environmental challenges, renewable energy also helps to ensure national energy security [13]. Over the last few decades, wave energy has garnered considerable attention due to its unique characteristics. For instance, it is non-intermittent, meaning that energy is available both day and night. Furthermore, since the working fluid is water, the energy density is high [7]. In this regard, it is estimated that wave

O. F. Farías Fuentes et al. (Eds.): CIBIM 2024, *Proceedings of the XVI Ibero-American Congress of Mechanical Engineering*, pp. 371–387, 2026.
https://doi.org/10.1007/978-3-032-22823-9_27

energy has a global potential of 2.11 ± 0.05 TW [4]. Chile, in particular, enjoys some of the most favorable conditions globally due to its high energy density in the central and southern regions of the country [7].

Wave energy converters, or WECs, can be classified according to their mode of operation. For example, a WEC that resembles a floating buoy, Fig. 1, is classified as an oscillating body or point absorber. This is one of the most common and prominent types [17]. Falnes [2] pointed out that, for a point-absorber WEC with a single degree of freedom, the maximum power can be obtained in the resonance state, where the frequency of the incoming regular waves matches the natural frequency of the converter. However, point-absorber WECs are characterized by high hydrostatic stiffness, resulting in a high natural frequency of the system [12]. Therefore, various active control strategies have been proposed to bring the system closer to its optimal state. These strategies include latching control, damping control, or reinforcement learning [1]. Another solution proposed by various authors is to use a mechanism with zones of negative stiffness to counteract the high hydrostatic stiffness, bringing the system closer to resonance. This is known as passive control [18].

The negative or nonlinear stiffness mechanism (NSM) is a structure capable of providing nonlinear restoring forces through specially configured components [14]. Initially, nonlinear stiffness mechanisms were used in vibration isolation and energy harvesting applications [8]. However, they have recently gained attention as a feasible, efficient, and simple solution for wave energy conversion [18]. Typically, NSMs are known as mechanical springs in compression (Fig. 1). In this field, a variety of mechanisms and components are currently used, such as electromagnets or pneumatic pistons [5,9,11,16]. Generally, the most studied system is the bistable one [18], as shown in Fig. 1.

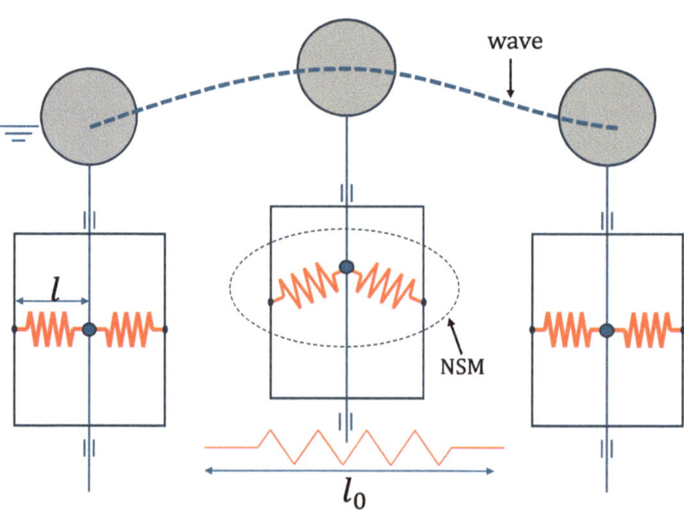

Fig. 1. Bistable negative spring mechanism.

Previously, Zhang et al. [17] proposed using compression springs to achieve the "snap-through" phenomenon with a bistable system. The purpose of this mechanism is to add a spring with negative stiffness to the equation of motion to counteract the high hydrostatic stiffness in the WEC. This reduces the natural frequency, bringing it closer to the wave excitation frequency, aiming for resonance. Zhang's numerical calculations and simulations showed a significant increase in the device's performance, as well as an increase in its capture bandwidth. Generally, there is a wide range of such studies, but they mainly focus on devices with vertical motion [18]. However, Tetû et al. [12] proposed using a similar mechanism presented by Zhang but for a pivoting WEC. They also performed experimental tests, concluding that the dominant physical effects were well modeled by their numerical simulation. Although the mechanism is quite similar to that of this study, the focus of this research is different. Tetû and his team solved the system's equation of motion using a custom MATLAB code. In contrast, this study presents a different methodology by conducting a simulation of a locally designed device using various ANSYS software modules.

The objective of this work is the numerical modeling of a scaled prototype of the Lafkenewen wave energy converter [6]. This 1:13 scale model is located in the laboratory of the Department of Mechanical Engineering at the University of Bío-Bío. It is worth mentioning that the study considers only regular wave conditions. The structure of this article is as follows: It begins by describing the scaled model and the corresponding wave tank in Sect. 2.1. The methodology section details the numerical simulation strategy and the equations developed in the software. Finally, Sect. 3 details an experimental validation of the model, the responses of simulations with and without the spring arrangement, and a comparison of the effects of using the nonlinear stiffness mechanism.

2 Methodology

First, the geometric model of the WEC is developed. This geometry is integrated into the ANSYS AQWA software to calculate the hydrodynamic coefficients and hydrostatic stiffness in the WEC Dynamics section. These parameters are vital for capturing the motion of the body in a mathematical model. Such parameters are exported to the Rigid Dynamics module, where the WEC dynamics are resolved. In the same module, the nonlinear stiffness spring is integrated, and its behaviour is evaluated. In addition, a damper is incorporated as a power extraction system. The motion and extracted power of the system with the nonlinear stiffness mechanism are compared to those of a WEC without NSM in the results section. This methodology is summarized in Fig. 2.

2.1 Geometric Model

The WEC under study is a 1:13 scaled model of the Lafkenewen wave energy converter. This device is located in a wave flume at the University of Bío-Bío,

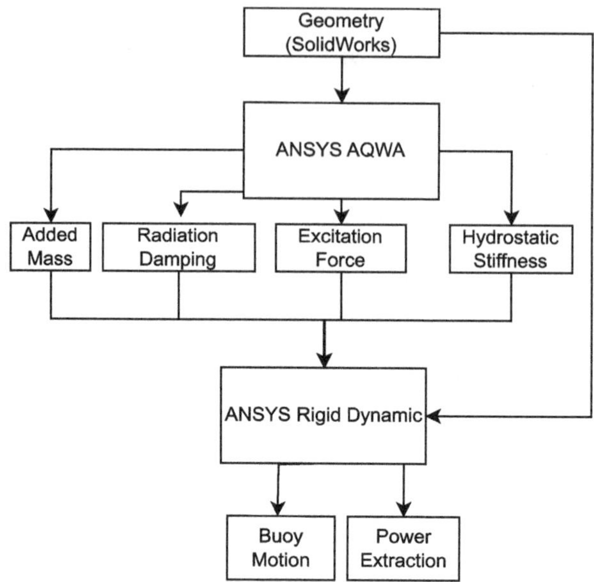

Fig. 2. Methodology for modelling the dynamic response of the WEC.

as shown in Fig. 3. The placement of the WEC in the wave flume and its dimensions are shown in Fig. 4 and Table 1. The general dimensions of the converter are presented in Table 2. Assuming that the WEC undergoes small motion amplitudes, the system in Fig. 3 can be represented as the typical mass-spring-damper system, as shown in Fig. 5. Here, k represents the hydrostatic stiffness, and c represents the radiative damping of the equivalent system, which models the interaction of the floating buoy with the water.

Table 1. Dimensions of the wave flume shown in Fig. 4.

Dimension	Measurement	Unit
l	5.4	m
W_t	0.45	m
h_m	0.65	m
h	0.20	m
l_d	1.40	m

Nonlinear Stiffness Mechanism. The configuration for adding the compression spring is shown in Fig. 6, and the dimensions are presented in Table 2. The main idea is that when the buoy is in its equilibrium position, the compression

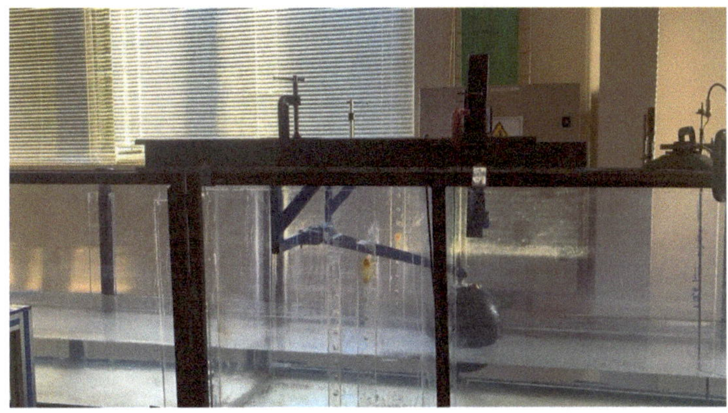

Fig. 3. Photograph of the WEC in the wave flume.

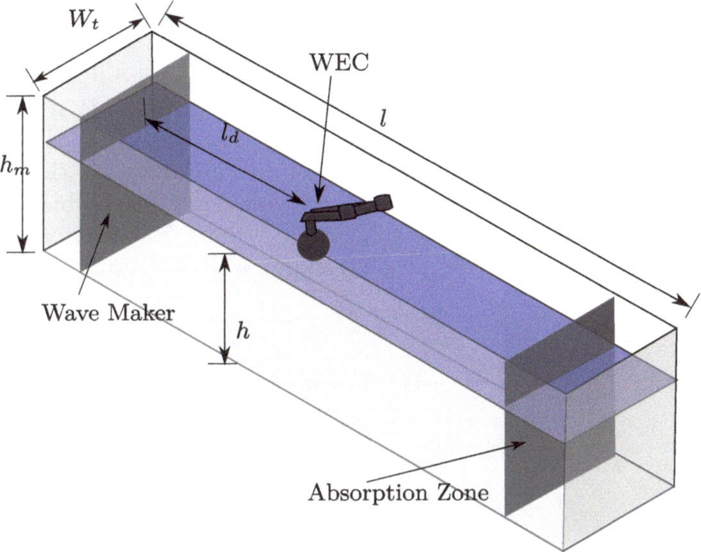

Fig. 4. Schematic and dimensions of the wave flume.

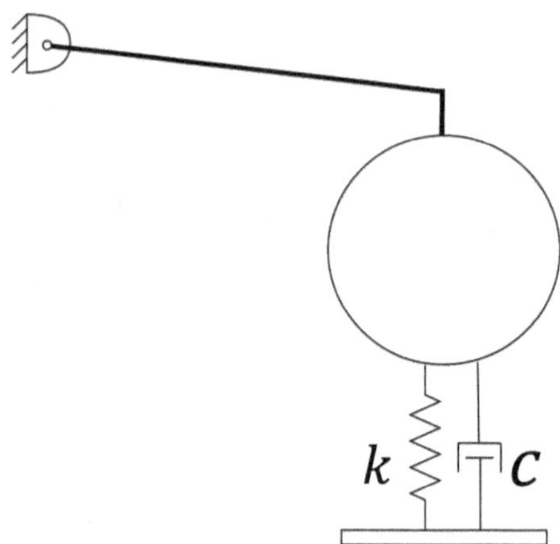

Fig. 5. Mass-spring-damper system, equivalent to the interaction of the floating buoy with water.

spring applies a force in the same direction as the arm. This ensures that at the equilibrium position, the spring does not generate torque. However, when an excitation force displaces the buoy from its equilibrium position, the spring generates torque and the "snap-through" phenomenon. The stiffness of the negative spring was determined using a heuristic approach, based on the stiffness used in [12] and evaluating which value generates the largest motion amplitude (see Table 3).

Table 2. Dimensions of the converter shown in Fig. 6.

Dimensions	Measurement	Unit
Buoy Mass	2.476	kg
Buoy Diameter	0.200	m
Arm Mass	0.804	kg
Arm Length	0.400	m
AD	0.272	m
AE	0.340	m
EC	0.060	m

2.2 WEC Dynamics

For regular waves of a given frequency, the equation governing the oscillation of the WEC is presented in Eq. (1), as described in [12]:

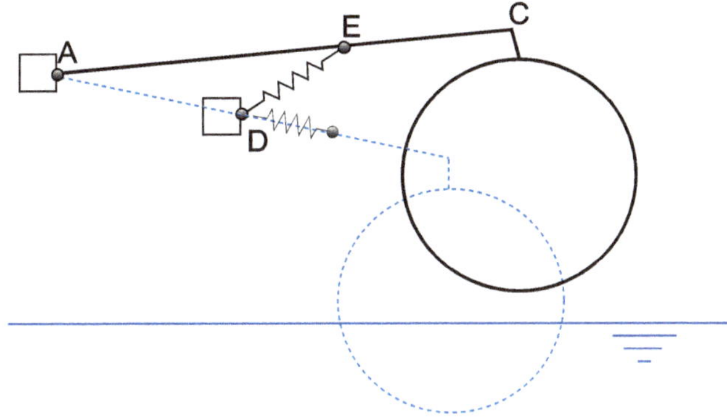

Fig. 6. Negative spring position.

$$(I + I_{\text{add}}(\omega))\ddot{\theta} + c(\omega)\dot{\theta} + k\theta = T_{\text{ext}}(\omega), \tag{1}$$

where:

Table 3. Angular Position and Related Parameters

θ	Angular position of the WEC.
ω	Wave excitation frequency
I	Moment of inertia of the WEC
I_{add}	Added moment of inertia, which represents an inertial increase due to the displaced water near the buoy during motion [3]
c	Radiative damping, representing a dissipative effect related to the energy transmitted to the water by the body's oscillations [3]
k	Hydrostatic stiffness
T_{ext}	Excitation moment due to incoming waves, normalized by the wave amplitude

As shown in the equation, there are three coefficients that depend on the excitation frequency or wave frequency ω: the added inertia I_{add}, the radiative damping c, and the excitation moment T_{ext}. To determine these hydrodynamic coefficients, ANSYS AQWA software was used, as described in Fig. 2. It is worth mentioning that a small wave amplitude of 25 mm was used in this study. This value ensures small motion amplitudes, allowing the system to behave linearly. Under this consideration, the hydrodynamic coefficients are calculated considering only the vertical motion of the system, as shown in Fig. 5.

Mathematical Modeling. The fluid-structure interaction is modeled based on potential flow theory, which considers the working fluid to be ideal, i.e., incompressible, non-viscous, and irrotacional. Under these conditions, the water velocity $\boldsymbol{u}$ can be expressed in terms of a scalar function called velocity potential Φ, which must satisfy the Laplace equation, (2) [3]:

$$\nabla^2 \Phi = 0 \tag{2}$$

In the context of wave energy, the flow must satisfy the Laplace Eq. (2) within the fluid domain, along with boundary conditions for a regular wave with a wavelength much greater than its amplitude. On the mean wetted surface of the body, the boundary condition is defined by Eq. (3). At the bottom surface of the wave flume, the condition is described by Eq. (4). Finally, at the free surface, the boundary condition is defined by Eq. (5).

$$\frac{\delta \Phi}{\delta n} = u_n \tag{3}$$

$$\frac{\delta \Phi}{\delta y} = 0 \tag{4}$$

$$-\omega^2 \Phi + g\frac{\delta \Phi}{\delta y} = 0. \tag{5}$$

where n is the normal direction vector, u_n is the velocity of the wetted body in the normal direction to its surface, y is the vertical position within the fluid domain, ω is the wave frequency, and g is the gravitational acceleration.

Except for simple geometries, it is not possible to analytically solve this boundary value problem. Therefore, numerical approaches are the most common method. The method used in this work is the Boundary Element Method (BEM), also known as the panel method [3]. All the equations and considerations mentioned above are incorporated into the AQWA module. It is important to note that, although BEM codes have several limitations, the first step in modeling WECs is almost universally a frequency-domain analysis using BEM codes [3].

ANSYS AQWA. As mentioned earlier, this ANSYS module is used to obtain the hydrodynamic and hydrostatic coefficients. First, the geometry of the WEC is integrated along with its mass and inertia values. The system is configured to pivot through a hinged connection. Subsequently, the buoys surface is discretized into boundary elements. The software automatically proposed an element size of approximately 8 mm, which was kept as it provides a sufficiently fine resolution for the present scaled geometry. It should be noted that, in Boundary Element Method (BEM) approaches, the exact element size is not as critical as in volumetric CFD meshes. What matters is achieving an adequate surface discretizations that captures the geometry without excessive computational cost, following the common practice reported in previous works [3,12]. Then, the "constant environment", representing the wave flume used by the software, is characterized. In this case, the water depth, length, and width of the flume, as well as the WEC location, are defined in the same way as presented earlier in Table 1. The simulation in this module is shown in Fig. 7.

Fig. 7. Simulation in AQWA.

ANSYS Rigid Dynamics. This ANSYS module is used because it can integrate the components required for the analysis. In this case, the mass-spring system shown in Fig. 5 is integrated into this module. First, the geometry is fixed so that it can only pivot, using a fixed rotational joint, similar to what is done in AQWA. Then, a spring and damper are added on the buoy in a vertical position. This spring-damper appears as a cylinder, as shown in Fig. 8, which is equivalent to what is shown in Fig. 5. The hydrostatic spring stiffness is imported from AQWA, and the same applies to the damping constant. Then, the added mass is assigned to the buoy as a point mass, Fig. 8. Additionally, a force acting on the buoy's surface is integrated. This force has a magnitude equal to the product of the wave excitation force and the wave amplitude. This force acts sinusoidally, with the previously defined excitation frequency.

Next, a spring is inserted parallel to the arm, Fig. 9. The position of this spring is detailed in Fig. 6 and Table 2. Subsequently, the damper, which acts as a power extraction system, is integrated, Fig. 9. To calculate the damping coefficient, it was considered that this damper must provide the same energy dissipation effect as the damping in the mass-spring system. This is due to the optimal damping must equal the radiation damping. In other words, and although it may seem paradoxical, absorbing a wave implies creating a wave [3].

Please note that an adjustment to the vertical position of the spring-damper, which simulates the effect of the hydrostatic spring, is needed. This occurs because, with the spring-damper in a vertical position, the restoring force does not have the desired direction when the buoy descends. If the buoy descends from its equilibrium position and the spring compresses, the movement will cause the

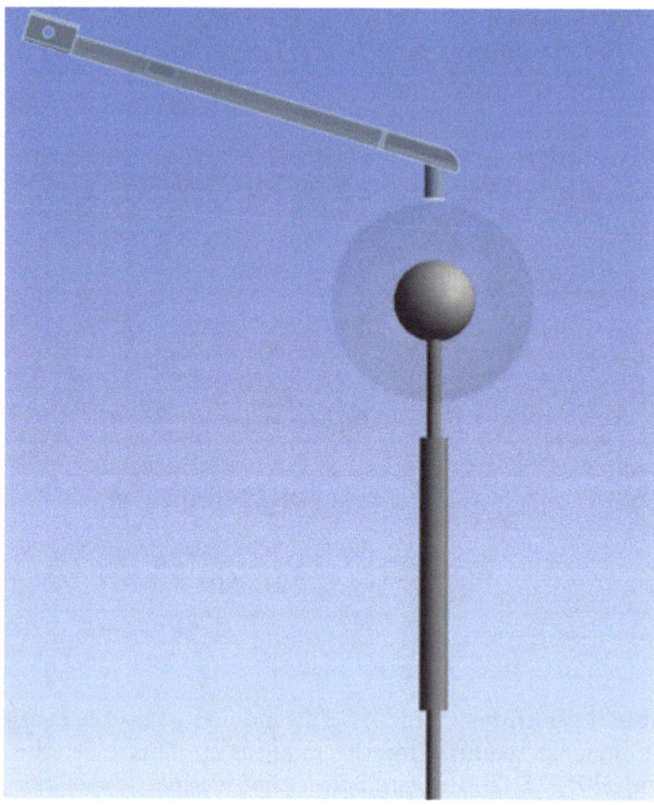

Fig. 8. Mass-spring-damper system in the ANSYS Rigid Dynamics module.

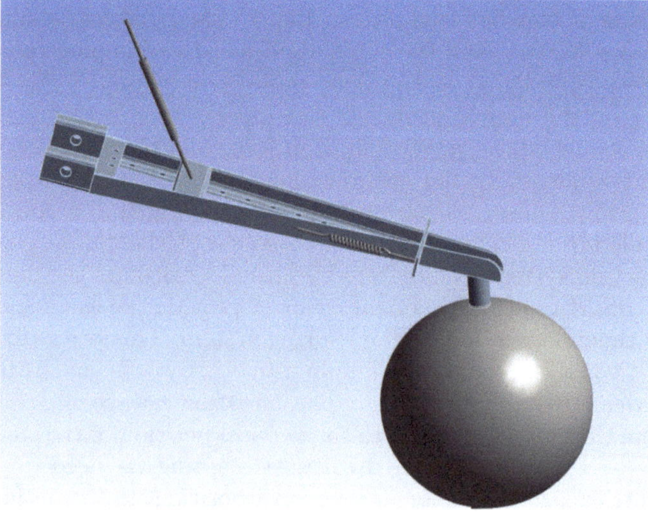

Fig. 9. System with Spring and Damper.

spring to tilt to the left. This tilt generates a restoring force in the radial direction, making the tangential component almost null and preventing the necessary torque for the buoy to ascend, resulting in an incorrect response.

To solve this problem, the spring-damper is tilted to the right, aligning it similarly to the tangent of the arm's trajectory with the buoy. However, to avoid significantly distorting the vertical stiffness value, the tilt angle must be small. To determine the optimal angle, an analysis is performed using different angles less than 15°, evaluating the buoy motion amplitude. This analysis determined that a tilt of 6ř generates optimal results. With this level of tilt, the variation in vertical stiffness is negligible, achieving results consistent with those simulated in AQWA, as it is shown in the results section. Nevertheless, this solution leaves room for future analyses on how to improve this aspect.

3 Results

3.1 BEM Results

The hydrostatic and hydrodynamic coefficients have been extracted from the ANSYS AQWA software. These values are shown in Table 4, considering that they are calculated based on an excitation frequency of 1 Hz, which is the frequency used for our subsequent analysis. On the other hand, Fig. 10 shows the buoy's motion relative to the wave height. It can be seen that the buoy is not in resonance, as the buoy's displacement amplitude is almost equal to the wave amplitude.

Table 4. Hydrodynamic coefficients.

Dimension	Value	Unit
Added Mass	1.39	kg
Radiative Damping	4.59	Ns/m
Excitation Force	188.00	N/m
Hydrostatic Stiffness	314.69	N/m

Experimental Validation. The model developed in ANSYS AQWA is validated using the wave tank. To do this, the amplitude of the experimentally generated waves is compared to the modeling in AQWA, which is 25 mm. Additionally, the motion of the WEC measured experimentally is compared to the computational modeling. A capacitive wave sensor was used to measure the wave amplitude, and a laser sensor was used to measure the WECs motion.

Although the excitation frequency used in this study's analysis is 1 Hz, due to technical limitations of the wave tank, the validation was carried out

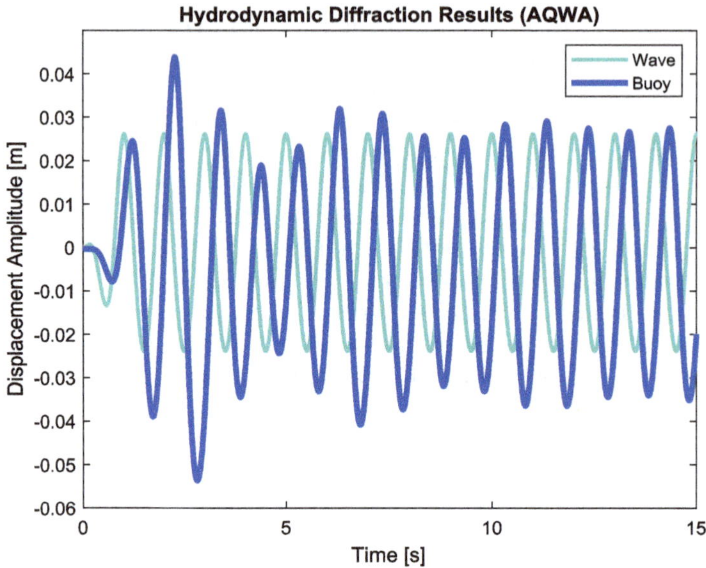

Fig. 10. Wave amplitude and buoy response of the BEM model.

at 1.3 Hz. The results of the wave amplitude and buoy response are presented in Figs. 11 and 12.

It can be observed that the model correctly predicts the WECs behavior for the studied conditions. In this regard, the root mean square error (RMSE) for the graph that experimentally validates the wave amplitude, Fig. 11, is 0.0076 m. This value is due to wave reflection within the wave tank. On the other hand, the root mean square error for the graph that experimentally validates the buoy's response, Fig. 12, is 0.0022 m. Therefore, now that the BEM model has been experimentally validated, the analysis of interest, which is the comparison of motion amplitude when applying the nonlinear stiffness mechanism, proceeds.

3.2 Rigid Dynamics Results

To validate the use of the Rigid Dynamics tool for studying spring mechanisms, the buoy's motion generated by AQWA is initially compared to that generated by Rigid Dynamics. Figure 13 shows the comparison of the buoy's motion at a frequency of 1 Hz, as initially mentioned. Here, the curves almost overlap, except at the beginning, since AQWA assumes the water is still before wave excitation.

The root mean square error of this comparison is 0.010051 m. This indicates that the physical components have been well incorporated into Rigid Dynamics. Once the physical model has been simulated in Rigid Dynamics, the nonlinear stiffness mechanism or compression spring is integrated. Using the method described in the methodology, the system stiffness has been set to 1600N/m, and the spring's initial force is set to 50 N. With these values, the mechanism

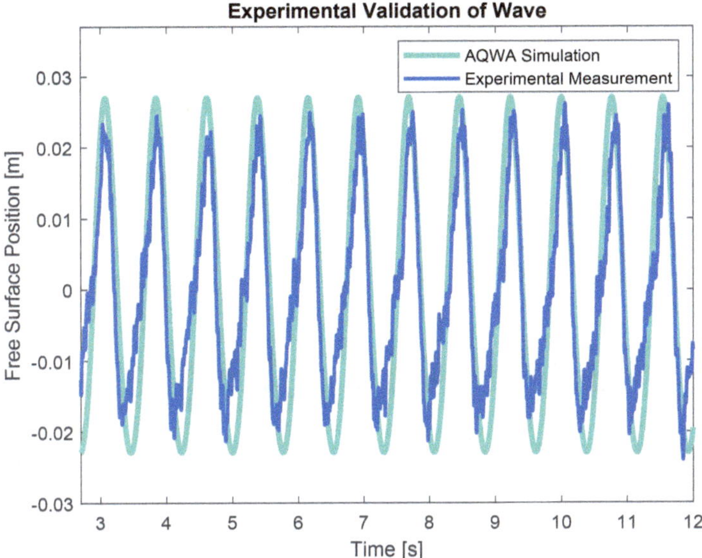

Fig. 11. Results of the BEM model and experimental data.

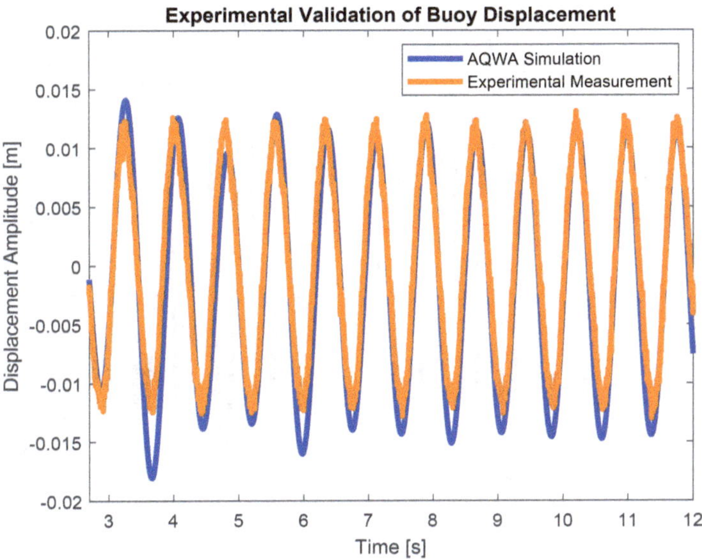

Fig. 12. Comparison of buoy displacement.

reduces the WECs natural frequency. Figure 14 shows the results of applying the spring to the system. As shown, the system clearly increases the motion amplitude. This is because the compression spring reduces the high hydrostatic stiffness, making the system more flexible and bringing it closer to resonance.

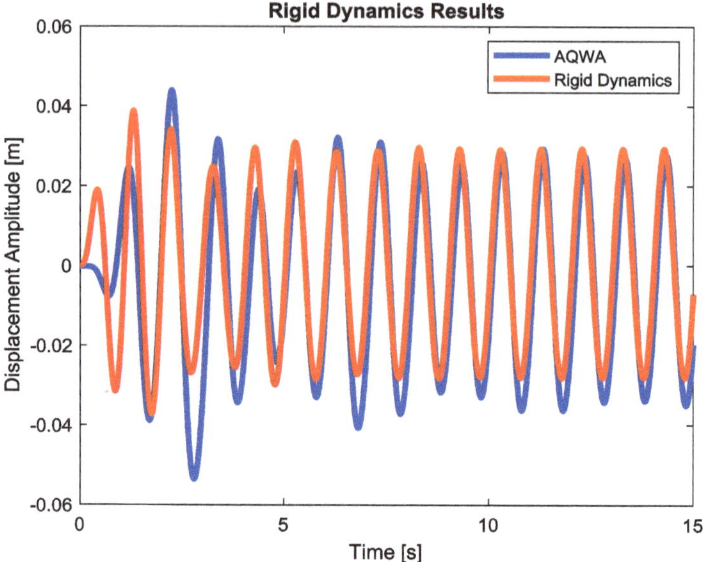

Fig. 13. AQWA-Rigid Dynamics comparison.

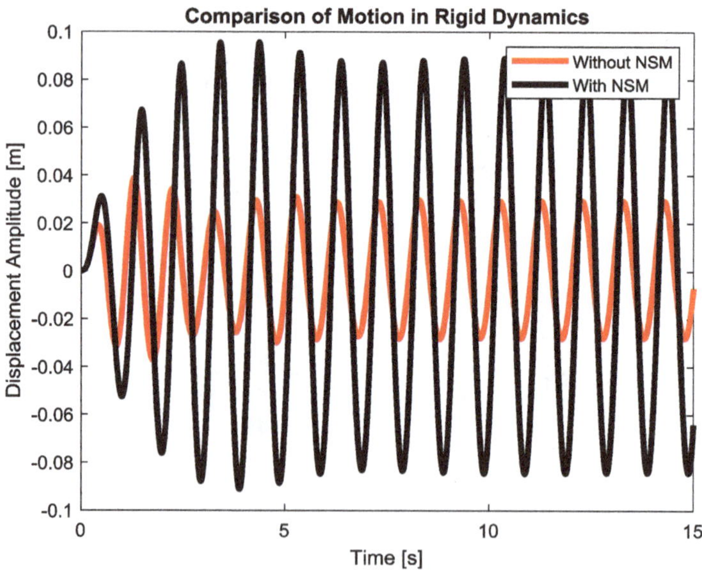

Fig. 14. Buoy displacement with nonlinear stiffness mechanism integration.

Using the motion amplitude data, the power generated by the damper, which represents the power extraction system, is calculated. When the system does not include the nonlinear stiffness mechanism (negative spring), the average power delivered is 41 mW. On the other hand, when the negative spring is integrated, the average power delivered is 158 mW. Thus, under ideal conditions, the power increases by 280%.

4 Conclusions

This article presents a simple methodology for modeling a nonlinear stiffness mechanism applied to the scaled version of the Lafkenewen wave energy converter. Using a spring with a nonlinear geometric configuration, negative stiffness has been generated, bringing the system closer to resonance. This resulted in a significant increase in motion amplitudes and a 280% increase in the average generated power.

The mathematical development was conducted using the ANSYS software, employing the AQWA and Rigid Dynamics modules. Unlike other studies that often use Computational Fluid Dynamics (CFD), this work opted for the Boundary Element Method (BEM) due to its lower computational costs and speed. BEM provides results in a matter of seconds. Additionally, BEM is universally used for modeling energy converters in the frequency domain [3]. Furthermore, the experimental validation demonstrated that BEM could adequately represent the behavior of the scaled converter for small oscillations with regular wave conditions. However, future work should incorporate CFD to obtain more accurate results when dealing with large oscillations and irregular wave conditions.

It is important to note that these models and simulations did not consider the effects of friction between mechanical components. Additionally, as described by various authors, there are practical challenges when using mechanical compression springs for this application. As reported by Wu et al. [15], achieving the required nonlinear stiffness demands springs with very long free lengths and large compressible lengths, which is very difficult to implement at prototype scale. Additionally, issues such as fatigue under cyclic loading, buckling, and corrosion in marine environments further limit their applicability. These constraints significantly reduce the potential performance gains when scaled up, which explains why alternative solutions, such as pneumatic cylinders, have been proposed in recent literature [18].

Despite these limitations, this study represents a valuable starting point for investigating different nonlinear stiffness mechanisms and exploring their advantages and applications in wave energy converters. These investigations will contribute to making these devices more robust and competitive, harnessing the tremendous energy potential available in Chile.

Acknowledgements. The authors would like to thank the support provided by University of Bío-Bío and RWTH Aachen University for making this research possible.

References

1. Anderlini, E.: Control of wave energy converters using machine learning strategies. Ph.D. thesis (2017)
2. Falnes, J.: Ocean Waves and Oscillating Systems. Cambridge University Press, Cambridge (2002)
3. Folley, M.: Numerical Modelling of Wave Energy Converters: State-of-the-art Techniques for Single Devices and Arrays. Academic Press (2016)
4. Gunn, K., Stock-Williams, C.: Quantifying the global wave power resource. Renew. Energy **44**, 296–304 (2012). https://doi.org/10.1016/j.renene.2012.01.101
5. Khasawneh, M.A.: Exploiting non-linearity to improve the performance of point wave energy absorbers. Ph.D. thesis, ProQuest Dissertations Publishing (2023)
6. Knobloch, T., Titulaer, H., Wolkenhauer, J., Hüsing, M., Corves, B., Villegas, C.: Leistungsoptimierung von wellenkraftwerken durch variation der eigenfrequenz performance optimization of wave energy converters by eigenfrequency variation. https://doi.org/10.17185/duepublico/82905
7. Pecher, A., Kofoed, J.P.: Handbook of Ocean Wave Energy. Springer (2016)
8. Pellegrini, S.P., Tolou, N., Schenk, M., Herder, J.L.: Bistable vibration energy harvesters: a review. J. Intell. Mater. Syst. Struct. **24**, 1303–1312 (2013). https://doi.org/10.1177/1045389X13489230
9. Qin, J., Zhang, Z., Song, X., Huang, S., Liu, Y., Xue, G.: Design and performance evaluation of an enclosed inertial wave energy converter with a nonlinear stiffness mechanism. J. Mar. Sci. Eng. **12** (2024). https://doi.org/10.3390/jmse12010191
10. Ram, M., Aghahosseini, A., Breyer, C.: Job creation during the global energy transition towards 100 forecasting and social change **151** (2020). https://doi.org/10.1016/j.techfore.2019.06.008
11. Todalshaug, J., et al.: Tank testing of an inherently phase controlled wave energy converter. In: Proceedings of the 11th European Wave and Tidal Energy Conference (EWTEC) (2015). https://doi.org/10.13140/RG.2.1.2211.8481
12. Têtu, A., Ferri, F., Kramer, M.B., Todalshaug, J.H.: Physical and mathematical modeling of a wave energy converter equipped with a negative spring mechanism for phase control. Energies **11** (2018). https://doi.org/10.3390/en11092362
13. Wang, C.N., Nhieu, N.L., Tran, H.V.: Wave energy site location optimizing in Chile: a fuzzy serial linear programming decision-making approach. Environ., Dev. Sustain. (2024). https://doi.org/10.1007/s10668-023-04320-8
14. Wu, Z.: Nonlinear stiffness concept in wave energy conversion. Ph.D. thesis, Universidade Federal do Rio de Janeiro (2019)
15. Wu, Z., Levi, C., Estefen, S.F.: Practical considerations on nonlinear stiffness system for wave energy converter. Appl. Ocean Res. **92** (2019). https://doi.org/10.1016/j.apor.2019.101935
16. Xi, R., Zhang, H., DaolinXu, Zhao, H., Mondal, R.: High-performance and robust bistable point absorber wave energy converter. Ocean Eng. **229** (2021). https://doi.org/10.1016/j.oceaneng.2021.108767
17. Xiao, L., Zhang, X., Yang, J.: Numerical study of an oscillating wave energy converter with nonlinear snap-through power-take- off systems in regular waves (2014). https://doi.org/10.13140/2.1.4414.4000
18. Zhang, X., Zhang, H., Zhou, X., Sun, Z.: Recent advances in wave energy converters based on nonlinear stiffness mechanisms. Appl. Math. Mech. (English Edition) **43**, 1081–1108 (2022). https://doi.org/10.1007/s10483-022-2864-6

Multiobjective Optimization of Centrifugal Fans Using CFD and NSGA-II Integrated with Factor Analysis

Matheus Costa Pereira[1,2,3](✉), Tiago Martins de Azevedo[1], Matheus Brendon Francisco[1], Anderson Paulo de Paiva[1], and Ronã Rinston Amaury Mendes[1]

[1] Federal University of Itajubá, Av. BPS, 1303, Pinheirinho, Itajubá, MG, Brazil
matheusc_pereira@hotmail.com, {matheus_brendon, andersonppaiva}@unifei.edu.br, rona.rinston@ifsuldeminas.edu.br
[2] University of Melbourne, 700 Swanston St, Carlton, Melbourne, VIC, Australia
[3] Universitat Politècnica de Catalunya, C/ Llorens i Artigas 4-6, Barcelona, Spain

Abstract. Centrifugal fans play a crucial role in various industries. This study investigates the multiobjective optimization of centrifugal fans using Computational Fluid Dynamics and the Non-dominated Sorting Genetic Algorithm II. Techniques such as Principal Component Analysis and Factor Analysis, combined to form Principal Components Factor Analysis, are applied to handle high-dimensional data, reducing complexity and facilitating multiobjective optimization. The genetic algorithm is employed to generate optimal non-dominated solutions. Once the solutions are obtained, they are evaluated using the Technique for Order of Preference by Similarity to Ideal Solution and Mahalanobis Distance. The fan is made of AISI 430 stainless steel and operates under high-temperature conditions, with modifications applied to variables related to the fan blades, resulting in responses such as mass, mass flow rate, torque, pressure, velocity, and turbulence. The multiobjective optimization results were replicated and demonstrated high reliability, with an error of less than 10%, contributing to the manufacturing of fans with improved performance and resource efficiency, without the need for extensive prototyping.

Keywords: Computational Fluid Dynamics · Multiobjective Optimization · Principal Components Factor Analysis · NSGA-II · Design of Experiments · Centrifugal Fan

1 Introduction

Centrifugal fans are devices used to move air or gases in ventilation, heating, and cooling systems. Their operation is based on centrifugal force: air is drawn axially into the fan housing and expelled radially outward through rotating blades. Centrifugal fans play a crucial role in numerous industries, such as engine cooling in hybrid vehicles [1], cooled

O. F. Farías Fuentes et al. (Eds.): CIBIM 2024, *Proceedings of the XVI Ibero-American Congress of Mechanical Engineering*, pp. 388–404, 2026.
https://doi.org/10.1007/978-3-032-22823-9_28

heat sinks [2], building ventilation [3], agricultural harvesters [4], wind power generators [5], HVAC systems [6], industrial film recovery machines [7], suction machines [8], solar heaters [9], and high-speed train systems [7], among others.

Despite their wide range of applications, many fan designs still require improvements. Multiobjective optimization (MOP) is directly related to finding the best design or operational parameters for centrifugal fans by evaluating multiple criteria or objectives simultaneously. MOP applied to centrifugal fan blades for ventilation system improvements is discussed by Zhou et al. [3, 10].

The use of Computational Fluid Dynamics (CFD), which employs numerical methods and algorithms to solve and analyze fluid flow problems, plays a crucial role in simulating centrifugal fans, especially when integrated with MOP [6].

When dealing with numerous response variables, data dimensionality can hinder the achievement of reliable results. Therefore, dimensionality reduction techniques such as Principal Component Analysis (PCA), commonly used in other industrial processes involving multi-response optimization [11–16], and Factor Analysis (FA) are applied. The combination of these techniques is known as Principal Components Factor Analysis (PCFA) [17–19]. Additionally, these methods help transform the variables into independent, uncorrelated components.

With reduced dimensionality, the MOP process becomes less computationally intensive, as the number of constraints decreases. For the optimization, the genetic algorithm (GA) known as Non-dominated Sorting Genetic Algorithm II (NSGA-II), developed by Deb et al. [20], is employed. A comprehensive review of optimization methods and algorithms for multi-objective problems in mechanical engineering is presented by Junho et al. [21].

To assess the quality of the solutions found, the Technique for Order of Preference by Similarity to Ideal Solution (TOPSIS), proposed by Hwang and Yoon [22], is used to aid decision-making based on the similarity of alternatives to an ideal solution. Additionally, Mahalanobis Distance (MD, proposed by Johnson and Wichern [23], serves as a multivariate metric that considers the variance-covariance matrix.

The experimental procedure is initially designed using a Design of Experiments (DOE) approach, followed by computational simulations. The results obtained with Ansys Fluent are then stored and subjected to dimensionality reduction. Subsequently, individual optimization of each variable and factor is performed, followed by MOP using NSGA-II. The results are then evaluated using the TOPSIS and MD metrics.

To apply this approach, a case study is conducted on a centrifugal fan made of AISI 430 stainless steel, operating under high-speed and high-temperature conditions. The objective is to optimize factors such as fan mass (M), outlet mass flow rate (OMF), required torque (T), average dynamic pressure magnitude (DP), average velocity magnitude (V), and average turbulence Ω (TΩ), which are subsequently divided into specific factors. The independent variables considered are the number of blades (BN), blade inlet angle (BA), blade outlet angle (BO), and blade length (BL). The DOE employs a Central Composite Design (CCD), resulting in 24 experiments.

The combination of these methods aims to simultaneously optimize multiple fan responses, improving performance with lower resource consumption and eliminating the need for extensive prototyping, all based on robust methodological approaches.

2 Methodology

Initially, the input and output data for the experiment are defined based on a DOE providing a solid foundation for the execution of the procedures. The steps of Computer-Aided Design (CAD), CFD, and result analysis are carried out using Ansys software, including SpaceClaim, Workbench, Ansys Fluent, and CFD-Post.

After obtaining the results, PCA is applied to determine the number of variables to be extracted. Simultaneously, FA with the Promax rotation method is employed to identify factors that exhibit similar behavior, enabling targeted optimizations within a single factor.

The eigenvalues (λ) and eigenvectors (v) of the covariance matrix (**C**) are calculated according to Eq. (1), while Eq. (2) describes the calculation of the explained variance (Var_{exp}) for the components.

$$\mathbf{C}v = \lambda v \tag{1}$$

$$Var_{exp} = \frac{\lambda_i}{\sum \lambda_i} \tag{2}$$

where: Each eigenvalue λ_i represents the variance explained between the i-th and j-th variables.

The factor loadings ($\mathbf{L_P^*}$) are obtained through Varimax rotation, and by raising the loadings to the power of κ, they are calculated using Eq. (3). The factor scores ($\mathbf{F_P}$) are determined using the structure matrix (**S**), as described in Eq. (4).

$$\mathbf{L_P^*} = \mathbf{L}|\mathbf{L}|^{\kappa-1} \tag{3}$$

$$\mathbf{F_P} = \mathbf{ZS} \tag{4}$$

where: **L** represents the factor loading matrix after orthogonal rotation (Varimax), **Z** is the standardized data matrix, and **S** is the correlation matrix between the factors and the original variables after applying the Promax rotation.

The first optimization phase focuses on individual optimization, evaluating the ideal conditions for each variable and factor. Then, a MOP is performed using the NSGA-II algorithm to find non-dominated solutions, that is, Pareto-efficient solutions.

The MOP results are subsequently evaluated using the MD metric and the TOPSIS method to rank the quality of the obtained solutions. The MD is calculated according to Eq. (5):

$$\mathrm{MD}(\mathbf{x}) = \sqrt{(\mathbf{x} - \mu)^{\mathrm{T}} \mathrm{S}_{\mathrm{cov}}^{-1} (\mathbf{x} - \mu)} \tag{5}$$

where: **x** represents the variable vector, μ is the mean vector of the variables, and $\mathbf{S_{cov}}$ is the covariance matrix of the variables.

The process to compute TOPSIS begins with data normalization and the weighting of the variables. Then, the Utopia and Nadir points for the variables are determined, followed by the calculation of the MD. Finally, the TOPSIS score is calculated considering the distances to the Nadir point (D_{Nadir}) and to the Utopia point (D_{Utopia}), as shown in Eq. (6).

$$TOPSIS = \frac{D_{Nadir}}{D_{Utopia} + D_{Nadir}} \tag{6}$$

To validate the best solutions obtained, new CAD models and CFD simulations are replicated to verify whether the results are consistent with the predictions from the NSGA-II.

3 Case Study

The case study illustrates the application of the approach to a centrifugal fan operating at high speeds and temperatures in an industrial environment. The fan is made of AISI 430, a ferritic stainless steel known for its corrosion resistance and robust mechanical properties. It has a density of 7.7 g/cm^3, maximum strength of 450 MPa, modulus of elasticity of 200 GPa, shear modulus of 77 GPa, Poisson's ratio of 0.27, thermal conductivity of 24.9 W/(mK), and specific heat of 460 J/(kg°C). The blade thickness, blade width, and the outer diameter of the volute are considered fixed variables, with no changes to these parameters.

The mesh plays a crucial role in obtaining the results; therefore, orthogonal quality in all experiments was kept above 0.14, with skewness below 0.95. To ensure the accuracy and reliability of the results, the convergence of iterations was monitored, verifying the continuity equation, the velocities in the three axes (x, y, and z), and the turbulence modules (k and Ω).

A total of 24 observations were carried out, each with a new geometry and different conditions based on the DOE. The independent variables are related to the blades: BN, BA, BO, and BL. The dependent variables are the CFD responses: FM, MFR, T, DP, VM, and TΩ.

Using Ansys software, it was possible to perform everything from the geometry setup to obtaining the experimental results. In addition to Ansys, Python programming was used for MOP with NSGA-II.

3.1 Experimental Data

Figure 1 shows the model of one of the fans, selecting the central point experiment, which includes 10 blades, an inlet angle of 15.0°, an outlet angle of 45.0°, and a blade length of 33.0 mm. The structure includes the volute and plates to provide a complete view of the assembly.

The fan mass is calculated excluding other plates without modifications. The subsequent measurements refer to the facets obtained through CFD. DP represents the pressure exerted by the airflow on the fan, while VM indicates the highest velocity magnitude of the flow. TΩ refers to the average turbulence of the airflow inside the fan.

MFR is calculated by the mass flow rate of the fluid passing through the cross-section over a period of time, according to Eq. (7):

$$\text{MFR} = \int \rho \mathbf{v} \cdot d\mathbf{A} = \sum_{i=1}^{n} \rho_i \mathbf{v}_i \boldsymbol{A}_i \tag{7}$$

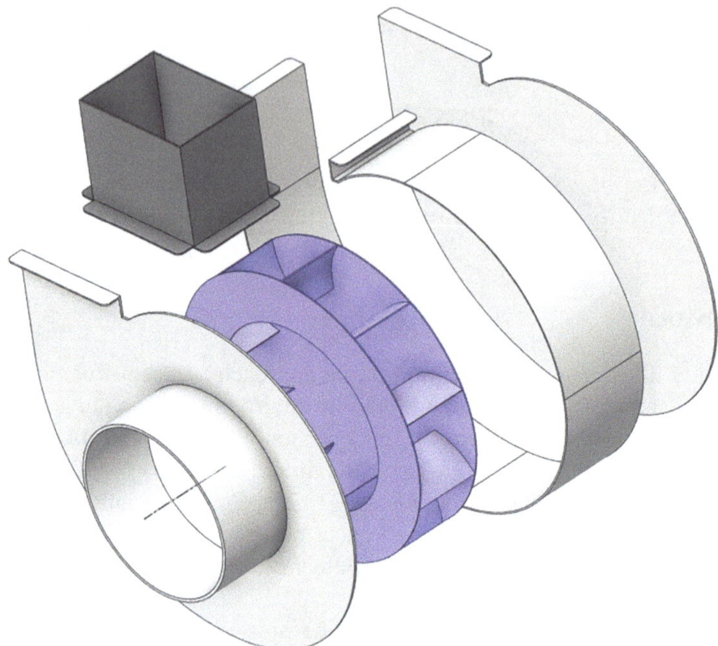

Fig. 1. Components of the centrifugal fan.

where: ρ is the fluid density, $\mathbf{v}$ is the fluid velocity, and $d\mathbf{A}$ is the area element.

T is the amount of rotational force required to rotate the fan blades and is calculated by:

$$\mathrm{T} = \mathbf{r}_{\mathbf{AB}}\mathbf{F}_{\mathbf{P}} + \mathbf{r}_{\mathbf{AB}}\mathbf{F}_{V} \tag{8}$$

where: **A** is the specific moment center, **B** is the origin of the force, $\mathbf{r}_{\mathbf{AB}}$ is the torque vector, $\mathbf{F}_{\mathbf{P}}$ is the pressure force vector, and $\mathbf{F}_{V}$ is the viscous force vector.

The combination of PCA and FA is fundamental for reducing dimensionality, facilitating the MOP process. The use of the Promax method, a variant of Varimax, provides an oblique transformation that simplifies the factor structure, allowing for a clear interpretation of FA results. The variance explained by the factors is illustrated in Fig. 2, while the FA loadings, both without rotation and with Promax rotation, are shown in Fig. 3. The first factor grouped FM, TΩ, DP, and VM, while the second grouped MFR and T.

Table 1 details the optimization direction of the variables and factors, along with their respective coefficients of determination (R2). Figure 4 shows the correlation among the responses, including both the original variables and the factors, represented by ellipses. Smaller ellipses indicate stronger correlations: ellipses close to the maximum value of + 1 (perfect positive correlation) are shown in light blue, while those close to −1 (perfect negative correlation) tend toward red. White ellipses represent no correlation (0).

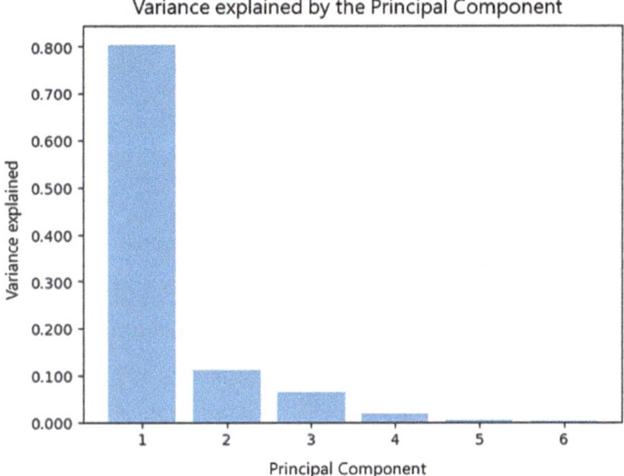

Fig. 2. Variance explained by the principal components.

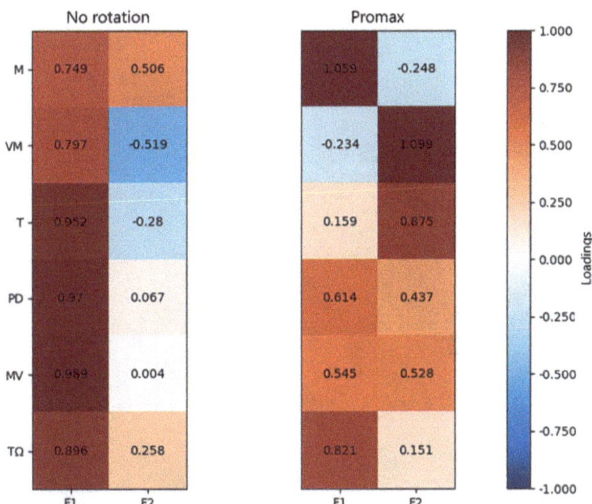

Fig. 3. Loadings without rotation and with Promax rotation.

Based on this information, Table 2 represents the original variables, divided into input variables (the first four columns) and the responses obtained from CFD (the following six columns).

The CCD analysis allows the evaluation of the effects of each input variable on each response. Table 3 presents the significant values for each response, considering only the original variables and using a significance level $\alpha = 0.05$. Interactions or quadratic variables are not shown. Additionally, the residuals of each result are analyzed, and possible outliers in the data are identified. Table 4 shows the coefficients of the original variables and the rotated factors using the Promax method.

Table 1. Optimization direction and coefficient of determination of the variables and rotated factors.

Variable	R^2	Direction
FM	0.993	Minimization
MFR	0.723	Maximization
T	0.802	Minimization
DP	0.894	Minimization
VM	0.877	Maximization
TΩ	0.972	Minimization
F_1	0.983	Minimization
F_2	0.747	Maximization

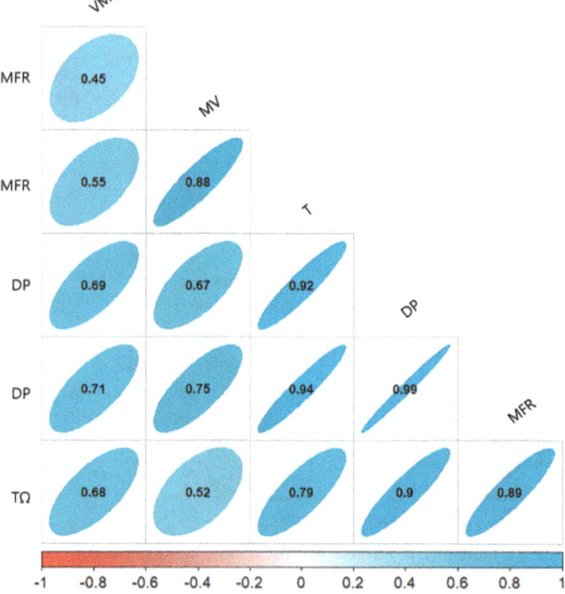

Fig. 4. Correlation between response variables and rotated factors.

Table 2. Input variables and output variables.

BN	BA	BO	BL	FM	MFR	T	DP	VM	TΩ
6	0.0	0.0	26.0	311.70	0.21	1.50	930.65	34.99	32,485.44
14	0.0	0.0	26.0	371.08	0.27	2.35	1,162.91	39.45	41,013.85

(*continued*)

Table 2. (*continued*)

BN	BA	BO	BL	FM	MFR	T	DP	VM	TΩ
6	30.0	0.0	26.0	322.36	0.22	2.16	1,150.97	37.56	37,535.55
14	30.0	0.0	26.0	395.95	0.26	3.51	1,842.95	47.03	51,205.75
6	0.0	90.0	26.0	316.63	0.22	1.68	981.04	35.66	32,880.31
14	0.0	90.0	26.0	382.58	0.31	3.31	1,463.40	44.13	46,250.06
6	30.0	90.0	26.0	328.48	0.23	2.41	1,179.40	38.32	38,107.64
14	30.0	90.0	26.0	410.21	0.34	5.18	2,279.15	53.10	52,456.55
6	0.0	0.0	40.0	370.59	0.24	1.85	943.41	35.86	32,268.89
14	0.0	0.0	40.0	461.93	0.28	2.41	1,207.25	40.70	42,095.26
6	30.0	0.0	40.0	395.49	0.25	2.87	1,502.91	42.34	41,536.15
14	30.0	0.0	40.0	520.05	0.21	2.39	1,647.11	44.39	50,456.45
6	0.0	90.0	40.0	378.17	0.26	2.25	1,118.70	38.46	33,053.57
14	0.0	90.0	40.0	479.62	0.32	3.42	1,656.60	47.32	46,983.50
6	30.0	90.0	40.0	405.84	0.26	3.16	1,504.42	43.34	40,741.25
14	30.0	90.0	40.0	544.19	0.39	5.79	2,686.47	57.63	54,986.01
18	15.0	45.0	33.0	467.72	0.35	4.72	1,894.58	49.39	50,731.61
10	−15.0	45.0	33.0	386.78	0.26	1.96	1,002.76	36.69	34,782.70
10	45.0	45.0	33.0	485.50	0.18	2.43	1,871.04	45.80	50,084.24
10	15.0	−45.0	33.0	388.58	0.23	2.17	1,168.07	38.92	38,706.01
10	15.0	135.0	33.0	408.54	0.18	1.28	878.08	33.95	37,655.31
10	15.0	45.0	19.0	304.01	0.21	2.18	1,264.67	39.82	45,087.67
10	15.0	45.0	47.0	463.71	0.27	3.09	1,748.64	47.87	42,854.63
10	15.0	45.0	33.0	387.72	0.27	3.36	1,592.95	44.74	43,079.98

Table 3. Significant variables.

Dependent Variable	Independent Variable
FM	BL, BN, BA, BO
MFR	BN
T	BN, BA
DP	BA, BN
VM	BN, BA
TΩ	BN, BA
F_1	BN, BA, BL, BO
F_2	BN

Table 4. Coefficients of the original variables and the rotated factors.

FM	MFR	T	DP	VM	TΩ	F_1	F_2
387.72	0.27	3.36	1,592.95	44.74	43,079.98	0.06	0.43
45.87	0.03	0.60	271.90	3.95	5,875.68	0.79	0.55
18.65	0.00	0.40	252.75	2.72	3,774.90	0.54	0.22
5.69	0.01	0.27	79.21	1.07	615.01	0.11	0.24
43.18	0.01	0.16	51	1.50	238.33	0.37	0.13
−2.64	0.01	0.15	−25.05	−0.31	−671.47	−0.11	0.16
12.25	−0.01	−0.24	−21.27	−0.62	15.11	0.07	−0.21
2.86	−0.01	−0.36	−124.72	−1.82	−1,048.10	−0.12	−0.33
−0.82	0.00	−0.13	−3.83	0.03	399.53	0.02	−0.07
6.26	0.00	0.13	100.10	0.87	345.61	0.14	0.08
2.41	0.02	0.37	123.11	1.60	934.25	0.13	0.36
10.94	−0.01	−0.17	−23.65	−0.45	−187.24	0.05	−0.16
0.82	0.01	0.19	33.12	0.41	−359.15	−0.01	0.17
6.27	0.00	−0.01	31.28	0.22	415.18	0.09	−0.01
1.43	0.01	0.13	53.13	0.71	122.10	0.05	0.14

4 Results

4.1 Individual Optimization

The individual optimization of the variables and the rotated factors related to each objective function is presented in Table 5, highlighting the best result for each item. It is important to note that some variables have opposite optimization directions and high correlation or conflicts with each other.

Table 5. Individual optima.

Y	x^{T*}	Y^*
FM	$\left[3\ 14.55\ 44.91\ 25.87\right]$	275.28
MFR	$\left[18\ 15.21\ 73.76\ 33.32\right]$	0.40
T	$\left[7\ 6.68\ 102.92\ 28.35\right]$	1.06
DP	$\left[10\ 10.62\ -35.96\ 33.80\right]$	856.95
VM	$\left[16\ 130.32\ 73.49\ 36.44\right]$	56.01

(*continued*)

Table 5. (*continued*)

Y	x^{T*}	Y^*
TΩ	$\begin{bmatrix} 3 & 5.31 & 661.25 & 32.53 \end{bmatrix}$	27,357.38
F_1	$\begin{bmatrix} 2 & 10.44 & 63.00 & 30.29 \end{bmatrix}$	−2.08
F_2	$\begin{bmatrix} 17 & 21.24 & 75.33 & 32.37 \end{bmatrix}$	2.60

The individual optimization performed with the GA and the construction of the Payoff Matrix include not only the Utopia of each variable and factor but also their pseudo-Nadir.

It is observed that all input parameters differ for each response, often resulting in an optimum for one response that is far from the optimum for others.

To visualize the behavior of the objective functions in relation to the most significant input variables, surface plots are used. Figure 5 illustrates the dependent variable F_1 in relation to BN and BA, while Fig. 6 illustrates F_2 in relation to BN and BO.

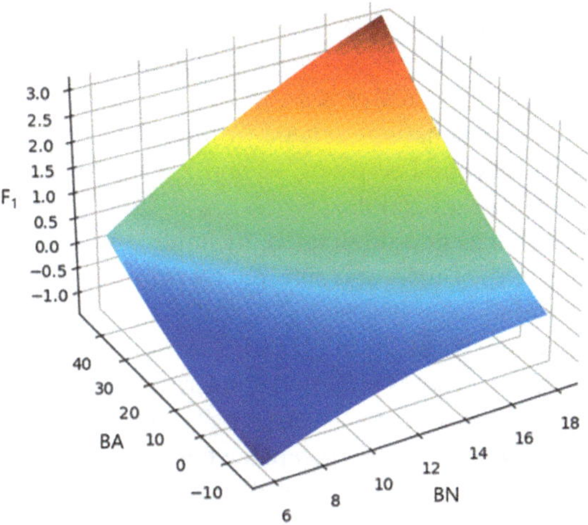

Fig. 5. Surface plot of F_1.

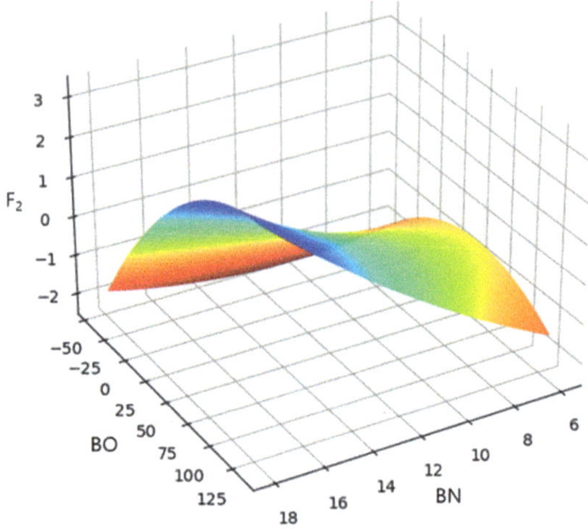

Fig. 6. Surface plot of F_2.

4.2 Multiobjective Optimization

The MOP performed with NSGA-II begins with the evaluation of candidate solutions by decision variable vectors, which are the four variables (BN, BA, BO, and BL). The initial population is randomly generated and evaluated against the utopias previously defined in Table 5. The population size is set to 100 individuals.

There are two objective functions related to the rotated factors (F_1 and F_2), with lower and upper bounds encoded between -2 and 2, as defined by the DOE. F_1 is for minimization and F_2 for maximization.

Solutions are classified as non-dominated, meaning a solution should not be worse than another in all objectives. The layers represent sets of solutions that are not dominated by any other in the population.

After this classification, a selection procedure based on the diversity of solutions throughout the optimization process is carried out. Crossover and mutation are applied to the "parents" to generate new solutions ("offspring"), aiming to find improved solutions based on the objectives. New populations replace the previous ones using a mechanism to avoid the premature loss of good solutions during the process.

This cycle of selection, crossover, mutation, and new population formation is repeated for several generations. The algorithm converges when it reaches the stopping criterion, defined at 100 generations.

The Pareto front is formed by the most relevant solutions, representing the non-dominated solutions that achieve the optimum between the objective functions. Figure 7 illustrates the Pareto front created for F_1 and F_2.

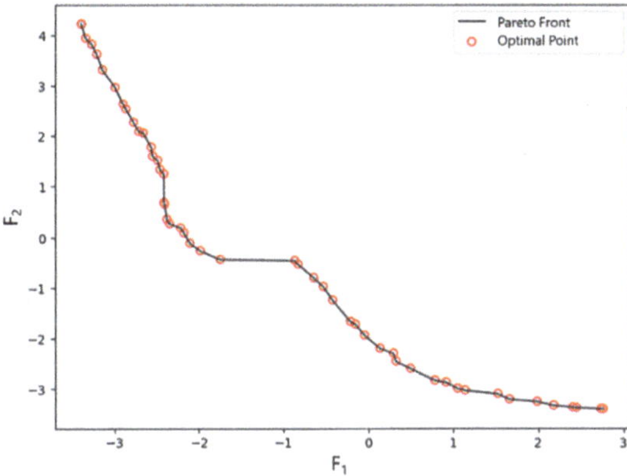

Fig. 7. Pareto front with respect to F_1 and F_2.

4.3 Evaluation Metrics

The validation metrics based on MD and TOPSIS aim to find the best solutions among the 100 possible generated by NSGA-II. The average MD result was 2.83, while the average TOPSIS result was 0.48. The best MD results (lowest values) were 0.59, 0.90, and 0.93, while the best TOPSIS values (highest values) were 0.99, 0.99, and 0.96, respectively. It is important to highlight that the best results concerning MD do not necessarily correspond to the best results regarding TOPSIS.

Analyzing the best results from each metric, it is observed that the values resulting in the lowest MD are represented by MD $\mathbf{x_{MD}} = \left[2\ 11.31\ 87.03\ 25.17\right]$, while the highest values in relation to TOPSIS are $\mathbf{x_{TOPSIS}} = \left[18\ 39{,}92\ 127{,}22\ 30{,}53\right]$. Table 6 presents the best results obtained with respect to the metrics.

Table 6. Independent variables in relation to the evaluation metrics.

Variable	Value (MD)	Value (TOPSIS)
FM	263.27	557.91
MFR	0.20	0.42
T	1.04	6.78
DP	454.77	2,938.02
VM	28.37	59.24
TΩ	25,715.50	59,315.31
F_1	−2.55	−1.59
F_2	2.76	3.39

MD has advantages in considering covariance, outlier detection, and working in multidimensional spaces. On the other hand, TOPSIS is advantageous due to its criteria balancing, flexibility in weighting, and ease of implementation. However, both have disadvantages: MD, although capable of detecting outliers, can be sensitive to them and requires higher computational resources due to inverse matrix calculations. TOPSIS can be sensitive to extremes and dependent on weight selection.

4.4 Validation with CFD Simulations

After obtaining the optimal results from the factor optimization process, it is necessary to perform new simulations to analyze and verify the results achieved by the MOP with NSGA-II. The entire process is conducted using Ansys software, with results extracted through CFD. The simulations are carried out under the same conditions and characteristics defined in the DOE.

Figure 8 illustrates the particle trajectory lines showing the fluid path within the fluid domain with respect to velocity, while Fig. 9 presents the dynamic pressure, and Fig. 10 shows the turbulence Ω. The colors indicate the corresponding values (not related to the optimization direction): closer to red indicates higher values, and closer to blue indicates lower values.

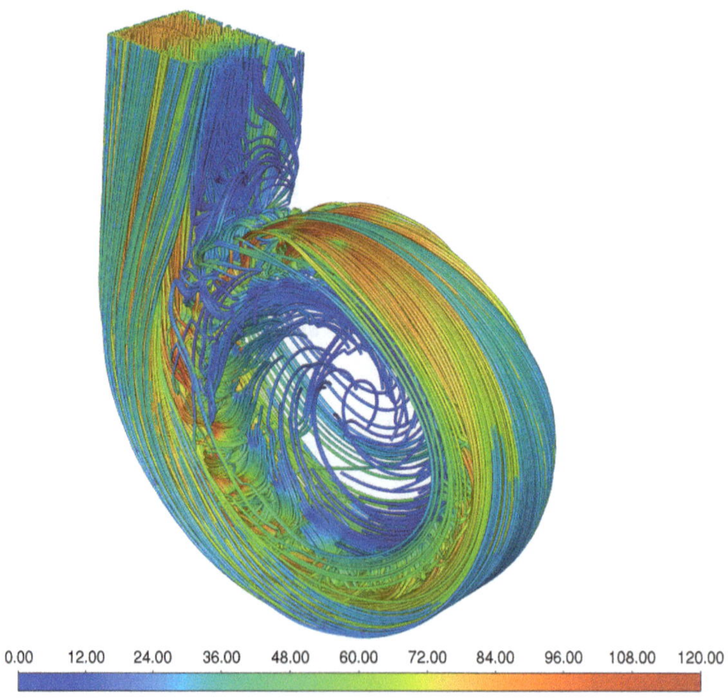

Fig. 8. Particle trajectory lines (velocity) for the model with the highest TOPSIS.

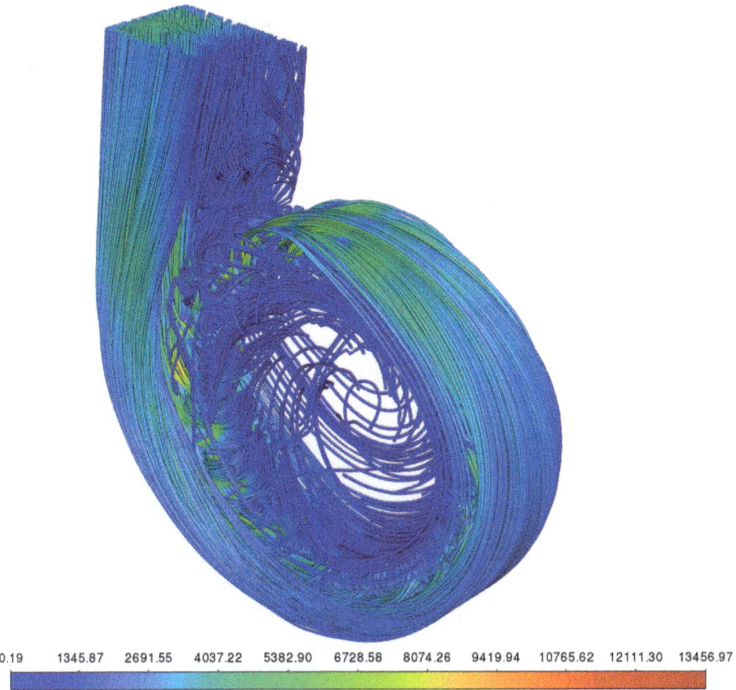

Fig. 9. Particle trajectory lines (dynamic pressure) for the model with the highest TOPSIS

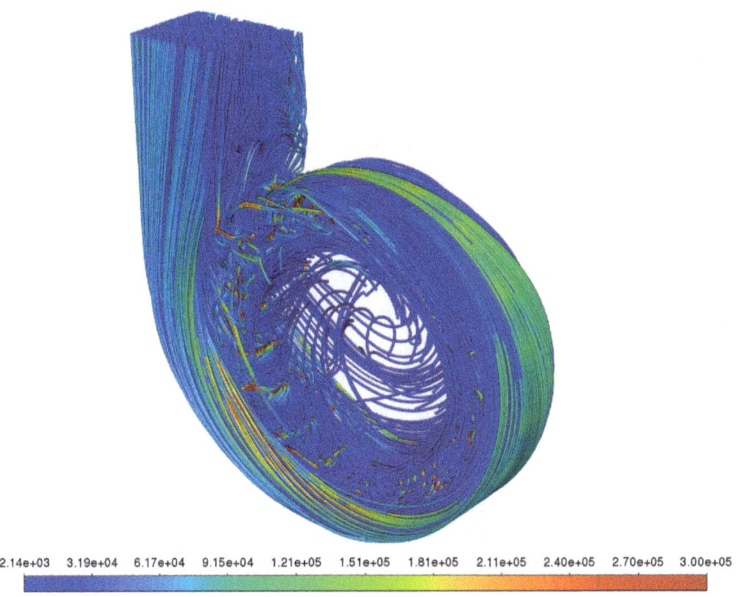

Fig. 10. Particle trajectory lines (turbulence Ω) for the model with the highest TOPSIS.

Performing an MOP with a greater number of variables, as would be the case with the original variables, can lead to several challenges, such as conflicts between multiple objectives, increased constraints, and contradictions in the optimization directions.

Furthermore, if a trial-and-error approach were adopted, the time required to simulate, obtain, and compare results would be significantly higher. Even using the proposed method, validating the results of the 100 possible solutions generated by NSGA-II would require high computational costs.

Finally, to determine the best result, it is crucial to use metrics that rank the responses, thus justifying the use of MD and TOPSIS. Validation was carried out through new simulations, which confirmed the effectiveness of the suggested approach. The results obtained by CFD showed a minimal difference, less than 10.00%, compared to the predictions from NSGA-II.

5 Conclusions

This paper aims to combine DOE techniques for planning and conducting experiments, MOP with the NSGA-II, and dimensionality reduction methods such as PCFA. These approaches are applied to fluid dynamics problems. With the use of CFD it is possible to carry out the simulations of the experiments as well as the confirmation simulations with the optimal solutions.

By integrating these methodologies, it was possible not only to identify the best design and operational parameters but also to simultaneously optimize multiple performance criteria, such as FM, MFR, T, DP, VM and TΩ. This was achieved not by working directly with the variables but through the scores of the rotated factors using the Promax rotation method, simplifying the complexity of the optimization problem.

The application of NSGA-II ensures the achievement of non-dominated solutions on the Pareto front, balancing the factors with a trade-off between the objectives. The evaluation of the results was carried out using metrics such as the TOPSIS and MD to qualify the solutions found, ensuring decisions based on solid criteria. It is important to note that the results presented by each metric were significantly different, with MD presenting control variables of $\mathbf{x_{MD}} = \begin{bmatrix} 2 & 11.31 & 87.03 & 25.17 \end{bmatrix}$, and TOPSIS of $\mathbf{x_{TOPSIS}} = \begin{bmatrix} 18 & 39.92 & 127.22 & 30.53 \end{bmatrix}$, to find the optimal results related to the response variables FM, MFR, T, DP, VM, and TΩ. For MD, the response values were $\mathbf{Y_{MD}} = \begin{bmatrix} 263.27 & 0.20 & 1.04 & 454.77 & 28.37 & 27{,}715.50 \end{bmatrix}$, while for the TOPSIS metric they were $\mathbf{Y_{TOPSIS}} = \begin{bmatrix} 557.91 & 0.42 & 6.78 & 2{,}938.02 & 59.24 & 59{,}315.31 \end{bmatrix}$.

The case study addresses an industrial centrifugal fan, where the entire proposed approach was applied, demonstrating performance improvements in the equipment without the need to carry out numerous simulations or create multiple physical prototypes. This represents resource savings, accelerates the development process, and provides valuable insights into the results obtained.

This study seeks not only to contribute to the advancement of centrifugal fan development but also to propose a systematic and effective method for engineering problems involving fluid dynamics systems.

In future work, a comparison with other MOP methods will be conducted. Additionally, the performance of the results obtained using original variables versus component and factor scores will be analyzed.

Acknowledgements. Acknowledgements are expressed to CAPES, CNPq, and FAPEMIG for the support provided to this work. This research was also made possible with the support of NOMATI-UNIFEI, which granted access to its laboratories, materials, and expertise.

References

1. Huang, J., et al.: A hybrid electric vehicle motor cooling system-design, model, and control. IEEE Trans. Veh. Technol. **68**(2019), 4467–4478 (2019). https://doi.org/10.1109/TVT.2019.2902135
2. Yeom, T., Huang, L., Zhang, M., Simon, T., Cui, T.: Heat transfer enhancement of air-cooled heat sink channel using a piezoelectric synthetic jet array. Int. J. Heat Mass Transf. **143**, 118484 (2019). https://doi.org/10.1016/j.ijheatmasstransfer.2019.118484
3. Zhou, S., Yang, K., Zhang, W., Zhang, K., Wang, C., Jin, W.: Optimization of multi-blade centrifugal fan blade design for ventilation and air-conditioning system based on disturbance CST function. Appl. Sci. **11**, 7784 (2021). https://doi.org/10.3390/app11177784
4. Shi, R., et al.: Design and experiments of crawler-type hilly and mountainous flax combine harvester. Nongye Gongcheng Xuebao/Trans. Chin. Soc. Agric. Eng. **37**, 59–67 (2021). https://doi.org/10.11975/j.issn.1002-6819.2021.05.007
5. Zhu, G., Liu, X., Li, L., Gao, S., Tong, W.: Design and analysis of the ventilation structure for a permanent magnet wind generator. Diangong Jishu Xuebao/Trans. China Electrotech. Soc. **34**, 946–953 (2019). https://doi.org/10.19595/j.cnki.1000-6753.tces.L80756
6. Zhou, S., Zhou, H., Yang, K., Dong, H., Gao, Z.: Research on blade design method of multi-blade centrifugal fan for building efficient ventilation based on Hicks-Henne function. Sustain. Energy Technol. Assess. **43**, 100971 (2021). https://doi.org/10.1016/j.seta.2020.100971
7. Guo, W., et al.: Development of a comb tooth loosening and pneumatic stripping plough layer residual film recovery machine. Nongye Gongcheng Xuebao/Trans. Chin. Soc. Agric. Eng. **36**, 1–10 (2020). https://doi.org/10.11975/j.issn.1002-6819.2020.18.001
8. Zhang, F., et al.: Design and field testing of the air-suction machine for picking up Chinese jujube fruits. J. Fruit Sci. **37**, 278–285 (2020). https://doi.org/10.13925/j.cnki.gsxb.20190424
9. Nowzari, R., Saygin, H., Aldabbagh, L.B.Y.: Evaluating the performance of a modified solar air heater with pierced cover and packed mesh layers. J. Sol. Energy Eng. Trans. ASME **143**, 011006 (2021). https://doi.org/10.1115/1.4047528
10. Zhou, S., Dong, H., Zhang, K., Zhou, H., Jin, W., Wang, C.: Optimal design of multi-blade centrifugal fan based on partial coherence analysis. Proc. Inst. Mech. Eng. Part C J. Mech. Eng. Sci. **236**, 894–907 (2022). https://doi.org/10.1177/0954406221999683
11. Gomes, J.H.F., Júnior, A.R.S., Paiva, A.P., Ferreira, J.R., Costa, S.C., Balestrassi, P.P.: Global criterion method based on principal components to the optimization of manufacturing processes with multiple responses. Stroj. Vestnik/J. Mech. Eng. **58**, 345–353 (2012). https://doi.org/10.5545/sv-jme.2011.136
12. Costa, D.M.D., Belinato, G., Brito, T.G., Paiva, A.P., Ferreira, J.R., Balestrassi, P.P.: Weighted principal component analysis combined with Taguchi's signal-to-noise ratio to the multiobjective optimization of dry end milling process: a comparative study. J. Brazilian Soc. Mech. Sci. Eng. **39**, 1663–1681 (2017). https://doi.org/10.1007/s40430-016-0614-7

13. Almeida, F.A., Santos, A.C.O., Paiva, A.P., Gomes, G.F., Gomes, J.H.F.: Multivariate Taguchi loss function optimization based on principal components analysis and normal boundary intersection. Eng. Comput. **38**, 1627–1643 (2022). https://doi.org/10.1007/s00366-020-01122-8
14. Rocha, L.C.S., Paiva, A.P., Paiva, E.J., Balestrassi, P.P.: Comparing DEA and principal component analysis in the multiobjective optimization of P-GMAW process. J. Brazilian Soc. Mech. Sci. Eng. **38**, 2513–2526 (2016). https://doi.org/10.1007/s40430-015-0355-z
15. Paiva, A.P., Gomes, J.H.G., Peruchi, R.S., Leme, R.C., Balestrassi, P.P.: A multivariate robust parameter optimization approach based on principal component analysis with combined arrays. Comput. Ind. Eng. **74**, 186–198 (2014). https://doi.org/10.1016/j.cie.2014.05.018
16. Paiva, A.P., Costa, S.C., Paiva, E.J., Balestrassi, P.P., Ferreira, J.R.: Multi-objective optimization of pulsed gas metal arc welding process based on weighted principal component scores. Int. J. Adv. Manuf. Technol. **50**, 113–125 (2010). https://doi.org/10.1007/s00170-009-2504-y
17. Pereira, M.C.: Otimização não-linear multiobjetivo das condições de eficiência usando uma abordagem híbrida CFD-DOE: uma aplicação prática em ventiladores centrífugos para fornos industriais. Dissertation, Universidade Federal de Itajubá (2025)
18. Pereira, M.C., Paiva, A.P.: Nonlinear multiobjective optimization of efficiency conditions using a CFD-DOE hybrid approach: a practical application in centrifugal fans for industrial ovens. Therm. Sci. Eng. Prog. 65, 103899 (2025). https://doi.org/10.1016/j.tsep.2025.103899
19. Pereira, M.C., Ribeiro, C.T., Mendes, R.R.A., Campos, P.H.S., Paiva, A.P.: A hybrid multivariate normal boundary intersection approach with post-optimization assisted by mixture design of experiments. Eng. Appl. Artif. Intell. 162(C), 112510 (2025). https://doi.org/10.1016/j.engappai.2025.112510
20. Deb, K., Agrawal, S., Pratap, A., Meyarivan, T.: A fast elitist non-dominated sorting genetic algorithm for multi-objective optimization: NSGA-II. In: Schoenauer, M., Deb, K., Rudolph, G., Yao, X., Lutton, E., Merelo, J.J., Schwefel, H.-P. (eds.) PPSN 2000. LNCS, vol. 1917, pp. 849–858. Springer, Heidelberg (2000). https://doi.org/10.1007/3-540-45356-3_83
21. Pereira, J.L.J., Oliver, G.A., Francisco, M.B., Cunha, S.S., Gomes, G.F.: A review of multi-objective optimization: methods and algorithms in mechanical engineering problems. Arch. Comput. Methods Eng. **29**, 2285–2308 (2022). https://doi.org/10.1007/s11831-021-09663-x
22. Hwang, C.-L., Yoon, K.: Multiple Attribute Decision Making. Springer, Berlin (1981). https://doi.org/10.1007/978-3-642-48318-9
23. Johnson, R.A., Wichern, D.W.: Applied Multivariate Statistical Analysis. Prentice Hall, Englewood Cliffs (1982)

Modeling of an Industrial Agglomerating Drum Using the Discrete Element Method

Carlos Henríquez N., Manuel Moncada M.(✉), and Cristian G. Rodríguez

Department of Mechanical Engineering, University of Concepción, Concepción, Chile
manuelmoncada@udec.cl

Abstract. Agglomerating drums are widely used in the mineral processing industry, particularly in heap leaching operations, to promote granulation by reducing fines and enhancing heap permeability. Despite its widespread use, there are limited studies proposing numerical models to delve into the optimization and improvement of these equipment. This work presents a comprehensive numerical study of an industrial-scale agglomerating drum—4.8 m in diameter and 16.3 m in length—operating at copper mine. Simulations were carried out using the Discrete Element Method (DEM). A liquid bridge contact model was employed to represent particle adhesion due to moisture. A parametric study was conducted considering different rotational speeds (5–7 rpm) and filling degrees (8%–12%). The results provide detailed insights into particle velocity profiles, power consumption, and moisture distribution under steady-state conditions. The simulations revealed that both increased rotational speed and filling degree significantly affect energy demand and particle mixing dynamics. Furthermore, higher angular velocities tend to increase particle segregation, whereas lower speeds result in a more uniform moisture distribution. These findings contribute to a better understanding of agglomerating drum performance and offer a numerical foundation for operational optimization and energy efficiency improvements in industrial applications.

Keywords: Agglomerating drum · DEM · liquid bridge

1 Introduction

Chile holds the largest copper reserves globally, with estimates exceeding 190 million tons, making it the largest copper producer in the world, with a 26.5% market share in 2022. The mining sector is one of Chile's most important industrial activities, representing 14.6% of the national gross domestic product and 58% of total exports in 2022 [1]. However, the progressive decline in ore grades has imposed increasing demands on the efficiency of extraction and processing methods [2].

For oxidized copper ores, leaching is one of the predominant extraction techniques. This hydrometallurgical process enables the recovery of high-purity copper through dissolution in aqueous solutions. A critical pre-treatment step to enhance leaching efficiency is agglomeration, which improves the percolation and uniform distribution of leaching

O. F. Farías Fuentes et al. (Eds.): CIBIM 2024, *Proceedings of the XVI Ibero-American Congress of Mechanical Engineering*, pp. 405–415, 2026.
https://doi.org/10.1007/978-3-032-22823-9_29

solutions throughout the ore bed. Agglomeration involves the formation of larger, more porous aggregates through the adhesion of fine particles onto coarser ones, facilitated by the addition of moisture [3]. According to research and industrial databases, the agglomeration process improves the permeability of the mineralized ore used in leaching by 10 to 100 times. This improvement reduces the time required for the process by between 33% and 50% and reduces the amount of sulfuric acid by between 20% and 30% [2]. Indeed, seeking new methods to optimize the leaching process could help reduce the industry's operational costs and improve energy efficiency [2].

Empirically determining the optimal operating parameters of a mineral processing agglomeration drum in the industry entails significant risk exposure, considering multiple operating parameters. The greatest risk involved in performing such an action is partial or total damage to the machinery. The main consequences for companies are downtime and unforeseen maintenance costs. Given the problems above, research is currently focused on validating a computational model capable of simulating the behavior of the particles in mineral processing equipment [3]. A model using the discrete element method (DEM) or computational fluid dynamics (CFD) allows for evaluating the forces, power, and torque to which the drum and its drive system are subjected. Analyzing and studying the results obtained using numerical methods makes it possible to determine the appropriate operating parameters. However, particle flow within the agglomerating drum has not been thoroughly studied [4, 5].

Several notable works have explored particle behavior in rotary drums. For instance, Yan et al. [6] developed a two-dimensional DEM model validated with Positron Emission Particle Tracking data, demonstrating high predictive accuracy. Yu et al. [5] conducted a three-dimensional DEM simulation of polydisperse particle flow in the rolling regime, providing insight into active and passive layer formation. Trung et al. [7] investigated the growth of a single granule within a dense flow of an initially homogeneous distribution of wet and dry particles using the discrete element method. Similarly, Pachón-Morales et al. [8] examined the flow of raw and torrefied biomass particles in a loose and dynamic condition in a rotating drum. DEM simulations were practical in studying the isolated effects of particle flow distribution and cohesion. Wang et al. [9] studied granulation in a rotating drum using a DEM model for application in the iron ore sintering process. They analyzed the small particles adhered to a central particle, showing that a low variation in water content impacts the adhesion of small particles.

This research aims to develop a numerical model using the discrete element method (DEM) using ANSYS Rocky software to model and analyze mineral behavior during agglomeration in a copper mineral processing agglomerating drum. The model incorporates adhesion forces through a liquid bridge model and employs real operational parameters. The main objective is to understand and analyze material behavior inside the drum and calculate the necessary power to operate this machine. The results are expected to resemble real operating conditions, enabling process optimization and important decision-making in the mining industry.

2 Methodology

The numerical model was developed using the discrete element method, a numerical technique used in engineering and applied sciences to study granular flows. DEM provides detailed insight into particle-scale interactions and is particularly effective for simulating dynamic processes such as those occurring within agglomerating drums.

The simulations were performed using Ansys Rocky, which allows for the implementation of advanced contact and cohesion models. The mechanical interactions between particles were modeled using the Hertz-Mindlin contact model for the normal and tangential force [10, 11], along with a Type-C rolling resistance model [12], which accounts for moment due to rolling friction.

To simulate the cohesive forces induced by liquid addition during agglomeration, a liquid bridge model was employed. This model does not resolve the full fluid dynamics of the liquid phase but effectively captures the capillary and viscous forces generated by interstitial liquid bridges between particles.

The capillary force F_c exerted by a liquid bridge between two particles is given by the following empirical expression [13]:

$$F_c = \pi \gamma r_g \left(e^{A\widehat{h}+B} + C \right) \tag{1}$$

where γ is the surface tension of the liquid, $r_g = \sqrt{r_i r_j}$ is the geometric mean of the radii of the two particles, $\widehat{h}$ is the dimensionless separation distance between the particles. In addition, A, B and C are coefficients that depend on the dimensionless volume of the liquid bridge $\widehat{V}$ and the contact angle θ, by the following expressions:

$$\begin{cases} A = -1.1\widehat{V}^{-0.53} \\ B = \left(-0.34\ln\widehat{V} + 0.96\right)\theta^2 - 0.019\ln\widehat{V} + 0.48 \\ C = 0.0042\ln\widehat{V} + 0.78 \end{cases} \tag{2}$$

In addition to capillary forces, the model includes viscous forces that resist the relative motion between particles or between particles and their boundary, connected by a bridge. These forces have both normal and tangential components, defined as [14]:

$$F_{\mu,n} = -6\pi \mu R * v_n^{rel}\hat{r} \tag{3}$$

$$F_{\mu,t} = -6\pi \mu R * v_\tau^{rel}\hat{r}(\frac{8}{15} ln\hat{r} + 0.9588) \tag{4}$$

where μ is the absolute viscosity of the liquid, R is the equivalent radius, v_n^{rel} and v_t^{rel} are the normal and tangential components of the relative velocity, and $\hat{r}$ is the dimensionless equivalent radius. Further details on the implementation and validation of the liquid bridge model can be found in the literature [15].

3 DEM Modeling

This study focuses on the modeling of an industrial agglomeration drum operating in a copper mining facility. The equipment has an internal diameter of 4.8 m, a length of 16.3 m, and a tilt angle of 6.5°, with a nominal processing capacity of 2,800 tons per hour. The rotational speed varies between 4 and 7 rpm.

Due to the high computational cost associated with simulating the full-scale geometry, a scaled-down segment equivalent to 1/163rd of the drum length (i.e., 0.1 m) was modeled, as shown in Fig. 1. This segment length ensures a reasonable computational load while adequately capturing the essential particle dynamics, given that the maximum particle diameter is 25 mm. The DEM domain includes a particle inlet zone through which material is continuously fed during the first two seconds of the simulation. Additionally, a liquid injection zone with dimensions of 0.1 m × 1.2 m × 0.1 m is implemented to replicate the spray bar system employed in the industrial setting (see Fig. 1(b)). As particles traverse this zone, a liquid film is numerically assigned to them, realistically simulating the wetting process critical for agglomeration.

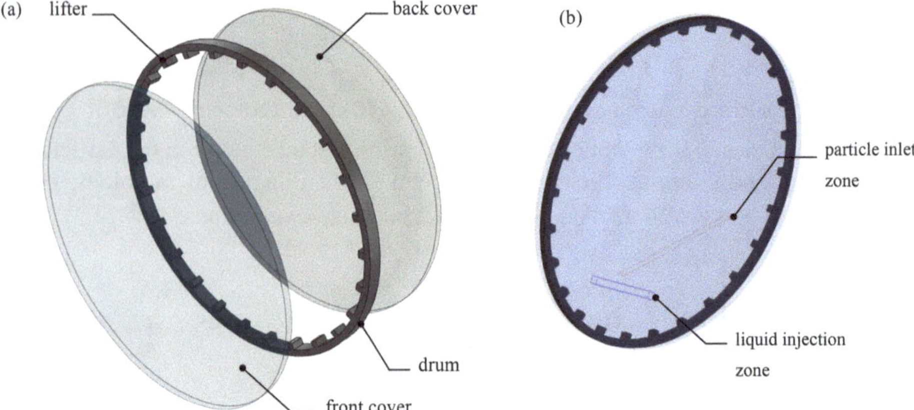

Fig. 1. Geometry of the simplified DEM model of the industrial agglomerating drum. (a) exploded view, (b) inlet and liquid injection zone configuration.

The copper ore was represented using spherical particles with the particle size distribution presented in Table 1. This distribution reflects the granulometry observed in industrial operations.

Table 1. Particle size distribution.

Particle size (mm)	Cumulated size (%)
25	100
15	95.2
10	85.8
6	67.3

The drum walls and internal lifters are modeled as steel surfaces, while the particles represent copper ore. The mechanical properties of both materials are presented in Table 2, and the properties of the liquid phase (water) are summarized in Table 3.

Table 2. Material properties of particles and walls.

Property	Particle (copper ore)	Wall (steel)
Density (kg/m^3)	1600	7850
Young's Modulus (MPa)	100	1000
Poisson's ratio	0.3	0.3

Table 3. Physical properties of the liquid.

Property	Value
Density (kg/m^3)	1000
Viscosity (Pa s)	0,001
Surface tension (N/m)	0.072
Maximum liquid fraction	0.01

To capture the interaction dynamics between particles and between particles and walls, a set of contact parameters must be defined. These include sliding and rolling friction coefficients, restitution coefficients, and liquid bridge parameters, as outlined in Table 4.

Table 4. Contact parameters of particles and walls.

Parameter	Particle-Particle	Particle-Wall
Sliding friction	0.6 [16]	0.5 [16]
Restitution coefficient	0.4 [17]	0.5 [16]
Rolling friction	0.33 [16]	0.33 [16]
Liquid bridge fraction	0.02–0.05 [9, 18]	0
Contact angle	20°–30° [9, 18]	30° [9, 18]

A total of nine simulations were conducted to evaluate the effects of rotational speed and filling degree on the internal dynamics of the drum. The selected operating conditions consisted of three rotational speeds (5, 6, and 7 rpm) and three filling degrees (8%, 10%, and 12%). A filling degree of 8% corresponds to a mass of 216.3 kg of particles within the simulated volume, with the other filling levels scaled accordingly.

The simulations were carried out using ANSYS Rocky 2023 R1.1 software on a high-performance workstation equipped with two NVIDIA A2000 GPUs. Each simulation spanned a physical time of 60 s, with total computational runtimes ranging from 10 to 15 days, depending on the specific simulation conditions.

4 Results

Figure 2 presents snapshots of the velocity profiles obtained from the DEM simulations of the industrial agglomerating drum. The particles are colored according to the velocity magnitude, with values ranging from blue (low velocity) to red (high velocity), thereby illustrating the internal flow behavior under different operating conditions. Nine simulation results are presented, where the drum rotation speed increases from left to right: 5, 6, and 7 rpm, while the filling degree increases from top to bottom: 8%, 10%, and 12%.

Each subfigure clearly distinguishes between the active and passive layers of particles. The active layer, characterized by a light blue to yellow hue, exhibits particle velocities between 2 and 6 m/s, and particles falling from the lifters can reach speeds close to 8 m/s. In contrast, the passive layer, in dark blue, maintains lower velocities, typically under 1 m/s.

As the drum's rotational speed increases, the regions of high velocity (light blue and yellow zones) become more prominent and widespread. This indicates greater energy input to the system, potentially improving particle mixing and agglomeration but also risking excessive dispersion. Regarding the effect of filling degree, an increase in the fill level leads to a thicker active layer. Additionally, greater fill levels result in increased energy dissipation due to enhanced interparticle contact and compaction, particularly from overlying particle layers.

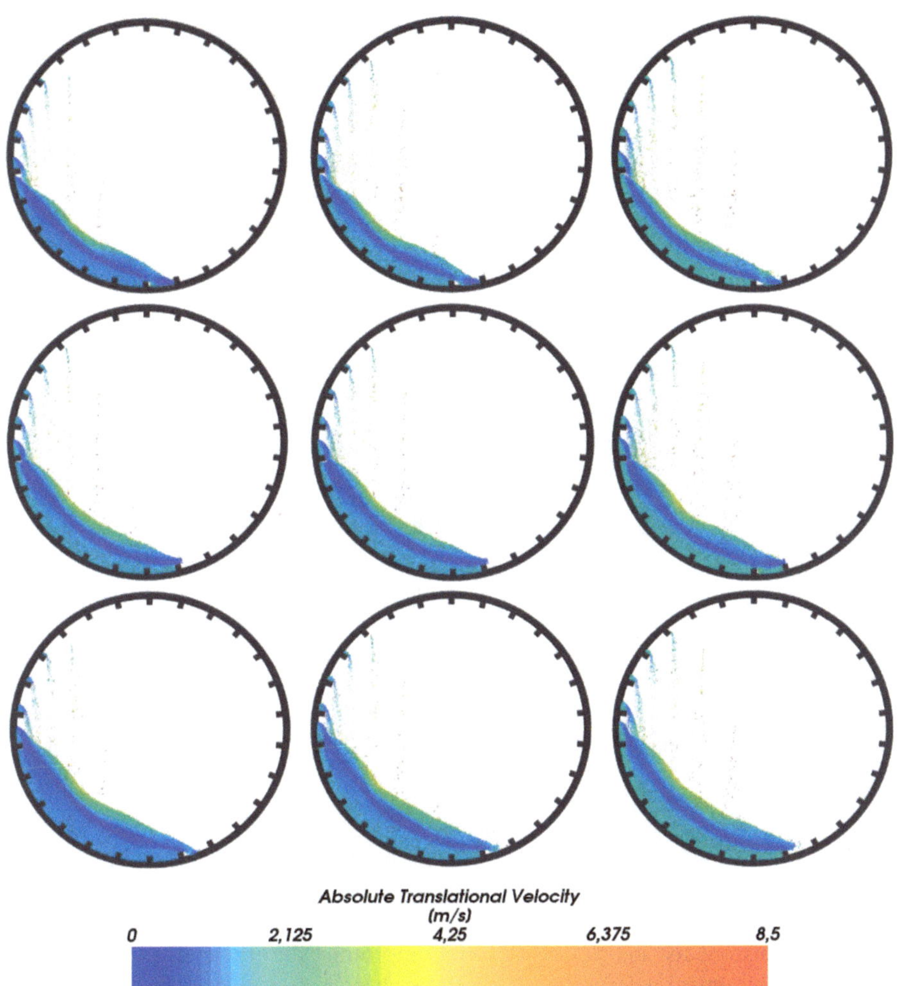

Fig. 2. Velocity profiles of the nine DEM simulations. From left to right, rotational speed increases (5, 6, and 7 rpm). From top to bottom, filling degree increases (8%, 10%, and 12%).

Figure 3 depicts the root mean square (RMS) power as a function of rotational speed for three degrees. As expected, both increasing the drum's angular velocity and fill level result in higher power consumption.. Increasing the drum fill level causes a shift in the interpolation curves in the vertical direction of the graph, i.e., a positive shift if the fill level increases, and vice versa. From the quadratic interpolation curves, the possible RMS power consumptions are extrapolated for rotation speeds of 4 and 8 rpm for the different fill levels. The power values for 4 rpm range from 155.11 to 213.96 kW, while for 8 rpm, the values range from 336.0 to 467.60 kW. Regarding energy demand, the least demanding condition corresponds to an 8% fill level and a rotation speed of 5 rpm, with a consumption of 197.15 watts. On the other hand, the most critical consumption is at a 12% fill level and 7 rpm, with a consumption of 401.17 watts, representing an increase of 203.5%. A 1 rpm increment leads to a power increase of approximately 18.92% to 22.75%, and a 1% increase in fill level results in an additional 15.83% to 20.10% power demand.

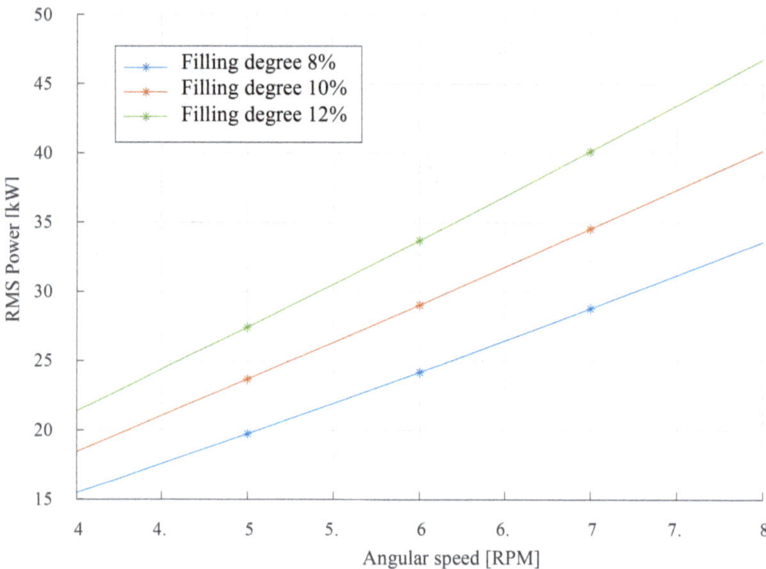

Fig. 3. RMS power values obtained from the DEM simulations under different rotational speeds and fill levels.

Moisture distribution analysis under steady-state conditions is presented in Fig. 4. These distributions highlight zones with uneven wetting inside the drum. A persistent central blue region with low liquid content is evident across several conditions, indicating suboptimal moisture distribution, which may compromise the efficiency of the agglomeration process. Interestingly, higher rotational speeds do not necessarily improve moisture distribution. In some cases, elevated speeds appear to intensify particle segregation, thus exacerbating non-uniform wetting.

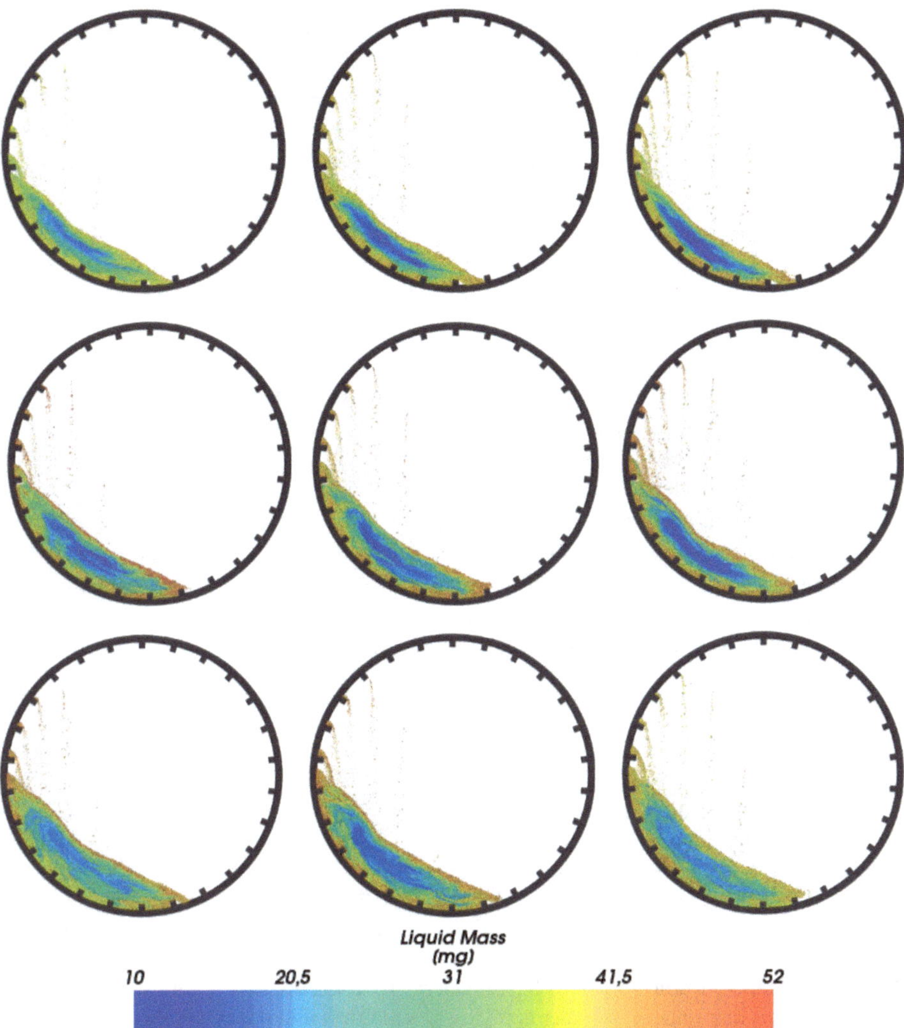

Fig. 4. Moisture distribution at 60 s of simulation. From left to right, rotational speed increases (5, 6, and 7 rpm). From top to bottom, filling degree increases (8%, 10%, and 12%).

5 Conclusions

This study developed and analyzed a numerical model based on the Discrete Element Method (DEM) to investigate the operational behavior of an industrial agglomerating drum used in copper mining. The drum features a diameter of 4.8 m and a length of 16.3 m, and the DEM model incorporated a liquid bridge contact model to account for the adhesive interactions caused by liquid addition during agglomeration.

A computational strategy was proposed to reduce the simulation domain, using a scaled section (1/163rd of the drum length) while maintaining accurate particle dynamics, thus making industrial-scale simulations computationally feasible. The simulation campaign evaluated the influence of two key operational parameters: rotational speed (5, 6, and 7 rpm) and fill level (8%, 10%, and 12%). The results obtained are power, particle velocity, and moisture distribution. The following can be highlighted from these results:

i. Power consumption: increasing the drum's rotational speed results in a nonlinear rise in RMS power consumption. The lowest consumption (197.15 W) was observed at 8% fill and 5 rpm, while the highest (401.17 W) occurred at 12% fill and 7 rpm—an increase of 203.5%. A 1 rpm increment produced a power increase between 18.92% and 22.75%, whereas a 1% increase in fill level led to a 15.83% to 20.10% rise in power.
ii. Velocity profiles and particle dynamics**:** Higher rotational speeds intensify particle movement within the active layer, promoting more vigorous mixing but potentially leading to excessive particle dispersion. Fill level also plays a critical role; higher levels result in a thicker active layer and greater energy dissipation due to compaction and interparticle interactions.
iii. Moisture distribution: uniform moisture distribution was not necessarily enhanced by increased drum speed. In fact, higher rotational velocities may increase particle segregation and hinder homogeneous wetting. Lower angular velocities were found to reduce the standard deviation of moisture content across the drum cross-section, suggesting better liquid distribution at moderate operational speeds.

References

1. Cifras actualizadas de la mineria (2023)
2. Wang, L., Yin, S., Wu, A.: Ore agglomeration behavior and its key controlling factors in heap leaching of low-grade copper minerals. J. Clean. Prod. **279**, 123705 (2021)
3. Toledo, P., Moncada, M., Ruiz, C., Betancourt, F., Rodríguez, C.G., Vicuna, C.: A review of the application of the discrete element method in comminution circuits. Powder Technol. **459**, 121027 (2025)
4. Dissanayake, S., Karunarathne, S.S., Lundberg, J., Tokheim, L.A.: CFD study of particle flow patterns in a rotating cylinder applying OpenFOAM and fluent. In: Proceedings of the 58th Conference on Simulation and Modelling (SIMS 58) Reykjavik, Iceland, 25th–27th September 2017, Norway (2017)
5. Yu, M., Zhang, H., Guo, J., Zhang, J., Han, Y.: Three-dimensional DEM simulation of polydisperse particle flow in rolling mode rotating drum. Powder Technol. **396**, 626–636 (2022)
6. Yang, R.Y., Zou, R.P., Yu, A.B.: Microdynamic analysis of particle flow in a horizontal rotating drum. Powder Technol. **130**, 138–146 (2003)
7. Trung Vo, T., et al.: Agglomeration of wet particles in dense granular flows. Eur. Phys. J. E **42**, 1–12 (2019)
8. Pachón-Morales, J., et al.: Potential of DEM for investigation of non-consolidated flow of cohesive and elongated biomass particles. Adv. Powder Technol. **31**, 1500–1515 (2020)
9. Wang, Y., Xu, J., He, S., Liu, S., Zhou, Z.: Numerical simulation of particle mixing and granulation performance in rotating drums during the iron ore sintering process. Powder Technol. **429**, 118890 (2023)
10. Hertz, H.: On the contact of elastic solids, pp. 156–171 (1881)

11. Mindlin, R.D., Deresiewicz, H.: Elastic spheres in contact under varying oblique force. Trans. ASME J. Appl. Mech. **20**, 327–344 (1953)
12. Wensrich, C., Katterfeld, A.: Rolling friction as a technique for modelling particle shape in DEM. Powder Technol. 409–417 (2012)
13. Mikami, T., Kamiya, H., Horio, M.: Numerical simulation of cohesive powder behavior in a fluidized bed. Chem. Eng. Sci. 53 (1927)
14. Nase, S.T., Vargas, W.L., Abatan, A.A., McCarthy, J.: Discrete chracterizacion tools for cohesive granular material. Powder Technol. **116**, 214–223 (2001)
15. ANSYS ROCKY: Liquid Bridge Model Module - Uptated for Rocky version 2024 R1.1 (2024)
16. Bouassale, N.-E., Sallaou, M., Aittaleb, A.: Development of a methodology for prediction and calibration of parameters of DEM simulations based on machine learning. Mech. Res. Commun. **141**, 104336 (2024)
17. Ho, C.A., Sommerfeld, M.: Modelling of micro-particle agglomeration in turbulent flows. Chem. Eng. Sci. **57**, 3073–3084 (2002)
18. Shi, X., Li, C., Wang, Q., Li, G., Zhang, W., Xue, Z.: Numerical study of the dynamic behaviour of iron ore particles during wet granulation process using discrete element method. Powder Technol. **401**, 117296 (2022)

Redefining Foot Point Trajectories in Theo Jansen and Klann Mechanisms Using Combined Cam-Linkage Mechanisms

Joan Puig-Ortiz(✉), Lluïsa Jordi Nebot, and Rosa Pàmies-Vilà

Mechanical Engineering Department, Universitat Politècnica de Catalunya-Barcelona Tech, Barcelona, Spain
joan.puig@upc.edu

Abstract. In mobile robotics, walking mechanisms offer distinct advantages over wheeled systems for moving over uneven terrain and avoiding obstacles such as stairs. This study presents a methodology for modifying the trajectory of a point representative of the foot in walking mechanisms by replacing a single rigid link with a combined cam-linkage mechanism. A systematic procedure for selecting the most suitable link for replacement is developed, based on the analysis of relative joint angles. The approach is applied to two classic mechanisms: Theo Jansen and Klann. In the Theo Jansen mechanism, the crank is replaced to maintain the original stride length while maximizing foot elevation. In the Klann mechanism, the crank is modified to extend the stride length. In both cases, cam profiles with continuous transitions are designed to achieve precise trajectory continuity. The results demonstrate the feasibility and versatility of the proposed method and highlight its potential to enhance the adaptability and performance of walking mechanisms in robotic applications.

Keywords: walking robots · combined cam-linkage mechanisms · kinematics · trajectory

1 Introduction

In mobile robotics, locomotion mechanisms must often adapt to complex and unpredictable environments. While wheeled systems are common due to their mechanical simplicity and suitability for flat surfaces, they encounter significant limitations on irregular terrain or when navigating obstacles such as stairs. In contrast, walking mechanisms –designed to mimic the movement of human or animal limbs– offer greater versatility and robustness in these challenging scenarios [1].

A common approach in the development of walking robots is the use of linkage mechanisms, where the trajectory of a representative point simulates the foot's motion. Several mechanisms have been proposed with this aim, including the well-known designs of Theo Jansen [2] or Klann [3]. Figure 1 illustrates both mechanisms, taking the robot body as the fixed frame to which, the walking mechanism is attached.

O. F. Farías Fuentes et al. (Eds.): CIBIM 2024, *Proceedings of the XVI Ibero-American Congress of Mechanical Engineering*, pp. 416–428, 2026.
https://doi.org/10.1007/978-3-032-22823-9_30

In applications where only, specific segments of the foot trajectory require adjustment –such as smoothing contact with the ground or increasing step height– modifying the length of a single link is a frequent strategy. However, this global change often alters the entire trajectory, making precise local control difficult.

To overcome this limitation, several authors [4–7] have introduced hybrid cam-linkage systems, combining rigid links with cam-actuated prismatic elements. In the aforementioned walking mechanisms, by replacing one of the links with a combined mechanism –consisting of two rigid bodies connected by a prismatic pair and a cam-follower mechanism [4]– it becomes possible to locally modify the trajectory while preserving the original motion elsewhere.

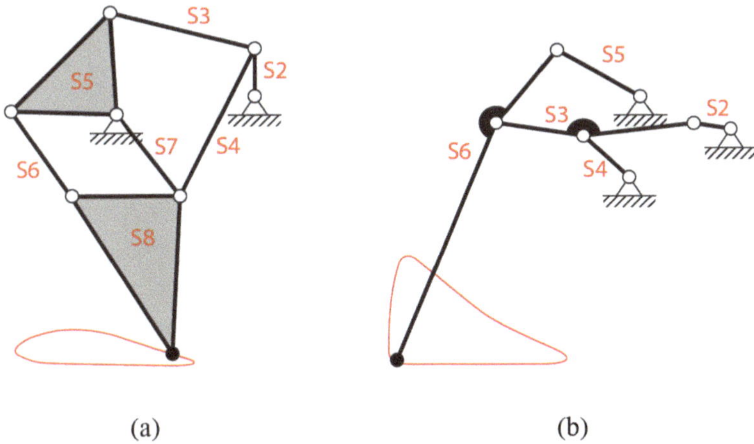

Fig. 1. (a) Theo Jansen mechanism and (b) Klann mechanism.

This study analyzes, through simulation, the possibilities for modifying walking mechanisms by replacing some of the links with combined cam-linkage mechanisms, specifically the Theo Jansen and Klann mechanisms, by incorporating combined cam-linkage mechanisms. In these cases, the replaced link becomes a variable-length element whose extension is governed by a cam profile attached to the component to which the original link was connected. This allows precise control of the link length as a function of its orientation, enabling localized trajectory modification without compromising the overall gait.

2 Methodology

To simulate the two walking mechanisms under study, the simulation software PAM [8] has been used.

The Theo Jansen mechanism comprises eight rigid bodies, including the frame as a fixed body, while the Klann mechanism consists of six rigid bodies, also including the fixed body. All joints between links in both mechanisms are modeled as revolute pairs.

The initial geometry and dimensions of the mechanisms are shown in Fig. 2, and Table 1 lists the coordinates of the fixed points attached to the frame of the both mechanisms.

In order to identify the most suitable bars of the mechanism to be replaced, the relative rotation angles between each pair of connected links are first analyzed. These angles directly influence the variation in distance between the endpoints of the replaced bar and therefore determine how effectively a cam-driven element can control the link's length.

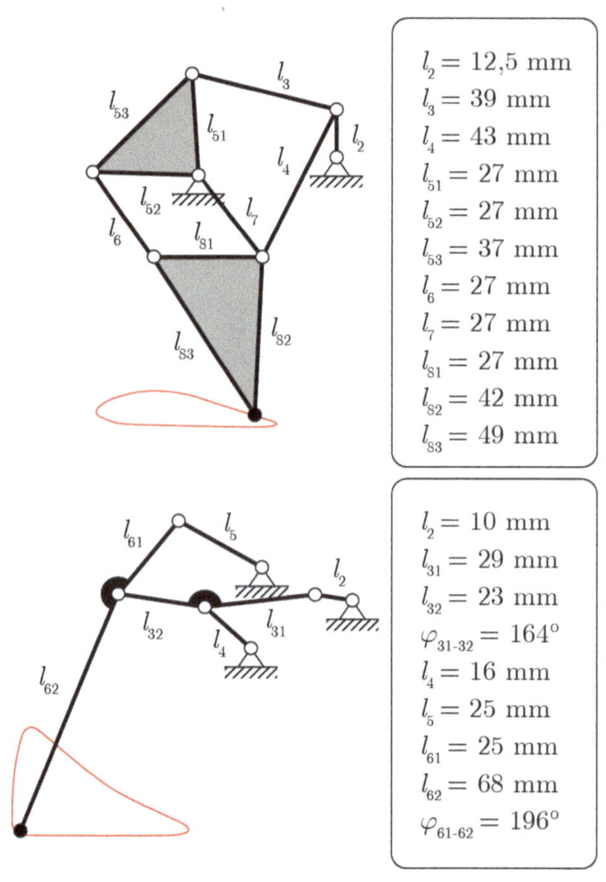

Fig. 2. Dimensions of Theo Jansen's and Klann's mechanisms.

To evaluate the potential for trajectory modification through link substitution, simulations were conducted for both the Theo Jansen and Klann mechanisms. For each mechanism and each candidate link for replacement, three simulations were carried out: one using the original link length, one with the length increased by 2.5 mm, and another with the length decreased by 2.5 mm. In each case, the replaced link was substituted with a combined cam-linkage mechanism, and the resulting trajectory of a representative foot point was obtained.

Table 1. Initial dimensions of the mechanisms.

Theo Jansen Mechanism	
Point	**(x, y)**
RB-2	(0 mm, 0 mm)
RB-5	(36 mm, 5 mm)
Klann Mechanism	
Point	**(x, y)**
RB-2	(0 mm, 0 mm)
RB-4	(−26 mm, −13 mm)
RB-5	(−23 mm, 8 mm)

This methodology enables the controlled alteration of the foot trajectory within a bounded region, defined by the maximum and minimum distances achievable through the variable-length link. The feasibility and flexibility of this approach are demonstrated through two application examples:

- A modified version of the Theo Jansen mechanism, designed to maintain the original stride length while maximizing the vertical displacement of the representative foot point.
- A modified version of the Klann mechanism, aimed at increasing the stride length beyond that of the original design.

For both cases the designs of the cam profiles are presented along with the resulting foot point trajectories.

3 Results

3.1 Selection of Links to Be Replaced

Figure 3 and Fig. 4 show the variation of the relative angles φ (i) between the different pair of links in the mechanisms. The index i indicates the revolute joint R between two connected bodies; for example, φ (RB-2) refers to the relative angle between the base B and the link 2. These relative angles are plotted as a function of the crank angle φ_2, taking as the initial configuration the configuration in which the crank is oriented vertically upwards.

As can be observed in the plots, both mechanisms exhibit two types of joint behavior: some joints undergo a continuous increase in relative angle from 0° to 360° over the course of a stride, while others show non-monotonic behavior with dead points –instances where the same relative angle occurs at two different positions during the stride. In the latter case, the distance between the endpoints of the link must be identical at both positions, limiting the effectiveness of using a variable-length element.

Although, in principle, any articulated link can be replaced by a combined cam-linkage mechanism, it is more advantageous to select links connected through joints with continuous 360° relative rotation. This ensures a one-to-one mapping between the crank angle and the distance to be controlled via the cam mechanism.

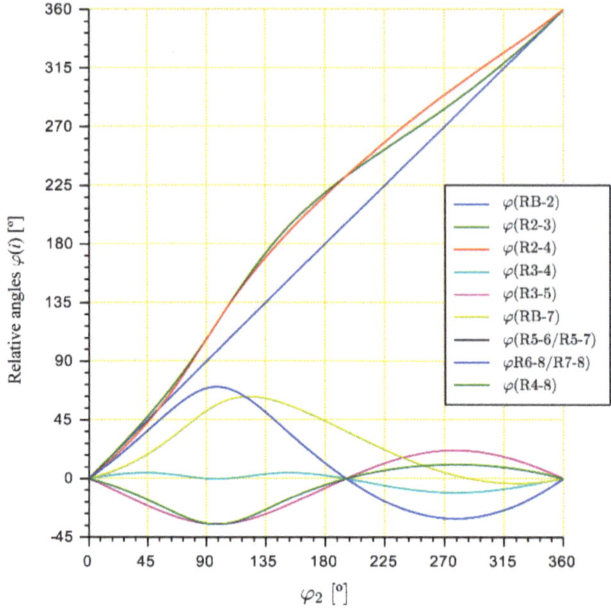

Fig. 3. Rotation angles, for one cycle of the Theo Jansen mechanism, of the joints studied.

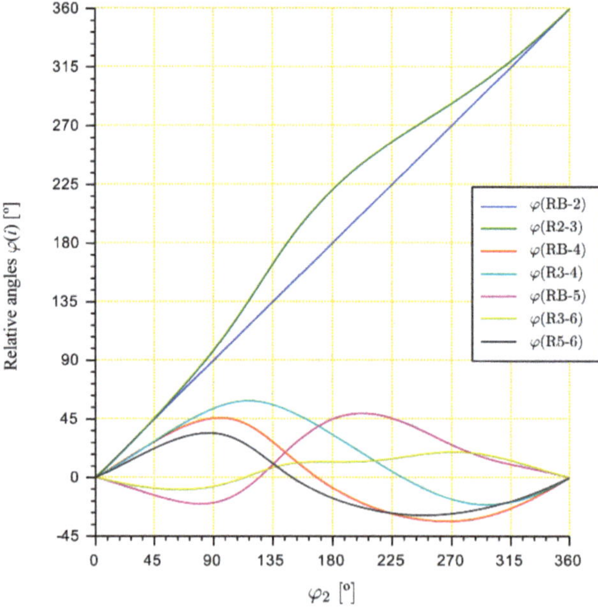

Fig. 4. Rotation angles, for one cycle of the Klann mechanism, of the joints studied.

In Theo Jansen's mechanism, the joints in which the relative angle between links reaches 360° are the joint between the base and link 2 (RB-2), the joint between link 2 and link 3 (R2-3) and the joint between link 2 and link 4 (R2-4). Among the bodies involved in these joints, the ones that are links are bodies 2, 3 and 4. If link 2 is selected for replacement, the control angle can be derived from joints RB-2, R2-3, or R2-4. If link 3 is replaced, the only valid control angle comes from joint R2-3. Similarly, if link 4 is replaced, control must be based on joint R2-4.

In the Klann mechanism, the joints where the relative angle between links reaches 360° are the joint between the base and link 2 (RB-2) and the joint between link 2 and link 3 (R2-3). In this case, the solids that are links are bodies 2 and 3. If link 2 is replaced, the control angle may be taken from either RB-2 or R2-3. If link 3 is replaced, control must be derived from joint R2-3.

3.2 Possible Trajectories

As described in the methodology section, three simulations were performed for each mechanism and for each candidate link to be replaced. Table 2 shows the modified lengths used in each of the analyses.

The objective of these simulations is to determine the range of achievable foot trajectories when substituting a standard link with a combined cam-linkage mechanism. The cam is assumed to provide a radial variation of ±2.5 mm relative to its base radius, enabling local trajectory control within a bounded region.

Table 2. Modified lengths in each of the analyses carried out.

Theo Jansen Mechanism	
Modification of link 2	Length S2 = 12.5 mm
	Length S2 = 15 mm
	Length S2 = 10 mm
Modification of link 3	Length S3 = 39 mm
	Length S3 = 41.5 mm
	Length S3 = 36.5 mm
Modification of link 4	Length S4 = 43 mm
	Length S4 = 45.5 mm
	Length S4 = 40.5 mm
Klann Mechanism	
Modification of link 2	Length S2 = 10 mm
	Length S2 = 12.5 mm
	Length S2 = 7.5 mm
Modification of link 3 (distance from R2-3 to R3-4)	Length S3 = 29 mm
	Length S3 = 31.5 mm
	Length S3 = 26.5 mm

3.3 Trajectories in Theo Jansen's Mechanism

Figure 5a shows the analysis of the modification of link 2 in the Theo Jansen mechanism. It displays the trajectories followed by the representative foot point for the three different lengths of link 2.

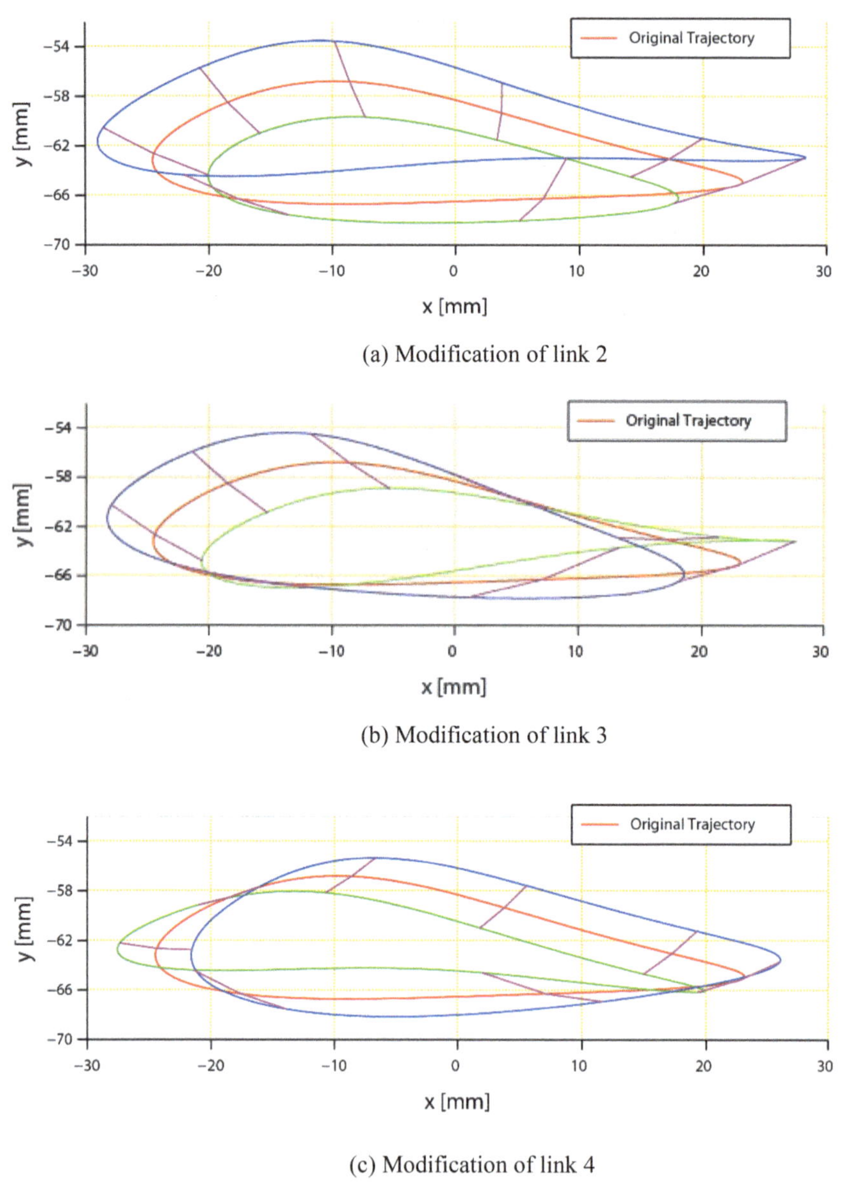

(a) Modification of link 2

(b) Modification of link 3

(c) Modification of link 4

Fig. 5. Region of possible foot point trajectories in the Theo Jansen mechanism resulting from variations in link length.

The red curve corresponds to the original trajectory with the mechanism's unmodified dimensions. The other two trajectories correspond to the cases where link 2 was lengthened or shortened by 2.5 mm. Magenta lines are drawn to connect equivalent crank positions across the three trajectories and are shown at crank angles that are multiples of 45°. The area enclosed by these connecting lines defines the region of possible trajectories achievable through this specific modification.

Figures 5b and 5c show the results of modifying links 3 and 4, respectively, in the Theo Jansen mechanism. These analyses are equivalent to the previous case.

Although all three links –2, 3, and 4– can theoretically be replaced with a combined cam-linkage mechanism, the modifications to links 3 and 4 result in relatively limited variations in the trajectory. Specifically, several points along the foot path remain nearly unchanged despite variations in link length. In contrast, modifying link 2 produces a broader and more uniform range of variation across the entire trajectory.

This analysis suggests that link 2 is the most effective candidate for achieving meaningful and controllable changes in the foot point trajectory within the Theo Jansen mechanism.

3.4 Trajectories in the Klann Mechanism

Figure 6 shows the trajectory modification analyses of links 2 and 3 in the Klann mechanism. As in the case of the Theo Jansen mechanism, modifying link 2 –which is the crank of the mechanism– results in significant trajectory variation across the entire foot path. In contrast, replacing link 3 leads to more limited influence on the trajectory. In particular, certain regions of the foot path remain largely unaffected by changes in link length, reducing the overall potential for meaningful trajectory modification in this case.

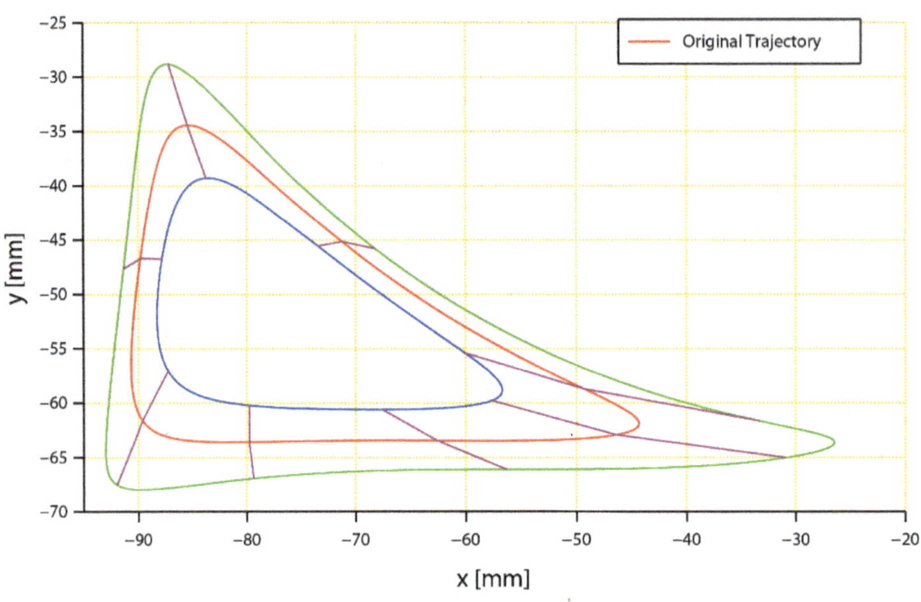

(a) Modification of link 2

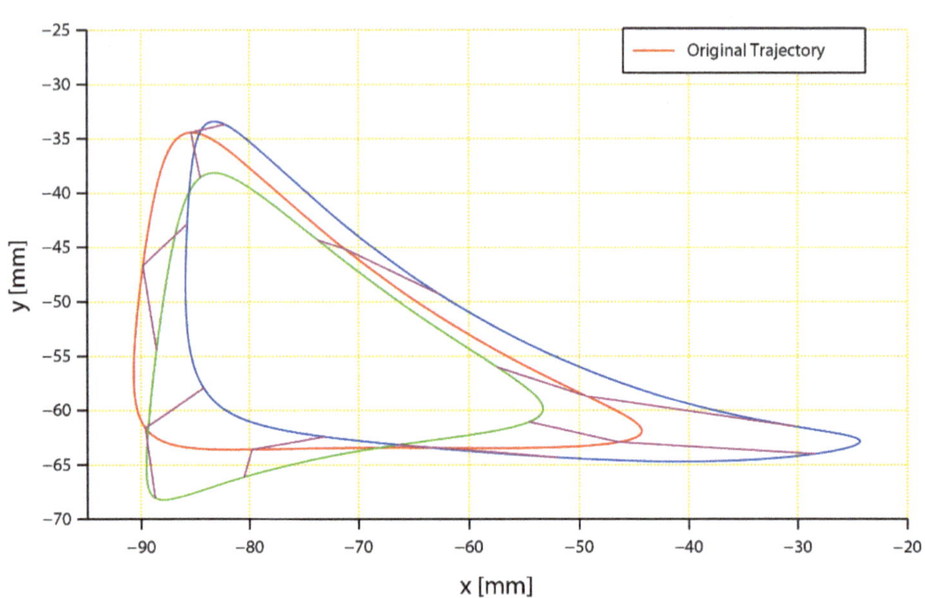

(b) Modification of link 3

Fig. 6. Region of possible foot point trajectories in the Klann mechanism resulting from variations in link lengths.

3.5 Example of Modification of Theo Jansen's Mechanism

To demonstrate the practical application of the proposed approach, a targeted modification of link 2 in the Theo Jansen mechanism was performed, using the original trajectory (shown in red in Fig. 5a) as a reference. The goal was to maximize the vertical displacement of the representative foot point. From the simulations in Fig. 5a, the maximum vertical coordinate was identified on the blue trajectory, corresponding to a link 2 length of 15 mm. This maximum point occurs when link 2 forms an angle of 354.6° with respect to the horizontal.

To implement this adjustment, a transition zone was defined as ±45° around the maximum point. Within this interval, a C2 continuous cam profile was designed to smoothly modify the effective length of the link. This length begins at 12.5 mm, starts to increase when the angle is 309.6°, reaches 15 mm at 354.6°, and then decreases back to 12.5 mm at 39.6° (i.e., 399.6°). This defines a single-dwell cam, whose radius evolution is shown in Fig. 7, where *R* represents the variable radius of the cam.

Figure 8 shows the trajectory obtained with the proposed modification for the Theo Jansen mechanism.

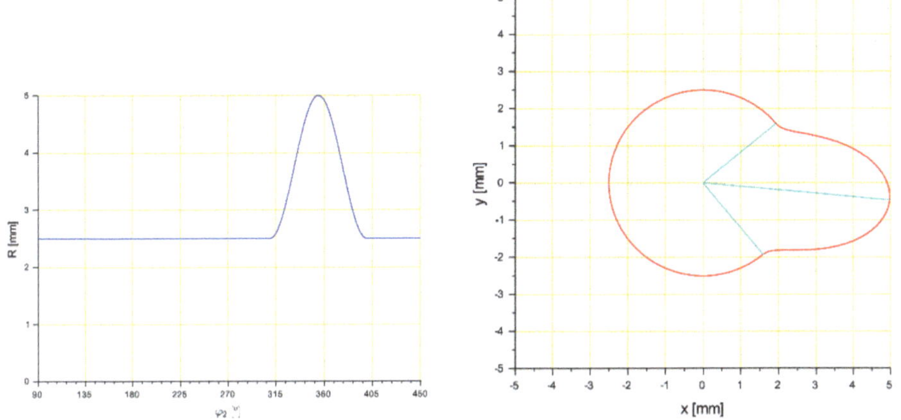

Fig. 7. Variation of the cam radius as a function of the crank angle. Cam profile used in the ex-ample modification of the Theo Jansen mechanism.

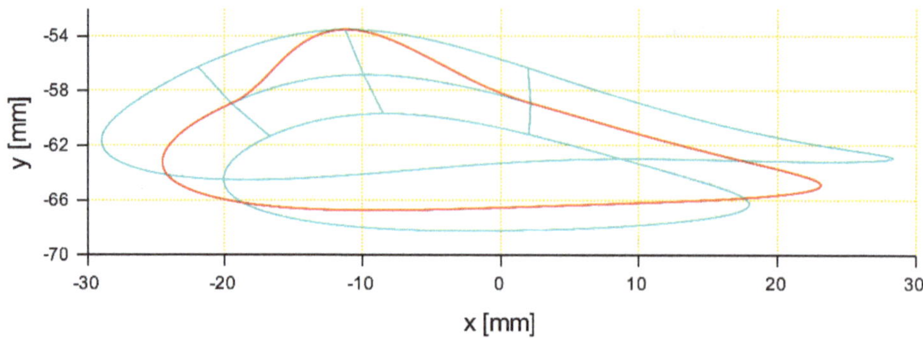

Fig. 8. Resulting foot point trajectory obtained after modifying the crank (link 2) in the Theo Jansen mechanism using the cam profile shown in Fig. 7.

3.6 Example of a Modification to the Klann Mechanism

A modification of the link 2 (the crank) in the Klann mechanism was performed, using the original foot point trajectory (shown in red in Fig. 6a) as a reference. In this case, the objective is to increase the stride length. The objective of this modification is to increase the stride length by selectively extending the lower, flat portion of the foot trajectory.

To achieve this, the cam controlling the variable-length link was designed to extend the link length from its nominal value of 10 mm to 12.5 mm during a specific portion of the crank's rotation. Specifically, the length increases as link 2 rotates from 0° to 45° with respect to the horizontal, remains constant at 12.5 mm between 45° and 135°, and then decreases back to 10 mm between 135° and 180°. For the remainder of the cycle, the length remains fixed at 10 mm.

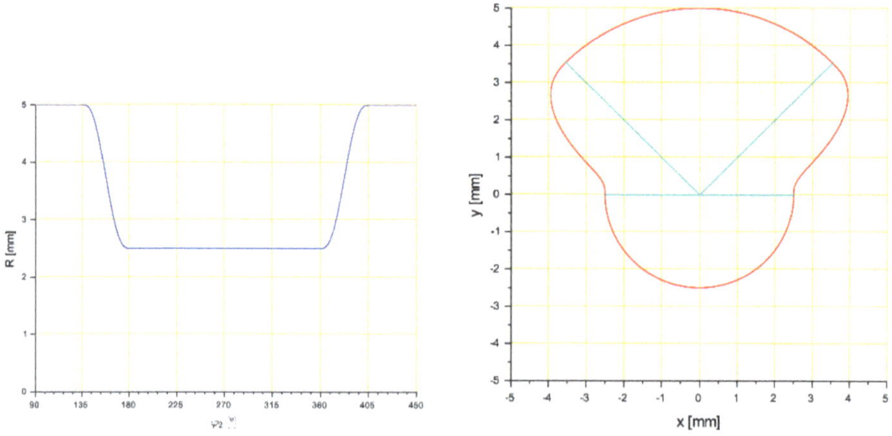

Fig. 9. Radius as a function of the angle and profiles of the cam used for the example of modification of the Klann mechanism.

A C2-continuous double-dwell cam profile was defined to implement this behavior. The corresponding cam radius R, plotted as a function of the crank angle, is shown in Fig. 9. The resulting foot point trajectory, obtained using this cam-driven modification, is presented in Fig. 10.

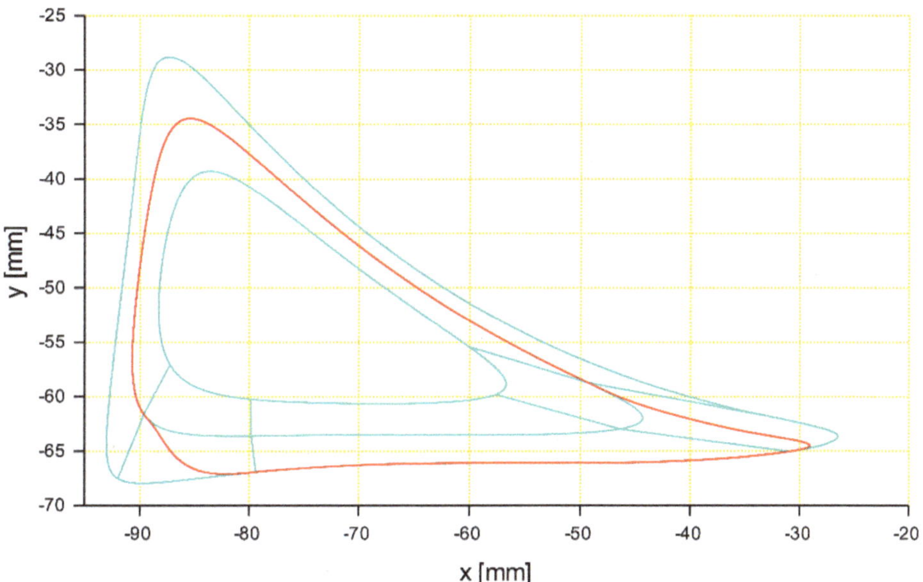

Fig. 10. Trajectory obtained from de modification of the crank of the Klann mechanism.

4 Conclusions

This work has explored the feasibility of enhancing two classic walking mechanisms –Theo Jansen and Klann– by replacing one of their rigid links with a combined cam-linkage mechanism. This approach introduces a variable-length element capable of selectively modifying the representative foot point trajectory while preserving the overall kinematic structure of the original mechanism.

Through systematic analysis of different link replacement options, the regions in which the representative foot point trajectory can be redefined were identified. Among the various possibilities, replacing the crank in both mechanisms proved to be the most effective strategy, offering the widest and most uniform range of trajectory variation.

To demonstrate the practical potential of the proposed methodology, two application examples were presented. The first, applied to the Theo Jansen mechanism, redefines the trajectory by imposing a specific elevated passage point. The second, applied to the Klann mechanism, transitions from the original foot path to an alternative extended stride along a targeted segment of the motion cycle. In both cases, a cam profile with C2 continuity was designed to achieve smooth, controllable variation.

These examples confirm that the proposed method allows for precise trajectory redefinition, enabling the generation of custom foot point paths within the predefined motion envelope. While the study focused on the Theo Jansen and Klann mechanisms, the methodology is generalizable and could be applied to other walking mechanisms, representing a promising direction for future research in adaptive robotic locomotion design.

References

1. Tenreiro Machado, J.A., Silva, M.: An overview of legged robots. In: International Symposium on Mathematical Methods in Engineering, pp. 1–40 (2006)
2. Nansai, S., Rojas, N., Elara, M.R., Sosa, R., Iwase, M.: On a Jansen leg with multiple gait patterns for reconfigurable walking platforms. Adv. Mech. Eng. **7**(3) (2015)
3. Kulandaidaasan Sheba, J., Elara, M.R., Martínez-García, E., Tan-Phuc. L.: Trajectory generation and stability analysis for reconfigurable klann mechanism based walking robot. Robotics **5**(3), 13 (2016)
4. Puig-Ortiz, J., Jordi Nebot, L., Zayas Figueras, E.E.: Análisis de los índices de prestaciones en mecanismos combinados de barras y levas. Actas del XXIII Congreso Nacional de Ingeniería Mecánica, Jaén (2021)
5. Soong, R.-C., Chang, S.-B.: Synthesis of function-generation mechanisms using variable length driving links. Mech. Mach. Theory **46**(11), 1696–1706 (2011). https://doi.org/10.1016/j.mechmachtheory.2011.06.011
6. Gatti, G., Mundo, D.: Optimal synthesis of six-bar cammed-linkages for exact rigid-body guidance. Mech. Mach. Theory **42**(9), 1069–1081 (2007). https://doi.org/10.1016/j.mechmachtheory.2006.09.006
7. Mundo, D., Liu, J.Y., Yan, H.S.: Optimal synthesis of cam-linkage mechanisms for precise path generation. J. Mech. Des. **128**(6), 1253–1260 (2006). https://doi.org/10.1115/1.2337317
8. Clos, D., Puig, J.: PAM, un programa de análisis de mecanismos planos de n grados de libertad enfocado a la docencia universitaria. Anales de Ingeniería Mecánica **15**(1), 757–765 (2004)

Automated Synthesis of Flexible Parallel Stages Using Screw Theory

Martín Pucheta[1,2](✉) and Alejandro Gallardo[1]

[1] Universidad Tecnológica Nacional, Facultad Regional Córdoba, Centro de Investigación en Informática para la Ingeniería (CIII), X5016ZAA Córdoba, Argentina
{mpucheta,agallardo}@frc.utn.edu.ar
[2] Consejo Nacional de Investigaciones Científicas y Técnicas (CONICET), Buenos Aires, Argentina

Abstract. Flexible parallel stages are used for high-precision positioning in various scientific, medical, and industrial applications. Screw Theory allows designers to synthesize this type of platform systematically. In this work, we address the problem of automatically locating the constraint beams for a flexible parallel platform starting from the desired three-dimensional motions, known as the freedom space, and additional spatial constraints. A new algorithm is proposed to compute the direction of the line screw that passes through a given point and is reciprocal to the desired freedom space. The points of the line screws to be analyzed in the resulting constraint space belong to the surface of the rigid body platform. The dimension of this space will indicate whether or not the mechanism can be physically implemented with beam flexures. This algorithm is used to synthesize a mechanism with one degree of freedom and another with three degrees of freedom.

Keywords: flexible parallel stages · Screw Theory · computational line geometry

1 Introduction

Flexible parallel stages (FPS) or platforms are used for high-precision positioning, laser light deflection, nanomanipulation robotics, and a variety of scientific, medical, and industrial applications. The Screw Theory formalism, due to its simplicity and mathematical soundness, is the most suitable to systematically design planar and three-dimensional FPSs with small deflections, to evaluate their vibration modes and frequencies [2], and to decouple the actuation increasing their controllability and precision [3]. One of the design stages that cannot be trivially automated is the conversion of constraint spaces into beams. Without computer assistance, the user must propose beams by following the well-known Blanding's Rules [5], which can become very tedious due to the complex three-dimensional geometry and/or combinatorial solutions. This conversion can be aided by predefined charts in Screw space atlases, such as those proposed by

O. F. Farías Fuentes et al. (Eds.): CIBIM 2024, *Proceedings of the XVI Ibero-American Congress of Mechanical Engineering*, pp. 429–440, 2026.
https://doi.org/10.1007/978-3-032-22823-9_31

Hopkins [1,4] followed by Blanding's Rules [5] or in an analytical way, as proposed by Su and Tari [6]. This last methodology, with a solid mathematical basis, is difficult to implement in an automated way.

Constraint-based design focuses on defining the motion of a mechanism by the position and orientation of its constraints. This approach was first introduced by Maxwell and formalized for precision machines by Blanding [4]. In this framework, slender beams act as ideal constraints, being stiff axially and flexible transversely, which allows simplified analysis for small displacements. A beam aligned with an axis restricts the motion along that axis while allowing transverse translations and rotations. Blanding's rules govern how constraints define allowable motions, notably stating that movement along a constraint line is prohibited, and that redundant constraints offer no added value. Degrees of freedom are described via "freedom lines" (axes of rotation), and Blanding's rule of complementary patterns states that every freedom line must intersect every constraint line, even if at infinity.

This work is the first to address the problem of automatically locating the constraint beams for a flexible parallel stage, given the desired three-dimensional motions and spatial restrictions. This geometric problem is very interesting because once it is solved, the designer will be able to explore a family of lines that satisfy the constraint space and focus on the tasks of selecting the most appropriate beams either by empirical rules or by computational optimization techniques.

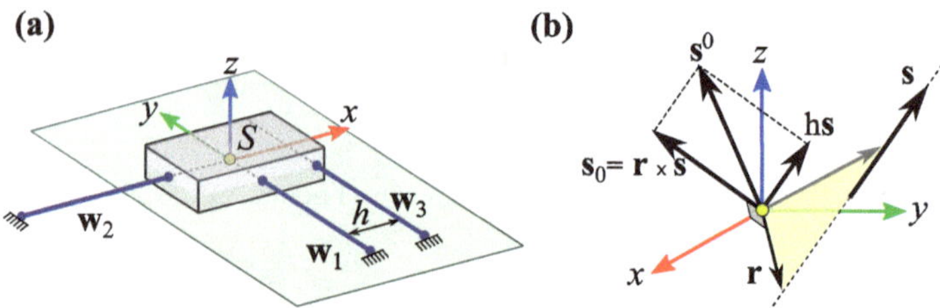

Fig. 1. (a) Motion of a flexible parallel stage constrained by three pure force wrenches; (b) Screw defined by a line $(\mathbf{s};\mathbf{s}_O)$ and pitch h.

The beams of a FPS can be represented as line screws, which are pure force wrenches; see Fig. 1(a). A line screw is defined by a point belonging to a line and by its direction vector. As initial data, the desired three-dimensional movements for the rigid body to be moved are expressed in terms of screws, which are used to define the desired freedom space. Additionally, spatial design restrictions can be considered, such as the attachment surface of the ends of the constraint beams. In terms of Screw Theory, the design problem of the FPS is to find the line screws that are reciprocal of the desired freedom space. A new algorithm is proposed to calculate the direction of the line screw that passes through a given point and is reciprocal to the freedom space. The points to be analyzed in the resulting space are those that belong to the lower surface of the rigid body, by constructing a constraint space made up only of line screws. The dimension of this space will indicate whether or not the mechanism can be physically implemented with beam flexures. This algorithm is applied to the synthesis of a mechanism with one degree of freedom and another with three degrees of freedom.

2 Screw Theory

A screw, denoted by \$, is a six-dimensional vector, composed of two three-dimensional vectors $\$ = (\mathbf{s}; \mathbf{s}^0)$. These vectors are shown in Fig. 1(b), where $\mathbf{s}$ is the direction vector of the screw axis and $\mathbf{s}^0$ is the sum of two vectors:

- $\mathbf{s}_0 = \mathbf{r} \times \mathbf{s}$ is the moment of the vector $\mathbf{s}$ with respect to the reference frame 0-xyz.
- $h\mathbf{s}$ is the product of the direction vector with the scalar h, called the *pitch* of the screw.

A screw is thus defined by three entities: (i) the direction vector s, (ii) the position vector r, and (iii) the scalar h called the pitch. When the pitch value is zero, the screw represents a straight line in space $(\mathbf{s}; \mathbf{s}_0)$, and this way of representing the straight line in space is defined as Plücker coordinates.

A screw can be used to represent the infinitesimal displacement (twist) of a body, through the combination of a translation along an axis and a rotation around the same axis. The twist is defined as

$$\mathbf{t} = \begin{bmatrix} \Delta\theta \\ \delta \end{bmatrix} = \begin{bmatrix} \Delta\theta \\ \mathbf{r} \times \Delta\theta + h\Delta\theta \end{bmatrix} \tag{1}$$

where $\Delta\theta$ is the rotational displacement vector pointing in the direction of the screw and δ is the linear or translational displacement, $\mathbf{r}$ is the location vector running from the origin to any point along the axis of the screw and h is the pitch of the screw which determines the relationship between rotation and displacement in motion. Force and moment, as well as rotation and translation, can be represented by a screw. The force/moment screw (wrench) is defined as:

$$\mathbf{w} = \begin{bmatrix} \mathbf{f} \\ \tau \end{bmatrix} = \begin{bmatrix} \mathbf{f} \\ \mathbf{c} \times \mathbf{f} + q\mathbf{f} \end{bmatrix} \tag{2}$$

where $\mathbf{f}$ is a force vector pointing in the direction of the screw and τ represents the moment of the screw about the origin of the coordinate system, $\mathbf{c}$ is the location vector running from the origin to any point along the axis of the screw and q is the pitch of the screw which determines the relationship between force and moment.

The work of a wrench $\mathbf{w}$ on a body with a twist $\mathbf{t}$ is defined by the reciprocal product of the wrench and the twist

$$\mathbf{w} \circ \mathbf{t} = \mathbf{f} \cdot \delta + \tau \cdot \Delta\theta \tag{3}$$

A twist is the reciprocal of a wrench when their reciprocal product is zero. Equation (3) can also be written as a product of vectors and matrices

$$\mathbf{w} \circ \mathbf{t} = \mathbf{w}^T \mathbf{Q} \, \mathbf{t} \tag{4}$$

where $\mathbf{Q}$ is the exchange matrix build as $\mathbf{Q} = \begin{bmatrix} \mathbf{O}_{3\times3} & \mathbf{I}_{3\times3} \\ \mathbf{I}_{3\times3} & \mathbf{O}_{3\times3} \end{bmatrix}$, where, $\mathbf{O}_{3\times3}$ and $\mathbf{I}_{3\times3}$ are, respectively, the 3×3 identity and zero matrices.

Blanding's first rule states that the motion of the body is perpendicular to the line of constraint, i.e., the work is zero. When the motion and the constraint are expressed in terms of screws, this means that they must be reciprocal. Furthermore, two screws are reciprocal if they intersect, either at a proper or improper point.

2.1 Freedom and Constraint Spaces

The motions that a body can undergo in space are represented by twists. The freedom space consists of all the linearly independent twists and is expressed in matrix form as

$$\mathbf{T} = \left[\mathbf{t}_1, \mathbf{t}_2, \ldots, \mathbf{t}_m\right], \tag{5}$$

where m is the dimension of the freedom space and the twists $\mathbf{t}_i$ are the basis vectors of the space of freedom. Thus, any allowed motion of the body can be expressed as a linear combination of the $\mathbf{t}_i$.

The constraint space consists of the linearly independent constraint wrenches that prevent certain body motions and, like the freedom space, is expressed in matrix form as

$$\mathbf{W} = \left[\mathbf{w}_1, \mathbf{w}_2, \ldots, \mathbf{w}_{6-m}\right], \tag{6}$$

The dimension of the constraint space depends on the dimension of the freedom space (or vice versa). Any constraint that can be expressed as a linear combination of the vectors $\mathbf{w}_i$ is a redundant constraint.

The definition of one screw being reciprocal to another one can be extended to systems of screws. A screw system $\mathbf{T}$ is reciprocal to another screw system $\mathbf{W}$ if the reciprocal product of each of its pair-wise taken screws is zero. For example, the constraint space $\mathbf{W}$ of body S in Fig. 1 is known, which consists of three pure force screws:

$$\mathbf{W} = \begin{bmatrix} 1 & 0 & 0 \\ 0 & 1 & 1 \\ 0 & 0 & 0 \\ 0 & 0 & 0 \\ 0 & 0 & 0 \\ 0 & 0 & h \end{bmatrix} \quad \rightarrow \quad \mathbf{T} = \text{null}(\mathbf{W}^T \mathbf{Q}) = \begin{bmatrix} 1 & 0 & 0 \\ 0 & 1 & 0 \\ 0 & 0 & 0 \\ 0 & 0 & 0 \\ 0 & 0 & 0 \\ 0 & 0 & 1 \end{bmatrix} \tag{7}$$

where, using the definition of the reciprocal product of Eq. (4), the freedom space $\mathbf{T}$ is the null space of $\mathbf{W}^T \mathbf{Q}$. The principles of constraint-based design can be expressed in the mathematical terms of Screw Theory.

3 Synthesis of Flexible Parallel Mechanisms

In the synthesis of parallel mechanisms, the starting point is the freedom space $\mathbf{T}$ and the dimensions of the rigid body of the mechanism to be moved. The objective is to determine the locations of the beam-like flexure elements that allow freedom space $\mathbf{T}$ and are connected to the rigid body.

Given a freedom space $\mathbf{T} = \left[\mathbf{t}_1, \mathbf{t}_2, \ldots, \mathbf{t}_m\right]_{6\times m}$, consisting of m linearly independent twists, and a pure-force screw $\mathbf{w} = \begin{bmatrix} \mathbf{w}_F \\ \mathbf{r} \times \mathbf{w}_F \end{bmatrix}$, which is a line screw that represents the axis of the beam-like flexure element. The force $\mathbf{w}$ belongs to the constraint space of $\mathbf{T}$ if the work of the constraint forces is null. In terms of screws, this is expressed by the nullity of the reciprocal product: $\mathbf{T} \circ \mathbf{w} = \mathbf{0}$. By expanding this equation, it is obtained

$$\mathbf{T}_\omega^T(\mathbf{r} \times \mathbf{w}_F) + \mathbf{T}_\delta^T \mathbf{w}_F = \mathbf{0} \tag{8}$$

where $\mathbf{T}_\omega$ is the $3 \times m$ matrix containing the direction vectors of the freedom space $\mathbf{T}$ and $\mathbf{T}_\delta$ is the $3 \times m$ matrix containing the translation vectors of $\mathbf{T}$. The vector product $(\mathbf{r} \times \mathbf{w}_F)$ can be replaced by a matrix product $(\tilde{\mathbf{r}}\mathbf{w}_F)$ by using the antisymmetric matrix $\tilde{\mathbf{r}} = \begin{bmatrix} 0 & -r_z & r_y \\ r_z & 0 & -r_x \\ -r_y & r_x & 0 \end{bmatrix}$ which is replaced in Eq. (8) to get $\mathbf{T}_\omega^T \tilde{\mathbf{r}}\, \mathbf{w}_F + \mathbf{T}_\delta^T \mathbf{w}_F = \mathbf{0}$ Then, by factoring out the line screw $\mathbf{w}_F$, we obtain

$$\begin{aligned} (\mathbf{T}_\omega^T \tilde{\mathbf{r}} + \mathbf{T}_\delta^T)\mathbf{w}_F &= \mathbf{0} \\ \mathbf{L}_{(m\times 3)} \mathbf{w}_F &= \mathbf{0} \end{aligned} \tag{9}$$

where $\mathbf{L}_{(m\times 3)}(\mathbf{T}, \mathbf{r})$ is the *Localization matrix* whose null space is the space of the direction vectors of $\mathbf{w}$ passing through point $\mathbf{r}$. The point $\mathbf{r}$ must be located within the space occupied by the rigid body of the mechanism. This ensures that the beam-type flexure element is correctly attached to the rigid body. Once the point $\mathbf{r}$ is defined, the direction vector $\mathbf{w}_F$ is obtained, and with these, the line screw $\mathbf{w}$ is built. The final step in defining whether the mechanism can be built with beam-type flexures is to find a set of line screws with dimensions of $6-m$. This step can be solved analytically or numerically with the new approach proposed here, which is algorithmically described in the following pseudocode:

Input : Freedom space $\mathbf{T} = [\mathbf{t}_1, \dots, \mathbf{t}_m] \in \mathbb{R}^{6\times m}$
Set of candidate points $\{\mathbf{r}_1, \mathbf{r}_2, \dots, \mathbf{r}_k\} \subset \mathbb{R}^3$ (within the rigid body)
Output: Set of line screws $\{\mathbf{w}_1, \dots, \mathbf{w}_n\}$ and their rank
Split $\mathbf{T}$ into rotational and translational parts:
$\mathbf{T}_\omega \in \mathbb{R}^{3\times m}$ (top 3 rows), $\mathbf{T}_\delta \in \mathbb{R}^{3\times m}$ (bottom 3 rows)
Initialize empty set $\mathbf{W} \leftarrow \emptyset$
foreach $\mathbf{r}_i \in \{\mathbf{r}_1, \dots, \mathbf{r}_k\}$ **do**
  Compute antisymmetric matrix $\tilde{\mathbf{r}}_i$:
  Compute linear constraint matrix: $\mathbf{L}_i = \mathbf{T}_\omega^T \tilde{\mathbf{r}}_i + \mathbf{T}_\delta^T$
  Compute null space of $\mathbf{L}_i$: solve $\mathbf{L}_i \mathbf{w}_F = \mathbf{0}$
  foreach *solution vector* $\mathbf{w}_F$ *in null space* **do**
    Construct constraint screw:

$$\mathbf{w} = \begin{bmatrix} \mathbf{w}_F \\ \mathbf{r}_i \times \mathbf{w}_F \end{bmatrix}$$

    Add $\mathbf{w}$ to set $\mathbf{W}$
  end
end
Compute rank of $\mathbf{W}$: $n = \text{rank}(\mathbf{W})$
if $n = 6 - m$ **then**
  Return: The mechanism can be built with beam-type flexures.
else
  Return: The mechanism cannot be built with beam-type flexures.
end

Algorithm 1: Numerical Synthesis of Parallel Mechanisms

4 Results

Next, the potential of the proposed algorithm will be illustrated with two examples.

4.1 One-Degree-of-Freedom Mechanism

When the freedom space $\mathbf{t} = [\mathbf{t}_\omega; \mathbf{t}_\delta]$ has a dimension of $m = 1$, the matrix $\mathbf{L}$ takes the form

$$\mathbf{L} = \mathbf{t}_\omega^T \tilde{\mathbf{r}} + \mathbf{t}_\delta^T. \tag{10}$$

Generally, the dimension of $\mathbf{L}$ will be $n = 1$, and the dimension of its null space will be $3-n$. This means that two linearly independent screws pass through any point $\mathbf{r}$. There is a special case when the degree of freedom is a rotation,

$$\mathbf{t}_R = \begin{bmatrix} \mathbf{t}_\omega \\ \mathbf{p}_1 \times \mathbf{t}_\omega \end{bmatrix} \tag{11}$$

and the matrix $\mathbf{L}$ is analyzed at the points on the rotation axis. Replacing Eq. (11) in (10) and working algebraically, it is obtained

$$\begin{aligned}\mathbf{L}_{(1\times3)} &= \mathbf{t}_{\omega}^{T}\tilde{\mathbf{p}} + (\mathbf{p}_1 \times \mathbf{t}_{\omega})^{T} \\ &= (-\mathbf{p} \times \mathbf{t}_{\omega})^{T} + (\mathbf{p}_1 \times \mathbf{t}_{\omega})^{T} \\ &= ((-\mathbf{p} + \mathbf{p}_1) \times \mathbf{t}_{\omega})^{T}\end{aligned} \tag{12}$$

where $\mathbf{p}$ and $\mathbf{p}_1$ are arbitrary points on the rotation axis, and their vector difference has the same direction as the rotation axis, that is, $-\mathbf{p} + \mathbf{p}_1 = \mathbf{t}_{\omega}$. Therefore, $\mathbf{L}_{(1\times3)} = \mathbf{0}$ will have dimension $n = 0$. This means that three linearly independent line screws pass through any point $\mathbf{r}$ located on the axis of rotation. Figure 2(a) illustrates a rigid body S for which the freedom space is to be obtained.

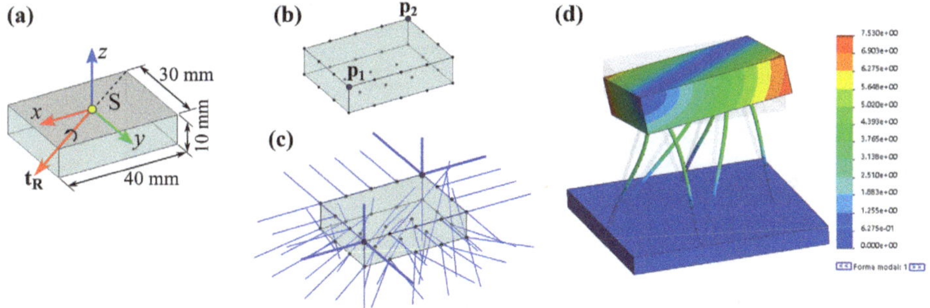

Fig. 2. Design of a 1-DOF parallel mechanism: (a) Dimensions of the rigid body about which a rotation $\mathbf{t}_R$ is desired. (b) Points of the body at which the ideal constraints will be analyzed.(b)-(c) Numerical solution for the synthesis, and (d) a FEM simulation example of chosen beams.

To solve the synthesis of the mechanism numerically, the null space of the matrix $\mathbf{L}_{(1\times3)}$ of Eq. (10) is determined for the set of points in Fig. 2(b). For graphical simplicity, points were selected only from the lateral and lower faces of body S, and not from all faces. The points are distributed on a 10 mm × 10 mm grid. Of all the points, only $\mathbf{p}_1$ and $\mathbf{p}_2$ are located on the axis of rotation. Figure 2(c) shows the set of line screws that belong to the constraint space of $\mathbf{t}_R$. Two line screws pass through each point, except for $\mathbf{p}_1$ and $\mathbf{p}_2$, where three are passing. This set of line screws has dimension 5. Therefore, the mechanism can be built with beam-type flexures.

This problem can also be solved analytically. The matrix $\mathbf{L}_{(1\times3)}$ in Eq. (10) depends on three variables $[r_x, r_y, r_z]$, and its null space will be determined using symbolic calculus in a Computer Algebra Software (CAS). To do this, it is necessary to assume two states for each variable: null and non-zero. The combination of these states results in 8 possible position vectors.

$$\begin{array}{llll} \mathbf{r}_1 = \begin{bmatrix} 0\,0\,0 \end{bmatrix}^T & \mathbf{r}_2 = \begin{bmatrix} r_x\,0\,0 \end{bmatrix}^T & \mathbf{r}_3 = \begin{bmatrix} 0\,r_y\,0 \end{bmatrix}^T & \mathbf{r}_4 = \begin{bmatrix} 0\,0\,r_z \end{bmatrix}^T \\ \mathbf{r}_5 = \begin{bmatrix} 0\,r_y\,r_z \end{bmatrix}^T & \mathbf{r}_6 = \begin{bmatrix} r_x\,0\,r_z \end{bmatrix}^T & \mathbf{r}_7 = \begin{bmatrix} r_x\,r_y\,0 \end{bmatrix}^T & \mathbf{r}_8 = \begin{bmatrix} r_x\,r_y\,r_Z \end{bmatrix}^T \end{array} \tag{13}$$

For each vector $\mathbf{r}_i$, at least two direction vectors will be obtained. Solving, we obtain the set of line screws $\mathbf{W}$ that belong to the constraint space of $\mathbf{t}_R$.

$$\mathbf{W} = \begin{bmatrix} 1 & 0\,0 & 0 & 0 & 0 \\ 0 & 1\,0 & 0 & 0 & 0 \\ 0 & 0\,1 & 0 & 0 & 0 \\ 1 & 0\,0 & 0 & 0 & 0 \\ 0 & 1\,0 & 0 & 0 & r_x \\ 1 & 0\,0 & 0 & 0 & -r_y \\ 0 & 1\,0 & 0 & 0 & 0 \\ 8/5 & 1\,0 & -r_z & (8r_z)/5 & 0 \\ 0 & 0\,1 & 0 & 0 & 0 \\ 8/5 & 1\,0 & -r_z & (8r_z)/5 & -(8r_y)/5 \\ -(8r_y)/(5r_z) & 0\,1 & r_y & -(8r_y)/5 & (8r_y^2)/(5r_z) \\ 8/5 & 1\,0 & -r_z & (8r_z)/5 & r_x \\ r_x/r_z & 0\,1 & 0 & 0 & 0 \\ 1 & 0\,0 & 0 & 0 & -r_y \\ 0 & 1\,0 & 0 & 0 & r_x \\ 8/5 & 1\,0 & -r_z & (8r_z)/5 & r_x-(8r_y)/5 \\ (5r_x-8r_y)/(5r_z) & 0\,1 & r_y & -(8r_y)/5 & -(r_y(5r_x-8r_y))/(5r_z) \end{bmatrix}^T \begin{matrix} \mathbf{r}_1 \\ \\ \\ \mathbf{r}_2 \\ \\ \mathbf{r}_3 \\ \\ \mathbf{r}_4 \\ \\ \mathbf{r}_5 \\ \\ \mathbf{r}_6 \\ \\ \mathbf{r}_7 \\ \\ \mathbf{r}_8 \\ \\ \end{matrix} \tag{14}$$

To the right of each line screw, the value of $\mathbf{r}_i$ with which it was calculated is indicated. The dimension of $\mathbf{W}$ is 5, indicating that the mechanism can be implemented physically.

4.2 Three-Degree-of-Freedom Mechanism

Given the 3-dimensional freedom space

$$\mathbf{T} = \begin{bmatrix} \begin{bmatrix} \mathbf{t}_{\omega 1} \\ \mathbf{t}_{\delta 1} \end{bmatrix} & \begin{bmatrix} \mathbf{t}_{\omega 2} \\ \mathbf{t}_{\delta 2} \end{bmatrix} & \begin{bmatrix} \mathbf{t}_{\omega 3} \\ \mathbf{t}_{\delta 3} \end{bmatrix} \end{bmatrix} \tag{15}$$

the matrix $\mathbf{L}$ takes the following form

$$\mathbf{L}_{(3\times 3)} = \begin{bmatrix} \mathbf{t}_{\omega 1}^T\ \tilde{\mathbf{p}} + \mathbf{t}_{\delta 1}^T \\ \mathbf{t}_{\omega 2}^T\ \tilde{\mathbf{p}} + \mathbf{t}_{\delta 2}^T \\ \mathbf{t}_{\omega 3}^T\ \tilde{\mathbf{p}} + \mathbf{t}_{\delta 3}^T \end{bmatrix} \tag{16}$$

The dimension of matrix $\mathbf{L}$ will be up to $n = 3$, and the dimension of its null space will be $3 - n$. From three degrees of freedom onwards, the row space of $\mathbf{L}$ can have a dimension equal to or greater than the column space of $\mathbf{L}$, which implies that its null space will consist solely of the null vector. In other words, there will be points in space where there are no line screws belonging to the constraint space of $\mathbf{T}$.

Among several possibilities, a particular 3-dof case is analyzed, whose freedom space is

$$\mathbf{T} = \begin{bmatrix} \mathbf{t}_{\mathrm{H}x} \ \mathbf{t}_{\mathrm{H}y} \ \mathbf{t}_{\mathrm{H}z} \end{bmatrix} = \begin{bmatrix} 1 & 0 & 0 & 1 & 0 & 0 \\ 0 & 1 & 0 & 0 & 1 & 0 \\ 0 & 0 & 1 & 0 & 0 & -1 \end{bmatrix}^T \tag{17}$$

where $\mathbf{t}_{\mathrm{H}i}$ are screw motions in direction $i \in \{x, y, z\}$. These movements are the desired ones on platform S, see Fig. 3(a).

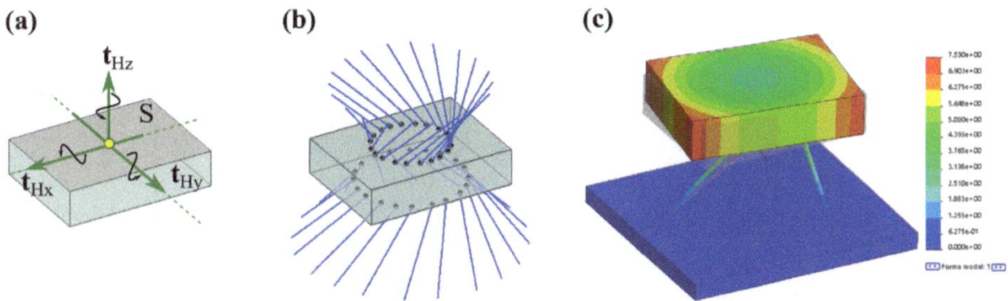

Fig. 3. Design of a 3-DOF parallel mechanism: (a) Freedom space, 3 mutually orthogonal screw motions $\mathbf{t}_{\mathrm{H}i}$. (b)-(c) Numerical solution for the synthesis, and (d) a FEM simulation example.

In this case, the matrix $\mathbf{L}$ takes the following form

$$\mathbf{L} = \begin{bmatrix} 1 & -r_z & r_y \\ r_z & 1 & -r_x \\ -r_y & r_x & -1 \end{bmatrix} \tag{18}$$

where $[r_x, r_y, r_z]$ are the coordinates of a generic point $\mathbf{p}$. For the numerical synthesis, a 10 mm × 10 mm grid is drawn on the faces of the rigid body, and at each point, the null space of the matrix $\mathbf{L}$ in Eq. (18) is determined. Figure 3(a) shows the set of line screws belonging to the constraint space. The dimension of this set of lines is 3; therefore, the mechanism can be constructed.

In the analytical synthesis, the determinant of the matrix $\mathbf{L}$ must be analyzed. For this example, the determinant of Eq. (18) is analyzed.

$$\det(\mathbf{L}) = r_x^2 + r_y^2 - r_z^2 - 1 \tag{19}$$

The $\det(\mathbf{L})$ must be zero; this ensures that the null space of $\mathbf{L}$ is at least 1-dimensional:

$$\begin{aligned} \det(\mathbf{L}) &= 0 \\ r_x^2 + r_y^2 - r_z^2 - 1 &= 0 \\ r_x^2 + r_y^2 - r_z^2 &= 1 \end{aligned} \tag{20}$$

The points whose coordinates satisfy Eq. (20) are located on a one-sheet hyperboloid; this can be seen in Fig. 3(b) as a graphical result of the numerical synthesis. Once the feasible solution space has been defined, it must be determined three points through which three linearly independent line screws pass. In this example, the solution is simple: three coplanar points must be chosen, each of which is located on the one-sheet hyperboloid. However, not all cases allow for a straightforward analysis. Therefore, once the determinant of **L** has been calculated, it must be evaluated for each of the position vectors in Eq. (13) and set equal to zero. The relationship between the coordinates of the points obtained will be substituted into the matrix **L**, and its null space will then be determined. The line screws obtained for this case are presented in Eq. (21), where two constants $a = r_z(-r_y^2 + r_z^2 + 1)^{1/2}$, and $b = (-r_y^2 + r_z^2 + 1)^{1/2}$ were defined for compactness.

$$\mathbf{W} = \left[\begin{array}{cccccc}
0 & -1 & 1 & 0 & 1 & 1 \\
0 & 1 & 1 & 0 & -1 & 1 \\
1 & 0 & 1 & -1 & 0 & 1 \\
-1 & 0 & 1 & 1 & 0 & 1 \\
-\frac{1}{(r_z^2+1)^{1/2}} & \frac{r_z}{(r_z^2+1)^{1/2}} & 1 & \frac{1}{(r_z^2+1)^{1/2}} & -\frac{r_z}{(r_z^2+1)^{1/2}} & 1 \\
\frac{1}{(r_z^2+1)^{1/2}} & -\frac{r_z}{(r_z^2+1)^{1/2}} & 1 & -\frac{1}{(r_z^2+1)^{1/2}} & \frac{r_z}{(r_z^2+1)^{1/2}} & 1 \\
-\frac{1}{r_y} & \frac{(r_y^2-1)^{1/2}}{r_y} & 1 & \frac{1}{r_y} & -\frac{(r_y^2-1)^{1/2}}{r_y} & 1 \\
-\frac{1}{r_y} & -\frac{(r_y^2-1)^{1/2}}{r_y} & 1 & \frac{1}{r_y} & \frac{(r_y^2-1)^{1/2}}{r_y} & 1 \\
\frac{r_z}{(r_z^2+1)^{1/2}} & \frac{1}{(r_z^2+1)^{1/2}} & 1 & -\frac{r_z}{(r_z^2+1)^{1/2}} & -\frac{1}{(r_z^2+1)^{1/2}} & 1 \\
-\frac{r_z}{(r_z^2+1)^{1/2}} & -\frac{1}{(r_z^2+1)^{1/2}} & 1 & \frac{r_z}{(r_z^2+1)^{1/2}} & \frac{1}{(r_z^2+1)^{1/2}} & 1 \\
\frac{(r_x^2-1)^{1/2}}{r_x} & \frac{1}{r_x} & 1 & -\frac{(r_x^2-1)^{1/2}}{r_x} & -\frac{1}{r_x} & 1 \\
-\frac{(r_x^2-1)^{1/2}}{r_x} & \frac{1}{r_x} & 1 & \frac{(r_x^2-1)^{1/2}}{r_x} & -\frac{1}{r_x} & 1 \\
-r_y & -(1-r_y^2)^{1/2} & 1 & r_y & (1-r_y^2)^{1/2} & 1 \\
-r_y & (1-r_y^2)^{1/2} & 1 & r_y & -(1-r_y^2)^{1/2} & 1 \\
(1-r_x^2)^{1/2} & r_x & 1 & -(1-r_x^2)^{1/2} & -r_x & 1 \\
-(1-r_x^2)^{1/2} & r_x & 1 & (1-r_x^2)^{1/2} & -r_x & 1 \\
-\frac{r_y+a}{r_z^2+1} & \frac{r_y r_z-b}{r_z^2+1} & 1 & \frac{r_y+a}{r_z^2+1} & -\frac{r_y r_z-b}{r_z^2+1} & 1 \\
-\frac{r_y-a}{r_z^2+1} & \frac{r_y r_z+b}{r_z^2+1} & 1 & \frac{r_y-a}{r_z^2+1} & -\frac{r_y r_z+b}{r_z^2+1} & 1
\end{array}\right]^T
\begin{array}{c}
\mathbf{r}_2 \\ \\ \mathbf{r}_3 \\ \\ \\ \\ \mathbf{r}_5 \\ \\ \\ \mathbf{r}_6 \\ \\ \\ \\ \mathbf{r}_7 \\ \\ \\ \mathbf{r}_8 \\ \\
\end{array} \qquad (21)$$

In this set of screws shown in **W**, the results for $\mathbf{r}_1$ and $\mathbf{r}_4$ are not included because there is no solution for those vectors. The dimension of **W** is 3, which confirms the result of the numerical synthesis.

5 Conclusions

Starting from the desired freedom space in a body, the constraint space consisting of beam-type flexures is determined. This was achieved by implementing Blanding's rules in terms of screws and interpreting the beam flexures as screw lines.

The proposed algorithm can be solved analytically (symbolically) and numerically. This new algorithm was implemented in the synthesis of a mechanism with one rotational degree of freedom and another mechanism with three degrees of freedom with screw motion. In the synthesis, the feasible space of beam flexures was obtained, allowing the construction of the mechanisms. These results are the first key steps toward achieving automatic design of parallel flexible mechanisms. Therefore, the algorithm can be included in graphical interfaces for computer-aided engineering systems.

Acknowledgment. The authors acknowledge the financial support from Universidad Tecnológica Nacional, Argentina, through project PID-UTN AMTCCO0008723TC.

References

1. Archer, N.C., Hopkins, J.B.: Analysis and synthesis of interconnected hybrid mechanisms using freedom and constraint topologies (FACT). Mech. Mach. Theory **200**, 105722 (2024). https://doi.org/10.1016/j.mechmachtheory.2024.105722
2. Pucheta, M., Gallardo, A.G.: Modal analysis of a 2R flexible platform using screw theory. In: Vizán Idoipe, A., García Prada, J.C. (eds.) IACME 2022, Proceedings of the XV Ibero-American Congress of Mechanical Engineering, pp. 50–56. Springer, Cham (2023). https://doi.org/10.1007/978-3-031-38563-6_8
3. Gallardo, A.G., Pucheta, M.A.: Synthesis of parallel flexure stages with decoupled actuators using sum, intersection, and difference of screw systems. Mech. Mach. Theory **192**, 105526 (2024). https://doi.org/10.1016/j.mechmachtheory.2023.105526
4. Hopkins, J.: Design of flexure-based motion stages for mechatronic systems via freedom, actuation and constraints topologies (FACT). PhD thesis, Massachusetts Institute of Technology, Massachusetts, U.S.A. (2010). http://hdl.handle.net/1721.1/62511
5. Blanding, D.L.: Exact Constraint: Machine Design Using Kinematic Principles. ASME Press, New York (1999)
6. Su, H., Tari, H.: On line screw systems and their application to flexure synthesis. J. Mech. Robot. **3**(1), 011009 (2011). https://doi.org/10.1115/1.4003078

Assessing the Use Potential of Calcined Volcanic Rock: An Experimental Approach to Strength Optimization and Mix Ratio Design

Pablo Cuello[1], Siva Avudaiappan[2], Cristian Canales[3](✉), Luis Felipe Montoya[1], María Gutierrez-Senepa[4], and Ramon Arrue[5]

[1] Department of Chemical Engineering, Universidad de Concepción, Bio Bio, Chile
pcuello2017@udec.cl
[2] Department of Construction Sciences, Universidad Tecnológica Metropolitana, Santiago, Chile
s.avudaiappan@utem.cl
[3] Department of Mechanical Engineering, Universidad de Concepción, Bio Bio, Chile
cristcanales@udec.cl
[4] Department of Metallurgical Engineering, Universidad de Concepción, Bio Bio, Chile
mariavgutierrez@udec.cl
[5] Department of Biological and Chemical Sciences, Universidad San Sebastián, Santiago, Chile
ramon.arrue@uss.cl

Abstract. Due to rapid urbanization and industrialization, the amount of natural resource exploration is increasing and causing carbon footprint, posing a significant challenge to environmental issues. For this effective management of natural resources and maintaining optimized utilization of natural resources, it is proposed to use calcined volcanic ash as a replacement for cement particles as an alternative material for construction industries. The study focuses on the mechanical (compressive and flexural) and thermal properties of concrete made from calcined volcanic ash as a replacement for cement particles at a ratio of 0% to 30% in 10% increments. Replacement of calcined volcanic ash by 30% cement has shown promising results that could improve the mechanical properties. Accordingly, to ensure the influential parameter of mechanical properties using calcined volcanic ash, microstructural investigations such as SEM (Scanning Electron Microscopy) and EDS (Energy Dispersive Spectroscopy) were conducted on volcanic rock samples. When compared to conventional concrete, it exhibits similar mechanical properties. The result of these results opens a new era of using calcined volcanic ash as a replacement for particles in cement, it also contributes to reducing the carbon footprint by using alternative materials in concrete production.

Keywords: Calcined cement · alternative construction materials · mechanical properties · thermal properties

O. F. Farías Fuentes et al. (Eds.): CIBIM 2024, *Proceedings of the XVI Ibero-American Congress of Mechanical Engineering*, pp. 441–452, 2026.
https://doi.org/10.1007/978-3-032-22823-9_32

1 Introduction

The production of Portland cement, an essential component in modern construction and the second most consumed material globally after water [1, 2], has reached the figure of 4.1 billion metric tons per year. This material, the basis of concrete, has been fundamental for the development of infrastructures and buildings around the world. However, its energy-intensive production and the consequent emission of carbon dioxide (CO2) have generated concern in the scientific community and in society in general [1]. Although concrete is considered more sustainable than other building materials such as glass and steel, its high consumption has eroded this advantage, making reducing its environmental impact an urgent challenge [1]. CO2 emissions associated with cement production account for approximately 8.6% of all anthropogenic emissions of this greenhouse gas, underscoring the need to look for more sustainable alternatives [3, 4].

The mechanical properties and durability of concrete are crucial indicators of its performance, and the microstructure plays a vital role in determining its overall performance [5]. Building materials have a very important influence on energy consumption. In a traditional building heat loss system, 70% of the building's heat loss is due to the surrounding structure. Therefore, improving the thermal performance of the building envelope is critical to achieving energy conservation. Among them, the rational use of heat-insulating materials is one of the most effective ways to improve energy utilization and conservation [6].

The search for solutions to mitigate the environmental impact of concrete has focused on various aspects, from the optimization of production processes to the modification of its composition. In particular, the partial replacement of Portland cement with supplementary cementitious materials has proven to be a promising strategy [7]. Previous research has explored the use of materials such as limestone, fly ash, blast furnace slag, and even calcined clay as partial cement substitutes, achieving significant reductions in CO2 emissions [8–16]. These materials, in addition to decreasing the demand for Portland cement, can improve the mechanical properties and durability of concrete, which contributes to the efficiency and sustainability of construction.

In this context, volcanic materials, such as calcined volcanic rock, have captured the attention of researchers and construction professionals [17, 18]. These materials, originating from the cooling and solidification of magma expelled during volcanic eruptions [19], have unique characteristics that make them especially attractive for use in construction. Their chemical composition, rich in silica and alumina, gives them pozzolanic properties, i.e. the ability to react with the free lime of the cement and form compounds that improve the strength and durability of concrete [21]. In addition, the availability of these materials in regions with volcanic activity makes them a local and sustainable option for cement and concrete production.

Chile, a country with vast volcanic activity and extensive deposits of volcanic rock, presents itself as a natural laboratory for the research and development of sustainable building materials based on these resources [20]. The presence of natural pozzolans, a by-product of past volcanic eruptions, offers a unique opportunity to harness this potential and contribute to the reduction of the construction industry's carbon footprint. The partial replacement of Portland cement with volcanic pozzolans not only represents a strategy to decrease CO2 emissions, but can also improve the properties of concrete, such as its compressive strength, bending and durability.

The present study focuses on evaluating the potential of hardened calcined volcanic rock as a partial substitute for cement in concrete mixtures. Through a rigorous experimental approach, the mechanical (compressive and flexural strength) and thermal properties of concretes with different proportions of cement substitution for calcined volcanic rock (0%, 10%, 20% and 30%) will be analyzed. The main objective is to determine the optimal substitution ratio that allows maximizing the performance of concrete in terms of strength, durability and thermal properties, while minimizing the environmental impact associated with cement production. The results of this research will provide valuable information for the construction industry, promoting the use of local and sustainable materials, and contributing to a more environmentally friendly future.

2 Materials and Methods

2.1 Volcanic Rock

Volcanic rocks for cement replacement were obtained by manual collection in the Andes mountain range area of the Araucanía region. In the preparation of the volcanic rock powder for use as a partial replacement for cement, it was subjected to a calcination process in a kiln at a temperature of 1000 °C, this calcination process aimed to improve the pozzolanic properties of the material and ensure its adequate chemical reactivity for the mixing process with the concrete components. Figure 1 shows scanning electron microscopy (SEM) analysis in order to investigate the microstructural characteristics of the material. Heterogeneous crystalline morphology and varied particle sizes are observed. Additionally, the mapping indicates a strong presence of elements such as oxygen, aluminum and silica, as also corroborated by Table 1 and Fig. 2 where the results of the X-ray energy dispersion (EDS) analysis are shown.

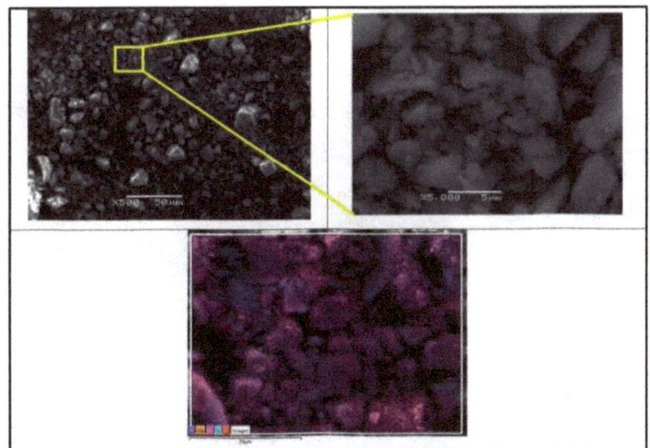

Fig. 1. SEM and mapping of calcined and pulverized volcanic rock.

Table 1. Chemical composition in percentage of volcanic rock.

Element	% weight	% atomic
Or	50.4	64.39
On	3.16	2.81
To the	7.78	5.9
Yes	33.12	24.11
K	3.26	1.70
Ca	0.84	0.43
Fe	1.05	0.38
Ti	0.11	0.05
Mg	0.28	0.24
Total	100	100

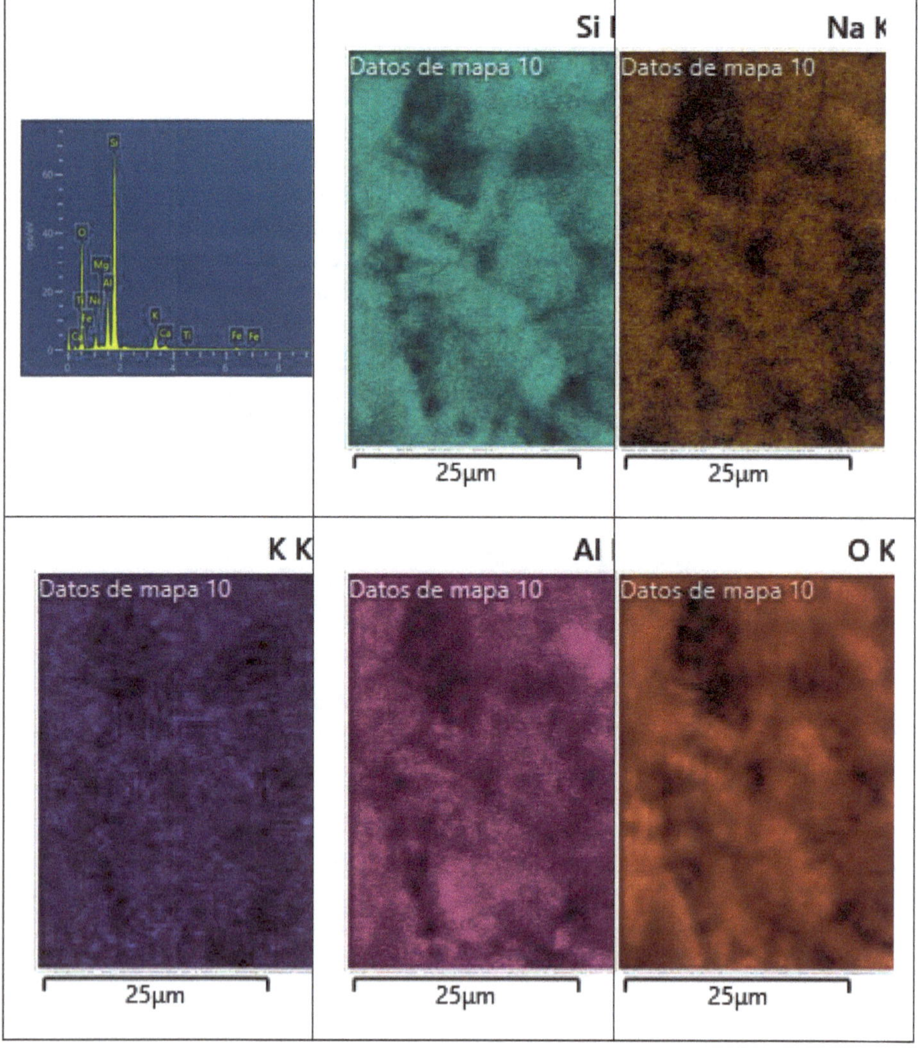

Fig. 2. EDS Images of Volcanic Rock Sample.

2.2 Fine Aggregate and Cement

The binder used in this study is a commercial product of the company Cementos Bio Bio (CBB), obtained by co-grinding clinker, pozzolan and gypsum. The product Cbb Cementos Especial is a cement obtained by co-grinding clinker, pozzolan and gypsum, current grade in accordance with the NCh 148 of68 standard [22]. The aggregate used was generic commercial fine sand. The physical properties are presented in Table 2.

Table 2. Physical properties of fine aggregate

Fine Aggregate Properties	Value
Water absorption	1,92%
Density	2663,9 kg/m^3
Material Content fine	1,8%

2.3 Mixing Ratio

This study aimed to investigate the mechanical performance of concrete based on a partial replacement of cement by calcined volcanic rock. Samples were prepared with different concentrations of pozzolanic cement replacement as a percentage of weight, in addition to a standard concentration. The proportion of materials used for concrete mixing is provided in Table 3.

Table 3. Proportions of mixed materials.

Naming of the mixture	Water/cement ratio	Cement/sand ratio	Percentage of cement replacement by volcanic rock (by weight)
Control	0,5	0,5	0%
10RV90			10%
20RV80			20%
30RV70			30%

The concrete mixing was done with the rotary mixing machine available in the laboratory. The mixing procedure of the concrete samples for this study was carried out strictly following the guidelines established by the ASTM C192 standard [23], thus ensuring the standardization and reproducibility of the results.

After being placed into the metallic prismatic molds, the specimens were left in the laboratory under ambient conditions for 24 h to harden. Following demolding, the samples were cured in water at room temperature (20 ± 2 °C) and a pH of 12 (±0.5) for 7 and 28 days prior to testing (Fig. 3).

Fig. 3. 10RV90 Samples

2.4 Test Methods

Physical Tests. To evaluate the consistency property of the mixtures, the concrete displacement test was performed with a flow rate according to the specifications of the ASTM C230 standard [24]. In addition, the density measurements of the mixtures were carried out, which in turn allowed the measurement of the percentage of water absorption and percentage of porosity of the samples.

Mechanical Properties Testing. To evaluate the influence of volcanic rock concentration on the mechanical properties of concrete at 7 and 28 days, mechanical tests were performed on flexural strength and prismatic samples of 40 × 40 × 160 mm and cubic samples of 50 × 50 × 50 mm. The bending tests were carried out in accordance with ASTM C78 [25], and the compression tests were carried out in accordance with ASTM C109 [26]. To determine the strengths, the averages of three samples were calculated in the case of bending and compression, all of them obtained from different mixtures of concrete with volcanic rock and curing times.

Thermal Conductivity Test. Thermal conductivity was measured using a model thermal property analyzer adopting the transient line heat source method in accordance with ASTM D5334 [27]. The instrument used (Fig. 4) is an advanced portable meter with a unique selection of model TLS-50 transient thermal conductivity sensors.

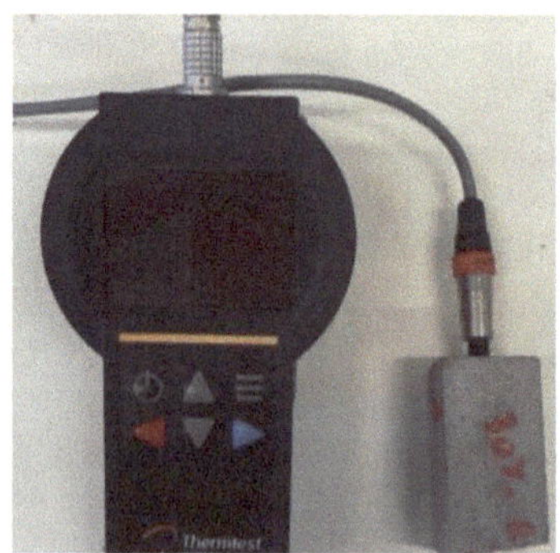

Fig. 4. TLS-50 Equipment

3 Results and Discussion

3.1 Workability and Physical Properties

Figure 5 shows the results of the concrete flow test. A higher flow indicates greater flowability, suggesting that the concrete spreads more easily, while a lower flow indicates a stiffer consistency.

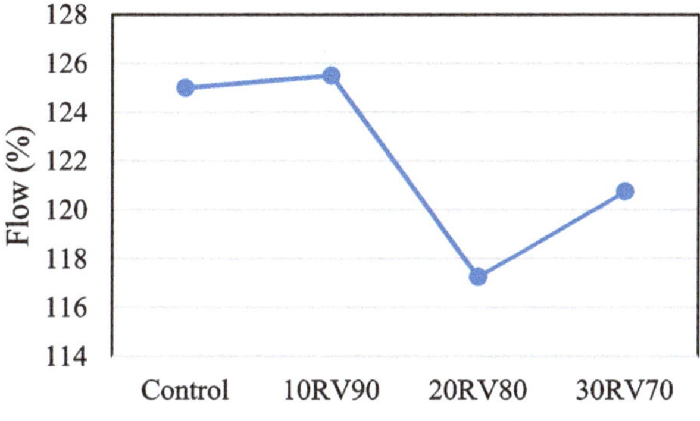

Fig. 5. Displacement test results.

These flow values indicate a significant degree of fluidity or workability of the concrete, suggesting that it was extended considerably during the test. In terms of practical applications, the results indicate that calcined volcanic rock has the potential to improve concrete workability at low replacement levels, but increasing the replacement percentage can decrease this workability.

Table 4. Physical properties of mixtures.

Mixture	Density (kg/m^3)	Absorption %	Porosity %
Control	2091.7	10.6	22.1
10RV90	2069.3	11.0	22.8
20RV80	2042.3	11.2	22.9
30RV70	2029.2	11.4	23.2

Table 4 shows the results of density, absorption and porosity tests in mixtures of concrete with different levels of cement replacement by volcanic rock dust.

A noticeable decrease in density is observed as the replacement percentage increases, suggesting that the inclusion of volcanic rock reduces the density of concrete. This reduction may be due to the lower density of volcanic rock compared to cement. On the other hand, water absorption and porosity increase significantly with higher replacement percentages. This increase in absorption and porosity indicates that mixtures with a higher content of volcanic rock are more likely to absorb water and have a more porous or less compact structure. These changes in absorption and porosity can have negative implications in terms of concrete durability, as more porous structures are more susceptible to deterioration processes such as freeze/thaw cycles and chemical attack.

3.2 Mechanical Testing

Flexion. Figure 6 shows the summary of the results obtained from the bending break tests, showing the average of three values for the tests at the ages of 7 and 28 days for the three different percentages of cement replacements, as well as the standard sample.

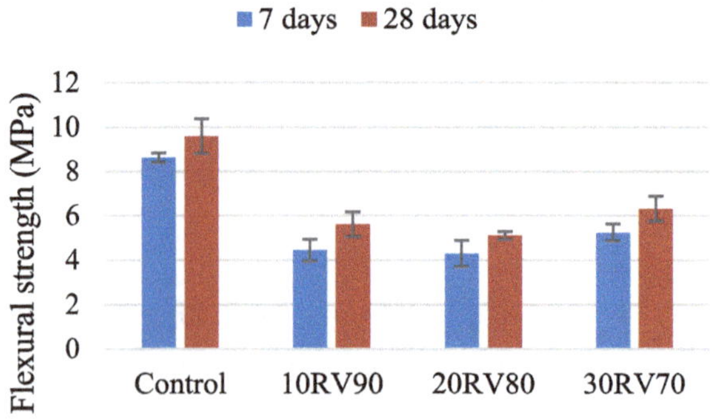

Fig. 6. Flexural Strength.

Analysis of the bending results reveals a clear trend in decreasing concrete strength by partially replacing cement with calcined and pulverized volcanic rock. In control samples, without any replacement, flexural strength of approximately 8 MPa is observed at 7 days, increasing to around 10 MPa at 28 days. This behavior is expected, as cement hydration continues to increase the strength of concrete over time.

In contrast, samples with 10% volcanic rock (10RV90) show significantly resistance reaching approximately 4.5 MPa at 7 days and about 6 MPa at 28 days. Similarly, samples with 20% replacement (20RV80) exhibit a resistance of about 5 MPa at 7 days and approximately 6.5 MPa at 28 days. These results indicate that the partial replacement of cement with volcanic rock reduces the initial flexural strength of concrete. However, samples with 30% replacement (30RV70) show a smaller decrease in strength compared to the other replacement mixtures. At 7 days, these samples reach about 5 MPa, and about 6.5 MPa at 28 days. Although these strengths are still lower than those of the control, the smaller reduction suggests that the calcined volcanic rock could be providing certain pozzolanic or filling benefits that improve strength as its percentage increases.

This behavior suggests that there is potential to optimize the percentage of cement replacement by calcined volcanic rock to maximize the strength of concrete. The results of 30% replacement indicate the possibility that higher percentages of replacement could offer better long-term strength.

Compression. The maximum resistance that a substance withstands under a compressive or crushing force before breaking or fracturing is called compressive strength. Figure 7 shows the summary of the results obtained from the compression break tests.

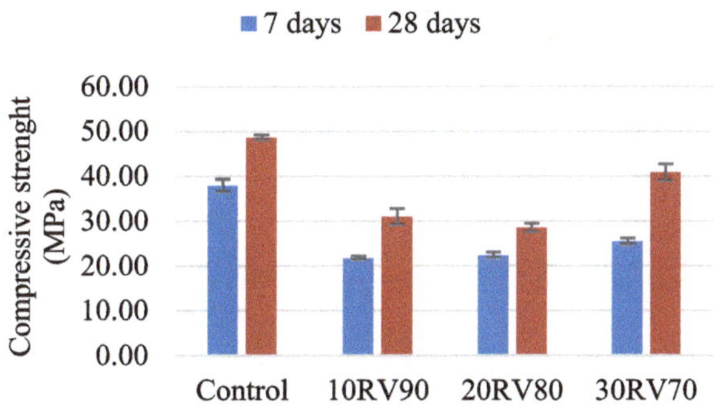

Fig. 7. Compressive strength

Analysis of the compression results reveals that the incorporation of calcined volcanic rock dust as a partial cement replacement significantly affects the strength of the concrete. The control samples, which do not contain volcanic rock, show a compressive strength of about 38 MPa at 7 days, increasing to about 50 MPa at 28 days.

In comparison, samples with 10% cement replacement for volcanic rock exhibit a markedly lower compressive strength, with approximately 21 MPa at 7 days and about 30 MPa at 28 days. Samples with 20% replacement also show a significant decrease, with resistances of approximately 23 MPa at 7 days and 29 MPa at 28 days. These results indicate that the partial substitution of cement by volcanic rock reduces the compressive strength of concrete. However, samples with 30% replacement show a smaller decrease in strength compared to the other replacement percentages. At 7 days, these samples have a resistance of approximately 27 MPa and reach about 41 MPa at 28 days. Although these strengths are still lower than those of the control, the smaller reduction observed suggests that calcined volcanic rock could provide benefits that improve strength as its percentage increases.

These findings suggest that although replacing cement with calcined volcanic rock powder generally decreases compressive strength, a 30% replacement appears to be more viable in terms of maintaining acceptable mechanical properties. This indicates the possibility that investigating replacement percentages greater than 30% can optimize compressive strength.

Thermal conductivity. Figure 8 shows the thermal conductivity results obtained by the TLS-50 at a current of 250 mA and a measurement interval of 60 s on samples of 28 days of curing.

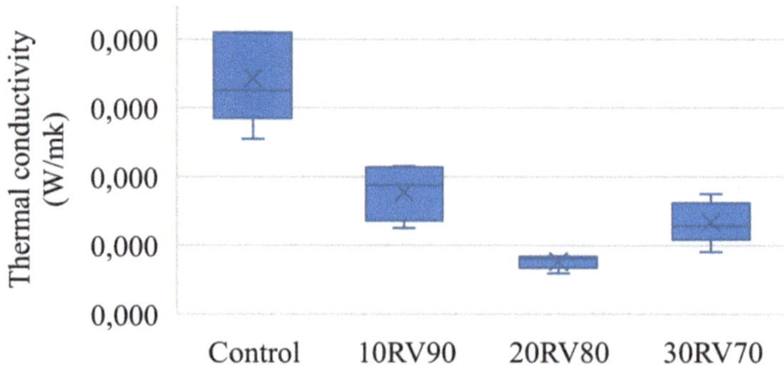

Fig. 8. Thermal conductivity at 28 days.

The results reveal that the incorporation of calcined volcanic rock as a partial replacement of cement produces a decrease in the thermal conductivity values in the concrete samples. This decrease is observed until a replacement of 30% where it increases slightly. The decrease in conductivity values is in accordance with the decrease in the density of the samples, on the other hand, in the case of a greater replacement (30%) it can be assumed that, due to an increase in the number of more ordered molecular structures, this due to the crystalline nature of the volcanic rock [28], the vibrations of the molecules can be transmitted more efficiently through the material allowing a better transfer of thermal energy [29].

To verify the impact of the use of volcanic rock, an Anova analysis of variance was carried out, which showed that there is a statistically significant difference between the replacement levels with a significance of 95%.

4 Conclusions

This study analyzed the influence of calcined volcanic rock dust on mechanical strength properties by a partial cement replacement to produce eco-friendly concrete. The most relevant results are presented below.

Concrete density was negatively affected by larger cement replacements, thus producing a lighter concrete. On the other hand, the values of water absorption and porosity increased. Although there is a reduction in flexural and compressive strengths, an increase in the values of both mechanical properties was observed with the highest percentage of cement replacement (30%), and a study was recommended for larger replacements. At low levels, the incorporation of volcanic rock can decrease the thermal conductivity of the material. From 30% onwards, an increase is perceived, which can be assumed by the more ordered molecular structure of the material.

The research demonstrated the feasibility of using calcined volcanic rock as a sustainable material in construction. However, this study recommends further studies to determine the effect of volcanic rock originating from different sources on other properties of concrete, as well as the viability of higher replacement values.

References

1. Adesina, A.: Recent advances in the concrete industry to reduce its carbon dioxide emissions. Environ. Challenges **1**, 100004 (2020)
2. Assi, L.N., Alsalman, A., Kareem, R.S., Carter, K., Ziehl, P., Alhamadani, Y.: Why sustainable concrete cannot penetrate concrete markets. J. Build. Eng. **91**, 109487 (2024)
3. Miller, S.A., Horvath, A., Monteiro, P.J.M.: Readily implementable techniques can cut annual CO_2 emissions from the production of concrete by over 20%. Environ. Res. Lett. **11**(7) (2016)
4. Kareem, R.S., Assi, L.N., Alsalman, A., Al-Manea, A.: Effect of supplementary cementitious materials on RC concrete piles. In: AIP Conference Proceedings (2021)
5. Huang, D., Feng, Y., Xia, Q., Tian, J., Li, X.: Research on mechanical properties and durability of early frozen concrete: A review. Constr. Build. Mater. **425**, 135988 (2024)
6. Zhang, H., Yang, J., Wu, H., Fu, P., Liu, Y., Yang, W.: Dynamic thermal performance of ultra-light and thermal-insulative aerogel foamed concrete for building energy efficiency. Solar Energy **204** (2020)
7. McLellan, B.C., Williams, R.P., Lay, J., Van Riessen, A., Corder, G.D.: Costs and carbon emissions for geopolymer pastes in comparison to ordinary Portland cement. J. Cleaner Prod. **19**(9–10) (2011)
8. Jungclaus, M.A., Williams, S.L., Arehart, J.H., Srubar, W.V.: Whole-life carbon emissions of concrete mixtures considering maximum CO_2 sequestration via carbonation. Resour. Conserv. Recycl. **206**, 107605 (2024)
9. Sinkhonde, D., Bezabih, T.: On the computational evaluation of carbon dioxide emissions of concrete mixes incorporating waste materials: a strength-based approach. Cleaner Waste Syst. **8**, 100149 (2024)
10. Mosaberpanah, M.A., Umar, S.A.: Utilizing rice husk ash as supplement to cementitious materials on performance of ultra-high-performance concrete: a review. Mater. Today Sustain. 7–8 (2020)
11. Liew, J.Y.R., Xiong, M.-X., Lai, B.-L.: Special considerations for high strength materials. In: Liew, J.Y.R., Xiong, M.-X., Lai, B.-L. (eds.) Design of Steel-Concrete Composite Structures Using High-Strength Materials, pp. 125–142. Woodhead Publishing, Cambridge (2021)
12. Tan, Y., Wu, C., Yu, H., Li, Y., Wen, J.: Review of reactive magnesia-based cementitious materials: current developments and potential applicability. J. Build. Eng. **40** (2021)
13. Zhou, H., Basarir, H., Poulet, T., Li, W., Kleiv, R.A., Karrech, A.: Life cycle assessment of recycling copper slags as cement replacement material in mine backfill. Resour. Conserv. Recycl. **205**, 107591 (2024)
14. Khankhaje, E., Kim, T., Jang, H., Kim, C.-S., Kim, J., Rafieizonooz, M.: A review of utilization of industrial waste materials as cement replacement in pervious concrete: An alternative approach to sustainable pervious concrete production. Heliyon **10**(4), e26188 (2024)
15. Gunasekaran, P.K., Chin, S.C.: Performance of bamboo biochar as partial cement replacement in mortar. In: Materials Today: Proceedings (2023)
16. Neeraja, P.G., Unnikrishnan, S., Varghese, A.: A comprehensive review of partial replacement of cement in concrete. In: Materials Today: Proceedings (2023)
17. Dahiru, D., Ibrahim, M., Gado, A.A.: Evaluation of the effect of volcanic ash on the properties of concrete. In: Proceedings (2019)
18. Gambo, S., Sanda, U.M., Ibrahim, A.G., Usman, J., Mohammad, U.H.: Strength properties of ordinary Portland cement concrete containing high volume recycled coarse aggregate and volcanic ash. In: Materials Today: Proceedings (2023)
19. Zou, C., et al.: Chapter 2 - Formation and Distribution of Volcanic Rock. In: Zou, C., et al. (eds.) Volcanic Reservoirs in Petroleum Exploration, pp. 11–29. Elsevier, Boston (2013)
20. W. Pablo, Chile: laboratorio natural para estudios de volcanes, Wolfy Pablo (2021)

21. Yoon, J., Jafari, K., Tokpatayeva, R., Peethamparan, S., Olek, J., Rajabipour, F.: Characterization and quantification of the pozzolanic reactivity of natural and non-conventional pozzolans. Cement Concr. Compos. **133** (2022)
22. Instituto Nacional de Normalización (INN): Cemento – Terminología, clasificación y especificaciones generales. NCh148.Of68 (1968)
23. ASTM: Standard Practice for Making and Curing Concrete Test Specimens in the Laboratory. ASTM C192 (2020)
24. ASTM: Standard Specification for Flow Table for Use in Tests of Hydraulic Cement. ASTM C230 (2023)
25. ASTM: Standard Test Method for Flexural Strength of Concrete (Using Simple Beam with Third-Point Loading). ASTM C78/C78M (2022)
26. ASTM: Standard Test Method for Compressive Strength of Hydraulic-Cement Mortars (Using Portions of Prisms Broken in Flexure). ASTM C349 (2018)
27. ASTM: Standard Test Method for Splitting Tensile Strength of Cylindrical Concrete Specimens. ASTM C496 (2017)
28. Scudder, N.A., Horgan, B.H.N., Rampe, E.B., Smith, R.J., Rutledge, A.M.: The effects of magmatic evolution, crystallinity, and microtexture on the visible/near-infrared and thermal-infrared spectra of volcanic rocks. Icarus **359** (2021)
29. Liu, Z., Chung, P.W.: Unusual thermal transport in molecular crystals. Materials Today Physics 36 (2023)

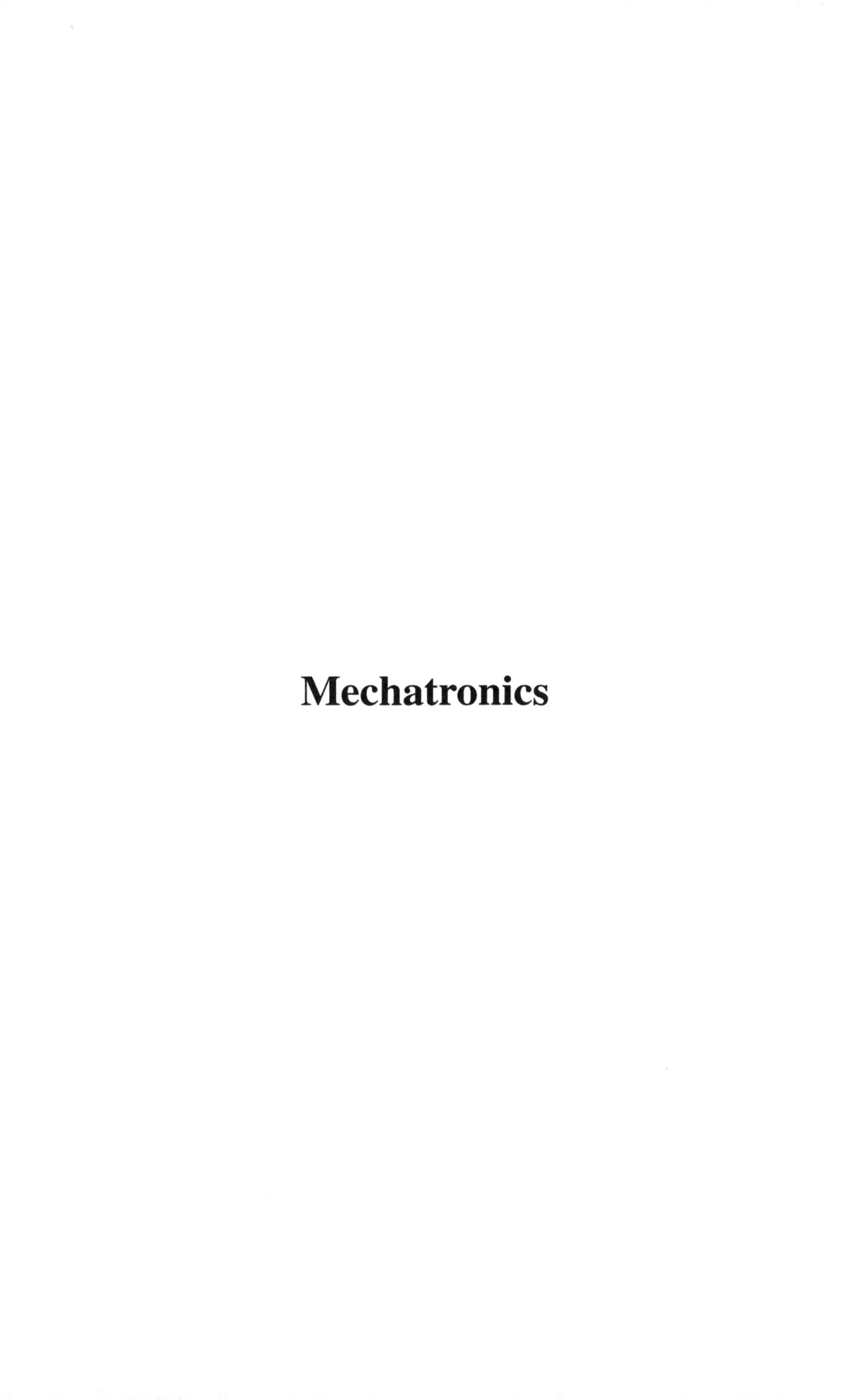

Mechatronics

Preliminary Design of a Mechatronic Device for Muscular Strength Training and Evaluation

Jorge González-Salazar(✉) and Brayan Álvarez-Gubelin

Mechanical Engineering Department, University of La Frontera, Francisco Salazar 01145, 4811230 Temuco, Chile
jorge.gonzalez@ufrontera.cl

Abstract. This study presents the preliminary design of a mechatronic device aimed at enhancing the training and evaluation of muscle strength. The motivation stems from the need to modernize traditional strength training methods by integrating computer-controlled variable resistance systems. The project draws from the fields of biomechanics, mechatronics, and product development, following a stage-gate model adapted from the U.S. industrial technology framework. This paper focuses on the first two stages: preliminary research and conceptual definition. The outcome establishes the initial design requirements and lead to the conceptual development of the proposed mechatronic device.

Keywords: Strength Training · Mechatronics · Preliminary Design · Biomechanics · Variable Resistance

1 Introduction

Muscular strength is the ability of muscle tissue to deform an object or alter its acceleration. Increased muscular strength is associated with enhanced performance in both general and sport-specific motor skills, as well as with a reduced risk of injury in athletes [1, 2]. Strength training contributes to the development of maximal or absolute strength, explosive strength, and muscle hypertrophy [3]. Moreover, it has proven effective in promoting both physical and mental health, leading to significant increases in lean body mass and basal metabolic rate. Additional benefits include the reduction of lower back pain, alleviation of arthritic discomfort, improved functional independence, enhanced gait control and speed, and its recognized role in the prevention of type 2 diabetes through improved glucose and insulin homeostasis [4]. Depending on the source of resistance, devices may include free weights or isoinertial equipment, chains, gravity-dependent machines, hydraulic and pneumatic systems, aquatic resistance, elastic resistance bands, and electronically controlled devices [1]. One such device is the isokinetic dynamometer, which integrates mechanical, electrical, electronic, and software components. Its primary function is to facilitate and evaluate muscular strength through electronically assisted exercises performed at specific speeds, torques, and joint positions. This is achieved using a servomotor with a closed-loop control system to

O. F. Farías Fuentes et al. (Eds.): CIBIM 2024, *Proceedings of the XVI Ibero-American Congress of Mechanical Engineering*, pp. 455–465, 2026.
https://doi.org/10.1007/978-3-032-22823-9_33

generate resistance [5]. The data collected by the dynamometer's sensors can be interpreted for early detection of injury risk factors, monitoring the functional recovery of injured individuals, and determining a patient's readiness to resume daily or athletic activities [6]. Recent technological advances have enabled the commercial development of computer-controlled, cable-driven training devices. These systems represent a significant improvement over traditional training methods by offering adjustable resistance and real-time performance monitoring, thereby enhancing the efficiency and precision of strength training and assessment.

This work aims to develop the conceptual design of a multimodal dynamometer, capable of supporting multiple training modes, for the assessment and enhancement of muscular strength in both sports and rehabilitation contexts. The design is envisioned to be adaptable in the future to meet the specific needs of the local population. Resistance will be generated by an electric servomotor and transmitted via a cable mechanism, like traditional pulley-based machines. The proposed dynamometer will support three primary training modalities: isometric, isotonic, and isokinetic modes [7]. The mechanical load or resistance can be programmed to follow different profiles, including the simulation of linear or nonlinear springs, viscous dampers, and dry friction. The system architecture will also allow for future expansion to support additional modes, enabling the exploration of novel training and rehabilitation strategies. It is also expected to be capable of capturing key performance metrics during exercise, including average and peak values of force, power, velocity, and range of motion. This will enable its use for both training and diagnostic evaluation.

2 Methodology

The development is based on an adapted Stage-Gate methodology from the Industrial Technologies Program, a framework employed by the U.S. Department of Energy for managing structured innovation projects [8] and tailored to the context of applied technology development in sports and physical activity. The present work focuses on the first two stages which are: Stage 1) preliminary research and analysis of relevant technologies, and Stage 2) conceptual definition of the device and the formulation of technical specifications. By the end of the development phase, the multimodal dynamometer will be ready to advance to Stage 3: conceptual development.

2.1 Stage 1: Preliminary Research and Analysis

At this stage, the current state of computer-controlled strength training equipment is investigated. This review enables an understanding of the most recent advancements in the field and helps identify the fundamental characteristics of such devices. In addition, a study of the dynamics of forces and movements during physical exercise was conducted. The activities carried out in this stage include:

External search: A comprehensive search was conducted to gather information on computer-controlled strength training devices. Sources included academic databases, specialized journals, commercial equipment websites, and available patents.

Product analysis: Leading commercial devices for computer-assisted strength training were examined and compared based on their key features. The following benchmarking methodology is applied:

1. Product selection for comparison: The leading devices were selected based on market share and user reviews.
2. Definition of relevant features: Key characteristics were identified, including variable resistance range, range of motion, power output, cost, additional functionalities, and the machine's size and weight.
3. Data collection: Information was obtained from technical specifications and user manuals available online.
4. Comparative analysis: A comparison table was used to evaluate the selected products based on the defined features.

Experimental assessment of dynamic requirements: Data on kinetic and kinematic characteristics were collected through the execution of two predefined exercises using a barbell and submaximal loads, that is, weights below the individual's one-repetition maximum (1RM). This experiment, along with the gathered data, supports the preliminary requirements for the conceptualization of the multimodal dynamometer. The study follows a protocol similar to that described in [9], enabling the extraction of dynamic variables related to barbell movement.

1. Participant: A young male athlete specializing in sprint events ranging from 100 m to 400 m. His physical condition, speed, strength, and familiarity with the exercises made him a suitable candidate for evaluating the design requirements of the device.
2. Protocol: A standardized warm-up routine aimed at minimizing the risk of injury by gradually increasing body temperature, enhancing blood flow to the working muscles, and improving joint mobility [10]. Subsequently, a set of 10–15 repetitions with light loads for each exercise under evaluation, focusing on proper technique and preparing specific muscles and joints for the main effort.
3. Instrumentation: Kinematic parameters were recorded using a smartphone mounted on a tripod, positioned 4 m from the athlete in the sagittal plane and 1.2 m above the ground, capturing video at 120 fps. A green marker was placed on one end of the barbell to enable automatic trajectory tracking using Tracker software [11], which tracks markers and performs automated calculations to extract spatial and temporal data.
4. Data Processing: The discrete data obtained from the tracking software—specifically, the reference point's horizontal x(t) and vertical y(t) positions —were exported to a spreadsheet with three columns. This data was used as input in MATLAB [12] for advanced data analysis. To reduce noise and better define the position signals, a moving average smoothing technique was applied. Velocity and acceleration in both directions were computed using forward finite difference approximations. From these calculations, numerical vectors of key dynamic parameters were obtained, including the resultant magnitude and angular orientation at each time point during the movement.

2.2 Stage 2: Concept Definition

The methodology applied in this stage is based on the product conceptual design principles proposed in [13]. The activities carried out include:
Problem Definition: A detailed analysis of the required functionality of the multimodal dynamometer was conducted, based on a functional diagram and the exercise modes the system must support.
Specification Setting: Preliminary design specifications were established, considering functionality, safety, ergonomics, and cost considerations.
Concept Generation: Concepts were developed based on a winch mechanism driven by a servomotor to generate resistance. The main subsystems were defined, and technical alternatives were explored to address the identified design challenges.
Concept Selection: Two system configurations were evaluated using a decision matrix, and the most suitable concept was selected, with justification provided for the choice.
Preliminary Design: A preliminary schematic of the selected concept was created, outlining the overall architecture and operation of the subsystems and components. In addition, a dynamic model of the dynamometer's resistance mechanism is presented.

3 Results

3.1 Stage 1: Preliminary Research and Analysis

Product Analysis: a comparison was made of the leading strength training devices currently available on the market. Five systems were selected for benchmarking purposes. Table 1 summarizes their key features.

Table 1. Comparison of Key Features of Strength Training Machines.

Characteristics	Model 1	Model 2	Model 3	Model 4	Model 5
Resistance range [kgf]	90	200	100	130	113
Dynamic modes	5	4	4	5	5
Cable Length [m]	182	300	-	-	-
Price [$USD]	3,995	2,900	2,899	1,709	3,999
Dimensions [cm]	129x55x13	12x117x52	185x35x71	19x120x58	207x115x203
Machine Weight [Kg]	68	38	85	45	158
Power output [W]	1440	1000	1600	1500	-

Dynamic Requirements Definition: To establish the preliminary dynamic requirements of the dynamometer, experimental tests were conducted with an athlete performing deadlift and squat exercises. Table 2 lists the 1RM and their closest percentages values within standard evaluation ranges [10], limited by the available weight plates. The kinematic and kinetic values from 2 squat repetitions are shown in Table 3. Additionally, Fig. 1 presents the 4 graphs obtained from the first repetition performed at 39% of 1RM in squats, depicting the movement dynamics during the exercise. It is evident that, although the exercises are primarily vertical in nature, horizontal displacement still occurs. Moreover, the force exerted by the athlete varies throughout the motion, at times falling below and at other times exceeding the barbell weight.

Table 2. 1RM and Corrected Workload Percentages for Deadlift and Squat.

	1RM	81%	69%	62%	42%
Deadlift [kg]	130	105	90	80	55
	1RM	83%	72%	61%	39%
Squat [Kg]	90	75	65	55	35

Table 3. Kinematic and Kinetic values for the first and second squat repetition.

Kinematic Characteristics	39%	61%	72%	83%
Max. Linear Velocity [m/s]	1,92/2,03	2,48/2,43	2,16/2,28	1,82/1,88
Avg. Linear Velocity [m/s]	0,65/0,73	0,86/0,9	0,75/0,84	0,73/0,65
Displacement [m]	0,63/0,62	0,83/0,87	0,79/0,81	0,69/0,71
Acceleration [m/s^2]	12,92/15,06	15,68/16,46	14,7/16,06	14,19/13,19
Kinetic Characteristics	39%	61%	72%	83%
Max. External Force [N]	561,84/623,73	1050,37/965,23	1113,39/1161,74	1318,91/1217,13
External Work [J]	437/431,16	901,9/1071,58	980,56/1059,33	976,03/995,82
Max. External Power [W]	945,31/1100,48	2221,43/2092,23	2068,2/2261,43	2004,05/2090,1
Avg. External Power [W]	224,76/261,17	510/517,18	508,72/582,55	557,24/495,15

Table 4 presents the kinematic and kinetic values from 2 repetitions of the deadlift exercise. Additionally, Fig. 2 displays the 4 graphs obtained from the first repetition performed at 42% of 1RM, illustrating the movement dynamics during the exercise. The results allow for the identification of the force and velocity ranges required for the design of the multimodal dynamometer, ensuring its capability to deliver accurate and reliable data during strength training and evaluation. These data are key for fine-tuning the device's control software, optimizing both performance and functionality.

Table 4. Kinematic and Kinetic values for the first and second deadlift repetition.

Kinematic Characteristics	42%	62%	69%	81%
Max. Linear Velocity [m/s]	1,34 / 1,44	1,04 / 1,03	0,88 / 0,95	0,82 / 0,85
Avg. Linear Velocity [m/s]	0,58 / 0,67	0,45 / 0,56	0,31 / 0,45	0,37 / 0,4
Displacement [m]	0,62 / 0,62	0,61 / 0,60	0,59 / 0,59	0,58 / 0,59
Acceleration [m/s^2]	8,42 / 10,29	10,58 / 9,45	6,2 / 8,3	5,43 / 5,66
Kinetic Characteristics	42%	62%	69%	81%
Max. External Force [N]	989,87/ 870,75	1245,9 / 1537,58	1440,85 / 1628,91	1217,48 / 1367,27
External Work [J]	668,74/ 663,74	925,25 / 902,08	1012,48 / 1020,6	1179,01 / 1124,65
Max. External Power [W]	929,41/ 840,23	901,32 / 1077,27	803.20 / 889,14	900,77 / 943,7
Avg. External Power [W]	311,72 / 347,33	338,57/425,33	272,02 / 396,57	378,08 / 391,14

Preliminary Dynamic Requirements: The dynamic analysis establishes two critical performance parameters: force range and linear velocity. An optimal point can be identified from an estimated force–velocity curve to maximize the power during exercise. The following were determined for the dynamometer design:

- Force range: up to 100 kgf (approximately 981 N) of maximum force.
- Linear velocity: up to 1.35 m/s of maximum linear speed.

3.2 Stage 2: Concept Definition

Based on the biomechanical requirements established in Stage 1, a systematic exploration of solution concepts was conducted. The following steps in concept definition were considered:

Problem Identification: The objective is to design a multimodal dynamometer, a mechatronic device intended for strength training, capable of providing variable resistance and delivering feedback on dynamic parameters during exercise. The resistance transmission mechanism will use a cable, which, together with attachment accessories, will enable the user to perform exercises. Resistance will be generated by an electric actuator and controlled electronically, distinguishing it from traditional machines that rely on weight stacks or inertial components. The four functional key subsystems, required for the

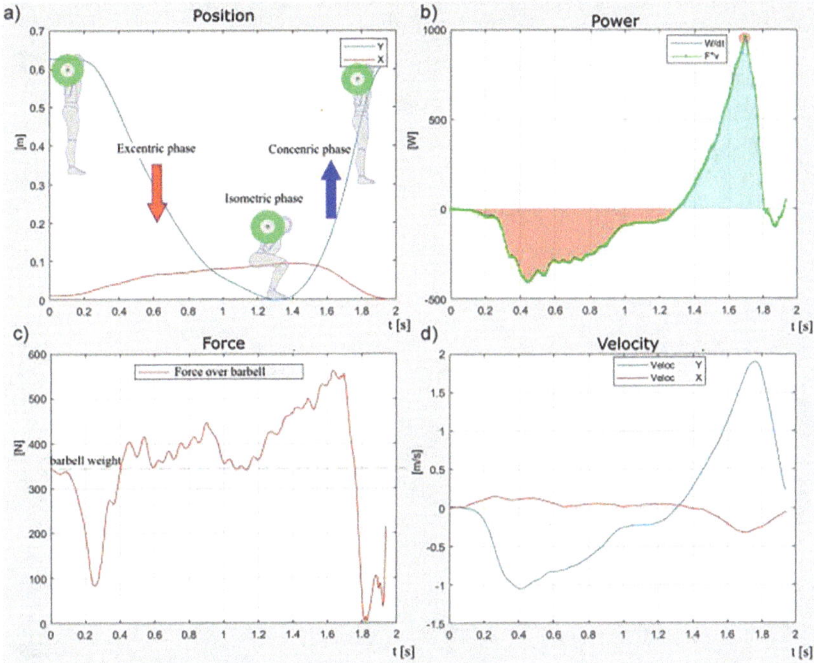

Fig. 1. Graphs from the first repetition at 39% of 1RM in squat. (a) Position [m], (b) Power [W], (c) External Force [N], (d) Velocity [m/s].

machine's operation, adapted from the operation of an isokinetic dynamometer, are as following:

1. *Actuator subsystem:* Consists of a servomotor and its controller. The motor will provide a resistance load based on the force exerted by the user.
2. *Control and command subsystem:* The human-machine interface allows the user to configure and operate the device. It includes all necessary electronic components to manage the system's functionality.
3. *Measurement subsystem:* Comprises the sensors required to capture dynamic exercise variables such as force, velocity, and position.
4. *Transmission subsystem:* Includes the mechanical elements responsible for transmitting the resistance generated by the actuator to the user's point of contact.

System Specifications the specifications of the multimodal dynamometer were defined based on the results obtained in the first stage and the characteristics observed in similar commercial equipment. The preliminary specifications include:

1. Force range: Up to 100 kgf or 981 N of maximum force.
2. Maximum linear velocity: Up to 1.35 m/s.
3. Maximum power: Approximately 1324 W, calculated by multiplying the maximum force and the maximum linear velocity.

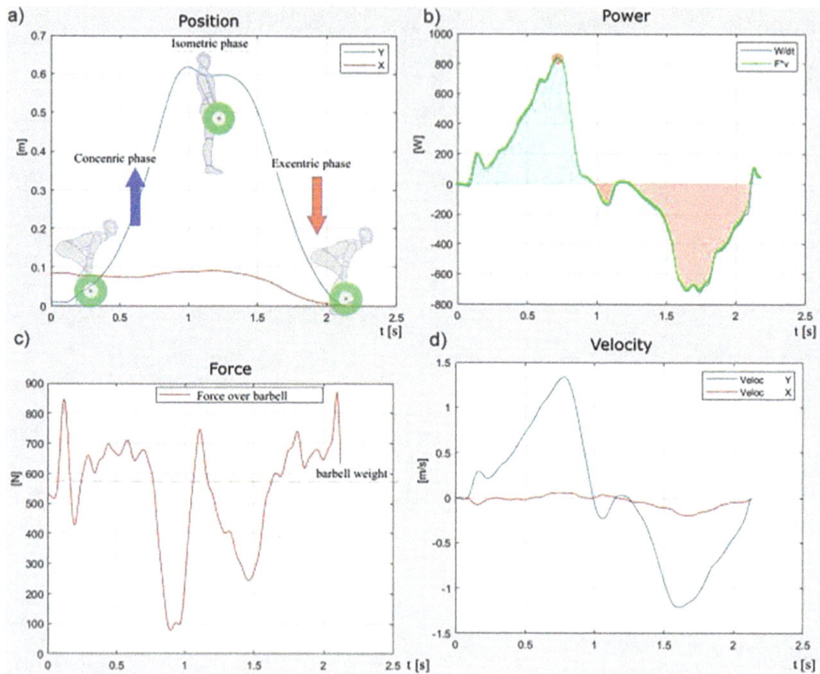

Fig. 2. Graphs from the first repetition at 42% of 1RM in deadlift. (a) Position [m], (b) Power [W], (c) External Force [N], (d) Velocity [m/s].

Solution Concepts. In [14], several design concepts for the multimodal dynamometer were proposed. One of the initial concepts involved a winch mechanism driven by a servomotor to generate resistance. This mechanism was divided into main subsystems, and key concepts were specified to address technical aspects of each. To select the most appropriate concept, two system configurations were outlined and evaluated using a decision matrix. This matrix considered factors such as functionality, safety, ergonomics, and costs. The most suitable configuration was selected, and the choice was justified based on the defined criteria.

Preliminary Design includes the definition of the main architecture and the overall functioning of the different subsystems and components. Figure 3 presents the dynamic model governing the resistance mechanism of the multimodal dynamometer, which will be needed in later stages of control system design.

The electromechanical equations that describe the system are given below:

$$V_t(t) = L_a \bullet \frac{di(t)}{dt} + R_a \bullet i(t) + K_e \bullet \dot{\theta}(t) \tag{1}$$

$$T_l(t) - T_m(t) = (J_m + J_l) \bullet \ddot{\theta}(t) + c \bullet \dot{\theta}(t) \tag{2}$$

where $T_m(t) = K_t \bullet i(t)$. By substituting the values of the state and the input vectors, and isolating the derivative of the state vector—which contains the parameters that depend

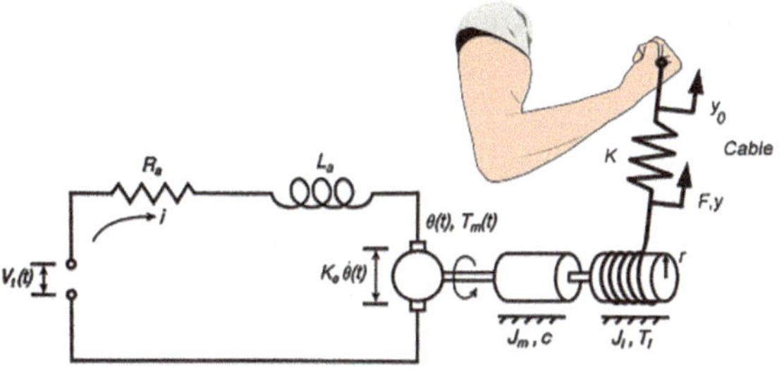

Fig. 3. Dynamic model of a DC motor applied to strength training.

on the physical characteristics of the motor and the load—it can be written as:

$$\begin{bmatrix} \dot{x}_1(t) \\ \dot{x}_2(t) \end{bmatrix} = \begin{bmatrix} -\frac{c}{J_t} & -\frac{K_t}{J_t} \\ -\frac{K_e}{L_a} & -\frac{R_a}{L_a} \end{bmatrix} \begin{bmatrix} x_1(t) \\ x_2(t) \end{bmatrix} + \begin{bmatrix} 0 & \frac{1}{J_t} \\ \frac{1}{L_a} & 0 \end{bmatrix} \begin{bmatrix} u_1(t) \\ u_2(t) \end{bmatrix} \quad (3)$$

This system has two degrees-of-freedom or two independent state-variables: the angular velocity of the rotor and the motor current. These two variables represent the independent quantities to describe the system's dynamic state. Finally, Fig. 4 depicts the conceptual design of the proposed multimode dynamometer.

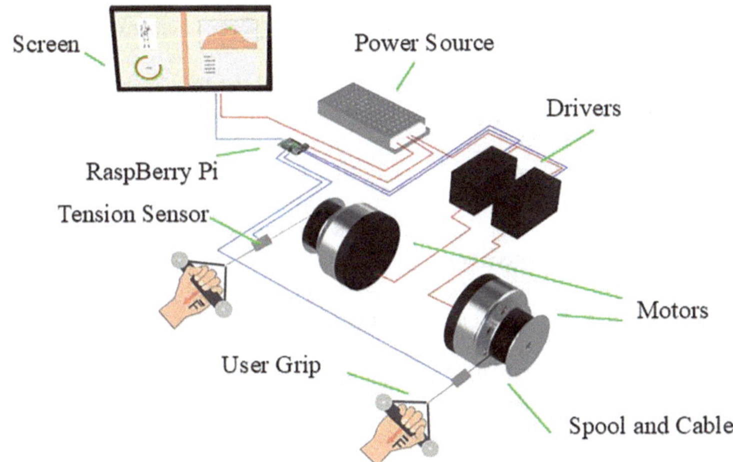

Fig. 4. Strength Training System Diagram with DC Motor and Raspberry Pi Controller.

4 Conclusions

This study presented the preliminary design process of a mechatronic device for muscular strength training and assessment. The use of the Stage-Gate methodology enabled a systematic approach to the project, covering the first two stages: preliminary research and concept definition. The external review of similar strength training devices revealed a growing trend toward the development of smarter and more adaptive machines. The dynamic analysis of deadlift and squat exercises performed at four relative loads of one-repetition maximum provided indicative ranges for average power output (224–582 W), average linear velocity (0.31–0.9 m/s), and maximum force (561–1628 N). These findings, along with insights from the comparative market analysis, informed us the initial requirements for the conceptualization of the multimodal dynamometer.

Further experimentation involving a broader range of test subjects and additional exercises is necessary to refine the machine's requirements before advancing to detailed design and prototype construction. A larger and more diverse participant pool—in terms of age, gender, and skill levels—would enhance the validity and representativeness of the experimental data used.

References

1. Pérez Soriano, P., Llana Belloch, S.: La Instrumentación en la Biomecánica Deportiva. J. Human Sport Exerc. **2**(2), 26–41 (2014)
2. Suchomel, T.J., Nimphius, S., Bellon, C.R., Stone, M.H.: The Importance of muscular strength: training considerations. Sports Med. **48**(4), 765–785 (2018)
3. Siff, M.C.: Biomechanical foundations of strength and power training. In: Zatsiorsky V.M. (eds.) Biomechanics in Sport, pp. 103–139. Wiley (2000)
4. Westcott, W.L.: Resistance training is medicine: effects of strength training on health. Curr. Sports Med. Rep. **11**(4), 209–216 (2012)
5. Ponce, D.A., Martins, D., Martin, C., Da Silva, F., De Mello, C., Ocampo, A.: Development of a scale prototype of isokinetic dynamometer. Ingeniare. Revista chilena de ingeniería **23**(2), 196–207 (2015)
6. Varas, E., González, E.: Determinación de la normalidad mediante evaluación isocinética de la musculatura del complejo articular del hombro. Revista Iberoamericana de Fisioterapia y Kinesiología **6**(2), 81–90 (2003)
7. Hamill, J., Knutzen, K., Derrick, T. Biomecánica: Bases del movimiento humano. 4th edn. Wolters Kluwer, Barcelona (2017)
8. U.S Department of Energy Homepage, https://www.energy.gov/eere/analysis/articles/stage-gate-review-guide-industrial-technologies-program, Accessed 30 May 2025
9. Balsalobre-Fernandez, C., Geiser, G., Krzyszkowski, J., Kipp, K.: Validity and reliability of a computer-vision-based smartphone app for measuring barbell trajectory during the snatch. J. Sports Sci. **38**(6), 710–716 (2020)
10. Woods, K., Bishop, P., Jones, E.: Warm-up and stretching in the prevention of muscular injury. Sports Med. **37**(12), 1089–1099 (2007)
11. Tracker Homepage, https://opensourcephysics.github.io/tracker-website, Accessed 30 May 2025
12. Matlab Homepage, https://matlab.mathworks.com/, Accessed 30 May 2025
13. Ulrich, K.T., Eppinger, S.D.: Product Design and Development. 6th edn. McGraw-Hill Education, New-York (2013)
14. Álvarez, B.: Diseño preliminar de un dispositivo mecatrónico para el entrenamiento y evaluación de fuerza muscular. Universidad de La Frontera, Temuco, Chile, Proyecto de título para optar al título de Ingeniero Civil Mecánico (2024)

Development of a Biomechanical Upper Body Model for Application in the Treatment of Stroke Patients

Juana Mayo[1](✉), Patricia Ferrand[2], María José Zarco[2], and Joaquín Ojeda[1]

[1] Universidad de Sevilla, Sevilla, Spain
juana@us.es
[2] Hospital Universitario Virgen del Rocío, Sevilla, Spain
patricia.ferrand.sspa@juntadeandalucia.es

Abstract. We present a biomechanical model of the upper limb based on protocols used in clinical practice. Its simplicity makes it suitable for use in patients with reduced mobility following a stroke. Motion capture analysis was performed in the Motion Analysis Laboratory of the Neurorehabilitation Unit at Virgen del Rocío Hospital in Seville, Spain. Joint angles and moments were computed using inverse kinematics and inverse dynamics, in both patients and healthy subjects performing a daily life task. The kinematic and kinetic results provide objective, quantitative data that complement information obtained using the Fugl-Meyer assessment scale in stroke patients.

Keywords: Multibody system dynamics · Stroke · Motion analysis · Biomechanic model · Upper extremity

1 Introduction

Cerebrovascular accident (CVA), commonly known as stroke, is a group of disorders that temporarily or permanently alter the functioning of one or more areas of the brain. Stroke occurs as a consequence of a cerebral circulatory disorder, either of blood vessels or of the quantity or quality of circulating blood, which causes affected cells not to receive oxygen and die [1]. It is estimated that 50% to 85% of patients who have suffered a stroke have an upper limb deficit during the acute phase that persists in approximately 60% of patients at 6 months after stroke. These deficits include muscle weakness, sensory deficits, loss of coordination, and spasticity, with a great impact on the functional capacity of these patients.

Within the framework of neurological rehabilitation, there are different therapeutic approaches to improve functional recovery of the upper extremity. However, in the studies carried out, the measuring instruments used, such as the Fugl-Meyer scale, which is specific for stroke, do not differentiate spontaneous recovery from the compensatory mechanisms performed by the patient. One

O. F. Farías Fuentes et al. (Eds.): CIBIM 2024, *Proceedings of the XVI Ibero-American Congress of Mechanical Engineering*, pp. 466–482, 2026.
https://doi.org/10.1007/978-3-032-22823-9_34

of the main reasons is that the measuring instruments used are based on the observational assessment of the individual, which is subject to various drawbacks, including the possible ceiling effect, or the subjectivity of the evaluator. In addition, the lack of standardization of the instruments in the assessment of the upper extremity makes it difficult to compare the results [2]. The Stroke Recovery and Rehabilitation Roundtable (SRRR), established in a consensus meeting the standardization of the different clinical measurement instruments, unified in COSMIN statements, based on the Classification of Functioning and Disability (CIF) promoted by the World Health Organization (WHO). However, SRRR considered that standard clinical measures do not provide insight into movement quality, are insensitive to the detection of the changes that occur, and mainly do not allow differentiating the changes that occur due to the existing recovery from the compensatory mechanisms that the patient puts into operation after stroke. Currently, biomechanical analysis is accepted as a tool that allows for an accurate, objective, and easily reproducible assessment of existing deficits. In particular, inverse dynamic analysis allows estimation of the forces involved in the movement under analysis.

However, there is little information in the literature on this type of analysis applied to stroke patients. The combination of the different kinematic and kinetic parameters is considered to provide a complete picture of the behavior of the upper limb of the patients. The challenge is to identify the kinematic and kinetic variables that best quantify upper limb movement in a reliable and accurate way, that discriminate between existing motor control, established deficit, and that differentiate the severity of the deficit and the compensatory movement performed [3]. Achieving this goal would allow for a better therapeutic approach and even a better prognosis for the patient [4].

The aim of this work has been to develop a biomechanical model of the upper body to obtain joint moments in stroke patients during the performance of daily living tasks.

2 Methodology

2.1 Biomechanical Model and Marker Protocol

An upper body biomechanical model was implemented consisting of five segments: right and left arms, right and left forearms, and trunk. The elbow joint has been modeled as a universal pair in which the varus-valgus motion has been restricted and in which the center of the joint has been defined at the distal humerus. The pronation-supination of the forearm has been modeled as a rotation around the axis connecting the center of the elbow to the distal ulna. The shoulder joint has been modeled as a spherical pair whose center of rotation is defined at the head of the humerus. The contribution of the scapula to shoulder motion has not been taken into account. The joint centers are assumed to be fixed in the local reference systems of the corresponding solids. The kinematics of the trunk has been calculated as the rigid solid motion of this segment with respect to the inertial reference system.

To reconstruct the position and orientation in space of each of the solids, a modification of the marker protocol proposed by Rab [5] has been used. A total of 16 markers located in the lateral zygomatic arches, the nasion, the acromions, the most proximal area of the sternum, the olecranons, the most distal area of the ulna and radius, the third metacarpophalangeal joint of the hand and the tip of the index finger were used (see Fig. 1).

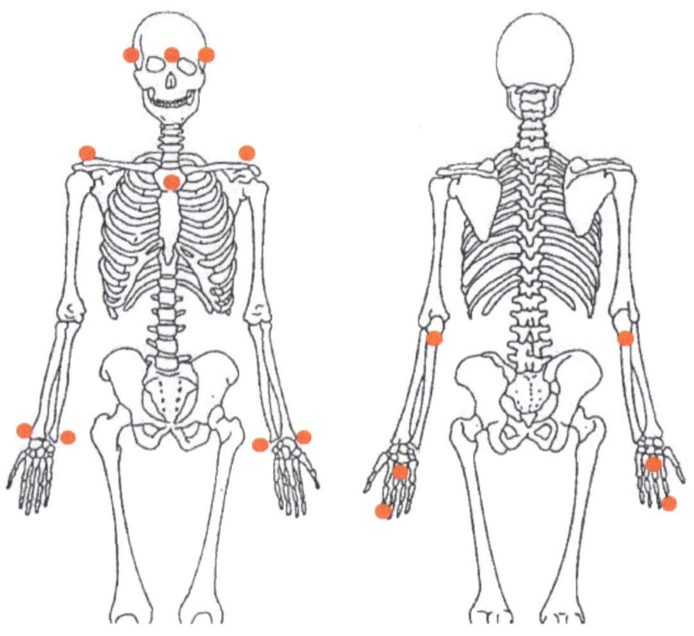

Fig. 1. Markers location. Source: Own elaboration.

2.2 Inverse Dynamic Problem

The inverse dynamic problem has been solved using the Newton-Euler equations using as input the kinematic data obtained in a previous work [6]. The solid-to-solid equilibrium equations were posed starting from the hand up to the arm. Mechanical model parameters were taken from the literature and scaled to each participant [7].

2.3 Study Design

The analized task was to bring the hand to the mouth. For this, the participant was seated with the hands on a table as shown in Fig. 2. During motion capture, the participant attempted to bring each hand alternately to their mouth.

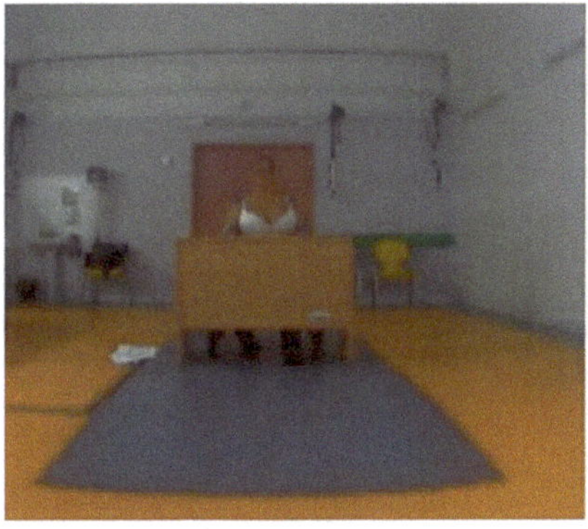
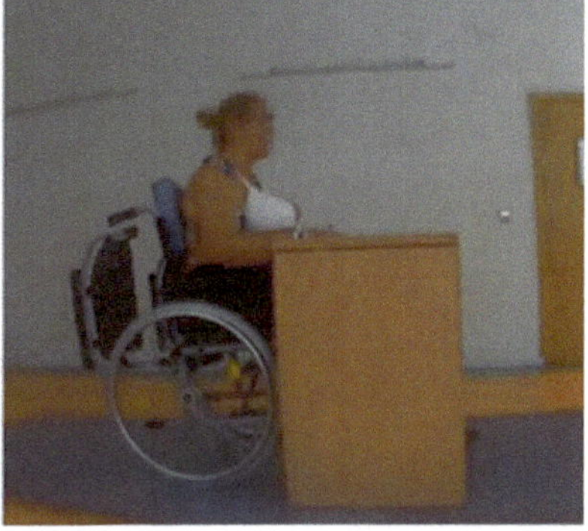

Fig. 2. Participant's position during motion capture. Source: Own elaboration.

2.4 Data Processing

A Matlab program was developed to perform inverse kinematics and dynamics. The marker coordinates were filtered using a two-way Butterworth low-pass filter using a cutoff frequency of 10 Hz. The cutoff frequency was selected using a residual analysis [8] to achieve a balance between signal distortion and the amount of noise allowed. The marker velocity of the right and left fingers was obtained by numerical differentiation.

Each movement was divided into three main phases: the reach phase (transport phase to the target), the approach phase (phase of the movement dedicated to accurately locate the target), and the return phase (transport phase back to the original resting position). The beginning of the reach phase was defined as the instant at which the modulus of the velocity of the finger marker exceeded a threshold value set at 50 mm/s [9]. The end of the reach phase coincides with the beginning of the approach phase and was defined as the instant at which the modulus of the finger marker velocity fell below 50 mm/s. The end of the approach phase, coinciding with the beginning of the return phase, was calculated as the instant at which the modulus of the finger marker velocity exceeded the set threshold. Finally, the end of the return phase was defined as the instant at which the velocity of the finger marker fell below the threshold.

3 Results and Discussion

The kinematic and dynamic results are shown for four subjects, one healthy and three with different scores on the Fugl-Meyer upper limb scale (FMA-UE). This scale between 0 and 66 assigns three degrees of impairment. It is considered mild between 58 and 66, moderate between 32 and 57, and severe below 31.

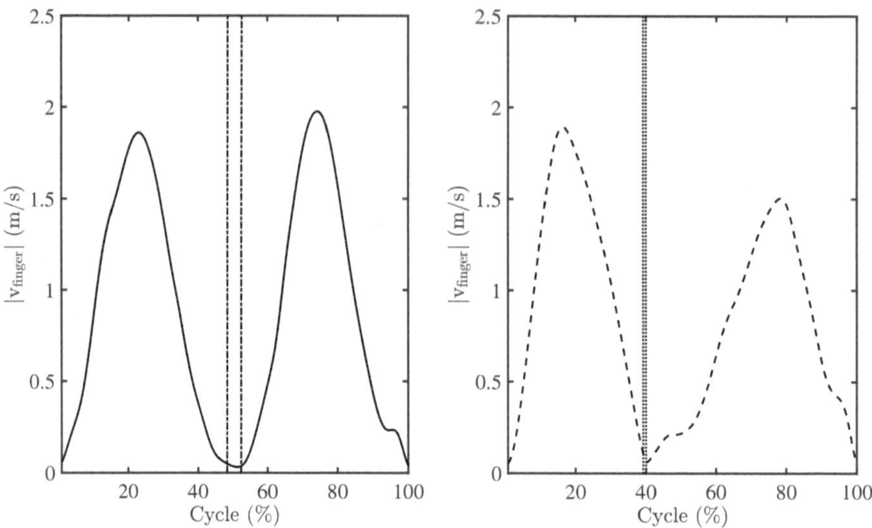

Fig. 3. Finger velocity modulus for a healthy subject. *Solid line* for the right, *dashed line* for the left. Source: Own elaboration.

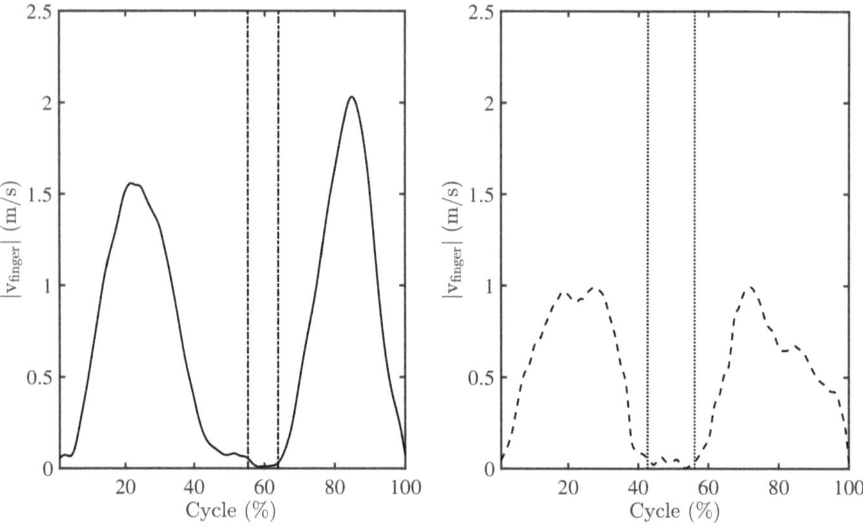

Fig. 4. Finger velocity modulus for a patient with mild pathology.*Solid line* for the right, *dashed line* for the left. Source: Own elaboration.

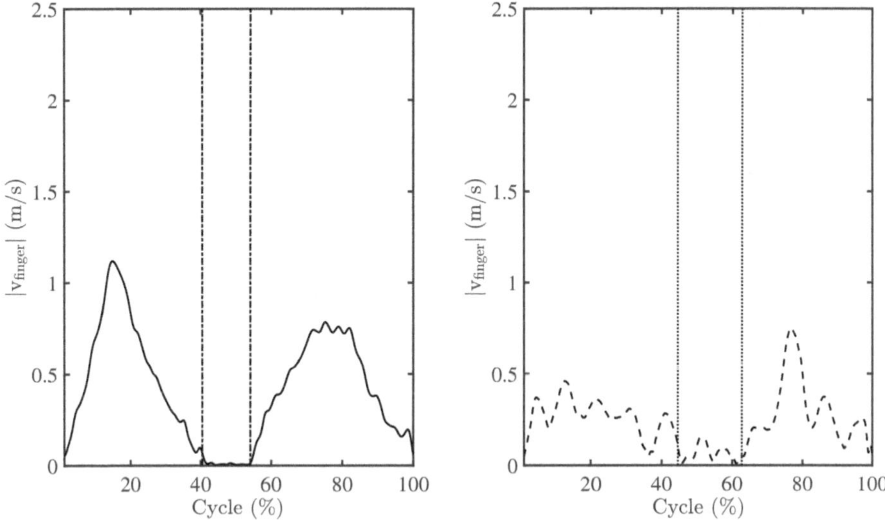

Fig. 5. Finger velocity modulus for a patient with moderate pathology. *Solid line* for the right, *dashed line* for the left. Source: Own elaboration.

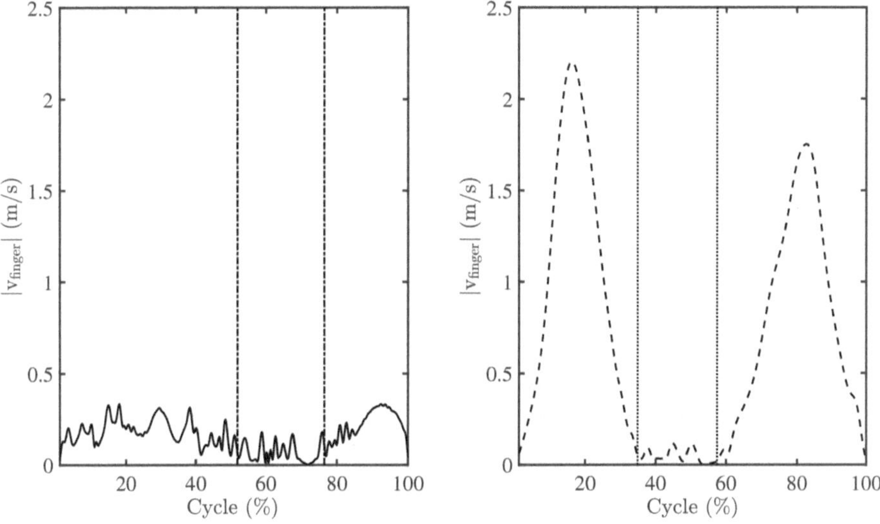

Fig. 6. Finger velocity modulus for a patient with severe pathology. *Solid line* for the right, *dashed line* for the left. Source: Own elaboration.

Four different figures are shown for each subject. The first one, Fig. 3 for the healthy subject, shows the modulus of the velocity of the finger marker. The left graph, in solid line, represents the right finger. The graph on the right, in dashed line, shows the evolution of the left finger. The beginning and end of the approach phase are also shown. A vertical dash and dot line has been used for the right limb and a vertical dots line for the left limb. This line pattern is maintained in the next three graphs, where the results for the right and left limbs are superimposed. The second figure, Fig. 7 for the healthy subject, shows the results for the three degrees of freedom of the shoulder. The third, Fig. 11 for the healthy subject, shows the results for the elbow and the fourth (Fig. 15) for the wrist. In these three figures, the left column corresponds to the joint angles while the right column corresponds to the joint moments. Figures 4, 8, 12 and 16 show a patient with mild involvement (59 FMA-UE) of the left limb. In Figs. 5, 9, 13 and 17, the FMA-UE scale rating is 42. It is a moderate pathology of the left limb. Finally, Figs. 6, 10, 14 and 18 show severe involvement (10 FMA-UE) of the right limb.

3.1 Finger Velocity Modulus

For a healthy subject, the curve shown in Fig. 3 has the shape of a double mountain with two well-defined maxima, the first corresponding to the reach phase and the second to the return phase. As the degree of impairment of the patient increases, an increase in the number of relative maxima is observed together with a decrease in the value of the relative maxima.

The velocity module of the right arm of the patient with mild impairment in Fig. 4 is similar to that of the healthy subject. However, for the affected limb, the maximum value decreases by approximately half, and although the double-peak shape is maintained, some relative maxima and minima begin to appear.

In Fig. 5 for the patient with moderate impairment in the left arm, eleven peaks overlap the double mountain profile. In addition, it can be seen that the movement of the unaffected limb is much slower than in the case of the healthy subject.

Finally, in the case of severe pathology (see Fig. 6), the double-mountain profile almost completely disappears in the affected limb, making it difficult to identify the different phases of movement. The number of peaks is more than twenty. For this patient, unlike the previous patient, the unaffected limb behaved very similarly to that of the healthy subject.

The number of relative maxima is an indicator of the smoothness of the movement. The smoothness of the movement was quantified using the number of movement units (NMU) [4,10]. An NMU is defined as the difference between a local minimum and the next maximum value of the hand velocity that exceeds a value of 20 mm/s, where the time between two consecutive peaks must be at least 150 [11,12]. In a healthy subject, the value of the NMU metric is 2. The further the metric deviates from this value, the less smooth the movement and therefore the greater the limb involvement.

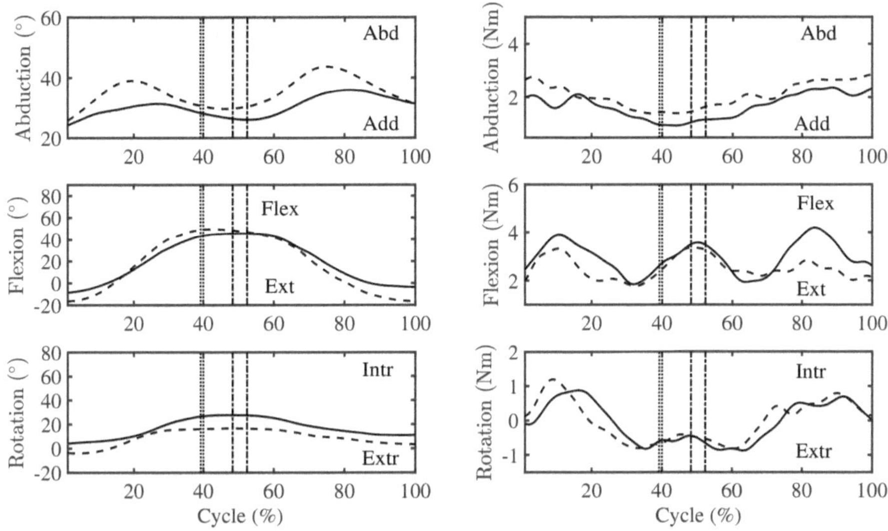

Fig. 7. Shoulder kinematic and kinetic results for a healthy subject. - for *Solid line* for the right, *dashed line* for the left. Source: Own elaboration.

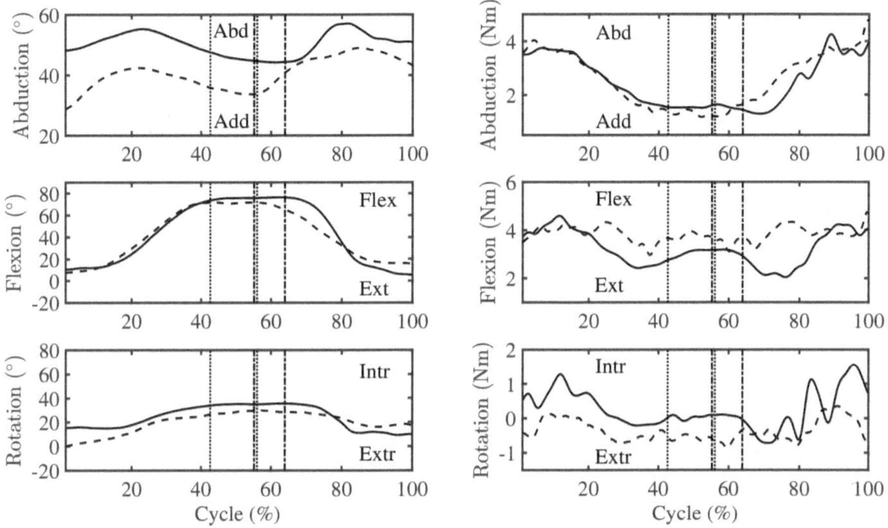

Fig. 8. Shoulder kinematic and kinetic results for a patient with mild pathology. *Solid line* for the right, *dashed line* for the left. Source: Own elaboration.

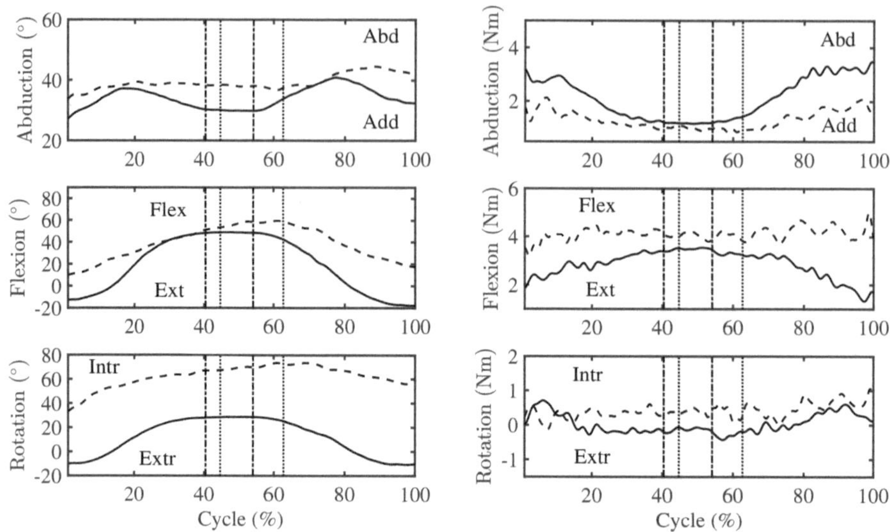

Fig. 9. Shoulder kinematic and kinetic results for a for a patient with moderate pathology. *Solid line* for the right, *dashed line* for the left. Source: Own elaboration.

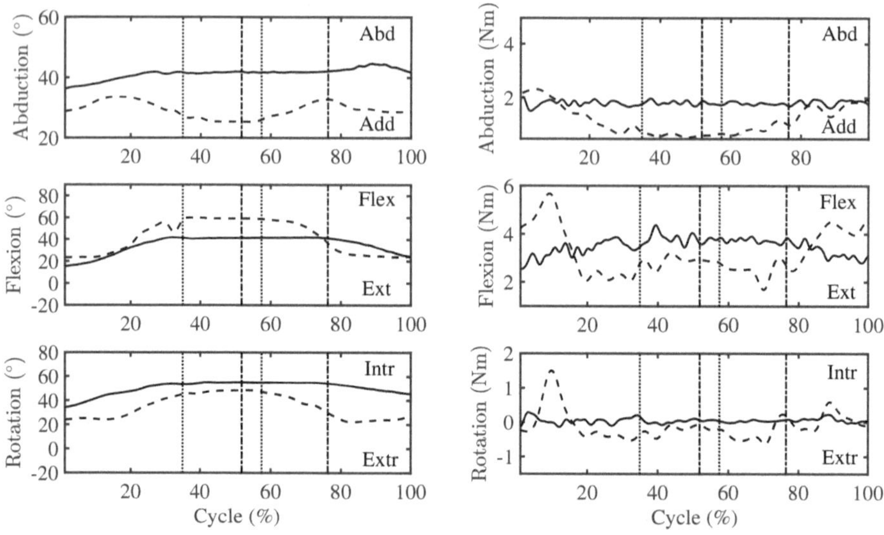

Fig. 10. Shoulder kinematic and kinetic results for a for a patient with severe pathology. *Solid line* for the right, *dashed line* for the left. Source: Own elaboration.

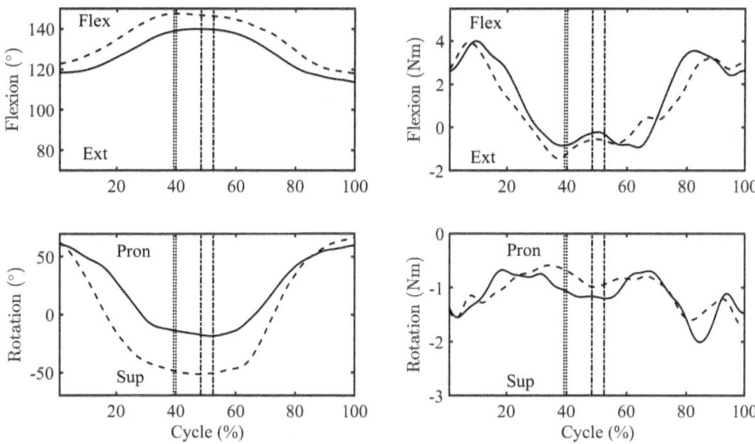

Fig. 11. Elbow kinematic and kinetic results for a healthy subject. *Solid line* for the right, *dashed line* for the left. Source: Own elaboration.

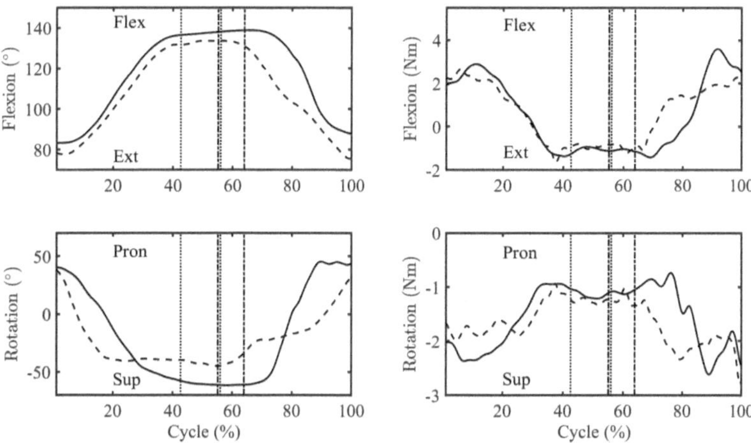

Fig. 12. Elbow kinematic and kinetic results for a patient with mild pathology. *Solid line* for the right, *dashed line* for the left. Source: Own elaboration.

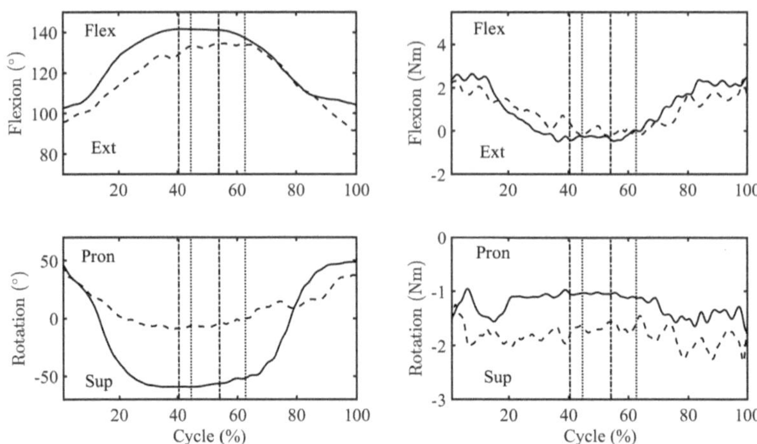

Fig. 13. Elbow kinematic and kinetic results for a patient with moderate pathology. *Solid line* for the right, *dashed line* for the left. Source: Own elaboration.

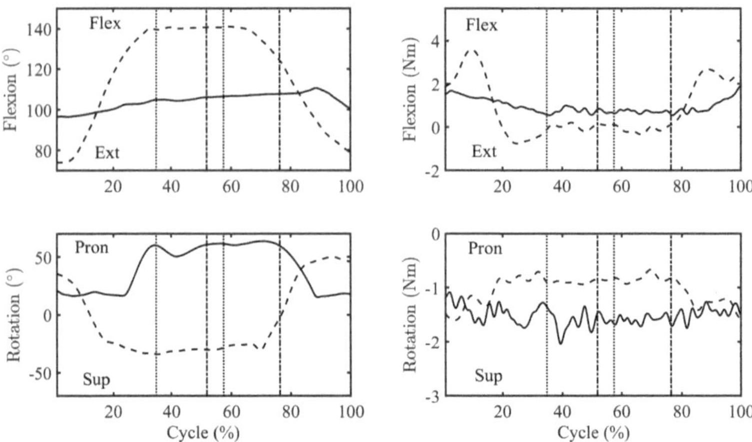

Fig. 14. Elbow kinematic and kinetic results for a patient with severe pathology. *Solid line* for the right, *dashed line* for the left. Source: Own elaboration.

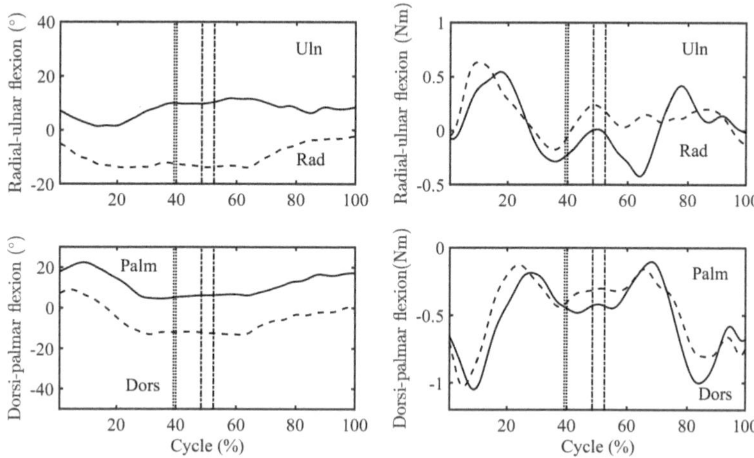

Fig. 15. Wrist kinematic and kinetic results for a healthy subject. *Solid line* for the right, *dashed line* or the left. Source: Own elaboration.

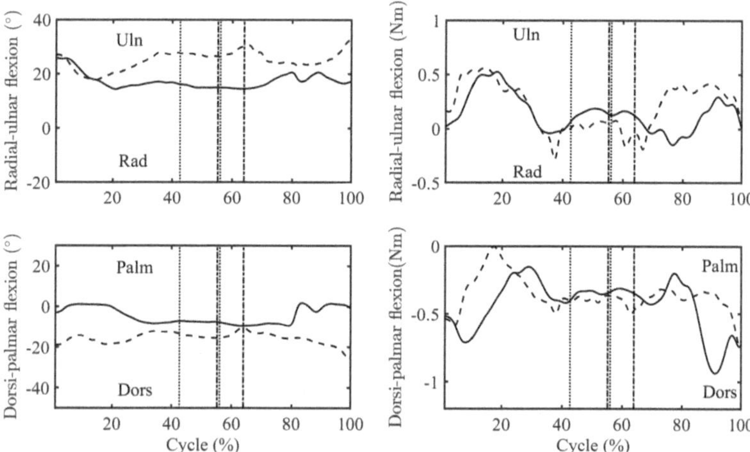

Fig. 16. Wrist kinematic and kinetic results for a patient with mild pathology. *Solid line* for the right, *dashed line* for the left. Source: Own elaboration.

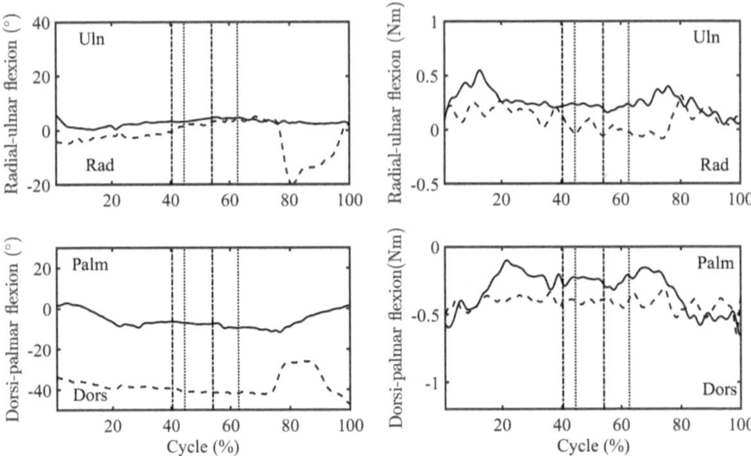

Fig. 17. Wrist kinematic and kinetic results for a patient with moderate pathology. *Solid line* for the right, *dashed line* for the left. Source: Own elaboration.

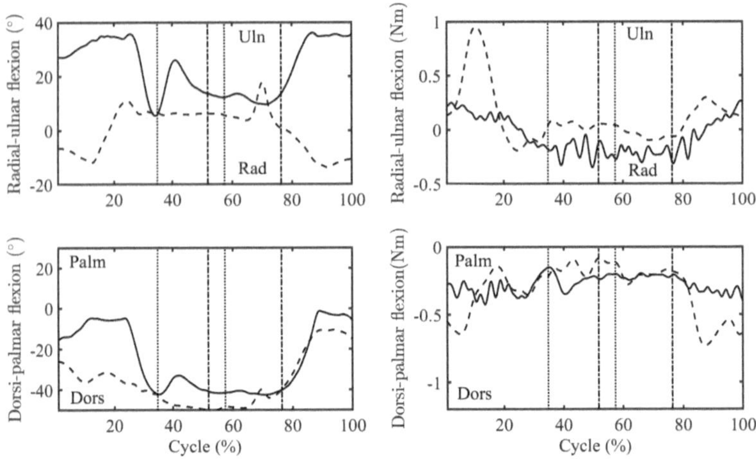

Fig. 18. Wrist kinematic and kinetic results for a patient with severe pathology. *Solid line* for the right, *dashed line* for the left. Source: Own elaboration.

3.2 Temporal Evolution of Joint Angles

When comparing the results for shoulder and elbow angles of the mildly affected patient (Figs. 8 and 12) with the healthy subject (Figs. 7 and 11), a significant increase in shoulder and elbow flexion range of motion (ROM) is clearly observed for the affected patient. The former goes from 53.33° to 64.6°. The second, more significant, goes from 21.1° to 56.2°. This increase is observed curiously in both limbs and not only in the affected one. However, a difference in the behavior of elbow rotation is observed between the right arm and the left arm of the patient with mild involvement. In the affected arm, rotation is performed abruptly almost at the beginning of the reach phase, remaining practically constant during most of the movement.

It is not possible to make a direct comparison with the kinematic results of patients with moderately and severely affected. In both cases, patients cannot touch the mouth with the hand of the affected limb. In the case of moderate involvement, approximately the height of the chin is reached, while in the case of severe involvement, only the height of the chest is reached.

The shoulder abduction curve in Figs. 9 and 10 shows a very different behavior between the two arms. The less affected arm shows a double mountain-shaped curve, similar to that in Figs. 7 and 8. As the hand approaches the mouth, the arm sticks to the body and detaches again when returning to the table. In patients with moderate and severe involvement, the upward movement of the affected limb is carried out keeping the arm separate from the body and the forearm practically horizontal, so that the abduction curve of the shoulder flattens and the intermediate valley disappears.

3.3 Temporal Evolution of Joint Moments

The joint moments must be able to balance the moment of the weights of the various segments that make up the upper body. But the joint moments must also be able to balance the moments associated with the inertia forces. Figures 7, 11 and 15 show for a healthy subject that the moment curves are practically symmetrical with respect to a central vertical axis. As the return path is similar to the reach path, so are the moment arms as well as the forces. This behavior is generally observed in all cases.

The accelerations appearing in HTM motion are not small. This makes the contribution to the joint moments of the inertia forces similar to the contribution of the weight. If, for example, we analyze the shoulder flexion moment in Fig. 7 corresponding to a healthy subject, we observe in the curve the existence of three maxima that cannot be explained by taking weight alone into account. The moment due to the weights would increase monotonically from the moment the hand leaves the table until it touches the mouth. However, on that same path there is first an acceleration followed by braking. This change of sign in acceleration is responsible for the valley between the first two maxima. A similar reasoning can be made for the second valley.

The decrease in smoothness of motion as the degree of limb involvement increases is not very visible in position, but is amplified by its derivatives. Thus, it was observed in the velocity modulus and is also observed in the joint moments that depend on the accelerations in Figures from 8 to 10 (shoulder), Figures from 12 to 14 (elbow) and Figures from 16 to 18 (wrist).

In previous figures, in case of patients with moderate and severe FMA-UE, when comparing the joint moments for the right and left extremities, it is generally observed that the curves become flatter and curlier for the affected extremity.

4 Conclusions

Joint moment curves have been obtained for stroke patients with different degrees of involvement and compared with those of a healthy subject. Kinetic results complement kinematic results and will allow a better assessment of the patient, help determine his or her treatment, and quantify his or her progress. The following conclusions should be validated by statistical analysis on a representative sample.

For one patient, the affected limb performs the HTM task significantly slower than the unaffected limb, with a higher NMU. For different patients, the affected limb moves slower and with a higher NMU the lower the value of the FMA-UE scale.

The unaffected limb does not always behave as a healthy one. In some patients, a reflection of the pathology is observed in the healthy limb.

The strong influence of accelerations on the kinetic results has been verified. For different patients, the curves of the joint moments of the affected limb show a higher curl the lower the value of the FMA-UE scale.

This study has some limitations. As suggested by other authors [13], the definition of phases (reach, approach, return) in the HTM task is complicated, especially for patients with low mobility. For this reason, it has been decided not to present the results by phases, although there are previous studies [14] that suggest potential different kinematic behaviors between the reach and return phases. However, only results from one capture per subject have been presented, instead of mean values. As seen in the figures, for each capture the time interval lasting the task is isolated and normalized to 100 points. Since in each capture the reach and return velocity and the approach time vary significantly, the calculation of the mean could affect points that do not correspond to the same position, falsifying the value of the mean. This is of particular concern in dynamics, where there are large oscillations associated with inertia forces. This problem can be addressed globally by nonlinear time interpolation [15] or locally by separating the captures into their three phases. These limitations do not affect the main objective of the work, which is to fill the gap in upper body kinetic results by developing a program to obtain joint moments in stroke patients during the performance of the hand-to-mouth task. The analysis is easily extendable to other tasks of daily life.

Acknowledgments. The authors would like to thank the Consejería de Transformación Económica, Industria, Conocimiento y Universidades (Junta de Andalucía) for funding the ProyExcel_00747 project, within which this study was carried out.

References

1. Plumacher, R., Ferrer-Ocando, O., Arteaga-Vizcaíno, M., Weir-Medina, J., Ferrer, O.: Enfermedades cerebrovasculares en pacientes con anemia falciforme. Investigaciones Clínicas **45**, 43–51 (2004)
2. Villepinte, C., Verma, A., Dimeglio, C., de Boissezon, X., Gasq, D.: Responsiveness of kinematic and clinical measures of upper-limb motor function after stroke: A systematic review and meta-analysis. Ann. Phys. Rehabil. Med. **64**, 101366 (2021)
3. Collins, K., Kennedy, N., Clark, A., Pomeroy, V.: Getting a kinematic handle on reach-to-grasp: a meta-analysis. Physiotherapy **104**, 153–166 (2018)
4. Schwarz, A., Kanzler, C., Lambercy, O., Luft, A., Veerbeek, J.: Systematic review on kinematic assessments of upper limb movements after stroke. Stroke **50**(3), 717–727 (2019)
5. Rab, G., Petuskey, K., Bagley, A.: A method for determination of upper extremity kinematics. Gait Posture **15**, 113–119 (2002)
6. Ojeda, J., Mayo, J., Ferrand, P., Martín, E., Zarco, M.J.: Influencia del procedimiento de reconstrucción en las métricas cinemáticas para valoración de pacientes de accidente cerebrovascular. Anales de Ingeniería Mecánica **23**, 110 (2023)
7. Ackermann, M.: Dynamics and Energetic of walking with prostheses, PhD Thesis, Stuttgart University, Germany (2007)
8. Winter, D.A.: Biomechanics and Motor Control of Human Movement, Fourth Edition. John Wiley & Sons, Inc (2009)
9. Topka, H., Konczak, J., Schneider, K., Boose, A., Dichgans, J.: Multijoint arm movements in cerebellar ataxia: abnormal control of movement dynamics. Exp. Brain Res. **119**, 493–503 (1998)
10. Schiefelbein, M.L., et al.: Upper-limb movement smoothness after stroke and its relationship with measures of body function/structure and activity – a cross-sectional study. J. Neurol. Sci. **401**, 75–78 (2019)
11. Alt Murphy, M., Willén, C., Sunnerhagen, K.S.: Responsiveness of upper extremity kinematic measures and clinical improvement during the first three months after stroke. Neurorehabil. Neural Repair **27**(9), 9 (2013)
12. Johansson, G.M., Grip, H., Levin, M.F., Häger, C.: The added value of kinematic evaluation of the timed finger-to-nose test in persons post-stroke. J. Neuroeng. Rehabil. **14**(1), 1 (2017)
13. Huang, X., Liao, O., Jiang, S., Li, J., Ma, X.: Kinematic analysis in post-stroke patients with moderate to severe upper limb paresis and non-disabled controls. Clin. Biomech. **113**, 106206 (2024)
14. Mesquita, I.A., Pinheiro, A.R.V., Velhote, M.F.P., Costa, C.I.: Methodological considerations for kinematic analysis of upper limbs in healthy and poststroke adults. part i: a systematic review of sampling and motor tasks. Top. Stroke Rehabil. **26**(2), 2 (2019)
15. Leo, K.H.: Establishing normative hand transport tangential velocity of a reach-to-grasp task: towards an objective assessment for UE stroke rehabilitation, PhD Thesis, Nanyang Technological University, Singapur (2014)

BY NC ND

Design and Validation of a Magnetic Bearing-Based Rotational Joint for Particulate-Free Cleanroom Robotics

G. Walter Torrestiana[1,2](✉), G. Leopoldo González[2], M. Emmanuel González[2], and Luis Bautista Cruz[2]

[1] Tecnológico de Estudios Superiores de Ecatepec (TESE), Ecatepec de Morelos, Mexico
walter@tese.edu.mx

[2] Universidad Nacional Autónoma de México (UNAM), Mexico City, Mexico

Abstract. This study investigates the integration of Active Magnetic Bearings as an enabling technology to reduce particulate contamination in ultra-clean environments, such as semiconductor fabrication and pharmaceutical cleanrooms. Traditional robotic joints rely on lubricated mechanical contact, which generates wear particles and outgassing compromising strict contamination control protocols. In contrast Active Magnetic Bearings enable contactless rotor levitation, inherently eliminating these contamination sources.

We present the design and implementation of a novel rotational kinematic pair, specifically developed for a Selective Compliance Assembly Robot Arm robotic joint, utilizing Active Magnetic Bearings to enable particle-free operation. The system incorporates a robust Linear Quadratic Gaussian controller to ensure precise and stable levitation and motion control. The control algorithm is implemented on a Field-Programmable Gate Array, enabling high-speed, real-time control required for Active Magnetic Bearing operation.

A custom mechanical structure was developed to integrate the Active Magnetic Bearings seamlessly into the robotic architecture, optimizing performance and minimizing particulate generation. Experimental validation was conducted in a controlled laminar flow hood, where key performance indicators including positioning accuracy, motion repeatability, and airborne particulate concentration were systematically evaluated using a laser particle counter.

The results demonstrate that the proposed Active Magnetic Bearing based system achieves sub-micron positioning accuracy and high repeatability while maintaining International Organization for Standardization Class 5 cleanliness levels during extended operation. These findings confirm the feasibility of Active Magnetic Bearings for cleanroom robotic applications, addressing a critical limitation of traditional actuation systems and enabling new use cases in contamination-sensitive environments.

Keywords: Active Magnetic Bearings · Linear Quadratic Gaussian Control · Cleanroom Robotics

List of Acronyms

ADRC Active Disturbance Rejection Control

O. F. Farías Fuentes et al. (Eds.): CIBIM 2024, *Proceedings of the XVI Ibero-American Congress of Mechanical Engineering*, pp. 483–503, 2026.
https://doi.org/10.1007/978-3-032-22823-9_35

AMB	Active Magnetic Bearing
CL	Cleanliness Level
DGAPA	General Directorate of Academic Personnel Affairs (from Spanish: Dirección General de Asuntos del Personal Académico)
FDM	Fused Deposition Modeling
FPGA	Field-Programmable Gate Array
ISO	International Organization for Standardization
LED	Light-Emitting Diode
LFH	Laminar Flow Hood
LQG	Linear Quadratic Gaussian
MOSFET	Metal-Oxide-Semiconductor Field-Effect Transistor
PAPIIT	Support Program for Research and Technological Innovation Projects (from SpanishPrograma de Apoyo a Proyectos de Investigación e Innovación Tecnológica)
PD	Proportional-Derivative
PID	Proportional-Integral-Derivative
PLA	Polylactic Acid
PWM	Pulse-Width Modulated
RPM	Revolutions Per Minute
SCARA	Selective Compliance Assembly Robot Arm
UNAM	National Autonomous University of Mexico (from Spanish: Universidad Nacional Autónoma de México)

1 Introduction

The use of magnetic fields to suspend metallic bodies and achieve contactless operation has been extensively studied across a wide range of engineering applications [1–3]. These systems operate by generating controlled electromagnetic fields to levitate a rotor within an Active Magnetic Bearing (AMB) [4]. The advantages of AMBs over conventional mechanical bearings are well established, chief among them being the elimination of physical contact between rotating components specifically, the rotor and the bearing. This absence of contact eliminates friction and the need for lubrication, thereby significantly reducing mechanical wear and extending component lifespan [5, 6].

In addition to wear reduction, AMBs offer several other inherent benefits, including high vibration tolerance, resilience to external disturbances, zero lubrication requirements, low maintenance, and high operational reliability [7–10]. Moreover, AMBs are capable of operating at high rotational speeds while maintaining precise positional control [11–13], making them well suited for demanding industrial applications.

However, AMBs are inherently nonlinear and unstable in open-loop configurations. Therefore, a closed-loop feedback control system is essential to ensure stable levitation. Such systems regulate the electromagnetic force used as the control input to maintain the rotor at its desired position [14].

AMB systems are complex mechatronic platforms comprising mechanical, electrical, and electronic subsystems, including position sensors, rotors, power amplifiers, data acquisition modules, embedded controllers, and electromagnets. In [15], a modeling approach for a two-degree-of-freedom AMB system is presented, using reduced-order modeling and a Proportional-Integral-Derivative (PID) controller to regulate the rotor position. However, the study focuses primarily on the mechanical model and does not incorporate the electronic subsystems.

Other studies have examined the impact of decentralized control strategies and how the nonlinear behavior of the AMB's radial force influences system performance [16, 17]. For instance, [18] employs a homopolar electromagnetic actuator and a Proportional-Derivative (PD) controller for rotor stabilization. A comparative analysis of Active Disturbance Rejection Control (ADRC) and PID control is reported in [19]. In [20], the authors explore the geometric interplay between AMBs and traditional bearings in hybrid configurations. Additionally, [21] proposes an optimized sensor placement strategy for AMB systems, which reduces cost and complexity while improving dynamic performance through enhanced sensor-actuator alignment.

This work proposes the use of an Active Magnetic Bearing in a rotational kinematic pair specifically designed for low-speed robotic applications. The prototype was fabricated using the Fused Deposition Modeling (FDM) additive manufacturing process with polylactic acid (PLA) material. The system's performance was evaluated through repeatability and cleanliness tests, conducted in accordance with current International Organization for Standardization (ISO) cleanroom standards.

2 System Description

In the proposed design, the axis of the kinematic pair is modeled as an inverted pendulum, whose position is actively regulated by an Active Magnetic Bearing (AMB) through magnetic forces generated by two pairs of electromagnets (Fig. 1).

The AMB is mounted on an aluminum frame that ensures proper alignment with a servomotor located at the base of the system. The shaft of the kinematic pair is positioned atop the servomotor shaft, transmitting rotational motion via a mechanical coupling.

At the upper end of the shaft, a radial extension acts as the end-effector. This component is used to conduct repeatability and imbalance tests, enabling evaluation of AMB performance under varying operational conditions.

A Linear Quadratic Gaussian (LQG) control strategy is employed in the prototype. This optimal control technique is particularly well suited for nonlinear systems that can be linearized around an operating point in this case, the rotor centered between the pairs of electromagnets. The electromagnetic actuators maintain this equilibrium position. Experimental validation confirms that the proposed control scheme successfully stabilizes the system and exhibits robust performance in the presence of external disturbances.

2.1 Mechanical Design

The prototype of the rotational kinematic pair consists of two mechanical subsystems: the shaft subsystem (Fig. 2), which transmits rotational motion from the servomotor to the end-effector, and the radial subsystem (Fig. 3), which houses the AMB components.

Fig. 1. Test bench for a rotational kinematic joint using an AMB.

All structural elements, with the exception of the shaft and electromagnets, were fabricated using additive manufacturing via the Fused Deposition Modeling (FDM) technique. Polylactic acid (PLA) was selected as the printing material due to its favorable mechanical, thermal, and environmental properties. Notably, studies such as [22] have demonstrated that PLA components can meet ISO Class 5 cleanroom cleanliness standards, validating their use in contamination-sensitive environments.

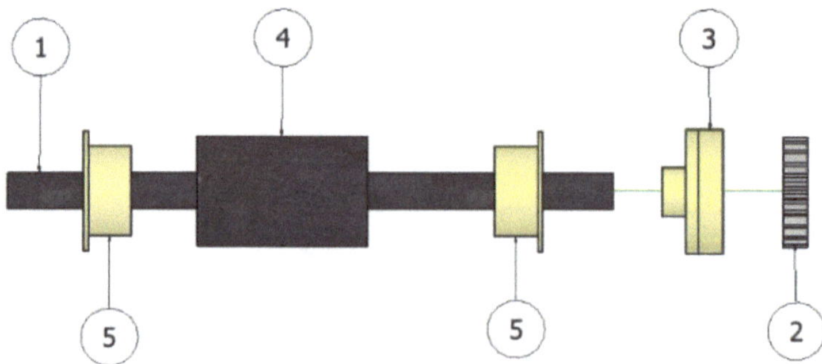

Fig. 2. Shaft subsystem.

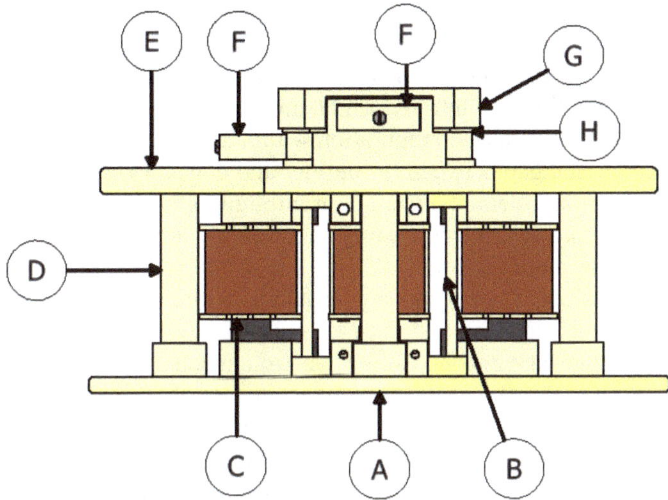

Fig. 3. Radial subsystem.

The shaft subsystem elements are presented in Table 1, and the radial subsystem elements are presented in Table 2.

Table 1. Shaft subsystem elements

Number	Element
1	Main Shaft
2	Servomotor shaft coupling
3	Primary shaft-servomotor coupling
4	Secondary shaft
5	Shaft couplings

Table 2. Radial subsystem elements

Letter	Element
A	Lower base
B	Front Support
C	Electromagnet
D	Axial Support
E	Upper base
F	Sensor holder
G	Sensor cover
H	Sensor holder base

2.2 Sensors and Actuators

The prototype employs optical position sensors, one per control axis of the AMB. Each sensor consists of a collimated infrared Light-Emitting Diode (LED) and a phototransistor arranged in a face-to-face configuration within a dedicated housing (Fig. 3). The shaft passes through the gap between the emitter and detector (Fig. 4). When the shaft undergoes lateral displacement, the amount of infrared light received by the phototransistor changes, resulting in a corresponding variation in output voltage. This voltage is processed to infer the shaft's position.

Experimental characterization confirms that the sensors operate effectively within a $\pm$ 1 mm range, with a resolution of 0.00074 mm.

Actuation is achieved via four C-core electromagnets (Fig. 5), whose geometry concentrates the magnetic flux, thereby enhancing the magnetic force applied to the shaft. Each electromagnet is driven by an amplifier stage based on a power Metal-Oxide-Semiconductor Field-Effect Transistor (MOSFET). These amplifiers are modulated using Pulse-Width Modulated (PWM) signals generated by a Field-Programmable Gate Array (FPGA) board, enabling high-precision control of the electromagnetic forces.

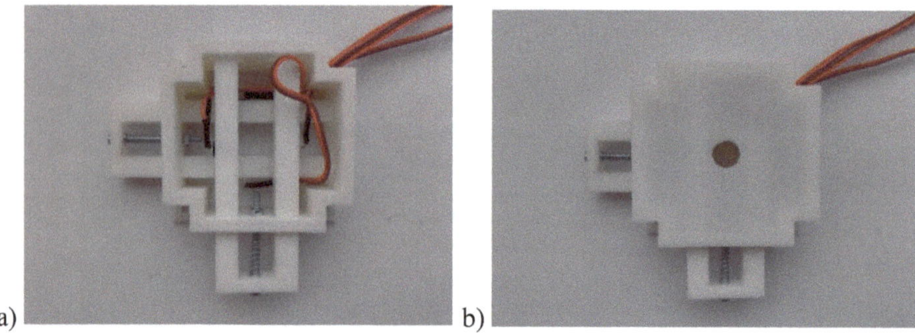

Fig. 4. Sensor base (Top view). a) Base without lid; b) Base with lid.

2.3 Control System

The AMB requires a closed-loop control system capable of dynamically modulating the magnetic forces generated by the electromagnets to maintain the shaft centered within the radial gap. To this end, a decentralized LQG controller was implemented, comprising two independent control loops arranged in a differential configuration one for each orthogonal control axis (Fig. 6). The system's equilibrium corresponds to the shaft being perfectly vertical and centered.

Fig. 5. Electromagnet used on the AMB test bench.

To design the controller, a nonlinear dynamic model of the system was first developed (Eq. 1) and subsequently linearized around the equilibrium point. The state vector was defined as shown in Eq. (2), including rotor position, velocity, and the current through the left and right electromagnets.

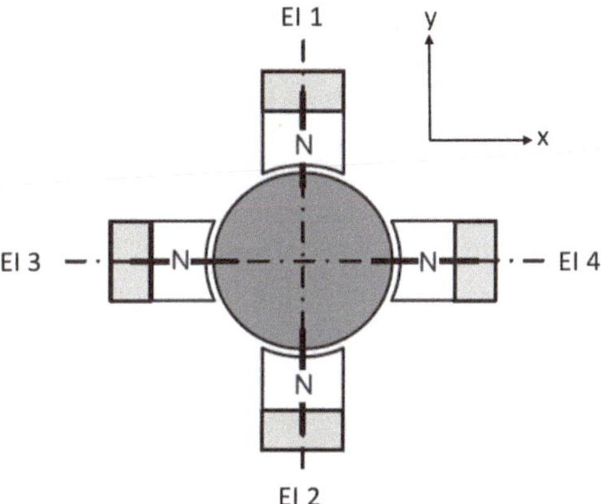

Fig. 6. AMB pole configuration.

The LQG controller was formulated to minimize a standard quadratic cost function (Eq. 3), with weighting matrices Q and R defined in Eqs. (4) and (5), respectively. These matrices were tuned to prioritize rotor position control while satisfying the system's dynamic response constraints.

$$\begin{bmatrix} \dot{x}_1 \\ \dot{x}_2 \\ \dot{x}_3 \\ \dot{x}_4 \end{bmatrix} = \begin{bmatrix} \dot{\theta} \\ \frac{b\dot{\theta} - glm(sen\theta) + \frac{c_L i_L^2 l cos\theta}{2(d_L + l sen\theta)^2} - \frac{c_R i_R^2 l cos\theta}{2(d_R - l sen\theta)^2}}{I + l^2 m (cos\theta)^2 + l^2 m (sen\theta)^2} \\ \frac{u_L - R_L i_L}{L_L} \\ \frac{u_R - R_R i_R}{L_R} \end{bmatrix} \tag{1}$$

$$x = \begin{bmatrix} \theta \\ \dot{\theta} \\ i_L \\ i_R \end{bmatrix} = \begin{bmatrix} x_1 \\ x_2 \\ x_3 \\ x_4 \end{bmatrix} \tag{2}$$

$$J = \int_{t0}^{tf} (x^T Q x + u^T R u) dt \tag{3}$$

$$Q = \begin{bmatrix} 100 & 0 & 0 & 0 \\ 0 & 100 & 0 & 0 \\ 0 & 0 & 1 & 0 \\ 0 & 0 & 0 & 1 \end{bmatrix} \tag{4}$$

$$R = 1 \tag{5}$$

A block diagram of the state-feedback control architecture, incorporating Kalman filter state estimation, is shown in Fig. 7. The control signal delivered to the electromagnet driver stage is a PWM waveform with variable duty cycle. The magnitude of the magnetic force is proportional to the duty cycle, enabling fine-grained control of rotor position.

To enhance rotor stiffness at the equilibrium position, a constant bias term is added to the LQG control signal. This bias generates a baseline magnetic force, the magnitude of which determines the maximum radial load the AMB can support.

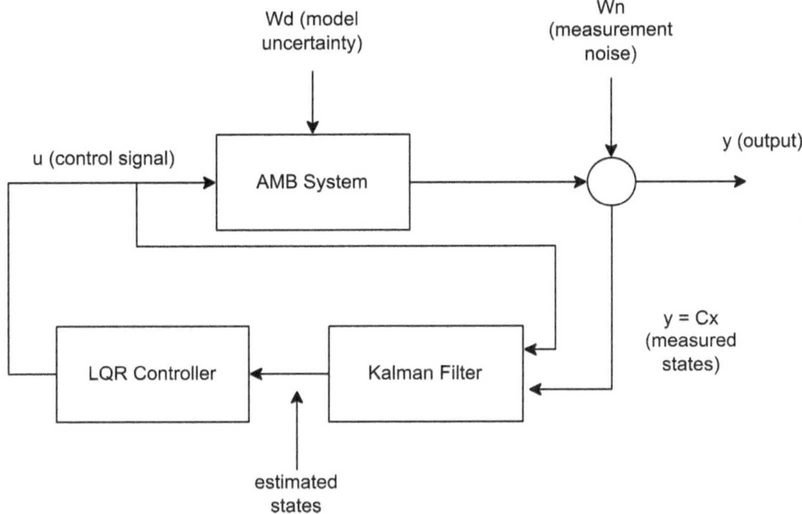

Fig. 7. Block diagram of the LQG controller

3 Repeatability Testing

To assess the performance of the Active Magnetic Bearing (AMB), a repeatability testing protocol was developed in accordance with ISO 9283, which provides the guidelines for characterizing the repeatability of robotic joints. The testing procedure consisted of three main phases: preparation, experimental testing, and analysis of results.

3.1 Preparation Phase

The preparation phase involved the following steps:

- Visual Inspection: The AMB was inspected for visible damage, loose connections, or other anomalies.
- Electrical Verification: All electrical connections were checked to ensure proper installation and secure fastening.
- Power Supply Setup: The AMB was connected to the designated power source.
- Initial Operational Check: The AMB was powered on and monitored for proper operation, verifying the absence of excessive vibration, abnormal noise, or overheating.
- Control Signal Response Test: The AMB's response to control inputs was evaluated to confirm expected behavior.
- Simulated Load Application: A representative load was applied to observe system performance under typical operating conditions.

3.2 Experimental Testing and Results

Workspace Characterization.
The operational workspace was defined as the three-dimensional volume in which the end-effector can move and perform tasks. This required establishing the AMB's reference coordinate systems. Figure 8 illustrates the reference frames, and Fig. 9 depicts the defined workspace.

Maximum Load Capacity.
Incremental weights of 10 g were applied at the origin of the local reference frames X_1, Y_1, and Z_1 on the end-effector. The resulting displacements were recorded to evaluate load sensitivity. The system showed a high sensitivity to applied loads. Results are summarized in Table 3.

Table 3. Weight and displacement

Weight [g]	Displacement [μm]
0	83.93
10	178.73
20	258.66
30	349.85
40	449.98
50	536.39
60	635.07
70	705.72
80	817.85
90	950.73
100	1029.67

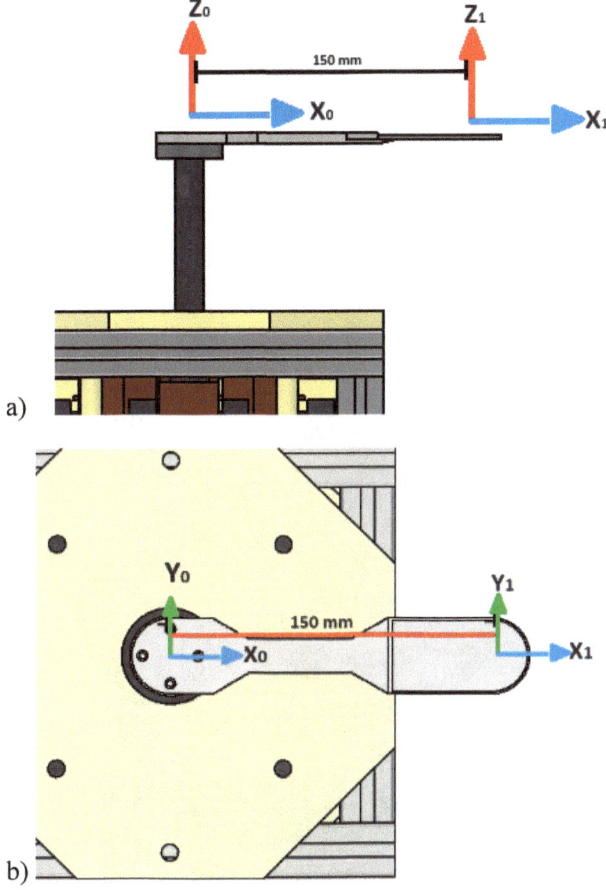

Fig. 8. AMB reference frameworks. a) Side view; b) Top view.

Maximum Operating Speed

The servomotor speed was increased in steps of 5 Revolutions Per Minute (RPM) while monitoring the shaft displacement using position sensors on the x and y axes. The maximum stable operating speed of the AMB was determined to be 20 RPM.

Repeatability Testing

Tests were conducted to evaluate the power-on and positioning repeatability of the end-effector, using a dial indicator and the control system's sensor array. For the power-on repeatability tests, the dial indicator was positioned as shown in Fig. 10 while the Active Magnetic Bearing (AMB) system was operating. The indicator was then set to zero, establishing this value as the reference point.

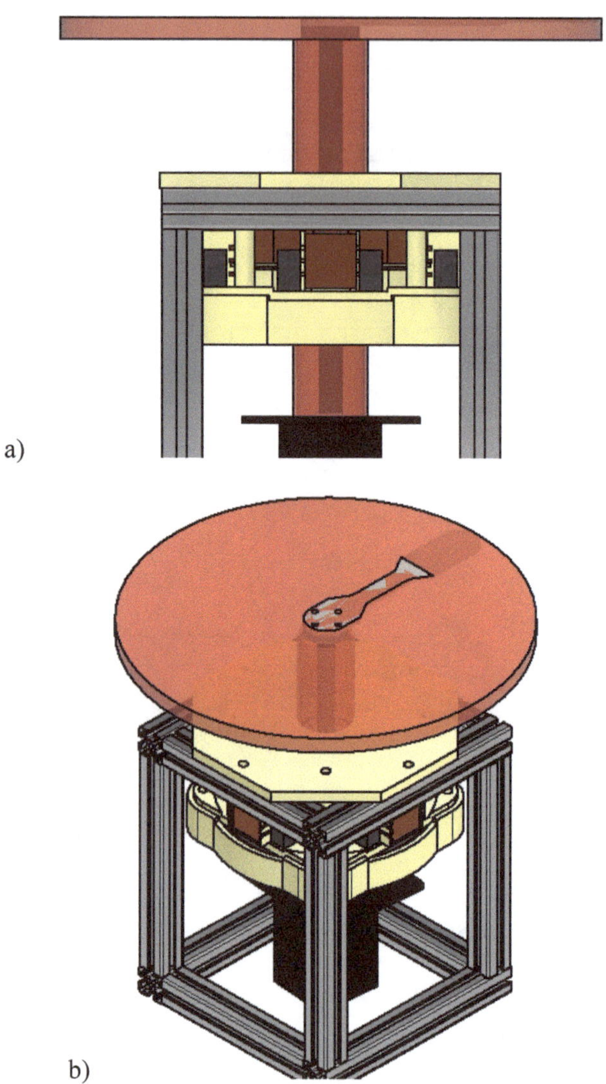

Fig. 9. AMB workspace. a) Side view; b) Isometric view.

The AMB control system was powered off and on for 30 cycles, in accordance with ISO 9283 standards for validating power-on repeatability. The results obtained are presented in Table 4.

For the positioning repeatability tests, the dial indicator was placed as shown in Fig. 11 while the AMB was powered on. This position was chosen because aligning the dial indicator with the Y_1 axis and adjusting it induced a moment that the AMB could not counteract, causing the shaft to deviate from the X_0, Y_0, and Z_0 reference origin. By selecting this position, the shaft was ensured to remain aligned with the origin. Once the new reference frame X_1, Y_1, and Z_1 was established, the shaft was rotated -90° about the Z_0 axis and then returned to the origin to collect data on the positioning repeatability of the end-effector.

Fig. 10. Dial indicator mounted on the end effector.

This procedure was repeated for 30 cycles. The positioning repeatability test was then performed again with a -180° rotation to validate the obtained results. The data are presented in Table 4.

To calculate the repeatability, Eqs. (6) to (10) were used, which are established in the ISO 9283 standard [23] and [24].

$$\overline{x} = \frac{1}{n}\sum_{j=1}^{n} x_j \tag{6}$$

$$l_{xj} = \left|x_j - \overline{x}\right| \tag{7}$$

$$\overline{l_x} = \frac{1}{n}\sum_{j=1}^{n} l_{xj} \tag{8}$$

$$S_x = \sqrt{\frac{\sum_{j=1}^{n}\left(l_{xj} - \overline{l_x}\right)^2}{n-1}} \tag{9}$$

$$RP_x = \overline{x} + 3S_x \tag{10}$$

Table 4. AMB ignition repeatability data and end effector positioning.

Cycle	Ignition repeatability			Positioning repeatability	
	X0	Y0	Z0	Y1 a 90°	Y1 a 180°
1	–182.435	543.540	0	0	2
2	–182.391	543.351	1	2	2
3	–180.708	543.600	0	3	2
4	–183.479	544.403	2	0	3
5	–185.592	541.861	0	0	3
6	–181.990	539.524	1	0	3
7	–181.572	540.693	1	0	2
8	–180.737	545.837	0	1	4
9	–181.672	540.460	1	0	3
10	–181.700	541.199	0	0	1
11	–182.925	543.536	0	1	2
12	–183.737	544.532	1	0	1
13	–179.576	541.909	0	0	3
14	–183.242	544.652	0	4	3
15	–184.415	544.448	0	1	4
16	–187.339	545.64	0	3	3
17	–184.961	542.978	0	4	1
18	–179.612	543.757	0	4	3
19	–183.974	540.171	0	0	3
20	–181.544	544.102	0	1	1
21	–181.311	540.291	0	1	3
22	–178.793	541.444	0	0	2
23	–180.210	544.436	0	0	3
24	–183.251	541.231	0	0	0
25	–183.801	540.215	0	0	0
26	–178.307	541.757	0	3	3
27	–181.961	543.062	2	4	2
28	–180.969	540.030	1	4	4
29	–180.267	544.468	0	1	1
30	–188.07	540.709	0	1	0

(*continued*)

Table 4. (*continued*)

Cycle	Ignition repeatability			Positioning repeatability	
	X0	Y0	Z0	Y1 a 90°	Y1 a 180°
Repeat-ability	6.888	5.575	1.819	5.923	5.728

4 Cleanroom Compatibility and Cleanliness Testing

To evaluate the cleanroom compatibility of the system, cleanliness tests were conducted in accordance with the ISO 14644–1 standard, which specifies the classification of air cleanliness by particle concentration in clean environments. The testing protocol began with determining the area of the Laminar Flow Hood (LFH), which served as the test environment. As shown in Fig. 12, the LFH has a base measuring 1.2 m × 0.6 m, yielding a total area of 0.72 m^2.

According to ISO 14644–1, for areas smaller than 1 m^2, a single sampling point is sufficient to determine the Cleanliness Level (CL). However, the standard also recommends that additional measurements be taken at locations deemed critical. In this study, particle concentration was assessed at three locations within the LFH, as illustrated in Fig. 12.

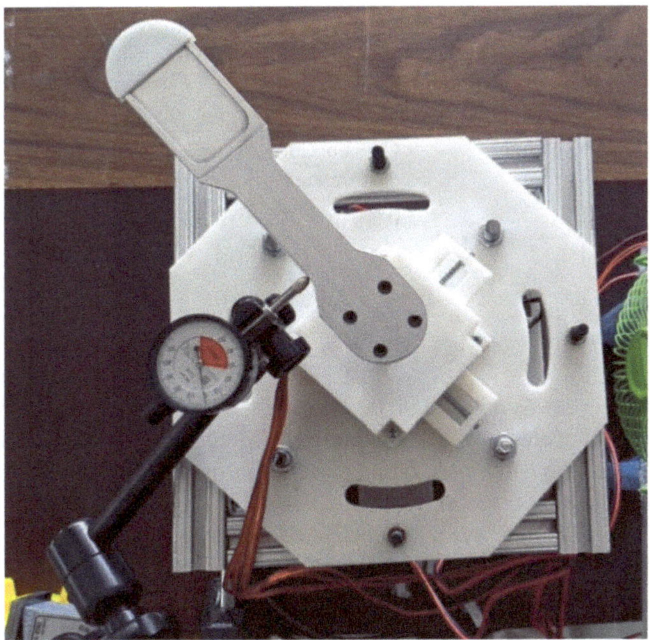

Fig. 11. Dial indicator positioned for positioning repeatability testing.

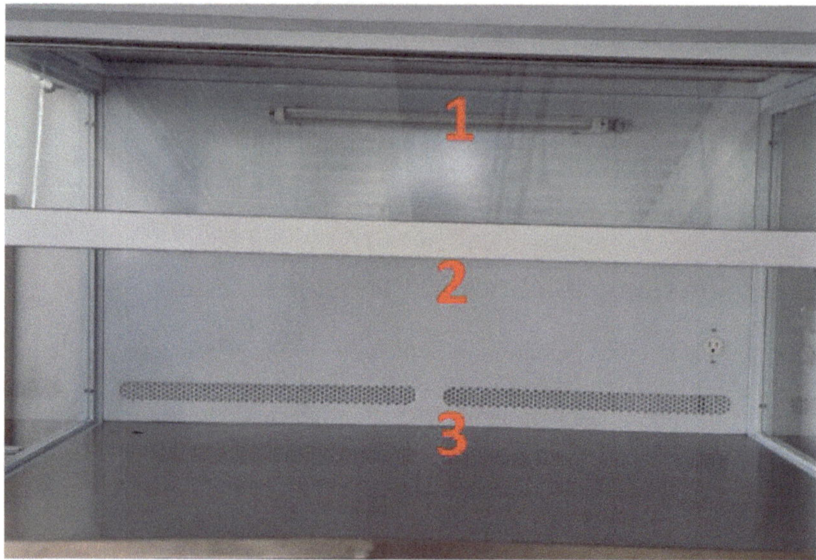

Fig. 12. Clean Level Sampling Locations.

The measurements obtained at all three locations confirmed that the interior of the LFH complied with the requirements for ISO Class 5 air cleanliness. These results are summarized in Table 5.

Table 5. Average concentration of particles inside the laminar flow hood.

Location	Concentration of particles per cubic meter	
	0.3 μm	0.5 μm
1	0	0
2	0	0
3	0	0

Prior to testing, the Active Magnetic Bearing (AMB) was cleaned using isopropyl alcohol and microfiber cloths, as recommended by Mathia [25]. After cleaning, the AMB was installed inside the LFH, and the preparation steps from the repeatability protocol were repeated to ensure correct operation.

The LFH's ventilation system was activated and left to run for 10 min to stabilize internal airflow and particle removal. After this stabilization period, particle concentration was measured at sampling location 3. Measurements confirmed zero particle counts for both 0.3 μm and 0.5 μm particles, thereby validating the internal cleanliness of the LFH prior to system operation.

Subsequently, dynamic cleanliness testing was conducted to assess particle generation during operation of the AMB. Sampling locations were selected as shown in

Fig. 13, 14 and 15. The drive system was then activated, rotating the end-effector while the particle counter recorded airborne contamination in real time.

a)
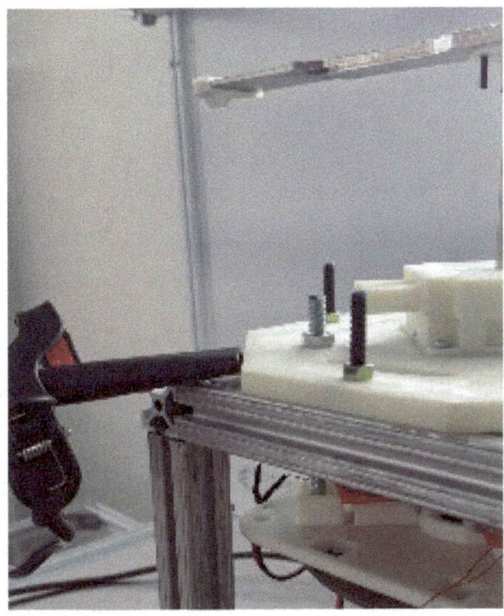

Fig. 13. Location 1 to measure particle concentration.

b)
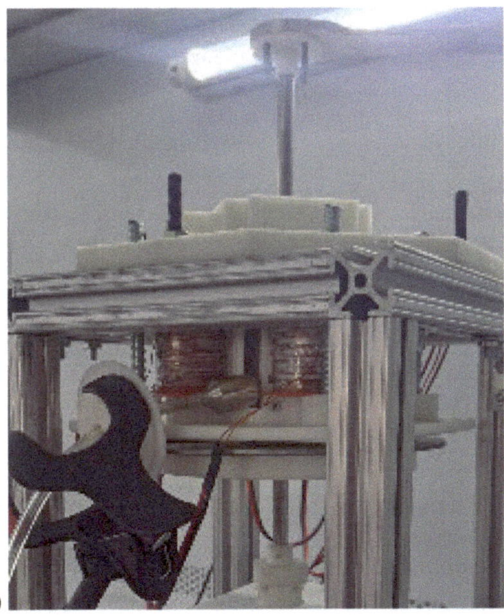

Fig. 14. Location 2 to measure particle concentration.

Fig. 15. Location 3 to measure particle concentration.

Measurements were taken at each location for one hour daily over a three-day testing period. The resulting particle concentration data, shown in Table 6, were used to verify the system's compliance with ISO Class 5 cleanliness requirements under operational conditions.

Table 6. Average particle concentration by location.

Location	Concentration of particles per cubic meter	
	0.3 µm	0.5 µm
1	45	0
2	239	18
3	61	14

5 Discussion of Experimental Results

The repeatability tests confirm that the proposed rotational kinematic pair achieves highly precise positioning performance. Experimental data reveal a power-on repeatability of less than 7 µm along each axis, underscoring the system's high degree of consistency across power cycles. It is worth noting that direct comparisons with existing Active Magnetic Bearing (AMB) systems are limited, as most AMB implementations are designed for turbomachinery or aerospace applications. Although research efforts toward robotic applications of AMBs exist, to date, no comparable prototypes have reported power-on repeatability metrics suitable for benchmarking.

The positioning repeatability of the end-effector was measured at 6 μm, which represents a high level of repeatability particularly advantageous in robotic systems where sub-millimeter precision is required.

In terms of accuracy, the absolute power-on accuracy of the AMB system was found to be 3.015 μm, while the absolute positioning accuracy reached 1.940 μm. For comparison, commercial Selective Compliance Assembly Robot Arm (SCARA) robots such as the Yamaha YK-TG Series typically offer an absolute accuracy of ± 20 μm. This contrast highlights the potential of AMBs to outperform conventional joint technologies in precision-demanding cleanroom robotics.

The results of the cleanliness tests further support the viability of the proposed system for use in particulate-sensitive environments. Components fabricated using additive manufacturing specifically, the Fused Deposition Modeling (FDM) technique with PLA material achieved compliance with ISO Class 5 cleanliness standards. The measured particle concentrations during AMB operation were 115 particles/m^3 for 0.3 μm particles and 10 particles/m^3 for 0.5 μm particles. These values are well below the maximum permissible limits for ISO Class 5 cleanrooms, which are 10,200 particles/m^3 and 3,520 particles/m^3, respectively.

After a total of 9 h of operation distributed over a 3-day testing period, the AMB system consistently maintained particle emissions within acceptable thresholds. These findings demonstrate the feasibility of integrating additively manufactured AMB components into robotic systems designed for cleanroom environments, combining high-precision motion control with minimal particulate generation.

6 Conclusions

The experimental results obtained from the various tests performed on the proposed magnetic bearing-based rotational joint prototype demonstrate high repeatability and positioning accuracy, both in power-on alignment and in end-effector placement. These outcomes were achieved by implementing a functional testing protocol based on ISO 9283 standards. The primary motivation for reducing the size of an active magnetic bearing (AMB) lies in its potential integration into robotic joints. Given the high repeatability achieved with the developed prototype, the proposed system can be considered a viable alternative to ultra-clean bearings currently employed in robotic joints designed for cleanroom applications.

It is worth noting that the prototype's overall dimensions can still be reduced, and further improvements to the AMB control system are possible to enhance its dynamic response under external loads, such as those applied by the end-effector.

The PLA-manufactured components showed favorable performance regarding particle emission. Additive manufacturing enabled the creation of complex geometries, and due to the cleanliness levels observed, this fabrication method proves promising for cleanroom-oriented applications. It offers advantages such as reduced production costs and rapid prototyping. To validate the cleanliness performance of the AMB, a particle emission test protocol was developed based on ISO 14644–1 requirements. The results confirmed compliance with ISO Class 5 cleanroom standards.

Acknowledgments. The authors gratefully acknowledge the support from DGAPA-UNAM through project PAPIIT IT101321, titled "Development of a Magnetic Bearing-Based Rotational Joint for Application in Cleanroom Robotics."

References

1. Arredondo, I., Jugo, J., Alonso-Quesada, S., Lizarraga, I., Etxebarria, V.: Modeling, analysis and control of active magnetic bearings systems. RIAI - Revista Iberoamericana de Automatica e Informatica Industrial **5**(4), pp. 1–13. Springer, Heidelberg (2016)
2. Binder, A., Sabirin, C., Popa, D., Craciunescu, A.: Modeling and digital control of an active magnetic bearing system. Revue Roumaine Des Sciences Techniques Serie Electrotechnique Et Energetique **52**(2), 157 (2007)
3. Huang, J., Wang, L., Huang, Y.: Continuous time model predictive control for a magnetic bearing system. PIERS Online **3**(2), 202–208 (2007)
4. Huang, T., Zheng, M., Zhang, G.: A review of active magnetic bearing control technology. In: Proceedings of the 31st Chinese Control and Decision Conference, CCDC 2019, pp. 2888–2893 (2019)
5. Hung, J.Y.: Magnetic bearing control using fuzzy logic. IEEE Trans. Ind. Appl. **31**(6), 1492–1497 (1995)
6. Husain, A.R., Ahmad, M.N., Yatim, A.H.M.: Deterministic models of an active magnetic bearing system. J. Comput. **2**(8), 9–17 (2007)
7. Ji, J.C., Hansen, C.H., Zander, A.C.: Nonlinear dynamics of magnetic bearing systems. J. Intell. Mater. Syst. Struct. **19**(12), 1471–1491 (2008)
8. Lindlau, J.D., Knospe, C.R.: Feedback linearization of an active magnetic bearing with voltage control. IEEE Trans. Control Syst. Technol. **10**(1), 21–31 (2002)
9. Meressi, T., Kao, M.C.: Modeling and control of a magnetic levitation system. In: Proceedings of the Eight IASTED International Conference on Control and Applications, Vol. 2006(2), pp. 48–51 (2006)
10. Raghunathan, P., Logashanmugam, E.: Position servo controller design and implementation using low cost eddy current sensor for single axis active magnetic bearing. J. Ambient Intell. Humanized Comput. (2018)
11. Rong, G., Gang, L., Yan, C.X.: Dynamic modeling and analysis of active magnetic bearings. Appl. Mech. Mater. 685–494–495.685 (2014)
12. Sun, Z., He, Y., Zhao, J., Shi, Z., Zhao, L., Yu, S.: Identification of active magnetic bearing system with a flexible rotor. Mech. Syst. Signal Process. **49**(1–2), 302–316 (2014)
13. Xu, Y., Shen, Q., Zhang, Y., Zhou, J., Jin, C.: Dynamic modeling of the active magnetic bearing system operating in base motion condition. IEEE Access **8**, 166003–166013 (2020)
14. Yaseen, M.H.A., Abd, H.J.: Modeling and control for a magnetic levitation system based on SIMLAB platform in real time. Res. Phys. **8**, 153–159 (2018)
15. Polajzer, B.: Modeling and control of horizontal shaft magnetic bearing system. In: ISIE' 99. Proceedings of the IEEE International Symposium on Industrial Electronics (Cat. No. 99TH8465), pp. 1051–1055. IEEE (1999)

16. Polajzer, B., Ritonja, J., Stumberger, G., Dolinar, D., Lecointe, J.-P.: Decentralized PI/PD position control for active magnetic bearings. Electr. Eng. **89**(1), 53–59 (2006)
17. POlajzer, B., Ritonja, J., DOlinar, D.: Impact of radial force nonlinearities on decentralized control of magnetic bearings. In: XIX International Conference on Electrical Machines-ICEM 2010, pp. 1–4. IEEE (2010)
18. Zhong, W., Palazzolo, A.: Magnetic bearing rotordynamic system optimization using multi-objective genetic algorithms. J. Dyn. Syst. Meas. Control **137**(2), 021012 (2015). https://doi.org/10.1115/1.4028401
19. Jin, C., Guo, K., Xu, Y., Cui, H., Xu, L.: Design of magnetic bearing control system based on active disturbance rejection theory. J. Vib. Acoust. **141**(1), 011009 (2019). https://doi.org/10.1115/1.4040837
20. Barbaraci, G., Mariotti, G.V., Piscopo, A.: Active magnetic bearing design study. J. Vib. Control **19**(16), 2491–2505 (2013)
21. Maslen, E.: Self-sensing magnetic bearings, pp. 435–459. Magnetic bearings Springer, Berlin (2009)
22. Pasanen, T., von Gastrow, G., Heikkinen, I., Vähänissi, V., Savin, H., Pearce, J.: Compatibility of 3-D printed devices in cleanroom environments for semiconductor processing. Mater. Sci. Semicond. Process. **89**, 59–67 (2019). https://doi.org/10.1016/j.mssp.2018.08.027
23. Mousavi, A., Akbarzadeh, A., Shariatee, M., Alimardani, S.: Repeatability analysis of a SCARA robot with planetary gearbox. In: 2015 3rd RSI International Conference on Robotics and Mechatronics (ICROM), pp. 640–644 (2015). https://doi.org/10.1109/ICRoM.2015.7367858
24. Mihelj, M., et al.: Accuracy and repeatability of industrial manipulators. Robotics 231–241 (2018). https://doi.org/10.1007/978-3-319-72911-4_15
25. Mathia, K.: Robotics for Electronics Manufacturing: Principles and Applications in Cleanroom Automation. Cambridge University Press (2010)

Implementation of a Collaborative Robot for Chemistry Lab Tasks

Isai Josue Rios Rosas[1(✉)], Francisco D. Calvo López[1], María Judith Percino Zacarías[2], Enrique Pérez-Gutiérrez[2], Manuel J. Heredia-Rios[2], David Pinto-Avendaño[3], and Humberto Salazar Ibargüen[4]

[1] Faculta de Electrónica, Decanato de Ingeniería, Universidad Popular Autónoma del Estado de Puebla, Puebla, México
isaijosue.rios@upaep.edu.mx, franciscodomingo.calvo@upaep.mx

[2] Unidad de Polímeros y Electrónica Orgánica, Instituto de Ciencias, Benemérita Universidad Autónoma de Puebla, Puebla, México
{judith.percino,enrique.pgutierrez}@correo.buap.mx, co1539773@colaborador.buab.mx

[3] Dirección de Innovación y Transferencia del Conocimiento, Benemérita Universidad Autónoma de Puebla, Puebla, México
david.pinto@correo.buap.mx

[4] Centro Interdisciplinario de Investigación y Enseñanza de La Ciencia, Benemérita Universidad Autónoma de Puebla, Puebla, México
humberto.salazar@correo.buap.mx

Abstract. The manuscript describes the use of an ABB YuMi IRB 14000 collaborative robot to recognize and handle common tools used in a chemistry lab. It details the selected tools, calibration procedures, and robot programming for handling and object recognition tasks. The results highlight the importance of correct workspace and tool calibration to ensure precision in object handling. The implications of these findings are discussed, and improvements for future applications of the robot in a chemistry lab are proposed.

Keywords: Collaborative · YuMi · Calibration · Handling · Recognition

1 Introduction

The implementation of collaborative robots in laboratory processes has gained significant relevance due to their ability to work safely alongside humans, thereby optimizing efficiency and precision in repetitive and delicate tasks.

This article reports the first attempts to integrate the ABB YuMi IRB 14000 collaborative robot into a chemistry lab environment specialized in polymer synthesis, to recognize and handle materials and items used for chemistry synthesis. The tools used, calibration procedures, and programming required to identify a workplace and objects used in the lab, as well as ensure safe and precise operation are described. Additionally, the results obtained are presented, and proposed improvements for future implementations are discussed.

O. F. Farías Fuentes et al. (Eds.): CIBIM 2024, *Proceedings of the XVI Ibero-American Congress of Mechanical Engineering*, pp. 504–523, 2026.
https://doi.org/10.1007/978-3-032-22823-9_36

2 ABB YuMi IRB 14000 Collaborative Robot

The ABB YuMi IRB 14000 is a dual-arm collaborative robot [1], specifically designed to work safely alongside humans in various environments (Fig. 1). Each of YuMi's arms has 7 degrees of freedom, allowing for complex movements with high precision. The arms are padded, and the robot features a compact and lightweight design that maximizes safety for personnel in direct contact with the system.

YuMi features include collision prediction and instant stop upon detecting any impact. This robot is ideal for handling small parts, delicate materials, and performing measurement and inspection tasks.

With its advanced safety and precision capabilities, the ABB YuMi IRB 14000 is ideally suited for laboratory environments that deal with sensitive materials, such as polymers. This section details the selected tools and their integration into the laboratory.

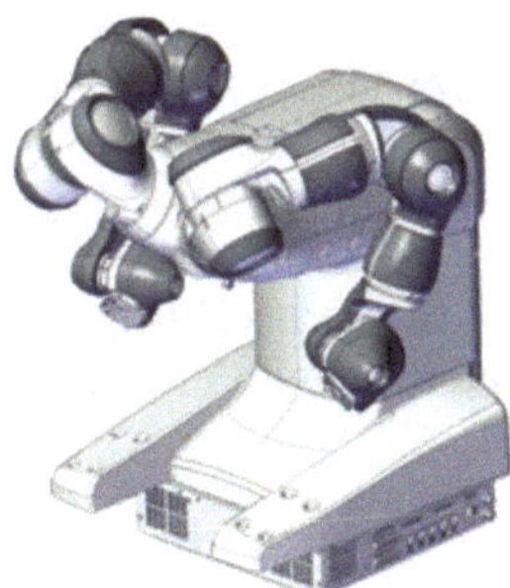

Fig. 1. General view of the ABB YuMi IRB 14000 robot

2.1 Tool Selection for the Robot

Gripper:

The gripper (Fig. 2) is used to grasp and manipulate objects of various shapes and sizes [2]. This tool is essential for the safe and precise handling of laboratory materials.

Fig. 2. General view of Smart Gripper ABB YuMi

Camera:
The camera enables object identification through computer vision, Fig. 3. This visual recognition capability is fundamental for identifying equipment and substances, as well as ensuring proper handling.

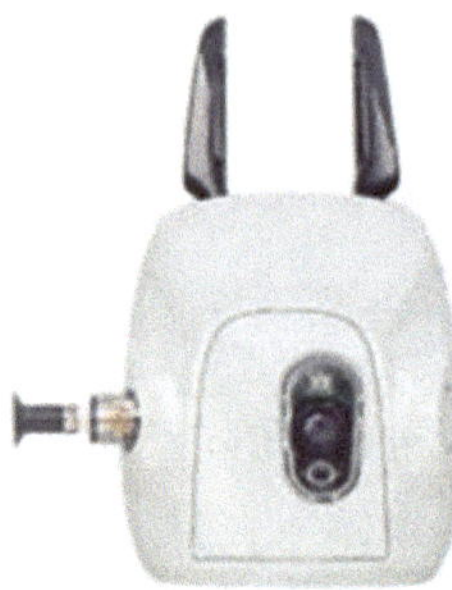

Fig. 3. ABB YuMi Smart Gripper with camera view

Anti-Reflective Mat:
The anti-reflective mat improves the camera's ability to recognize objects by eliminating unwanted reflections that could interfere with the vision system.

2.2 Identification of Laboratory Materials to Be Used

To carry out the synthesis process and other lab tasks, a variety of laboratory materials will be utilized, including test tubes, flasks, and specialized containers for storing and handling chemical reagents. To ensure process integrity, it is to test each tool and container for compatibility and resistance with the selected robotic tool.

3 Preparation of the Robot, Tools, and Workspace

The process begins with calibrating the robot and its tools using RobotStudio software [3]. First, the robot model is imported, and the Smart Gripper, the camera, and the workspace are configured. These steps are crucial to ensure safe and efficient operation.

3.1 Calibrations

To program the robot correctly, it is necessary to calibrate each tools being used; otherwise, collisions with the workspace or potential harm to laboratory personnel could occur.

3.1.1 Robot Calibration

Initial Positioning: Both arms of the YuMi must be placed in a reference position. This is typically a predefined configuration that aligns the arms in a neutral posture.
Execute Calibration: Running the calibration procedure, which generally includes updating the "Revolution Counter."
Validation and Adjustment: Validate that the positions reached by the arms match the predefined positions. If necessary, make fine adjustments until the arms are correctly calibrated.
Save Configuration: Once both arms are calibrated, save the configuration to the robot's system to ensure the calibration is maintained during operations.

3.1.2 Camera Calibration

It is important to have an initial position and reference point for the camera to acquire the image properly, Fig. 4.

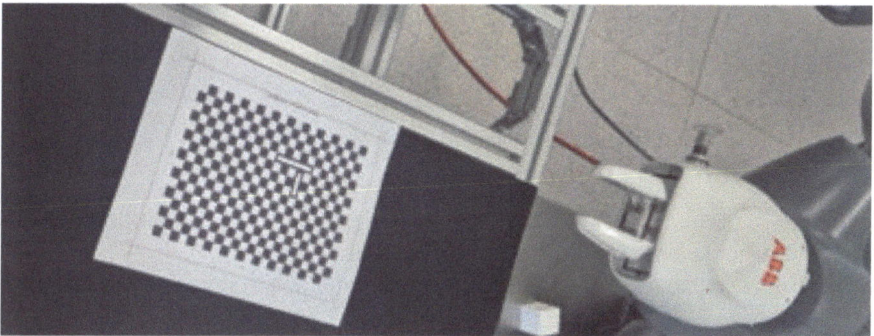

Fig. 4. Positioning for image acquisition

A calibration grid, as shown in Fig. 5, is placed within the workspace.

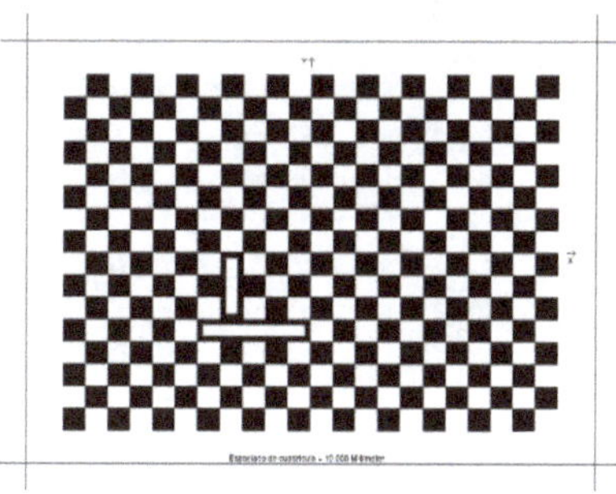

Fig. 5. Calibration grid

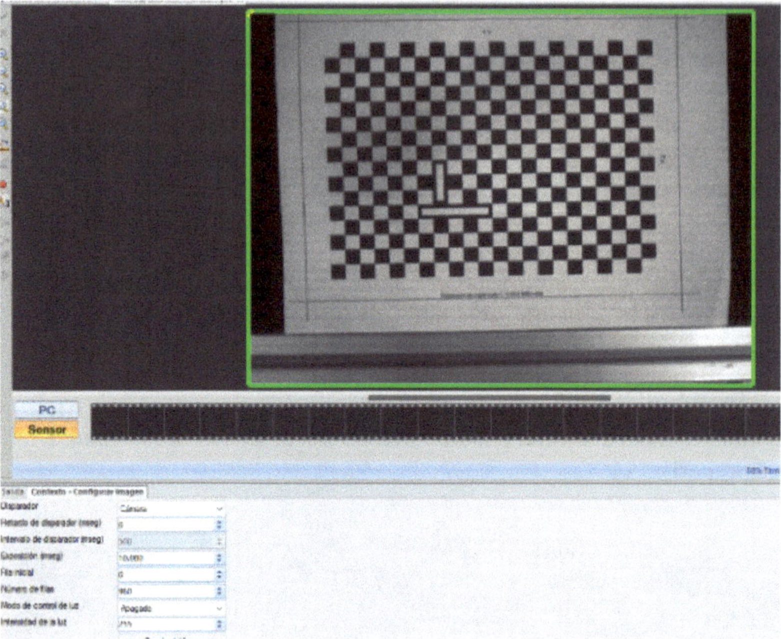

Fig. 6. Image acquisition configuration in software

An image is acquired, and then the parameters are adjusted within the interface to ensure the image quality is adequate for further processing, Figs. 6, 7.

A new image is acquired, and the camera is calibrated according to the manufacturer's instructions.

Fig. 7. Recognition of the calibration sheet in the software

The results and the reliability of the workspace calibration are obtained.

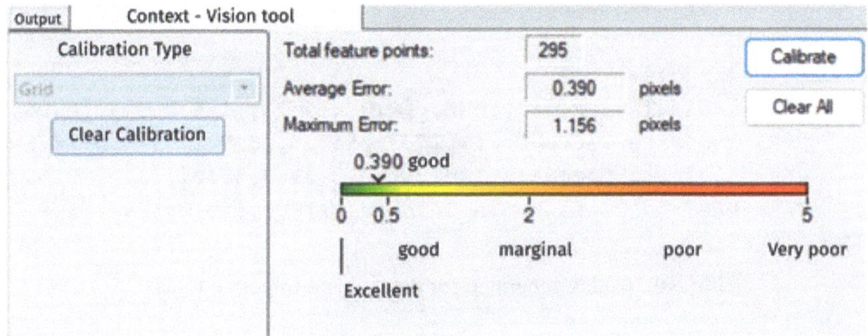

Fig. 8. Calibration results

With this, the camera calibration relative to the workspace is completed, Fig. 8.

3.1.3 Workspace Calibration

From the software interface, a new Tool Center Point (TCP) is created—this will be the tip of the gripper, which will be used as a probe or reference point. The Z-distance is set to 135.00 mm, Fig. 9.

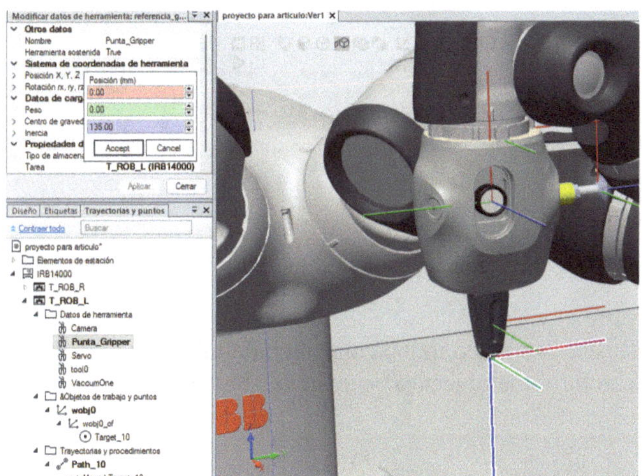

Fig. 9. Tool creation in the software workspace

The work cell is then synchronized with RAPID to retrieve the values of the tools being used. These are subsequently registered in the FlexPendant under the “Calib Data” file (Fig. 10).

```
1  MODULE CalibData
2      PERS tooldata Servo:=[TRUE,[[0,0,114.2],[1,0,0,0]],[0.24,[8.
3      PERS tooldata VaccumOne:=[TRUE,[[63.5,18.5,37.5],[0.70710678
4      PERS tooldata Camera:=[TRUE,[[-7.3,28.3,35.1],[0.5,-0.5,0.5,
5      TASK PERS tooldata Punta_Gripper:=[TRUE,[[0,0,135],[1,0,0,0]
6  ENDMODULE
```

Fig. 10. Codes generated for created and imported tools

Within the “Jogging” tab, the loaded tool can be found and selected to assist in the calibration process, as shown in Fig. 11.

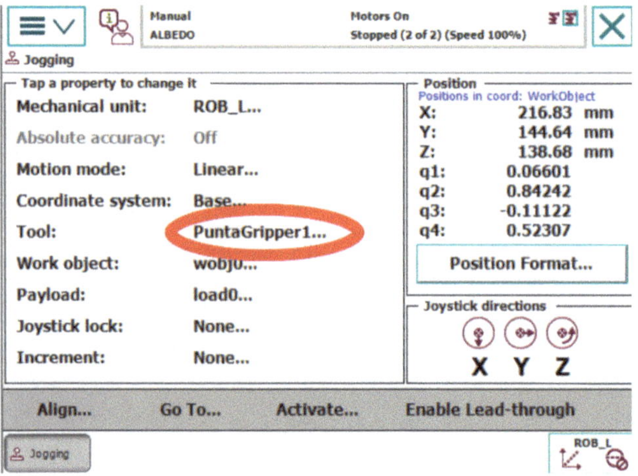

Fig. 11. “Jogging” window in the physical controller

A new workspace is created in the controller, defining the boundaries within which the robot will search for the object to be detected, Fig. 12.

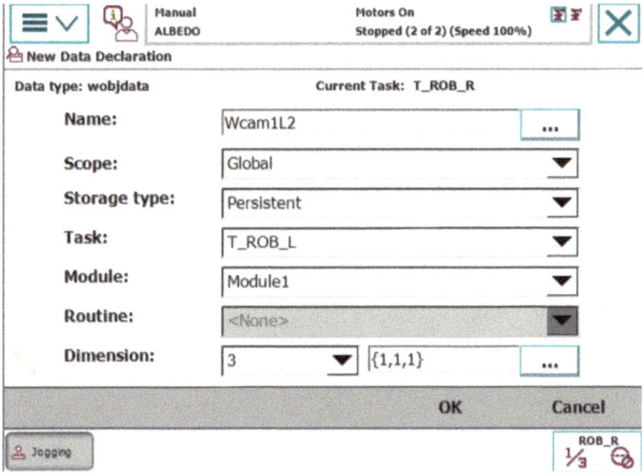

Fig. 12. Workspace creation in the physical controller

The workspace is defined using three reference points: center, X-axis limit, and Y-axis limit, as shown in Fig. 13.

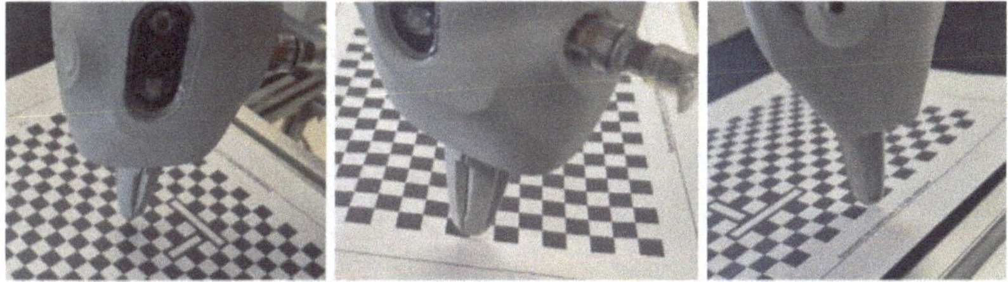

Fig. 13. Tool positioning at workspace references: center, X-end, and Y-end, respectively

The corresponding positions are then modified in the controller and accepted. The system then calculates the new workspace limits (Fig. 14).

To verify the setup, the robot arm can be manually moved while observing how the X, Y, and Z values change according to the tool's position (Fig. 15).

The system is then returned to the Home position, and the calibration grid is removed.

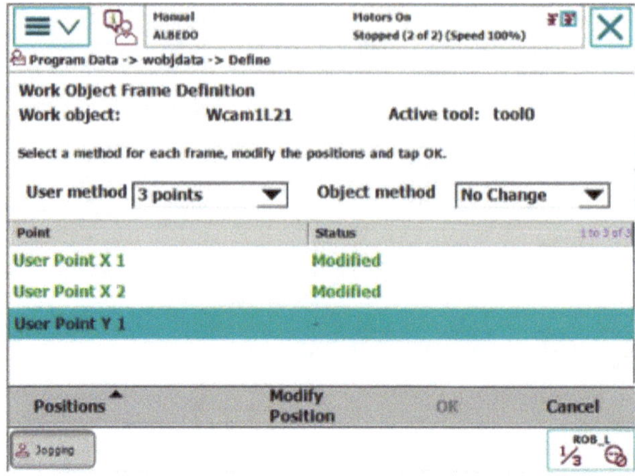

Fig. 14. Point acquisition for the workspace

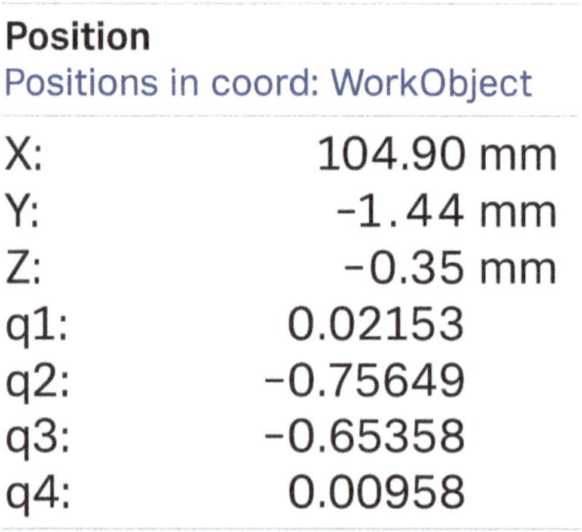

Position

Positions in coord: WorkObject

X: 104.90 mm
Y: -1.44 mm
Z: -0.35 mm
q1: 0.02153
q2: -0.75649
q3: -0.65358
q4: 0.00958

Fig. 15. X, Y, and Z coordinates of the workspace

3.1.4 Gripper (Servomotor) Calibration

To calibrate the gripper, navigate to the “Smart Gripper” tab and set the gripper to its minimum position before starting the calibration process, as shown in Fig. 16.

A test program was created to verify the correct gripping of an object (Fig. 17). For this test, a glass vial was used. The gripper was initialized with a grip force of 10 N and a speed of 10 mm/s, and then a holding force of 7 N was applied to ensure the container´s integrity (Fig. 18).

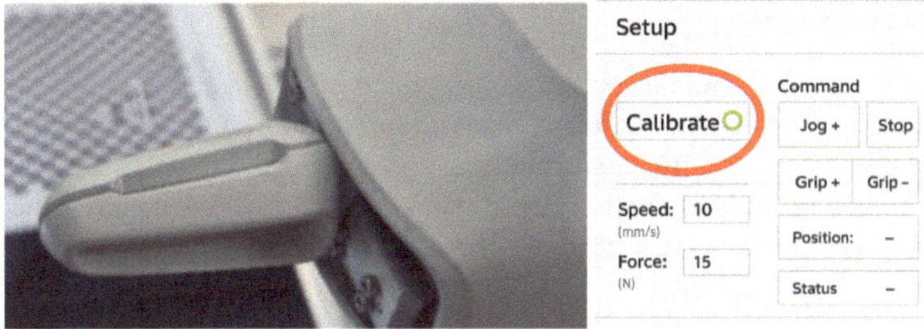

Fig. 16. Gripper calibration position and calibration window

```
PROC Pick_obj()
    !inicializa el gripper con valores default
    g_Init \maxSpd := 10, \holdForce := 10;
    WaitTime 5; !Tiempo de espera
    !Agarra el objeto con 7N de fuerza
    g_GripIn \holdForce := 7;
    WaitTime 5; !Tiempo de espera
    !Solta el objeto
    g_GripOut \holdForce := 7;
ENDPROC
```

Fig. 17. Sample code demonstrating gripper functionality

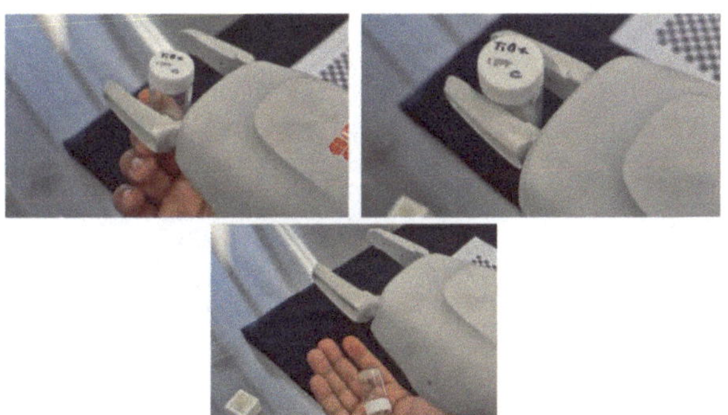

Fig. 18. Physical tests of the gripper picking up and releasing the object

4 Robot Programming

To program the ABB YuMi IRB 14000 robot, RobotStudio software is used to configure the camera and the necessary tools for object handling and visual identification within the workspace. The process includes defining precise trajectories to avoid collisions and ensure proper object manipulation.

The camera configuration enables image acquisition and pattern recognition to identify the laboratory items that the robot will handle. Additionally, auxiliary positions are programmed to guide the robot safely to the image capture point, ensuring efficient and safe operation.

4.1 Robot Positioning

The goal is to reach a specific point using defined coordinates to capture the image correctly. Auxiliary trajectories are also used to guide the robot to the desired position while avoiding collisions with its surroundings or with itself.

The reference point is created at coordinates [400, 0, 400], located at the center of the workspace (Figs. 19, 20 and 21).

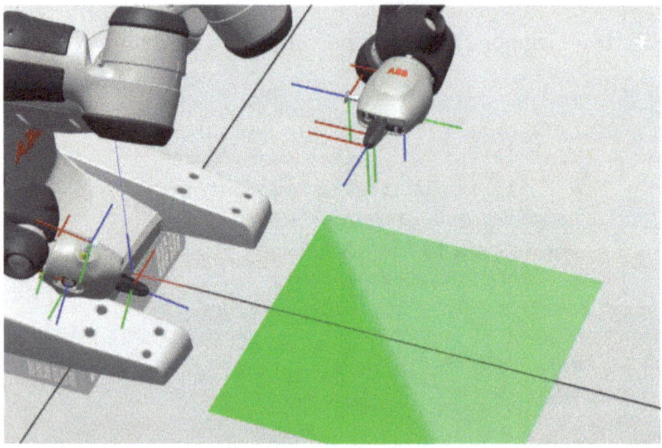

Fig. 19. Image acquisition position relative to the work surface in the simulator

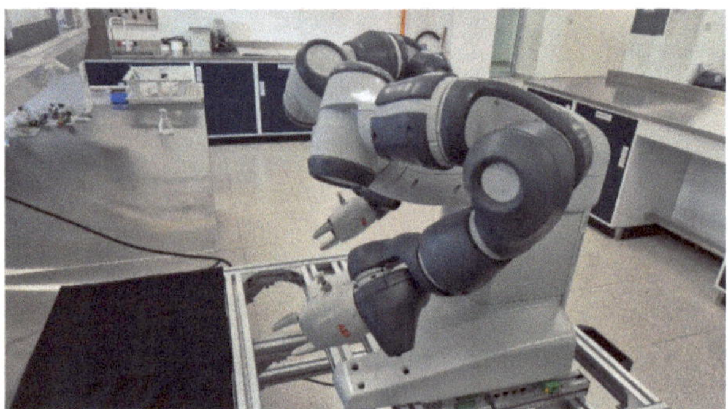

Fig. 20. Robot auxiliary position

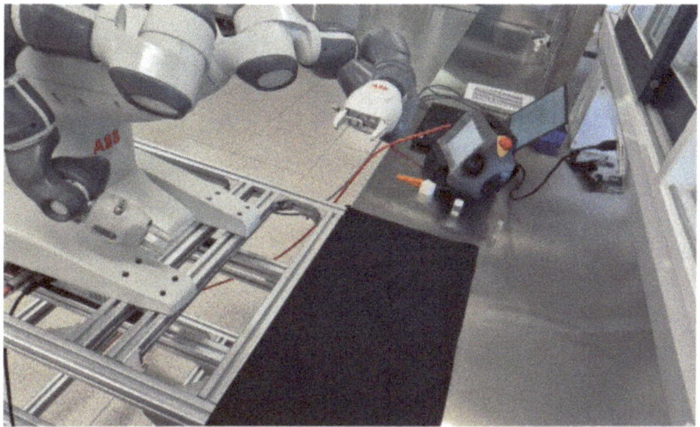

Fig. 21. Final image capture position

Once the robot reaches the target position, the image is captured to enable object recognition programming.

4.2 General Camera Programming

The object to be identified by the robot is a glass vial. An image of the workspace is acquired with the target object placed within the camera's field of view, Fig. 22.

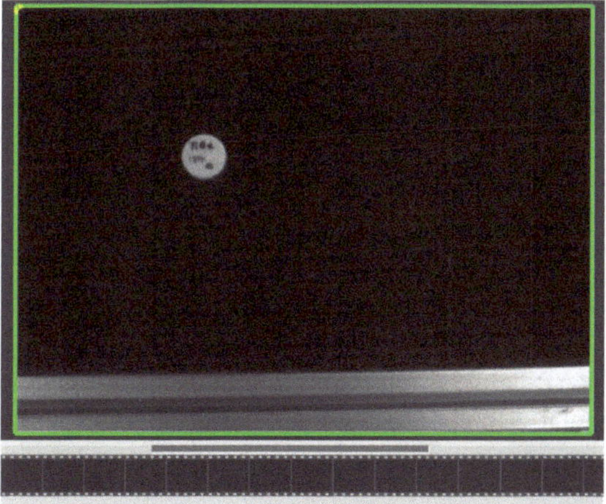

Fig. 22. Image acquisition in the software

Using the options provided by the software, the identification and pattern recognition tools are selected, as shown in Fig. 23.

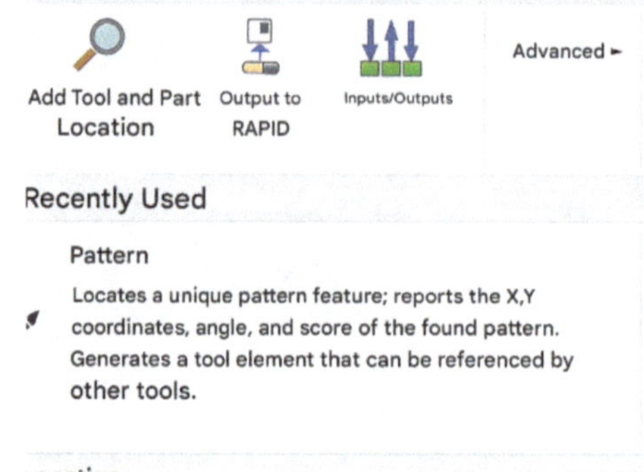

Fig. 23. Tool selection in the software

Within the "model" region, the specific part to be identified is selected—in this case, the cap of the vial (Fig. 24).

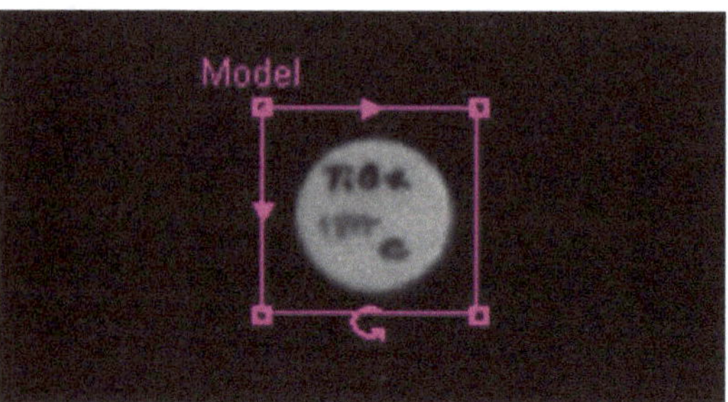

Fig. 24. Region of the model to be trained

After confirming the selection, the identification tool is ready to detect the object and determine its center coordinates.

Finally, these coordinates must be extracted from the camera and sent to the robot controller, so that the robot knows where to move to pick up the object.

The "Output to RAPID" function is selected to transmit the detected object's X, Y coordinates, and angular orientation, as shown in Fig. 25 and Fig. 26.

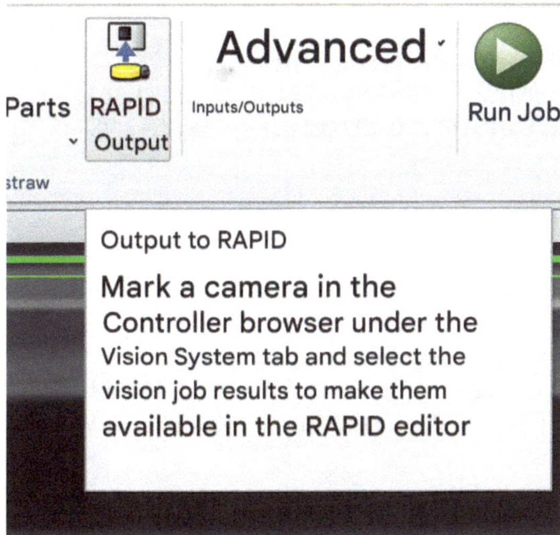

Fig. 25. Data output tool

Correlación de datos de la cámara para la pieza 'Tapa_Frasco'

Componente	Grupo	Resultado
Position x	identificacion_tapa	Fixture.X
Position y	identificacion_tapa	Fixture.Y
Rotation z	identificacion_tapa	Fixture.Angle
Value 1	Constant	0
Value 2	Acquisition	0
	Circle_1	
	identificacion_tapa	
	Inputs	
	Job	

Mostrar todos los ... cámara inclui

Fig. 26. Coordinates to be sent to the controller

With this, the camera calibration for object identification is complete.

4.3 Operation Code

!define the object frame for the vision system

```
PERS wobjdata Wcam1L2 := [FALSE, TRUE, "", [[383.361, -12.2866, -0.704813],
[0.71137, 0.0153852, 0.0151982, 0.702485]], [[4.05847, -3.98945, 0], [0.967522, 0, 0,
0.252787]]];
```

!assign the name of the vision job

```
CONST string myjob := "calibracion_articulo.job";
VAR cameratarget mycameratarget;
```

!define auxiliary object position before picking

```
CONST robtarget objeto_auxiliar_antes := [[-5, -14.52, 100.44], [0.355457, -0.673058,
-0.549159, 0.345062], [-1, -1, 2, 5], [142.013, 9E+09, 9E+09, 9E+09, 9E+09, 9E+09]];
```

!define object position to pick

```
CONST robtarget objeto := [[-28.29, -10.78, 25.56], [0.355467, -0.673066, -0.549144,
0.345059], [-1, -1, 2, 5], [142.013, 9E+09, 9E+09, 9E+09, 9E+09, 9E+09]];
```

!move to image acquisition position

```
PROC GoFotoL()
   MoveJ [[125, 210, 230], [0.0644092, 0.843286, -0.105959, 0.522965], [0, -1, 1, 4],
[141.536, 9E+09, 9E+09, 9E+09, 9E+09, 9E+09]], v1000, z100, tool0;
   MoveJ [[355, 205, 270], [0.469667, -0.443146, 0.480795, -0.593186], [-1, -1, 2, 4],
[142.664, 9E+09, 9E+09, 9E+09, 9E+09, 9E+09]], v1000, z100, tool0;
   MoveJ [[375, -7, 290], [0.0609033, 0.78477, 0.616734, 0.00809459], [-1, 0, 0, 4],
[161.611, 9E+09, 9E+09, 9E+09, 9E+09, 9E+09]], v1000, z100, Camera;
   MoveJ [[400, 0, 400], [0.0, 0.707106781, 0.707106781, 0.0], [-1, 0, 0, 4], [161.611,
9E+09, 9E+09, 9E+09, 9E+09, 9E+09]], v1000, z100, Camera;
ENDPROC
```

!detect object with vision system

```
PROC DetectedObject()
   CamSetProgramMode cam_izq;
   CamLoadJob cam_izq, myjob;
   CamSetRunMode cam_izq;

   WaitTime 7;
   CamReqImage cam_izq;
   CamGetResult cam_izq, mycameratarget;
   Wcam1L2.oframe := mycameratarget.cframe;
ENDPROC
```

```
!move to detected object using coordinates from vision system
PROC MoveToDetectedObject()
  MoveJ objeto_auxiliar_antes, v100, fine, Servo \WObj := Wcam1L2;
  MoveJ objeto, v100, fine, Servo \WObj := Wcam1L2;
  Pick_obj;
  MoveJ objeto_auxiliar_antes, v100, fine, Servo \WObj := Wcam1L2;
ENDPROC

!move robot to Home position
PROC GoHomeL()
  MoveAbsJ [[0, -130, 30, 0, 40, 0], [135, 9E9, 9E9, 9E9, 9E9, 9E9]] \NoEOffs, v1000,
fine, tool0;
ENDPROC

!pick up the object
PROC Pick_obj()
  g_GripIn \holdForce := 7;
ENDPROC

!initialize gripper
PROC open_gripp()
  g_Init \maxSpd := 10, \holdForce := 10;
  g_GripOut \holdForce := 7;
ENDPROC

!main program
PROC main()
  GoHomeL;
  open_gripp;
  WaitTime 5;
  GoFotoL;
  DetectedObject;
  MoveToDetectedObject;

  !move to release position
  MoveJ [[400, 0, 400], [0.0, 0.707106781, 0.707106781, 0.0], [-1, 0, 0, 4], [161.611,
9E+09, 9E+09, 9E+09, 9E+09, 9E+09]], v1000, z100, Camera;

  WaitTime 5;
  open_gripp;
  WaitTime 5;
  WaitTime 20;
ENDPROC
```

4.4 First Pick and Place

Once the program has been completed, execution proceeds up to the **DetectedObject()** function (Fig. 27), since it is necessary to define the grasping position manually.

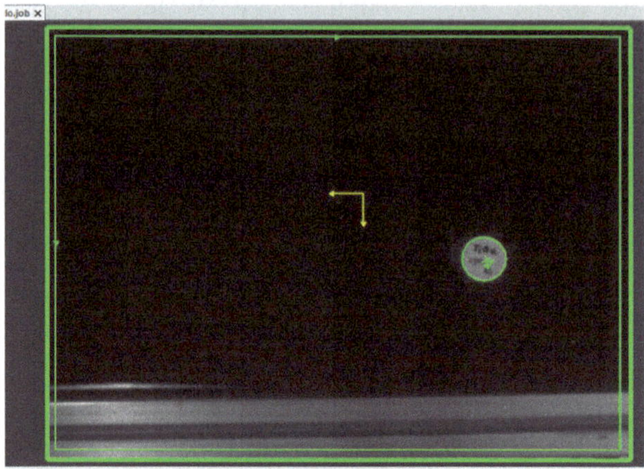

Fig. 27. Object automatically identified in the software

After the robot completes that function, it must be manually guided to the exact grasping position as desired (Fig. 28).

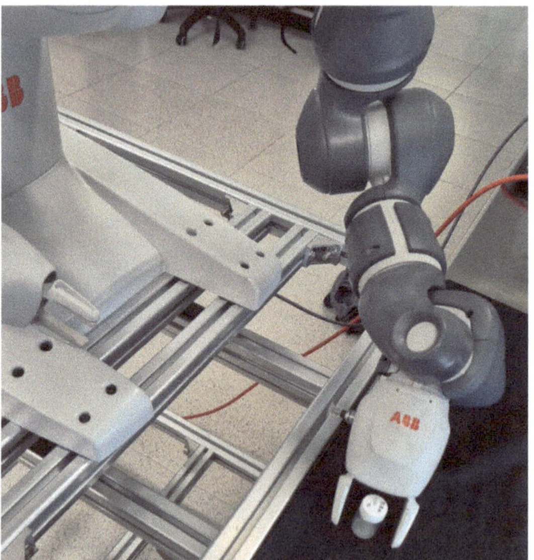

Fig. 28. Manual positioning of the robot at the pickup location

Once this is done, the robot is returned to the Home position, and the full program is executed.

5 Results

The results obtained for the implementation of the ABB YuMi IRB 14000 collaborative robot to identify and handle basic items used in a chemistry lab are described as follows.

5.1 Workspace and Calibration

One of the main findings was the importance of having a workspace with specific and well-defined measurements. During the experiments, it was observed that the robot's accuracy in picking objects could be affected by variations of 2 to 6 mm if the workspace and the probing tool were not correctly calibrated.

To address this issue, the following actions were proposed:

- Establish a specific workspace with precise dimensions, ensuring that all necessary elements for the robot's operations are located within this area.
- Use a dedicated probing tool for calibration, which helps minimize deviations and ensures consistent precision in the robot's tasks.

5.2 Impact on Precision

The data analysis revealed that correcting positioning deviations is very difficult without a well-defined workspace and a specific calibration tool. The robot's accuracy in object manipulation did not meet the standards required for laboratory operations.

5.3 Practical Examples

Specific tests were conducted to evaluate the impact of improper calibration. In one test, the robot was tasked with picking up a glass vial placed in the workspace. The results showed that the robot was able to pick up thevial; however, a deviation of the 2- to 6-mm was recorded. This error posed a risk of potential damage or operational error.

Demonstration video available as supplementary at:

https://drive.google.com/drive/folders/1HRT-eKiJI2Jx32LE9qgdxKoUK9LX LiHB?usp=drive_link.

6 Future Implementation

In the future, the goal is to correct the positioning deviations identified during the object pickup stage, thereby expanding the use of the YuMi robot to move objects and collect other types of containers or items used in a chemistry lab. This includes automating various laboratory tasks such as sample handling, test preparation, and material testing. The integration will also be extended to precise and repetitive processes, such as substance recognition and interaction with measurement equipment. In addition, it is expected that pattern recognition and Data Matrix code reading will be used to ensure optimal inventory management and process traceability. Future developments will also address scalability and safety challenges, proposing innovative solutions to maintain system integrity and efficiency in the laboratory environment.

7 Conclusions

The recognition and handling of a glass vial, commonly used in a chemistry lab, by the ABB YuMi IRB 14000 collaborative robot has proven to be a first step in implementing the robot for the automation of delicate and repetitive tasks in the lab. The main conclusions drawn from this first phase of implementation are as follows:

Importance of Calibration:

Precision in object manipulation critically depends on proper calibration of the workspace and the tools used. Slight deviations in calibration can significantly affect the robot's effectiveness in pickup and handling tasks.

Tool Selection:

Selecting the appropriate tools, such as the gripper and the camera, is essential to ensure efficient and safe operation. These tools enable the robot to handle objects of various shapes and sizes, as well as perform visual recognition tasks with high accuracy.

Programming Procedures:

Detailed programming of the robot, including the definition of trajectories and auxiliary positions, is crucial to prevent collisions and ensure correct object manipulation.

The camera setup for image acquisition and recognition is also crucial for the success of the robot's assigned tasks.

Impact on Research:

Integrating the YuMi robot into the laboratory presents new opportunities for automating research processes in lab chemistry, improving efficiency, and reducing the risk of human error. Additionally, the robot's ability to work safely alongside researchers allows for greater collaboration and resource optimization within the lab.

Future Improvements:

For future implementation phases, it is necessary to address the positioning deviations identified during this first stage. Continued data collection and refinement of calibration and programming procedures will help improve the robot's accuracy and effectiveness in more complex tasks.

References

1. ABB Robotics: Product Manual – IRB 14000 YuMi. ABB AB. https://library.e.abb.com/public/fc0d5c31604f48ed998229f626f72036/3HAC052983%20PM%20IRB%2014000-es.pdf, Accessed 10 Oct 2024
2. ABB Robotics: IRB 14000 Gripper Manual – YuMi. ABB AB. https://library.e.abb.com/public/6c35d74e5be34fcb93bf6810c449a2bd/3HAC054949%20PM%20IRB%2014000%20Gripper-en.pdf, Accessed 10 Oct 2024
3. ABB Robotics: Operator Manual – RobotStudio. ABB Group. https://library.e.abb.com/public/6aeb483836740e11c1257b4b0052375b/3HAC032104-005_revE_es.pdf, Accessed 10 Oct 2024
4. ABB Robotics: Operating Manual – IRB 14000 YuMi. ABB AB. https://cdn.pennair.com/media/2023/11/3HAC052986-OM-IRB-14000-en.pdf, Accessed 10 Oct 2024

Author Index

O. F. Farías Fuentes et al. (Eds.): CIBIM 2024, *Proceedings of the XVI Ibero-American Congress of Mechanical Engineering*, pp. 525–527, 2026.
https://doi.org/10.1007/978-3-032-22823-9

BY NC ND

Zeitfracht Medien GmbH
Ferdinand-Jühlke-Straße 7
99095 Erfurt, Deutschland
produktsicherheit@kolibri360.de